Evolution of the Monitor Lizards

This book is a comprehensive review of the entire lineage of monitor lizards (Varanidae), living and extinct. It fills a gap in our understanding of the relationships and role of environmental and ecological diversity and their effects on evolution and ecology. The evolutionary history of varanids stretches back some 70 million years. These lizards, so often described as "morphologically uniform" differing mainly in size, are incredibly diverse. This variation comprises numerous small features adding up to 80-plus distinctly different species. Synthesizing the latest research, plus a half-century of original observation and research by the author, this text presents a wide-ranging account of the broad biology and evolution of a successful group of vertebrates adapted to environments from the harshest deserts to lushest forests and even tropical beaches. Monitors are shown to be far from a homogeneous group, a lineage in which subtle variations have led to profoundly divergent organisms.

Robert George Sprackland is a herpetologist who has been interested in zoology since early childhood. He studied herpetology and paleontology at the University of Kansas (BA), San José State University (MA), and University College London (Ph.D.), where he conducted the bulk of his research at the Natural History Museum, London. Robert began studying varanids while an undergraduate and has pursued his interest in the lizards across Australia, Europe, and New Guinea for over 45 years. He has published three books and over 40 papers about monitors, plus dozens of other zoological papers. Robert spent most of his academic career teaching college biology, anatomy, and physiology, and has been conducting research in the Department of Vertebrate Zoology at the Smithsonian Institution since 2015.

Evolution of the Monitor Lizards

Systematics, Morphology, and Biogeography of the Varanids

Robert George Sprackland

CRC Press
Taylor & Francis Group
Boca Raton London New York

CRC Press is an imprint of the
Taylor & Francis Group, an **informa** business

Designed cover image: Robert George Sprackland

First edition published 2027
by CRC Press
2385 NW Executive Center Drive, Suite 320, Boca Raton FL 33431

and by CRC Press
4 Park Square, Milton Park, Abingdon, Oxon, OX14 4RN

CRC Press is an imprint of Taylor & Francis Group, LLC

Library of Congress Cataloging-in-Publication Data
Names: Sprackland, Robert G. author
Title: Evolution of the monitor lizards : systematics, morphology, and biogeography of the varanids / Robert Sprackland.
Description: First edition. | Boca Raton, FL : CRC Press, 2026. | Includes bibliographical references and index.
Identifiers: LCCN 2026000222 (print) | LCCN 2026000223 (ebook) | ISBN 9781032730318 hbk | ISBN 9781032732954 pbk | ISBN 9781003463528 ebk
Subjects: LCSH: Monitor lizards--Evolution | Monitor lizards--Morphology | Monitor lizards--Ecology
Classification: LCC QL666.L29 S685 2026 (print) | LCC QL666.L29 (ebook)
LC record available at https://lccn.loc.gov/2026000222
LC ebook record available at https://lccn.loc.gov/2026000223

ISBN: 978-1-032-73031-8 (hbk)
ISBN: 978-1-032-73295-4 (pbk)
ISBN: 978-1-003-46352-8 (ebk)

DOI: 10.1201/9781003463528

Typeset in Times
by KnowledgeWorks Global Ltd.

This book that represents so many years of work is dedicated with
thanks and love to my dear wife,
Teri.
On the road from Genoa to Cairns, there could not have been a better wife,
friend, colleague, and emotional supporter.

Contents

Preface xi
Acknowledgments xiii

PART I Overview and History

Chapter 1 Introduction to Monitor Lizards 3

1.1 Why Study Varanids? 3
1.2 Plan of the Book 6
1.3 Materials and Methods 6
1.4 Taxonomic Considerations 8

Chapter 2 The Herpetologists and the History of Monitor Lizard Research 10

2.1 Introduction 10
2.2 The Beginning of Varanid Research 10
2.3 Historical Paleontological Studies 10
2.4 Modern Herpetological Studies of Extant Varanids 12
2.5 Robert F. Mertens 14
2.6 From Morphology to Ecology 18

PART II Morphology

Chapter 3 Characters and Character States of the Varanidae 23

3.1 Introduction 23
3.2 Phylogenetics and Cladistics 23
3.3 Characters of Taxonomic Significance 25

Chapter 4 Monitor Anatomy I: External Morphology 34

4.1 Introduction 34
4.2 General Features of the Body 35
4.3 The Head 35
4.4 The Body 42
4.5 The Tail 52
4.6 Limbs 56
4.7 Coloration and Pattern 57
4.8 Ontogenetic Variation and Sexual Patterns 63
4.9 Conclusions 65

Chapter 5 Monitor Anatomy II: Internal – the Skeleton 68

5.1 Introduction to Varanid Skeletons 68
5.2 Shared Characters of The Skulls of *Varanus* 71
5.3 The Cranium 75
5.4 The Mandible 104
5.5 Cranial Foramina 108
5.6 Osteological Convergences of the Skulls within the Varanidae 109
5.7 The Vertebrae 111

Chapter 6 Monitor Anatomy III: Internal – the Soft Anatomy ... 113

6.1 General Features of the Soft Anatomy ... 113
6.2 Cranial Anatomy ... 113
6.3 Post Cranial Soft Anatomy ... 125
6.4 Variation and Ecological Considerations ... 140

PART III Ecology and Behavior

Chapter 7 Behavior ... 145

7.1 Introduction ... 145
7.2 Intelligence ... 145
7.3 Locomotion and Posture ... 151
7.4 Foraging and Feeding ... 153
7.5 Thermoregulation and Ultraviolet Radiation Regulation ... 155
7.6 Agonistic Behavior ... 158
7.7 Patterns of Behavior ... 161

Chapter 8 Monitor Lizard Ecology ... 164

8.1 Introduction ... 164
8.2 Habitat and Environment ... 165
8.3 Contemporary Biogeography ... 166
8.4 Metabolism and Energy ... 167
8.5 Respiration ... 169
8.6 Water Dynamics ... 169
8.7 Food and Energy Expenditure ... 170
8.8 Ultraviolet Radiation and Vitamin D ... 172
8.9 Competition ... 174
8.10 Island Variants ... 175

Chapter 9 Mating and Reproduction ... 178

9.1 Introduction ... 178
9.2 Mating ... 179
9.3 Eggs and Incubation ... 182
9.4 Hatchlings ... 189
9.5 Sex Determination ... 190
9.6 Hybrids and Parthenogenesis ... 192

PART IV Systematics and Evolution

Chapter 10 Origin, Evolution, and Prehistoric Biogeography ... 199

10.1 Introduction ... 199
10.2 Origins ... 200
10.3 Evolution and Speciation ... 201
10.4 Distribution over Time ... 203
10.5 Tertiary Varanids ... 208
10.6 Varanid Zoogeography and Taxonomic Considerations ... 212
10.7 Other Anatomical Features ... 215

Chapter 11 Relationships: The Systematics of Monitor Lizards 218

11.1 Systematic Review of the Varanidae 218
11.2 Macro-Compared with Micromorphological Taxonomies 219
11.3 Systematic Review of the Genera of the Varanidae 222
11.4 Unresolved Species-Complexes 242
11.5 Resolved Higher-Level Relationships (Hopefully) 249
11.6 Phylogenetic Summary 250

Appendix I 251

Appendix II 253

Bibliography 257

Index 285

Preface

Humans have known monitor lizards for as long as we have been on Earth. They predate our species by at least 70 million years (my). Large, conspicuous monitors had been common in Asia during the days of the last dinosaurs during the Late Cretaceous, 70–65 mya (million years ago), in North America and Europe since at least the Eocene (55–33 mya), and eastern Africa since the Miocene, or about 23 my before the first humans foraged in Olduvai Gorge. As people spread across the Old World, they encountered more species of monitors, in the Middle East, southern and southeast Asia, the Indo-Philippine islands, New Guinea, Australia, and numerous Pacific islands, some of which may have been introduced by humans. Relicts of the Oligocene pre-*Varanus* monitors of Europe may have lingered during our ancestors' earliest explorations as far west as Spain (Augé 2005, Estes 1983a, b, Fejérváry 1935, Georgalis et al. 2017). Many species encountered would be one- to two-meter-long giants of their family, used by our ancestors for food, leather, and medicine. Though not as well represented in the earliest tales as more conspicuous and dangerous animals, some of our ancestors considered them important because they thought that the presence of monitors indicated there were crocodiles in the area, dubbing the lizards "salvaguardia," or "safeguards," or an equivalent name. Australia had giant monitors, too, including the largest known land-living lizard (Molnar 2004, Nopcsa 1912, Owen 1859, Rich and Hall 1979), but there were also many much smaller species that still thrive there. Today, monitors that grow to more than one meter (three feet) in length are conspicuous even to people who might be oblivious to more numerous smaller lizards. Consequently, folklore, myths, and reliable observations pertaining to these lizards predate the origins of modern scientific practices and of course vary in reliability.

I began my interest in monitors around 1969, following discussions with Samuel B. McDowell, Jr., about their generally overlooked variety. In the intervening years, I have had opportunity to study a lot of monitors in the field and in collections on three continents, and I have researched varanids for my master's and doctoral degrees. There has been so much written about monitors, largely taxonomic works (Bayless and Sprackland 2000a, b, Böhme et al. 1994, Boulenger 1885, Gaulke 1991b, Koch et al. 2007, Mertens 1942a–c, 1959, Sprackland 1991a, b, 1993a, b, 1995), regional fauna guides (Cogger 1992, De Rooij 1915, McCoy 1980, 2006, Philipp and Philipp 2007, Pianka and King 2004, Sprackland 2010a, Storr 1980, Storr et al. 1983, Wilson and Knowles 1988, Worrell 1963), and in sections of broader herpetological texts (King and Green 1999, Neugebauer 1976, Pianka 2003, Pianka and King 2004, Porter 1972). Several species have been exceptions as subjects of comprehensive monographs (Ariefiandy et al. 2017, Auffenberg 1981b, 1988, 1994, Auffenberg et al. 1988, 1989, Cowles 1930, Eidenmüller 2016, Gaulke 1989, Murphy et al. 2002, Pfeffer 1965, Surahya 1989; see also Auliya 2006). They are well represented in varanid-only books (Bennett 1998, Burden 1927b, Eidenmüller 1997, 2007, 2021, Eidenmüller and Philippen 2008, Green and King 1993, Houston 1998, King and Green 1999, Philippen and Eidenmüller 2007, Pianka and King 2004, Rotter 1963, Schmida 2017, Suzuki 2006), with many titles focused on captive varanids (Berghof 2009, Brown 2012, Camina 2019, Eidenmüller 2016, Sprackland 1992, 2009, Vincent and Wilson 1999). I have no intention of repeating the literature more than necessary and I have no illusion that my bibliography is exhaustive (though the website https://repfocus.dk/Varanus_bibliography.html may come close), nor was it my plan to write such a document. Instead, I give a brief overview and cite references to pertinent earlier sources that, in turn, will yield more extensive citations.

As one of my recent projects over the past eight years, I have been part of the team inventorying several large bequests from retired colleagues for our museum's division library. We have catalogued thousands of books, theses, and collected works during the first three years. I created the inventory database to have it accessible as an online database. One advantage of this work is that as I wander through the shelved materials, I get an excellent education in the types of literature that exist – and, more important to this story, do not exist – for herpetology. Field guides and treatises predominate by a large margin, followed by regional accounts. Missing was the type of book I would have found extremely useful as a university student as well as later for my research, a solid overview of extant squamate anatomy. Chapters in many books gave nodding attention to the subject. The obvious go-to reference, Romer's *Osteology of the Reptiles* (Romer 1956), is overwhelmingly devoted to fossil taxa. Hoffmann (1890) provided the most complete single anatomical review of living reptiles, but only one species of *Varanus* was used. As is the case in most anatomical overviews of lizards (and Squamata and Reptilia), the range of taxa presented is vast but the focus at family level is small. The mammoth series, *Biology of the Reptilia* (see Gans 1969–2010), covers topics from morphology and physiology to ecology and behavior in 22 volumes, with contributions from leading subject specialists. Its only serious drawbacks are that there are few sections that are taxon-specific below the ordinal level and, naturally, it doesn't cover everything. The authors graciously left considerable material for others to study. A considerable body of character states and phylogenetic hypotheses have been provided by Conrad (2008), Dong et al. (2022), Estes (1983a, b), Estes et al. (1988a, b), Evans (1984, 1988), Head et al. (2009), Hocknull et al. (2009), Lee (1997, 2005), McDowell and Bogert (1954), Norell and Gao (1997), Reippel (1980), and others. Gauthier et al. (2012) produced a staggeringly detailed and exhaustive review of character states across the Squamata that has been of monumental importance to morphologists and systematists since its publication and has been of inestimable value to the present volume. Bucklitsch et al. (2016) have compiled a family-wide description of varanid scale morphology, especially useful for covering a previously neglected bit of anatomy and doing so in great depth. There is simply no single comprehensive treatise on lizard anatomy (but see Harris 1963). This book is, however, a monograph that brings together in an easily referenced single volume a broad range of topics relating to the family Varanidae, particularly the living genus *Varanus*, that are represented in some two centuries of widely dispersed publications or are new contributions.

Biological journals are filled with reports on the anatomy of what I think of as *regional* anatomy, that is, muscles of the forelimb, the hearts of one or two species, the vertebral structure in a genus or family, or the relationships of a limited representation of taxa. It is not uncommon for studies to use a very small sample size and extrapolate to the entirety of the species, to the family, or to all reptiles. For example, among these valuable contributions is the taxonomic limitation of the number of species covered, and then only as regards that single anatomical region (e.g., Bahl 1937, Bellairs 1949, Benedict 1932, De Meester et al. 2019, Jollie 1960, Oelrich 1956). It is my opinion that the field could use a few works that cover the range of biological information within a family and of more than a few species. Monitor lizards now include over 80 named species with many others waiting for discovery or description (as I write this, material I have seen makes me think the number, so far, will be over 100 species). As I hope to repeatedly make clear, monitors are not a nearly uniform group that varies chiefly by size. For so many important references, I often needed to seek assistance from specialist librarians and colleagues who very kindly went in search on my behalf. Part of the reason for this book is to highlight those resources for other varanophiles to let them know they exist and to serve as an introduction to what has already been done.

To study the Varanidae is to study a continuum of evolutionary diversification. This book is offered as an overview of that entire family, living and extinct, and will be useful not just to varanid or squamate specialists but to other zoologists as well. It is not an attempt to cover every aspect of varanid biology, and certainly not to replace or simply update the many excellent works cited above, but I hope it will serve as a source of broad information about what we know and what we have yet to learn. The comparative length or brevity of different sections of this book attempts to reflect some gauge of which areas have been well studied and which remain intellectual orphans. We have many volumes at hand on skeletal morphology and variation, a topic that takes the most space of any chapter in this book. Natural history, the business of being alive, takes three chapters here. Sections of soft anatomy are patchy, with some areas much better known than others. Where I have omitted presentation, it is because there is relatively little varanid-specific literature or, for example, regarding the subject of pathogens, the information is not about the lizards themselves. Finally, the number of people interested in varanids, from enthusiasts to professional keepers to university and museum researchers, continues to increase and with that so does the flood of new information about life histories, reproduction, husbandry, systematics, and every other aspect of their biology from molecules to evolutionary history. Frankly, some of what I have written for this preface will be outdated before the entire manuscript is written, which I find an exciting reality. Without question, then, this author has left considerable material for others to study.

Robert George Sprackland
Washington, D.C.
2026

Acknowledgments

Over five decades have passed since I first began my observations on the monitor lizards, and this report therefore encompasses assistance from innumerable institutions and people, many of whom have become good friends and too many no longer with us. In total I have seen over 5,000 specimens, which has given me an incredible opportunity to study variation and its effects on phenotype and natural history. My thanks to the army of people who made my work possible: Susan Evans (UCL), Colin McCarthy, Barry Clarke, and Angela Milner (BMNH), Geoffrey Swinney (NMS), Roy McDiarmid, Carole Baldwin, Kate Bemis, Esther Langhan, Steve Gotte, Kevin de Queiroz, George Zug, Polly Lasker, Sue Zwicker, Tom Munroe, and visiting researchers Arthur de Lima and Paulo Pinheiro (NMNH), Marinus Hoogmoed (RMNH), Wolfgang Böhme (ZFMK), Giuliano Doria (MSNG), Ross Sadlier and Gerry Swan (AM), Harry Greene (MVZ), Konrad Klemmer, Günther Kohler, Maren Gaulke, and Frau Monika Laudahn (SMF), Ulrich Gruber and Dieter Fuchs (ZMM), Alan Resetar (FMNH), Patrick Couper (QM), Jens Vindum (CAS), Robert Hanson (CSU), Rainer Günther (ZMB), Heinz Grillitch, Anita Cabella, and Franz Tiedermann (NMW), Iliah Bigilali (NMPNG), and Michel Thireau and M. LeRoi (MNHN) for allowing me to examine specimens in their care and for their numerous kindnesses on my several visits to the collections in their care. Thanks also for the many contributions by Mark O'Shea, O.B.E.

I am indebted to several mentors, friends, and colleagues who are no longer with us: E.N. Arnold, Walter Auffenberg, Mark Bayless, Joseph T. Collins, Herndon Dowling, Henry Fitch, George Foley, Ulrich Gruber, Hans-Gerg Horn, Steve Irwin, Larry Martin, Hymen Marx, James Oliver, Russell Rak, Edward Taylor, Garth Underwood, and Richard Zweifel. I offer most special thanks to the mentor, friend, and professor who taught me so much and directed my interest toward monitor lizards, Samuel B. McDowell, Jr.

The preserved and skeletal materials I have personally examined are every varanid in the following 23 collections (as of given dates): Australian Museum (2003), British Museum (the Natural History Museum) of Natural History (1996), California Academy of Sciences (1991), Field Museum of Natural History (2002), Museum of Comparative Zoology (1997), Museum für Naturkunde der Humboldt Universität zu Berlin (1995), Museum National Histoire Natural (1997), Museum of Vertebrate Zoology (1992), National Museum and Art Gallery of Papua New Guinea (2003), Museum Storia Natural Genoa (1991), National Museum of Natural History, National Museums of Scotland (1997), Naturhistorishes Museum Wien (1991), Queensland Museum (2003), Rijksmuseum van Natuurlijke Historie (1994), San Diego Natural History Society (1998), San José State University Vertebrate Museum (1990), Forschungsinstitut und Natur Museum Senckenberg (1997); University College London (1995), Zoologisches Forschungsinstitut und Museum Alexander König (1997), University of Amsterdam Natural History Museum (1994), Zoologisches Staatsmuseum München (1995). I have also examined significant parts of the varanid collections of the American Museum of Natural History, US National Museum of Natural History, Florida State Museum, Museo Civico di Storia Naturale de Milano, and the University of Kansas Museum of Natural History.

Among the zoological parks where I have examined, photographed, and videotaped live monitors are Australian Reptile Park, Australia Zoo, Aquarium Berlin, Arnhem Zoo, Cairns Night Zoo, Cincinnati Zoo, Clyde Peeling's Reptiland, Cologne Zoo, Columbus Zoo, Denver Zoo, Drachenfels Zoo (Bonn), Frankfurt Zoo, Fred Kundert's Reptile Zoo (Basel), London Zoo, New York Zoological Society, Oklahoma City Zoo, Philadelphia Zoo, Parque Zoológico de París, Prague Zoo, Reptilienzoo Königswinter, Reptilia Zoo (Toronto), Rotterdam Zoo, San Diego Zoo, Schönbrunn (Vienna) Zoo, Silver springs Reptile Institute, Smithsonian National Zoo, Taronga Zoo, Tierpark Berlin, Topeka Zoo, Wilhelma Stuttgart, and Woodland Park Zoo.

Gaining access to every described (and a few undescribed) species of *Varanus*, plus information on breeding, young, and variation in behavior and coloring, is a herculean undertaking, and it would not have been possible for me to complete this task without the extensive and generous support of an army of people in the private sector, to each of whom I am indebted. My sincere and heartfelt thanks to John Adragna, Jr., Chris Applin, Robert Balan, Richard Biffin, Gin Black, Brittany Bonfante, Allen Both, Kathryn Brown, Katy Bruce, Tom Crutchfield, Bernd Eidenmüller, Scott Eipper, Jim Elliot, Julian Firth, Maurice Fischer, Lee Gardella, Ian Gniazdowski, Jack Hartley, Kenneth Jensen, †Gunther Köhler, Codey Kopp, Peter Kuhn, Fred Kundert, Nick Lillis, Helmut Magdefrau, Brian Manning, Guy Masters, Keith McPeak, Benjamin Miles, Savannah Noland (for providing non-varanid comparative materials), Roy Payne, Kai Phan, Kai Philip, Dante Rodriguez, Ben Siegel, Nathan Souza, Michael Stefani, Rainer Thissen, Ralf Vitt, Bill Walsh, Rowan Wolf, and Przemysław Zdunek. Of these wonderful people I give special thanks to Michael Stefani for acting as unofficial liaison between the varanid breeding community and myself, accomplishing in days what may have taken me months; Chris Applin for providing me a huge and nearly unmanageable pile of field data from Indonesia, including photographs and ecological notes on species awaiting formal description; and Przemysław Zdunek who provided literature so obscure that I could not find them in my museum's extraordinarily extensive library; I shall still be trying to properly sort them long after this book is published. Half a century is a long time; if I have omitted anyone, I beg your forgiveness.

The extent and length of travel my work has entailed would not have been possible without the extraordinary kindness of Rainer Thissen, Hans-Georg Horn, Helmut Magdefrau, Ulrich Gruber, Bernd Eidenmüller, and Ursula Stilger and their families, who kindly provided lodging for my wife and me during most of our time in Germany; Gerry Swan and Steve and Terri Irwin provided accommodations in Australia; Michel Thireau in France; and Iliah Bigilali in Papua New Guinea. Wolfgang Böhme and Rainer Gunther also provided living accommodations at their institutions during my travels, as did the Queensland Department of Wildlife and Fitzroy Island Resort.

Some of my work was funded by an Overseas Research Scholarship and grants from the Graduate School (University College London), University of London Central Research Fund, the Linnean Society, the San Diego Herpetological Society, and the Shoestring Foundation, to whom I extend my gratitude.

Above all I thank my dear wife, Teri, a journalist by profession who has learned a lot about zoology over our 50 years together. She made this work possible, providing logistical, emotional, and cheerleading support with love. She has also been my primary editor, art critic, translator for Dutch and French works, traveling companion, and field assistant. I truly could not have done this research without her love and encouragement. Without doubt this work is dedicated to Teri.

Part I

Overview and History

Varanids have been on Earth far longer than humans, so it is unsurprising that people have known about them for the entirety of their history. The lizards of the Afro-Asian realm, home to the earliest humans, included large species that were used for food, sources of leather, and supposed harbingers of the presence of crocodiles. Yet, despite this long association, varanids remained largely neglected by inquiring humans until the middle 18th century and the beginning of the European-centric Great Age of Exploration. The known species were joined by larger – sometimes much larger – eastern Asian monitors and the great range of species, including many dwarves, in Australia. It is perhaps the discovery of the largest living lizard, the Komodo monitor, in the early 20th century that set the stage for the ever-increasing interest in monitors.

Varanid research has, until the late 20th century, been characterized by two dominant trends. First, the aspects studied were markedly defined by their times, going from early descriptive, then more taxonomic accounts, followed by extensive morphological work, and then focusing on the habits and reproduction of living lizards. Since the 21st century began, there has been considerable integration and collaborative efforts, including the addition of molecular systematics techniques and CT scan morphological work.

The second trend is the historical national nature of varanid research and publication. England and France (where, arguably, herpetology was born and matured) produced numerous accounts of new species and detailed descriptions of genera and the family. France and Hungary took the lead in describing the meagre fossil materials, though the material was much increased by the early 20th century. The mainly descriptive phase became more morphology and systematics-driven in the early to mid-20th century, with Robert Mertens having such a large influence that his work is discussed at length here. For most of that century, German researchers dominated varanid studies, but herpetologists in Australia and the United States quickly grew in number from about 1970 onward. The major studies explored the life histories of monitors and included extensive observations on wild and captive animals.

This section explores three topics: first, a quick introduction to monitor lizards. Second, it asks why we study varanids and offers the scope of this book. Finally, I present a brief description of the major varanid researchers and their contributions to the study of varanid biology.

DOI: 10.1201/9781003463528-1

1 Introduction to Monitor Lizards

1.1 WHY STUDY VARANIDS?

What is so special about monitor lizards? Monitors – a commonly used term for the many varanid lizards, along with the Australian goanna, Asian biawak, Sri Lankan kabaragoya, and South African leguaan – have used their seemingly limited morphological toolbox to evolve into an extraordinary group of active predators that have thrived in most Old World tropical and subtropical habitats for over 70 million years. All of them are active foraging predators that sit near or atop their local food pyramid. Their regular prey may include deer, crocodiles, highly venomous snakes, and scorpions, though most species consume clams, insects, crabs, other arthropods, small eggs, and small vertebrates such as lizards and rodents. They perform intricate displays to threaten rivals and potential predators while lacking any of the spines, crests, bright color changes, or frills of lizards such as horned lizards, basilisks, and chameleons. All are expert diggers, most are excellent swimmers, though none has webbed digits or other major adaptations for a semi-aquatic life, and a large number live in trees. And their feats can be astounding: there are monitors that excavate 3 m vertical spiral burrows in which they communally lay their eggs; dig into concrete-hard ant and termite mounds, also to deposit their eggs; swim at sea dozens of miles from land; remain submerged for an hour on a single breath of air, chase and tackle prey at surprising speed and with great accuracy; and climb trees with the speed and facility of any squirrel. The range of colors borne by monitors runs from drab grays and browns to bright orange, yellow, green, blue, or some combination. A monitor's tail may be about the same length as the head and body (snout–vent length, SVL) combined and strongly keeled, or more than twice the SVL and glossy. And they are smart: monitors, like Jurassic Park's fictionalized *Velociraptors*, routinely explore their enclosures, looking for weaknesses. Some test doors and screens, learn how and where food will be presented (and terraria opened), and they remember. Some monitors are alleged to be able to count (Pianka and Vitt 2003), engage with behavioral enhancement objects ("toys" to most of us), and can even pick the lock on terrarium doors (pers. obs.). The largest can weigh 81.5 kg (180 lb on an empty belly) and measure 3.13 m (10.27 ft) or more in length, while the smallest only weighs 0.016 kg (0.04 lb) and reaches 0.23 m (0.75 ft) – which means the largest may be over 2,000 times the mass and 13 times the length of the smallest, a disparity not seen in any other genus of vertebrates.

Most people who have heard of monitors tend to think of the Komodo Dragons of Indonesia. "Oh, like Komodo lizards," is the most common reply from people who ask what I study. For decades that seemed natural, given the frequent appearance of Komodo Dragons on nature documentaries, plus the incredible popularity and interest in dinosaurs, which have been featured for decades as huge lizard-like creatures. It is, honestly, a large part of what got me interested in the family over half a century ago. But that same mass media has subsequently introduced people to more than just Komodo Dragons, and the popularity of monitors has grown exponentially since the 1980s. At that time, I proposed a book about the world's monitors, about which two editors said that there just wouldn't be the interest, and so instead my first idea morphed into *Giant Lizards* (Sprackland 1992, 2009).

However, a rapidly increasing interest in monitors did take place. Books, magazines, and television popularized the lizards, as did the increasing numbers being displayed in zoos and kept by private individuals. Prior to the 1980s, monitor literature was overwhelmingly in the form of scientific papers and monographs, often in journals not readily available at the public library. In fact, until 1988 there were no books specifically about monitors in English (though Jiri Rotter's (1963) German book was an excellent starting point), and the books and most popular articles that were published were about a single species (surprise), the Komodo Dragon (e.g., Auffenberg 1981, Burden 1927, Murphy et al. 2002, Pfeffer 1965, Surahya 1989). From that point forward, though, the books have increased in number, languages, and scope of subject material. Auffenberg's 1981 book on Komodo Dragons was to be followed by his books on Gray's (1988) and Bengal (1994) monitors, and these were subsequently joined by *Giant Lizards* (Sprackland 1992, 2009), *Goanna: The Biology of Varanid Lizards* (Green and King 1993, King and Green 1999), *Faszination Warane: Pflege und Zucht* (Kirschner et al. 1996), *Monitor Lizards: Natural History, Biology & Husbandry* (Bennett 1998), *Varanoid Lizards of the World* (Pianka and King 2004), *Monitor Lizards: Natural History, Captive Care and Breeding* (Eidenmüller 2007), *TERRALOG: Varanoid Lizards, Monitor Lizards* (Eidenmüller and Philippen 2008), *Australian Monitors* (Schmida 2017), *Varanos* (Camina 2019), and many more (Fig. 1.1). As in all fields, the books range from poor to excellent, and interest in monitors, especially as evidenced by certain websites (which also range from poor to excellent), is still growing.

So, why am I writing another book to join the large field of published volumes? What more can be said, indeed, what more is there to know? Let me preface the answer by stating that this is not another book of species accounts; there are plenty of those already available (e.g., Auliya and Koch 2020, Bennett 1998, Brown 2012, Camina 2019, Eidenmüller 2007, Houston 1998, Mertens 1942a–c, Pianka and King 2004, Storr et al. 1983, Vincent and Wilson 1999).

DOI: 10.1201/9781003463528-2

FIGURE 1.1 An idea of the amount of published material specifically about monitor lizards can be somewhat appreciated by these views of the author's library on the subject. Not shown are the numerous other books, periodicals such as *Varanus*-specific *Biawak* and *Monitor*, general herpetological and zoological journals with varanid papers, books kept in my museum office, and those donated to the division library, and about two-thirds of the author's notebooks.

At the rate that new varanids are being discovered and named, a guide to species would be outdated before publication. For example, while editing the proofs for this book, three more species were named. So first, and putting aside the Komodo Dragon, the myriad other species of monitors are generally described as a conservative group. Translated from Biology-speak, that phrase suggests that all monitors look alike and are essentially similar, barring their most conspicuous variation, size. However, the dwarves are not simply miniatures of the giants (Sprackland 2001b), nor is one giant species necessarily much like another (Auffenberg 1981b, 1988, 1994, Christian et al. 1995, Shine 1986). When we look at the animals more closely, we see details that show us a world of variety that has been little studied, and because variation is the raw material upon which evolution works its effects, I have chosen to write this book about that diversity among monitors and how that diversity affects their lives.

Second, monitors are versatile in most tropical and subtropical environments. Though each species favors a particular place, climate, food, and lifestyle, almost all species can adapt when environments change. There are 2-plus-meter giants that live in forest canopies, and desert dwarves that may never see a live tree. Some are averse to entering water, and others can stay submerged for over an hour. The amazing thing about the 90 -odd described species in this family is that they have become so ecologically and behaviorally diverse while evolving seemingly very minor variations in their phenotypes, which mirrors the same model as seen in Darwin's finches, where seemingly insignificant alterations in phenotype may have important effects that may lead to success for one phenotype and extirpation for another (Grant and Grant 2014). To understand their great ecological and behavioral variability, it is important to understand their structure – the morphology – in greater detail. The history of varanid morphology research is similar to the story of the blind men and the elephant: each man got to feel one part of the elephant, leading him to form an incorrect description of the entire animal until all their observations were combined. In this book I take both information from others, which scientists call "the literature," and original observations I have been making since 1969. Unlike the blind men, we do pretty much know what a monitor is, but only recently to the detailed level now possible. This book is intended as a compendium of research and observations that have given us that detailed picture and, hopefully, provide background data to researchers embarking on new studies.

Third, the trend of the past 30 years has been to separate the fields of study so much that there is seemingly less integration of research in ecology with biogeography, paleontology with behavior, or genetics with morphology. This book's central theme is morphology, the shape and structure of organisms, and the links morphology has with other branches of herpetology. At a time when genetic and molecular studies have taken predominance in zoology, morphology remains an essential field of investigation. After all, morphology is the direct study of the phenotype, and as demonstrated by the Grants (Grant and Grant 2014) with Darwin's finches, and by Losos and colleagues (Losos 2011) with anoles, morphology can dramatically show us the stepwise progression of biological adaptation in the face of environmental change.

Finally, the range and breadth of published information about the Varanidae is truly staggering with the result that information from the minute to expansive is distributed in several languages (predominantly English, German, Latin, French, and Italian) in a seemingly unknowable number of publications (Fig. 1.2). Knowing that something has been studied and published is one thing; finding the reference has often become something far more difficult. This book is partially an indication of the level of my exasperation at trying to find, confirm, and keep track of that Himalayan pile of data and to make the subsequent climb less of a burden to others coming along the path. However, the fact that some topic is not, however superficially, mentioned here is not indicative that the information does not exist. Rather, I found it irrelevant to the overall subject of the book.

This book is a herpetological monograph, but my chosen writing style will, I hope, make it accessible, and thus useful, to non-herpetologists, college students, keepers, and other readers with a serious interest in natural history. One of my mentors told me that "we don't just need publications that only six other specialists will read." I agree, and have

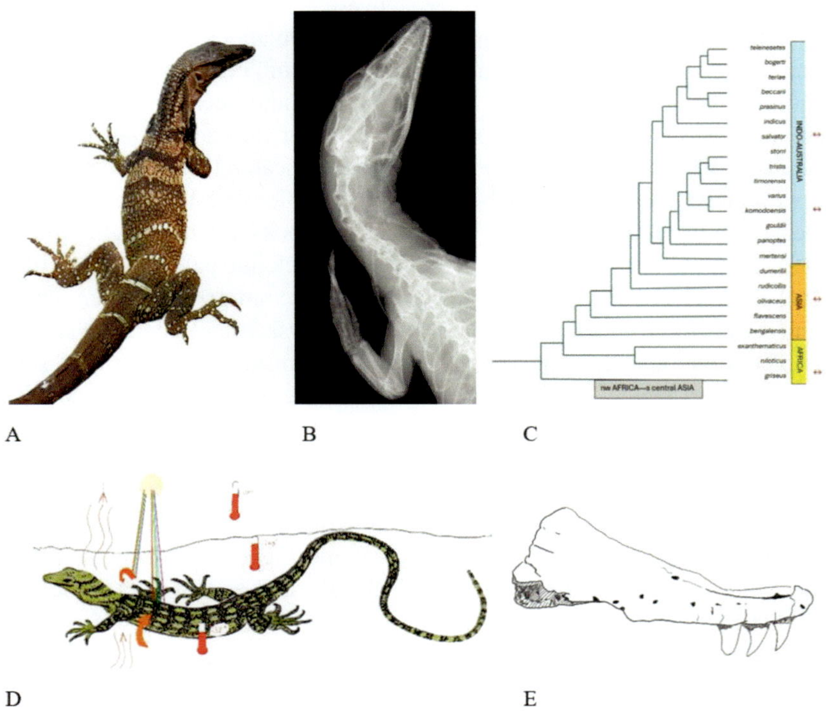

FIGURE 1.2 Varanids (A) are frequently studied animals for numerous biological disciplines, including (B) morphology, (C) phylogenetics and evolution, (D) ecology, behavior, and physiology, and (E) paleontology. A and B, *Varanus rudicollis* (B is USNM 259133), D is *Varanus kordensis*, E is the extinct European †*Varanus marathonensis*.

therefore written this volume to reach at least a few dozen other readers. I hope it serves you well.

1.2 PLAN OF THE BOOK

This monograph presents varanid biology in four parts. **Part I** is the **general overview** and introduction to the lizards, the history of varanid research, and a discussion of the major trends that the research has covered. In this chapter, I introduce monitors and the intention and plan of the book. Because a relatively small number of researchers have produced the greatest quantity of studies on the monitors between 1800 and 1975, I include a semi-biographical account of the contributions of each of those herpetologists and paleontologists in Chapter 2. I start with a history of varanids in the literature and a critique of the various publications, mainly from 1758 through 1975. The first writings were primarily assignations of names to specimens, followed by a period of increasingly more detailed written descriptions. After 1900, descriptions continued to appear – especially between 1990 and 2018 – but there was considerably more emphasis on taxonomy and systematics. By the late 1900s, there was also a large body of research published on varanid physiology and reproduction. The focus, however, is not on the foibles, hobbies, or adventures of the herpetologists, but specifically on their contributions to our better understanding of monitors. Several herpetologists who study goannas are, at the time of this writing, extant and are friends or colleagues of mine, and following tradition, I do not cover them, though their works are well represented in the text and literature cited. Each has started a path toward understanding that began in a different place from the others, such that several independent and only slightly overlapping studies resulted, giving us a wide variety of topics pertaining to varanid biology.

Part II covers the **morphology of monitors** inside and out and represents the bulk of this work. Chapter 3 reviews the alleged synapomorphies of the Varanidae as established by the most recent researchers (Conrad et al. 2008, Estes et al. 1998, Gauthier et al. 2012, Rieppel and Grande 2007, Simões and Pyron 2021, Sprackland 1991, 1995), along with my analyses and comments on the characters. Chapter 4 covers the broad range of external anatomy and variation. Because my research has focused largely on the variation of phenotypes within this taxon, I give considerable space to those topics. However, I have not tried to rewrite the volumes by Mertens (1942a–c) nor give a point-by-point comparison between varanid variation and the large character suites given by Estes (1983) or Gauthier et al. (2012). What I present is information either not presented elsewhere or that needs revision. The discussion of external anatomy follows on my earlier work with the *Varanus prasinus* (1991) and *Varanus indicus* complexes (1995). To those works I have herein added new characters and new analyses of the characters, notably as they pertain to ontogeny, sex, and coloration. My review of cranial anatomy is based in small part on articulated skulls, a subject thoroughly covered in Mertens's second volume (1942b) in which he gave species-by-species descriptions of a small suite of skeletal characters that used a set of measurement ratios (Mertens 1942b:122–123). My use of ratios is minimal. Instead, I present, where available, taxon-specific characters, and most of my data come from "exploded" – that is, isolated bones not attached to others – cranial material, so my descriptions will emphasize three-dimensional morphology and its variation or uniformity.

Because the post-cranial skeleton of varanoids has been very well covered elsewhere (Bochaton et al. 2019, Cieri 2018, Hoffstetter and Gasc 1969, Islam 2023, Muzammil 1949), I restrict my review to those elements that display considerable variation or that have an important bearing on evolution within the family.

My coverage of viscera is less extensive, and few publications include a specific section on varanid internal organs. Literature on several aspects of individual organs is widely scattered in journals (e.g., Ali et al. 1982 on bile; Bartholomew and Tucker 1964, Burggren and Johansen 1982, Galli et al. 2006, Hanemaaijer et al. 2019, and White 1969, on the heart; Becker 1991, Becker et al. 1989 on lung morphology; Yeager 1973 on liver vascularization; Khoirunnisa 2020 on the kidney; and Hoffmann 1890 for an excellent anatomical overview). Even so, there is not much prior information on any single organ or system, so my presentation focuses strictly on varanid-specific features of soft anatomy.

Part III is a blend of topics that can be summarized as "**natural history**," subjects that meld very smoothly one with another: where do monitors live, why can they live there, how do they live there, and how do they keep their populations going? In more formal terms, that covers distribution, ecology and physiology, behavior, and reproduction, and because those are some of the best studied aspects of varanid biology, my section is more an overview of the research than a contribution of new material than in other sections. Part III is essential for understanding **Part IV**, which looks at the **origin and evolution of monitors**, going back into paleontological time. Of course, from those earliest varanids came many changes that represent both their physical and geographical evolution. That eons-long journey resulted in some taxa staying in place instead of spreading across the globe. Some stayed put until extinction; others dispersed and founded successful populations with unique lineages in places such as Africa and Australia. Understanding that story is important to developing a full systematic account of monitors. I conclude by reviewing the varanid higher taxonomy, relationships, and their current global status as survivors in the Anthropocene.

1.3 MATERIALS AND METHODS

Because my research notes for this book go back to 1969, it is important to describe where and how the studies were conducted. I have studied monitors in every kind of place that has them. I examined every *Varanus* held in the respective museum collections of 11 nations as of the date I was there (see Appendix I: List of Institutions), numbering several thousand wet, dry, and skeletal materials. I was able to make field observations during three visits to Australia and two to Papua New Guinea. Most live lizards were examined in zoos and private collections, and members of the private sector often made important donations of deceased monitors, skulls, and shed skins for my research. Fifty-five years of varanid research have been influenced by different techniques and individual project goals, leading to some inconsistencies in data collection methods. For example, at the point I became interested in varanids' ability to stay submerged, I had not collected similar data for other taxa during the prior decades. Although my early work was among the first to use "total evidence" data analysis (Sprackland 1991a, b, 1995b), I did not incorporate genetic/molecular information, and so the molecular, microcomplement, and chromosome information I did use was all based on the published literature. I have personally studied behavior, morphology, taxonomy, and zoogeography extensively, and, to a lesser extent, physiology. The illustrations came in several different formats, from rough sketches in my notebooks to photographs of varying quality to video frame-grabs. Photos came from slides, negatives, and iPhone cameras. Many photos were converted into drawings for clarity, and in some cases, I present a photo and drawing together to highlight certain features of a specimen. I have striven to make the illustrations clear and comprehensible. Except where noted, all illustrations are my original work, facilitated by use of camera lucida, Da Vinci Eye: AR Trace & Draw (Da Vinci Eye LLC), and ProCreate® (Savage Interactive Pty Ltd.) applications and saved as JPEGs.

Live and preserved lizards were examined for characters relating to scalation, body proportions, coloration and pattern, osteology, sex, and gross morphology. Only specimens with reliable locality data were included in phylogenetic analyses and assigned species identifications, and when possible, literature citations were cross-checked to ascertain identification by examining the actual specimen. Where specimens were listed in the literature but were unavailable for examination, I omitted them from the study, and only specimens that I personally examined were chosen for analyses and descriptions. All measurements are in millimeters, and I reexamined specimens at random to compare the objectivity of my interpretation of character states. Measurements were taken with a measuring tape, plastic rulers, and calipers.

States that were not unequivocally discrete were eliminated from some, but not all, analyses (e.g., differing numbers of supraocular scales). Sex was determined for lizards over 160-mm snout-vent length (SVL) by following the technique proposed by Arnold (1984), in which a longitudinal incision is made posterior to the cloaca and examination made for the presence (male) or absence (female) of the transversus penis muscle. In many cases, it was possible to examine lizards for gonads or hemipenes, and the presence of post-cloacal hemipenial bulges or spiny tufts helped confirm the sex of males. Juveniles (specimens of less than 160-mm SVL) are highly variable in coloration and pattern and may differ considerably from conspecific adults, for which reason they were not included in analyses with adults, though details of pattern and color variation of young are provided. Osteological nomenclature principally follows McDowell and Bogert (1954) and Evans (2008), with consideration of Bahl (1937), Fejérváry (1935), Gauthier et al. (2012), and Reippel (1980). I followed the recommendations of Wüster et al. (1992) in initially grouping samples into operational taxonomic units (OTUs), based on shared phenotypes, in preparing data sets for statistical analysis. Composite data for each phenotype and locality were used in analyses of species and species-group relationships.

Scale, bone, and organ structure were microscopically examined with a Nikon SMZ25 microscope and 16.25-megapixel DS-RI2 Color CMOS Camera with associated NIS Elements in the laboratory of the Division of Ichthyology at the National Museum of Natural History (USNM), Washington, DC. Images were saved in JPEG format. Additional examination was performed with a 5X hand lens and from images photographed with an iPhone 12. Scales were examined from intact skin (live and preserved specimens) and shed skins, and when possible, both exterior and interior views of skin were examined, and cross-sections of scales were made from intact skin using scalpels. Shed skin was first observed dry, followed by examination of samples immersed in water, the latter providing some amplification of microstructure. Skin samples were taken from *Varanus* and representatives of additional families for comparison. The specimens (sample size in parentheses) included: **Crotaphytidae:** *Crotaphytus collaris* (2); **Teiidae:** *Crocodilurus amazonicus* (2); *Dracaena guianensis* (3); **Anguidae:** *Abronia smithi* (1); *Abronia aurita* (4); **Anniellidae:** *Anniella pulchra* (2); **Helodermatidae:** *Heloderma horridum* (1); **Lanthanotidae:** *Lanthanotus borneensis* (6); **Varanidae:** *Varanus acanthurus* (2), *Varanus albigularis* (2); *Varanus auffenbergi* (3); *Varanus bogerti* (3); *Varanus caspius* (3); *Varanus caudolineatus* (2); *Varanus chlorostigma* (6), *Varanus cumingi* (4); *Varanus doreanus* (3); *Varanus dumerilii* (4); *Varanus exanthematicus* (4); *Varanus flavescens* (2); *Varanus flavirufus* (2); *Varanus giganteus* (2); *Varanus gilleni* (2); *Varanus glauerti* (5); *Varanus gouldii* (3); *Varanus griseus* (3); *V. indicus* (5); *Varanus jobiensis* (2); *Varanus keithhornei* (2); *Varanus komodoensis* (3); *Varanus kordensis* (3); *Varanus mertensi* (2); *Varanus mitchelli* (2); *Varanus niloticus* (4); *Varanus olivaceus* (3); *Varanus orientalis* (6); *Varanus panoptes* (5); *Varanus melinus* (3); *V. prasinus* (4); *Varanus primordius* (1); *Varanus rainerguentheri* (2); *Varanus rudicollis* (5); *Varanus salvadorii* (1); *Varanus salvator* (2); *Varanus spinulosus* (2); *Varanus storri* (4); *Varanus timorensis* (2); *Varanus togianus* (3); *Varanus tristis* (2); *Varanus varius* (2); *Varanus* sp. (4).

A list of complete specimens examined is given in Appendix II.

1.4 TAXONOMIC CONSIDERATIONS

For almost every formally named species of monitor, there are many synonyms, ranging from typographical errors (= lapsus), authors unaware that a species had already been named (junior synonyms), changes due to priority, and other causes. Most of the species have well-supported scientific names and are much less likely to be changed than others. However, it is not always the long-used familiar names that will remain stable, and I give a few examples to show how taxonomic confusion might result.

Southeast Asia is home to monitors that are second only to the Komodo Dragon in size and have long been known as *V. salvator*, the Water Monitors. For over a century, the name *V. salvator* had been associated with this large species, with populations between Sri Lanka and the Andaman Islands to the East Indies and north into the Philippines divided into subspecies, such as *Varanus s. nuchalis* and *Varanus s. bivitattus*. In 2007, a team of herpetologists reviewed specimens of water monitors from 41 localities and elevated several subspecies to species status (as they had originally been named). The absence of a holotype for *V. s. salvator*, a specimen illustrated by Seba (1735) but subsequently lost, left the question of which population should be designated as *V. salvator*. By comparing Seba's illustrations with known populations, Koch et al. (2007) established Sri Lanka as the type locality and designated a neotype from there. At present, then, the more familiar and widespread water monitor is known as *Varanus macromaculatus* (or as Koch et al. 2007 published, *V. salvator macromaculatus*). Taxonomically, that is a sound decision, but it presents some practical concerns. Prior to 2007, most reports about *V. salvator* may, in fact, have been based on entirely different species. Where locality data are available, the species can usually be fixed; water monitors from Java and Flores are classified as *V. bivitattus*. More often, though, the specimens lack such data, calling into question which species had been referenced. A formidable volume of literature – arguably the vast majority – about "*Varanus salvator*" must now be reevaluated, much of it applying to *V. macromaculatus* or *V. bivitattus*. A very similar case took place regarding *V. gouldii* and *V. panoptes* (International Commission of Zoological Nomenclature/ICZN 2000, Sprackland et al. 1997), the latter having originally been described as a subspecies of the former. This led to legal implications in the early 1990s, resulting in the need to petition the ICZN for nomenclatural clarity. The Commission designated a neotype (BMNH 1997.1, formerly WAM R131792, Commission of Zoological Nomenclature 2000) to resolve the potential status of *V. panoptes* as a junior synonym for *V. gouldii, and the two taxa are classified as distinct species, supported by morphological and ecological parameters.*

More complicated is the taxonomic situation regarding Pacific Mangrove Monitors, which have become known as the *V. indicus* complex. Prior to 1991, there was one, and by 2000, there were eight recognized species (Sprackland 1995, Table 1.1). Between 1999 and 2024, another 14 species were recognized (Böhme and Ziegler 1997, Weijola and Sweet 2010, Weijola et al. 2016, Ziegler et al. 1999), and I am aware of several more species awaiting formal taxonomic description. Three of the post-2000 species had been delegated to junior synonym status under *V. indicus* (*chlorostigma, douarrha, and tsukamotoi*). In part, this is the result of having a relatively small sample of specimens; just a few hundred spread among major natural history museums in eight countries. That may sound like a lot, but the *V. indicus* group has a tremendous distribution, from Halmahera east to the eastern Solomons, and Micronesia south to far northern Australia (a total area larger than the continent of Australia and its islands), and after 200 years of collecting, there had been very few specimens taken from any given location. When I was studying the *V. indicus* group for my doctoral thesis, the number of specimens with locality data (N=186, sometimes simply "western New Guinea") across several museums in eight countries tallied *V.* "*indicus*" 153, *V. doreanus* 12, *V. jobiensis* 18, and *Varanus finschi* 3. For the latter three species, there were 2, 3, and 1 localities, respectively. As for the "*indicus*" specimens, multivariate analyses split them into several groups across 28 localities, but without more specimens, I did not think any further resolution would be adequately supported. I left *V. indicus* as a metaspecies (*Varanus "indicus"**), which roughly translates into "*V. indicus et alia*." Later work by researchers who sampled live animals at several localities did, in fact, resurrect several species from "*indicus*" (Weijola 2010, Weijola and Sweet 2010, 2015, Weijola et al. 2016). These species, though morphologically cryptic, nonetheless have phenotypic differences that justify their species status (Table 1.1). They also differ behaviorally, as anyone who has worked with them can attest, and have an assortment of habitat preferences. The

TABLE 1.1
Recognized Species in the *Varanus indicus* Group, Subgenus *Euprepiosaurus*, as of March 2025

1. ***Varanus* sp. nov.** (presently being described)	12. *Varanus juxtindicus*
2. *Varanus bennetti*	13. *Varanus lirungensis*
3. ***Varanus caerulivirens***	14. *Varanus louisiadensis*
4. ***Varanus cerambonensis*** (now a jr. syn. of *Varanus indicus*)	15. ***Varanus melinus***
5. *Varanus chlorostigma**	16. *Varanus obor*
6. *Varanus colei*	17. *Varanus rainerguentheri*
7. ***Varanus doreanus***	18. *Varanus semotus*
8. *Varanus douarrha**	19. *Varanus tanimbar*
9. ***Varanus finschi***	20. *Varanus tsukamotoi**
10. ***Varanus indicus***	21. ***Varanus yuwonoi***
11. ***Varanus jobiensis***	22. *Varanus zugorum*

Note: **Additional species are known that are awaiting description. Species in bold were recognized prior to 2000. Those with an asterisk (*) had been in junior synonymy to V. indicus until after 2010.**

taxonomic clarity came with literature confusion. While some earlier studies gave locality data that allowed us to know which of these species was involved (Hediger 1934, Rodda et al. 1991, Uchida 1966, 1967, Wikramanayake and Dryden 1988, Wiles et al. 1989), many others did not. Was an author examining the eastern *V. finschi* or the far western *V. indicus* proper? Unfortunately, this leaves some important data in a limbo in which information and species cannot be linked. Taxonomic issues are covered in more detail in Chapters 10 and 11.

Due to the innumerable cases of multiple synonyms, incorrect and missing locality data, I had to establish a consistent way of presenting information to be accurate and not further muddy the waters. My six criteria are:

1. Names used are based on the most recent taxonomy at the time of this writing.
2. Where taxonomy has changed more than twice since 2000, I have placed the name used in a paper in parentheses following the current usage, e.g., *V. chlorostigma* (= *V. indicus*). If I choose to make an exception, my reasons will be presented following the name in parentheses.
3. Where I cannot confidently align an old published name with current taxonomy, I use the name given to the species in the publication. Where there is serious doubt that the published name is accurate, I added (?) after the name.
4. Varanids have long been divided into subgenera (Mertens 1942a). The membership in subgenera has also been quite fluid, perhaps most stable in the subgenus *Odatria*, and the category is without doubt useful for taxonomists and ecologists. I chose to use the designation only when discussing particular taxonomic issues, with the following exception:
5. The subgenus *Odatria* has been used in such a stable manner for so long and refers to so many species that I often use it alone when discussing the small varanids of Australia, along with the term odatrian.
6. I do not use subspecific epithets except when recognizing a subspecies at the species level. There are sound arguments for and against using that designation, but it is sufficiently common to see subspecies elevated to species rank that I take the route taxonomists call "a splitter" (see Burbrink 2001, Burbrink et al. 2000, 2022).

2 The Herpetologists and the History of Monitor Lizard Research

2.1 INTRODUCTION

The large monitors would always have been conspicuous, but they never became the objects of concern as did crocodiles, large mammals, and edible birds, so consequently there had been little written about these marvelous lizards for most of human history. It wasn't until the discovery by Europeans (for local peoples had long known of the lizards) of the great dragons of Komodo and nearby islands in the early 20th century that any special interest in the varanids was sparked (Burden 1927, Dunn and Burden 1927, Pfeffer 1965). The dragons themselves were amazing (Auffenberg 1981), much larger than any other living lizards, reportedly and incorrectly claimed to reach 15 ft in length (de Rooij 1915); what could their relatives tell us about them? Were there other, undiscovered dragons out there? Prior to discovery on Komodo, monitors were mainly important to some local peoples (still so, as food, leather, medicine, and "messengers"), and a few scientists who had a convenient large lizard upon which they could write monographs and comparative anatomy texts (Bahl 1937, Bellairs 1949, Benedict 1932, Brongersma 1958, De Jong and Brongersma 1927, Fejérváry 1918, 1935, Nopcsa 1903a, b). Papers about varanids consisted of those few, though often detailed, monospecific anatomical accounts of the largest species, descriptions of new species, or inclusions on lists of species encountered during the years that were the pinnacle of western exploration of the Earth (Table 2.1). A collection of all the dedicated publications from before 1970 barely spreads across more than 180 mm of my bookshelf. At the time I began writing this manuscript (2019), each of two bookshelves was longer than an adult Komodo Dragon (Fig. 1.1).

2.2 THE BEGINNING OF VARANID RESEARCH

The first varanid named according to Linnaean taxonomy was the Nile monitor (originally published as *Lacerta monitor* by **Carolus Linnaeus** (1707–1778) himself, in 1758, but changed to *L. nilotica* in 1766). Nine other species were subsequently named between 1758 and 1820 (*salvator* [1768], *varia* [1790], *exanthematica* [1792], *albigularis* [1802], *indicus* [1802], *bengalensis* [1802], *griseus* [1803], *ornatus* [1803], and *bivittatus* [1820]) (Table 2.1). Fourteen researchers have named three or more species representing 71 of presently named monitors (Table 2.2). It was **Blasius Merrem** (1761–1824) who proposed the genus name *Varanus* in 1820, and the name has been in use ever since but has not always been the only name used. All ten of these pre-1821 species remain valid today (with minor amendments to the names to bring them into the proper gender, e.g., *varia* became *varius*) and represent lizards that exceed 3 ft total adult length. It wasn't until 1831 that John Gray (see below) named the first of the "dwarf" monitors, *Varanus timorensis*. As of 2023, 90 species and subspecies had been described and named, ranging in size from Australia's dwarf, 8-in *Odatria brevicauda* and recently described *Odatria sparnus* (Doughty et al. 2014), to the 10-ft Indonesian Komodo Dragon, *Varanus komodoensis*. Several other species from Asia and Australasia are currently awaiting formal description. As for the genus names used after 1820, see the section on John Gray below.

The number of monitor species gives us several excellent opportunities to observe evolution, adaptation, ethology, and behavior in comparative ways not often available in other animal groups. It is perhaps the relatively "conservative" body plan of these animals, despite wide-ranging and species-specific habitat preferences, that make them so attractive to comparative zoologists. The detailed study of monitors exposes a formerly ignored breadth of variation that, though equaled or excelled in many animal groups, is nearly unique in that it stays within very well-circumscribed limits. Where an observer might never mistake a savannah or emerald monitor for anything but a monitor, he would be hard pressed to imagine the former (a stocky ground dweller) in the habitat of the latter (a slim, graceful tree dweller). This discrepancy between relative sameness and distinct difference is a major cause of the attention they have received in recent years (Böhme 1988, Gray et al. 2019, Green and King 1993, Sprackland 2009). As different as some taxa may be from each other, an observer might imagine that among similar species the variation is minimal. This would be an error. Even within the various "complexes" (a term that has largely been synonymous with "subgenera," prior to subgeneric name publication) of seemingly near-identical species, the number of characters that display notable variation is extensive. These groups and their diagnostic ambiguities are covered in Chapters 4, 10, and 11.

The rest of this chapter reviews the people who made major contributions to our understanding of the Varanidae. Additional biographical information may be found in the works of Adler (2014) and Adler et al. (2007, 2012).

2.3 HISTORICAL PALEONTOLOGICAL STUDIES

The possible links between varanoids, mosasaurs, and snakes were presented by American zoologist **Edward Drinker Cope** (1840–1897), who divided those reptiles into taxa he named Pythonomorpha and Streptosauria

DOI: 10.1201/9781003463528-3

TABLE 2.1
Graph Depicting the Number of Species of *Varanus* Described by Year, Beginning in 1762 Shortly after Linnaeus' 10th Edition of *Systema Naturae* Was Published, through 2016

The recognition and describing of monitor lizards follow historical lines that I dub the Linnaean, Mertensian, and fieldwork. The Linnaean begins with the acceptance and use of Linnaeus' taxonomic plan that almost exclusively utilized specimens based on published illustrations (primarily from Seba), specimens preserved in alcohol, and stuffed skins (see Fig. 2.7). The Mertensian covers a period with larger numbers of alcoholic specimens, most of which had been discovered in jars on museum shelves. Unlike the Linnaean period, the Mertensian presented much more detailed and taxonomically useful descriptions that included designated holotypes, and describers who had done no personal collecting of specimens. Both periods shared having the majority of specimens having been collected decades prior to description and the specimens were in museums and private cabinets of curiosity. The fieldwork period contains two overlapping trends, the commercial and ecological activities. Commercial collectors have discovered many new species, either via specimens or information where taxonomists could go collect them, an ongoing process that has contributed in large part to why so many taxa have been named during this time. Fieldworkers from scientific institutions who were conducting (usually) ecological or physiological studies in the field discovered and collected specimens, particularly in Australia. In both cases specimens were much more recently collected and better preserved than in earlier periods, and the descriptions typically include information about the species' life history.

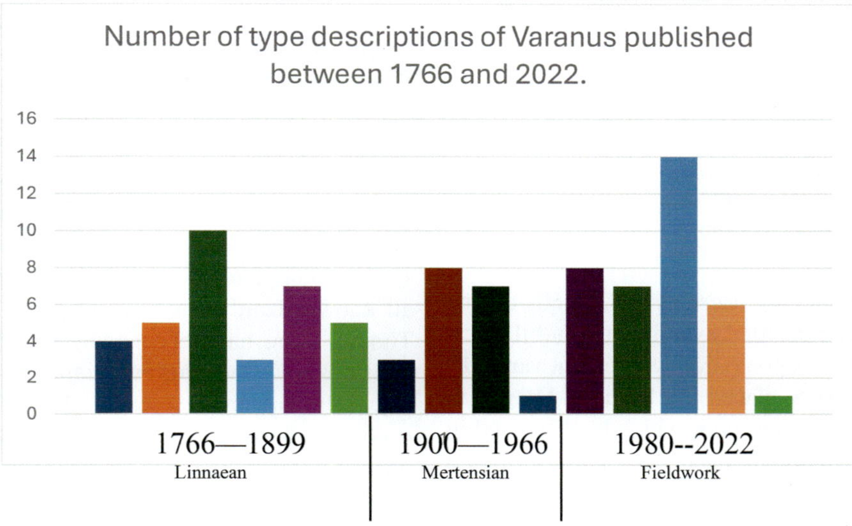

Note: Most species named since 1932 are from Oceania–Australia (34), followed by the Philippines (8), and western Asia (2). Excepting the Komodo Dragon discovered in 1912 in Indonesia, the last varanid species to be described outside those regions was Varanus nebulosus in 1831, from the Indian subcontinent and adjacent countries.

TABLE 2.2
The 14 Authors Who Have Described Three or More Species of *Varanus*

Describer and Date of Publication	Number of Species Described
Daudin (1802–1803)	5
Gray (1838–1845)	6
Schlegel (1839–1844)	4
Peters (1869–1878)	3
Boulenger (1885–1898)	3
Mertens (1841–1966)	9
Deraniyagala (1944–1947)	4
Storr (1980)	4
Böhme (1989–2025)	14
Ziegler (1994–2007)	8
Gaulke (2001–2010)	3
Philipp (1999–2007)	3
Koch (2009–2025)	6
Welton (2010–2014)	3

(Cope 1869) (Figs. 2.1 and 2.2). Cope was predominantly known as a dinosaur paleontologist whose 1,400-plus writings encompass herpetology, ichthyology, and theoretical biology, but little to do with fossil lizards. The first notable works concerned primarily with fossil varanids were written by paleontologists, beginning with the Hungarian **Baron Franz Nopcsa von Felső-Szilvás** (1877–1933, Fig. 2.1A). Nopcsa was interested in dinosaurs and other large extinct reptiles, notably mosasaurs. He was one of the first to note the probable relationship between mosasaurs, aigialosaurs, dolichosaurs, and extant varanids and wrote several important papers (Nopcsa 1903a, b, 1912, 1923) on the subject. His work influenced another Hungarian zoologist **Baron Géza Fejérváry de Komlóskeresztes** (1833–1914, Fig. 2.1B), cited as "Fejérváry," a paleontologist who, in his colorful career, was a Hungarian army general and later prime minister. Fejérváry produced two important monographs on varanids (1918, 1935), comparing available fossils with extant species. However, he used a limited number of specimens, representing the largest species, and got some

FIGURE 2.1 The early students of varanids. (A) Baron Franz Nopcsa von Felső-Szilvás (1877–1933). (B) Baron Géza Fejérváry de Komlóskeresztes (1833–1914). (C) Bernard-Germain-Étienne de La Ville-sur-Illon, comte de Lacépède (1756–1825). (D) André Marie Constant Duméril (1774–1860). (E) Gabriel Bibron (1805–1848). (F) John Edward Gray (1800–1875).

of his information wrong (see Mertens 1942a). His contributions include the first detailed examination and illustration of osteology of fossils compared with extant varanids. In a 1927 paper, Fejérváry placed the Komodo Dragon in the monotypic genus *Placovaranus*, which he discussed in detail in his 1935 posthumous paper, and he described a new fossil varanid, *Varanus Bolkay* (sic). He also named

FIGURE 2.2 The 20th-century students of varanids. (A) George Albert Boulenger (1858–1937). (B) Charles Lewis Camp (1893–1975). (C) Charles W. Gilmore (1874–1945). (D) Robert Friedrich Mertens (1894–1975). (E) Charles Mitchill Bogert (1908–1992). (F) Samuel Booker McDowell, Jr. (1928–2014). (G) Walter Auffenberg (1928–2004).

three distinct morphotypes based on the angle and shape of the dorsal surface of the maxilla, which he termed hypsiprosopic, platyprosopic, and mesoprosopic. Though useful in describing a complex feature, they are of limited taxonomic use, having been eschewed by Mertens (1942a).

Two American paleoherpetologists made major phylogenetic advances by routinely including all lizards, living and extinct, in their analyses. **Charles Lewis Camp** (1893–1975, Fig. 2.2B) published the cornerstone work for lizard relationships that became one of the earliest robust phylogenies of lizards. Most of his characters were osteological, but he included many features of musculature for living taxa (Camp 1923). The classifications of Boulenger (see below) and Camp would serve as the baseline model for lizard phylogenetics until the 1950s. Camp's analysis of the monitors was brief but important as a foundation for later phylogenetic studies.

Leading the charge of those later studies was **Richard Estes** (1932–1990), who employed the cladistic methods that were, at the time, the predominant way to study phylogeny. He was, appropriately, an editor and contributor to a volume on lizard phylogenetics published as a commemoration of Camp (Estes et al. 1988). Estes used paleontology to understand the evolution of lizard distributions (Estes 1983b, 1984), and he published a comprehensive review of all known lizard taxa down to genus level (Estes 1983a). Though there have been many new taxa and taxonomic revisions since then, his encyclopedia stands as an essential adjunct to squamate research. **Jack Conrad** (1976–2016) was a paleontologist who made several important contributions to varanid evolution studies (e.g., Conrad 2008, 2015, Conrad et al. 2006, 2008, 2010).

2.4 MODERN HERPETOLOGICAL STUDIES OF EXTANT VARANIDS

Linnaeus and his students of the late 18th century did little more than name species, many of which were based on illustrations in Seba's *Thesaurus*, a set of volumes with drawings of Seba's massive personal collection (Seba 1735–1765, Thireau et al. 1998) (Figs. 2.1–2.7, Table 2.1). It wasn't until the very end of the 18th and the beginning of the 19th centuries that there evolved literature of a descriptive nature. Animals already named, plus new species were given morphological and geographic descriptions. The most important volumes of this early literature were written by French zoologists, and it can be argued that modern herpetology began in France in the first decade of the 1800s (Sprackland 1995c). **François Marie Daudin** (1776–1803, no known illustration) wrote his eight-volume *Histoire naturelle, générale et particulière des reptiles* (Daudin 1802), with its many descriptions of new species, is an essential reference for herpetological taxonomy. Daudin provided lengthy accounts of ten *Varanus* (as *Tupinambis*), five of which were original descriptions (Fig. 2.3 and Tables 2.1 and 2.2). Daudin used the collection of the new National Museum of Natural History in Paris, where specimens were not typically accompanied by detailed collecting data, or data were

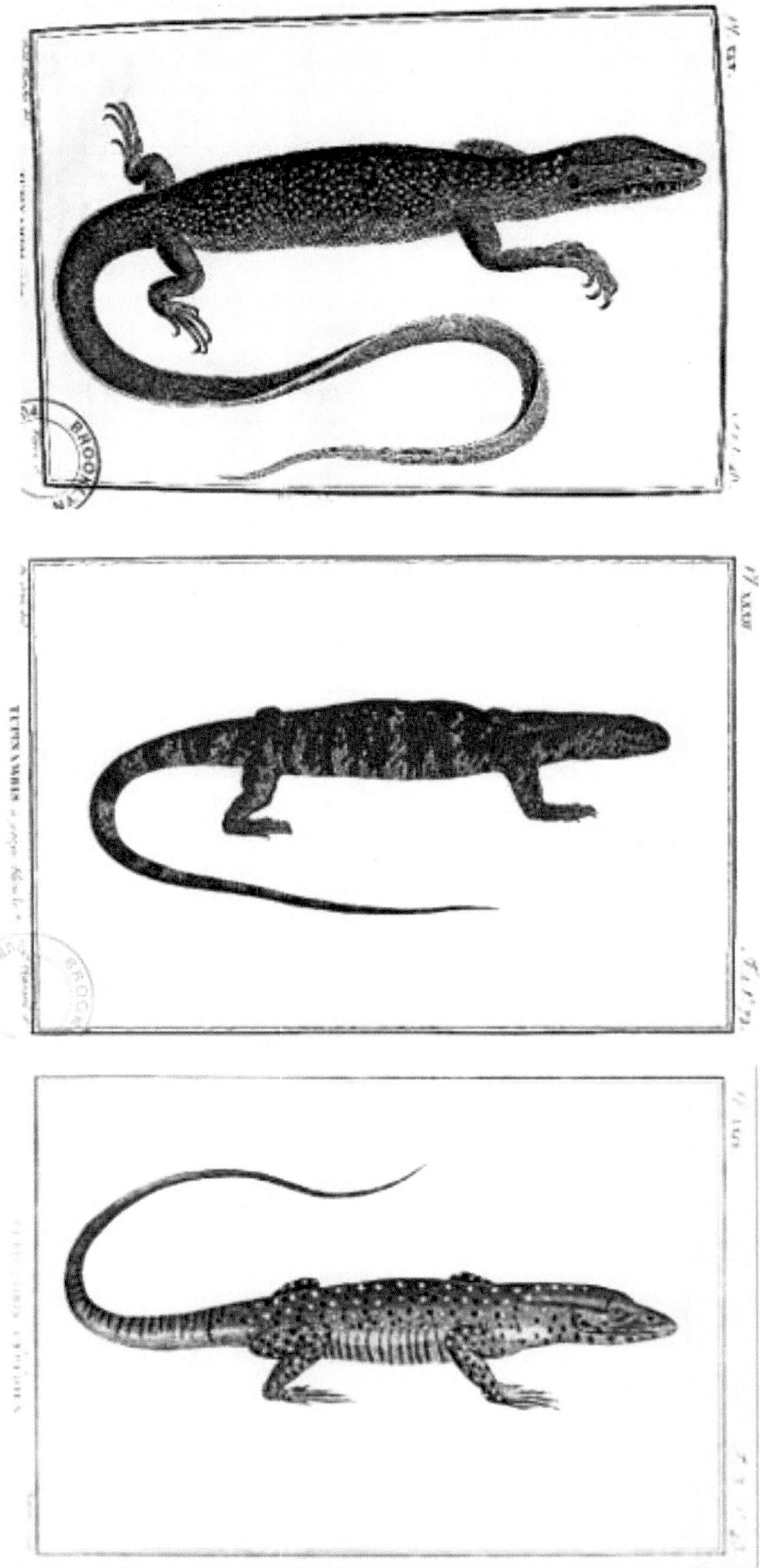

FIGURE 2.3 Three of the illustrations from Daudin (1802), by his wife, Adel, depicting species that he described. From top to bottom: *Tupinambis indicus* (= *Varanus indicus*), *Tupinambis albigularis* (= *Varanus albigularis*), *Tupinambis cepedianus* (= *Varanus bengalensis*).

incorrect or just missing (Thireau et al. 1998). Daudin's earlier contemporary, **Bernard-Germain-Étienne de La Ville-sur-Illon, comte de Lacépède** (1756–1825, Fig. 2.1C), generally known simply as Lacépède, also produced a multi-volume work, *Histoire naturelle des quadrupèdes ovipares et des serpens* (Lacépède 1788, 1789). Though similar in format and scope to Daudin's work, Lacépède provided fewer species accounts, no new species descriptions, and his writing focused more on the natural history than had Daudin. Neither author provided illustrations of taxonomically useful quality. The first important work that provided descriptions and a detailed classification of varanids was that of **André Marie Constant Duméril** (1774–1860, Fig. 2.1D) and **Gabriel Bibron** (1805–1848, Fig. 2.1E) in their nine-volume global herpetological review, *Erpétologie Générale* (1834–1854), though it allied distantly related species based on incomplete character analyses. In total, they recognized 12 species, bringing all together into a single genus, *Varanus*, for the first time. Included in their account was the first phylogenetic tree for the family (Duméril and Bibron 1836:470; Fig. 2.4), based simply on a small number of morphological characters.

As French herpetology was ascending, the British were gradually establishing their own zoological collections (Sprackland 1995b). Prolific British zoologist **John Edward Gray** (1800–1875, Fig. 2.1F) published the first catalogue of lizards in the British Museum (Gray 1845:6–13), a small volume with sparse descriptions of specimens in the collection. Grey recognized 22 species of monitors, seven of which were new, and seven new monitor genera (including "*Uaranus,*" likely a typographical error, or lapsus, for Merrem's *Varanus*). Gray was enthusiastic, but rejected a single genus, erecting seven new genera. (Arguably he only established six, because his *Uaranus*, which he attributes to Merrem (1820), must be an error in transcription for Merrems' *Varanus*.) It was largely from Gray's generic names that Mertens would get some of his subgeneric names, e.g., *Psammosaurus*, *Empagusia*, and *Odatria*. One of Gray's successors at the British Museum, the Belgian **George Albert Boulenger** (1858–1937, Fig. 2.2A) wrote the second, and three volume, edition of the *Catalogue of the Lizards* (1885), following his French predecessors, Duméril and Bibron, both in overall writing style and in his varanid classification. He used the single genus *Varanus*, relegated Gray's names to junior synonyms, recognized 27 valid species and added a new dwarf Australian species. His *Catalogue* (often referred to by herpetologists as "THE Catalogue") was a book of descriptions in systematic order and he introduced the practice (usually) of listing specimens in the museum and assigning accession numbers, practices that did not become standard until the mid-20th century. The number of named lizard species has roughly quadrupled since 1885, but Boulenger's *Catalogue* is still considered an essential inclusion in any herpetological taxonomy library.

2.5 ROBERT F. MERTENS

Unquestionably the most influential and prolific authority on varanids in the 20th century was **Robert Friedrich Mertens** (1894–1975, Fig. 2.2D), long-time curator and

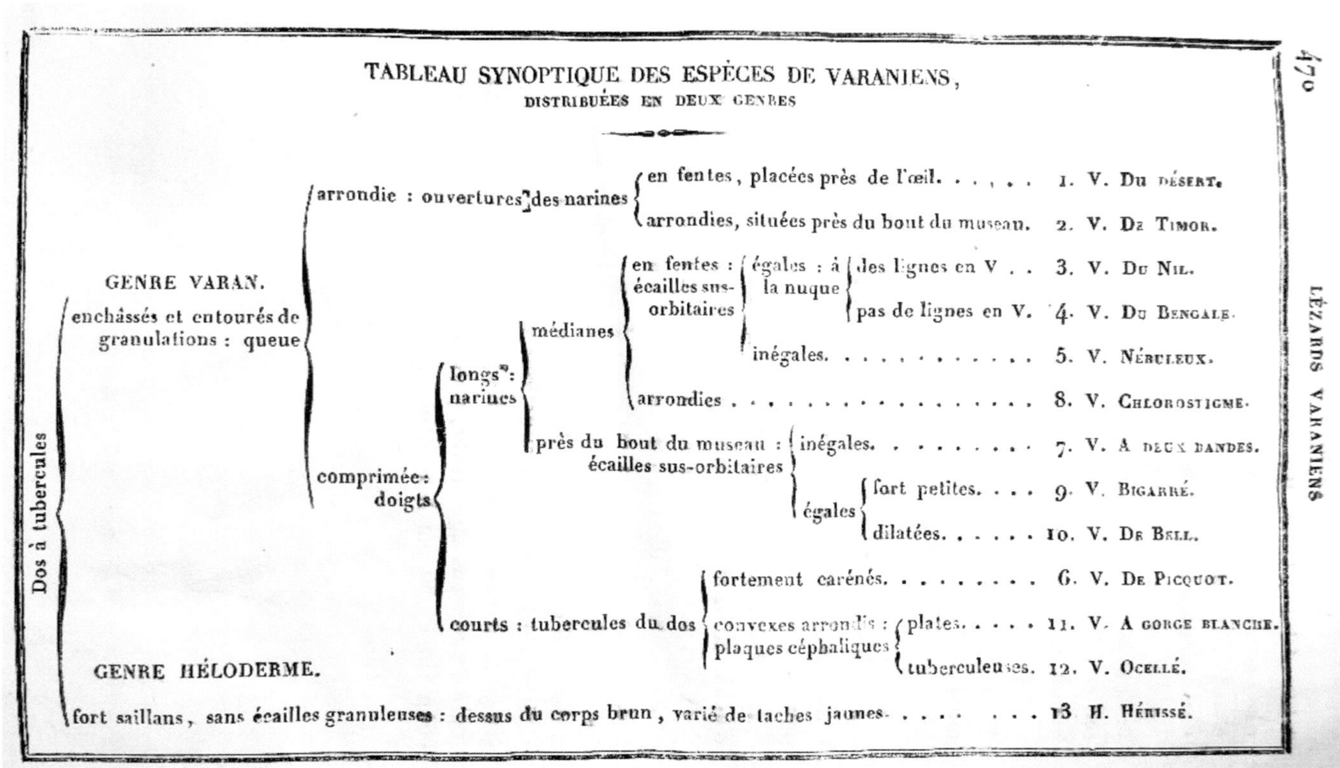

FIGURE 2.4 The phylogeny of varanids as presented by (Duméril and Bibron 1836:470). (Photographed from the author's library.)

director of Frankfurt's Senckenberg Museum, who produced taxonomic accounts of varanids between 1926 and his death in 1975. Incredibly, Mertens produced a major review and classification of varanids during World War II (Mertens 19421–c) in a monumental three-volume monograph, *Die Familie der Warane*. It remains an extensive overview, with descriptions and taxonomy of the entire family as it was then known. Supporting his authoritative text were numerous photos of preserved specimens and their skulls for almost every known species (by Hannes Rosenberg, for whom Mertens would later name the Australian *Varanus rosenbergi*). Mertens established eight subgenera from what would later also known as "species complexes" but maintained a single extant genus. Eventually, he increased his subgenera to ten, all but two (*Varanus* and *Odatria*) being monotypic. Departing from the practice of earlier herpetologists, Mertens named no new species in his monograph, though he named 12 species during his pre- and post-war years (Mertens 1926, 1941, 1958, 1959). His work was limited by access only to specimens in German, Austrian, and Swiss museums but was later greatly expanded in a series of emendations (e.g., Mertens 1951, 1958, 1959). After the war he visited major varanid collections at museums in Australia, Britain, and the United States. His first contribution on new Australian monitors (Mertens 1958) would be his largest, followed by papers describing new species from Australian and American museums, plus specimens that overseas colleagues sent to him in Germany. He named new species of monitors for Australian curators who had given him access to their collections, including Ludwig Glauert (*Varanus glauerti*) and Glenn Storr (*Varanus storri*); in turn, Glauert named a large aquatic monitor for Mertens (*Varanus mertensi*). He and his contemporaries went on to name numerous new species after the war, and by the time of Mertens' death in 1975, the number of named taxa had nearly tripled, and it is still growing. Mertens' original volumes went out of print several decades ago, the first volume being particularly difficult to find outside libraries, but the entire set was re-issued, with annotations and some updates, in a single volume 22 years after Mertens' death (Mertens 1997).

Mertens, however, never dealt with either fossil varanids or the larger issue of varanid phylogeny.

It is worth a brief description of the monograph for four reasons. First, it remains an excellent resource because it presents clear and useful information about the morphology of all of the species Mertens had available, especially his documentation of meristic characters. For most species the data remain relevant for modern researchers and remain the starting point for much of ongoing varanid studies. Second, Mertens presents the collected data for every specimen that has it, though too many are "*ohne fundort*" (= without data). The lack of such data is not Mertens' fault for he merely reported whatever information had been supplied, often decades earlier, by collectors. It is invaluable distribution information in our times of widespread species decline. Third, the information about skulls is also comprehensive, representing nearly every species he includes in the monograph, far more complete than all prior descriptions combined, and presenting the Australian dwarf species for the first time. The descriptions of the exterior and those parts clearly visible without an "exploded" skull are very pertinent to neontologists and paleontologists equally. Fourth, the monograph includes several topics, even if briefly, that had barely been addressed earlier. Notable among these are taxonomic history, biogeography, hypotheses linking skull form and function, and consideration of the actual variety of the genus to establish subgenera.

The first volume of *Die Familie der Warane*, "*Allgemeines*," presents a broad overview of extant *Varanus*, including a history of the literature, especially the taxonomic work. The six chapters include (1) history and nomenclature, (2) general morphology, (3) ecology, (4) distribution, (5) phylogenetic remarks, and (6) evolution remarks. Prior to the 1950s, most ecological studies or fieldwork with varanids had been limited to a small number of large species (Burden 1927a, b, Cowles 1930, Dunn and Burden 1927, Hediger 1934, Stirling 1912), and so Mertens' review of ecology spanned only 14 pages. He presented a useful guide to the nomenclature used with varanids and extensive (though not complete) lists of synonyms, which remain pertinent for contemporary taxonomists. This volume contains 20 black-and-white plates, only the first three depicting live lizards. Plate 4 reproduces illustrations from a few early publications. Plates 5 through 12 show entire specimens, 13–15 heads, and 16 tails, scalation, cloacae, and toes. The remaining plates depict numerous varanid skulls, which represent a still unequaled collection (Fig. 2.5).

The second volume is entirely dedicated to monitor skulls. The overview gives brief descriptions of the individual bones and notes any examples that are exceptions to the rule. Teeth are described in two pages, followed by the effects of ontogeny on skull morphology. The longest chapter (53 pp.) provides short descriptions of the skulls of each species and subspecies, particularly regarding any species-level variations, and a variety of comparative proportional measurements. Because the observable range of variation depends on having large numbers of each species to examine, the sections on well-represented (and, generally, large) taxa get the most coverage. The final chapter provides Mertens' taxonomic and ecological results based on his examination of the skulls. The volume concludes with plates 21–34, all of skulls in different views. This volume remains the most extensive review of varanid skulls to date.

Volume three is Mertens' taxonomic overview of species and subspecies. In of the 12 chapters Mertens first establishes subgenera within *Varanus*, and species allocated to them. The presentation is a catalogue of extant Varanidae, providing synonymies, lists of specimens examined, and detailed physical descriptions of each taxon. The bibliography is just short of 11 pages, complete for its time and incredibly short by today's standards.

It was not until Mertens (1942a–1942c, 1959, 1963) that a serious attempt was made to examine intrageneric relationships among varanids. Mertens recognized that varanids are not as morphologically conservative as many

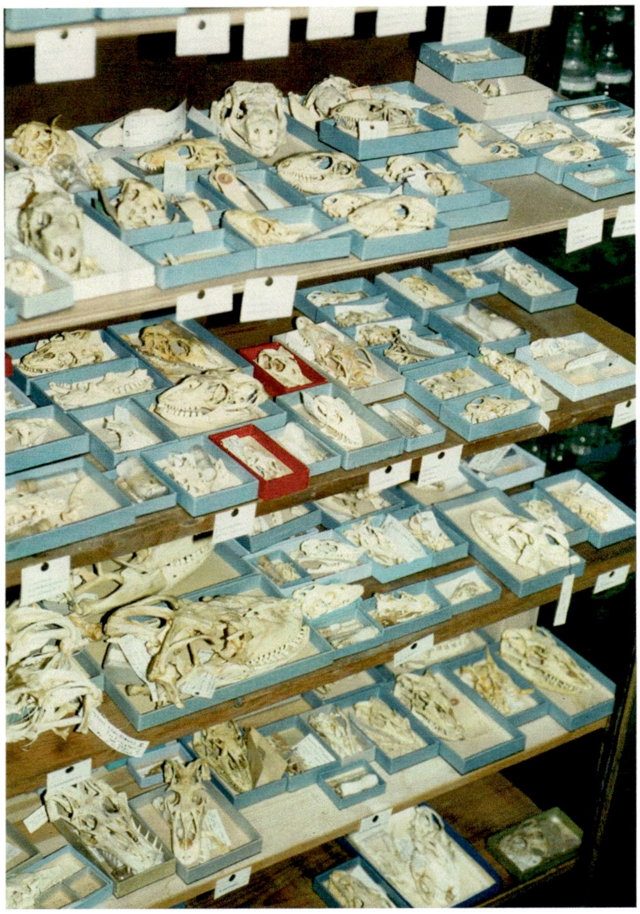

FIGURE 2.5 The collection of *Varanus* skulls used by Robert Mertens for his *Die Familie der Warane*. It remains housed in the herpetology department of Frankfurt's Senckenberg Museum.

herpetologists claimed and saw "natural" groupings that he termed subgenera. It is not clear whether Mertens used this category to sidestep the possible taxonomic confusion that erecting several genera might produce or to indicate that further resolution was necessary before such a step might be valuable. In light of the subgeneric shuffling that has taken place since 1942 (Böhme 1988, Dong et al. 2022, King and King 1975, King et al. 1991), Mertens' taxonomy served the interests of nomenclatural stability well at the alpha (= species descriptions) level. Estes (1983) extended Mertens' decision to retain *Varanus* by suggesting that either all varanids (including some fossil taxa) be combined in the genus *Varanus*, or all subgenera be elevated to generic rank. Though subgeneric designations are still widely used, their status and membership remain controversial (Dong et al. 2022, Sprackland 1991b).

Thus, in 391 pages, 33 plates, and numerous figures, Mertens produced a monumental and classic piece of morphological and taxonomic work that is frequently cited as a primary source of information. He had recognized 23 species of *Varanus* in his monograph (Mertens 1942a–c, table 2.3) though much of the taxonomy has changed. The fact that he managed to compile and publish such a formidable herpetological review in Germany during World War II is remarkable; Germany was frequently and increasingly being bombed, the zoological collections of the museum had been moved to a safer location out of town, and visiting museums was limited to those within the Third Reich and occupied countries.

Mertens' work requires significant updating because of our rapidly increasing knowledge of these lizards, and new methods available to examine herpetological specimens, both of which were unavailable to him. Some of these deficiencies include the following:

1. Mertens had limited access to specimens and literature from other countries during World War II. While comparing his work with actual museum specimens during 1991, I was aware that some specimens, though cited by Mertens, could not have been examined by him (e.g., it is doubtful he would really have confused *Varanus rudicollis* for *Varanus dumerilii*, or *Varanus varius* for *Varanus salvator*), and his inclusion of such specimen numbers in his publications may reflect examination of museum registers and earlier literature, rather than specimens. Mertens noted the extreme variability within "*Varanus indicus*" and claimed he could not tell the locality from which specimens came without recourse to labels (Mertens 1942c:262). Schmidt (1932) was unable to use Mertens' (1926) descriptions to identify two subspecies of the specimens of *Varanus indicus* collected from the Solomon Islands. In 1991, using the actual specimens, I could identify lizards to locality with greater than 90% accuracy (Sprackland 1995), leading me to believe that Schmidt had a limited number of specimens and filled in details from published accounts. Mertens later (1958, 1959, 1962, 1963) appended his three-volume monograph but never incorporated the use of diagnostic apomorphies, nor did he clarify relationships among cryptic taxa.
2. Though Mertens' literature review is extensive for the time, it is deficient in many places, a fact he acknowledged (Mertens 1942a). Additionally, cited works are not always listed in the bibliography, or references are incomplete (e.g., the citation for Kubary [1872] in text, though not especially important, was not in the literature cited). The war prevented Mertens from examining specimens outside occupied Europe, making the important holotypes in the British Museum (Natural History) out of reach, so his information was restricted to published data that did not always apply to the species, or specimens, in question. Even when he subsequently examined Australian and British museum collections, Mertens (1958, 1959) often did not have types available and his examinations were hurried, resulting in further errors and few

clarifications. It is also common to find literature accounts that represent two or more species (see Sprackland 1991b).

3. Mertens' taxonomic scheme is based on relatively few, predominantly plesiomorphic (= ancestral) or ecotypic (= environmentally versus genetically influenced) characters. Features that may be interpreted as convergent (e.g., compressed or round tails) are used as both primary diagnostics and as derived characters for assigning relationships. These characters are useful for discriminating among and diagnosing species groups ("subgenera") but often fail to separate similar species or record intraspecific variation; in some monitors tail shape varies depending on whether or not the lizard is obtaining sufficient food (pers. obs.). In particular, Mertens failed to associate nostril position and shape, important morphological characters, with ontogeny (Auffenberg 1994).
4. Mertens was limited in his use of living material (about 150 specimens, Mertens 1942a), causing him to omit features such as color in life, color pattern, male–male combat behavior, nest building/excavation, and other agonistic behaviors. His work far preceded the era of captive breeding, making the examination of ontogenetic changes that can be done today impossible in the early 1940s. In addition, species that have only recently become frequent imports for zoos and collectors, including *V. indicus* and its allies, were rarely, if ever, to be seen in captivity in the 1930s and 1940s. Of those species represented at the time in zoos, most were single specimens in "postage-stamp" collections, and of limited taxonomic usefulness. While obtaining extensive preserved samples showing ontogenetic change is possible, such series were rare in the 1940s and are still not widely available today. In at least one case, Mertens proposed a subspecies of mangrove monitor (*V. indicus rouxi*, Mertens 1926) based on juvenile coloration. As will be demonstrated below, juvenile color and pattern are the most variable intraspecific elements of mangrove monitor morphology. Additionally, even with such series, much basic taxonomic research must still be considered limited without access to live samples, and omission of color data can delay detection of cryptic taxa (Sprackland 1995).
5. For many taxa there were very few (N = 1–3) specimens in museum collections. In such cases, Mertens had only literature or single specimens from which to draw his conclusions. For example, where Mertens (1942c) had 105 specimens from the *indicus* group for his monograph, I had nearly 500 for my dissertation (Sprackland 1995).
6. Mertens reviewed the scant literature on fossil varanids but did little to incorporate fossil data into a larger scheme of varanid phylogeny. Much of his terminology was taken from Fejérváry (1935), whose paleontological work is in grammatically convoluted English, thus confusing many points as they emerge from Hungarian to "English" to German. Errors are more likely to occur as information goes through multiple translations. Further compounding this problem is Fejérváry's limited use of comparative material from extant varanids, making many of his conclusions erroneous or ambiguous. For example, as noted by Mertens (1942a), Fejérváry (1935) placed *V. komodoensis* into a monotypic genus *Placovaranus* because he believed the osteoderms of *komodoensis* were unique among "varanians"; they are not. The net result for Mertens was dismissing most of Fejérváry's observations on extant varanids and further omitting much of the fossil data of Fejérváry and prior paleontologists (Cope 1869, Gilmore 1922, 1928, Nopcsa 1903a, b, 1912, 1923). Estes (1983) noted that the differences between living and extinct varanids are either so minor as to warrant either including all post-Mesozoic taxa in a single genus *Varanus*, or significant enough to elevate all living subgenera to genera. This concern is a valid issue from both the taxonomic and biodiversity perspectives and is worth further consideration.
7. Given both the similarities among, and differences within the mangrove monitor complex, for example, Mertens may have been misled in diagnosing valid taxa by including juveniles in his sample (see point 4) and assigning them subspecific status. Ontogenetic changes are extreme within many monitors and could easily cloud morphological patterns until populations of the same species are studied. For example, the relative position of the nostril is more posterior in juveniles than adults for *Varanus jobiensis, Varanus doreanus*, and some *V. indicus*. Not surprisingly, Mertens named (1926) and removed to junior synonymy (1942c, 1963) his own subspecies *rouxi*, which was based on a juvenile from the Aru Islands. Nor is this juvenile pattern variation peculiar to Pacific varanids; the widespread African *Varanus albigularis* has at least five distinct juvenile color morphs but only two relatively similar adult patterns (Bayless and Sprackland 2000a, b).
8. Taxonomic analysis now uses many approaches that were simply unavailable in Mertens' time, including computer programs capable of dealing with large numbers of taxa and characters, molecular characters, and zoogeographical theory well founded in plate tectonics. Mertens was prevented from discovering the diversity and relationships among monitors for the same reason that genetic principles eluded so many workers before Gregor Mendel: he tried to find reliable and predictable

patterns from too many characters at one time. Though Mertens (1942c) seemed close to accepting the proposition that he might be dealing with sympatric, cryptic (= sibling) species, he did not ultimately arrive at that conclusion (Mertens 1942c:261–262). Perhaps the most significant systematic advance since 1942 was the development of Hennigian phylogenetics, or cladistics (Ax 1987, De Queiroz 2007, Forey 1992, Hennig 1967, Wiley 1980). While cladistic methods are not a panacea for discovering relationships or species boundaries, they do provide an objective analysis of multiple characters that was almost impossible prior to the development of phylogenetic computer programs, if only for the vast number of characters and their character states that can be used. Additionally, systematists now have complex statistical programs that can sort huge clusters of data into species groups. Neither of these approaches was available to Mertens.

Mertens remained interested in monitors until the end of his life and revised his monograph by producing a few additional papers describing new taxa (Mertens, 1951, 1958, 1959), including new subgenera (Mertens 1958, 1962). However, he never addressed the broader questions of evolution, systematics, or variation within the family in depth. He skirted the question of relationships by deferring to the need for paleontological data, while leaving the only substantial paleontological contributions of the day, with its shortcomings, to Fejérváry (1935).

Though he was unquestionably an academic herpetologist, Mertens both recognized and supported the contributions to herpetology made by "amateurs," people not employed in academia. He helped build a strong bridge between the groups in the form of the *Deutsche Gesellschaft fur Herpetologie und Terrarienkunde* (DGHT), which still publishes the respected journal *Salamandra*, and the special subject sister publication *Mertensiella* (see below).

2.6 FROM MORPHOLOGY TO ECOLOGY

Arguably one of the most cited and criticized papers regarding varanid morphology and systematics was produced by **Samuel Booker McDowell, Jr.** (1928–2014, Fig. 2.2E) and **Charles Mitchill Bogert** (1908–1992, Fig. 2.2F) working at the American Museum of Natural History. They provided detailed morphological descriptions (McDowell and Bogert 1954) of all the families within the *Anguimorpha* (their "Anguinomorpha"), many of which were known from very few specimens (e.g., *Shinisaurus* and the main subject, *Lanthanotus*), and made strong arguments that the ancestors to snakes were among the varanids and close relatives, as did Cope (1869), and sought to properly ally the *Mosasauridae* within the *Anguimorpha*. In an odd departure from their goal to examine anguimorphan systematics, McDowell and Bogert went on to review numerous other squamate groups, especially the snakes. It remains an important contribution and is one of the most cited works in publications about squamates. The most specific critique was by Underwood (1957), though his corrections were minor.

Studies of varanid biology had been numerous and largely limited to anatomy and taxonomy until the 1960s. **Walter Auffenberg** (1928–2004, Fig. 2.2G) undertook field studies of several large varanid species in their native habitats over several years that culminated in four epic publications. His years on Komodo Island resulted in several papers and a monograph (Auffenberg 1981 and references therein). After "rediscovering" the rare Philippine *Varanus olivaceus* in a drawer in the Smithsonian's Museum of Natural History (Auffenberg 1976; Fig. 2.7), Auffenberg went hunting for and found the 6-ft lizard alive and well on Luzon Island (Auffenberg 1979a). Again, he conducted fieldwork, where he discovered that the species ate large tree snails and fruits. This was the first known frugivorous varanid, though two additional species would be discovered several years later. Auffenberg made observations on captive lizards and with his field data published the second of his major monographs (Auffenberg 1988). This was followed by work in Pakistan and India studying the Bengal Monitor (Auffenberg 1979b, 1994), a familiar, widespread southeast Asian lizard about which little was known. The fourth monograph, an ecological study of the Indian Yellow Monitor, *Varanus flavescens* (Auffenberg et al. 1989), published in a journal rather than as a book. All of the monitors he studied are on endangered species lists, including CITES and are used to guide conservation practices.

Auffenberg deserves special recognition for including a natural history version of combined data set research by incorporating morphology, evolutionary history, taxonomy, and biogeography into his work on levels not often seen in other monotypic ecological monographs. Each of his volumes contains much of value to non-ecologists and non-herpetologists, providing detailed geological, cultural, and habitat data while also leaving the avenues for further exploration clearly presented. This gives his work a value-added feature that has intellectually prodded later workers, notably those who continued his studies with Philippine herpetofauna. His influence has inspired other herpetologists to do field studies of the unusual frugivorous varanids, at least two species unknown in Auffenberg's lifetime, the other, *V. olivaceus,* benefitting from attention of several sources in the bush and in captivity (Bennett 2014, Gaulke et al. 2007, Law et al. 2016, 2017, Recchio n.d.). Like Mertens, Auffenberg has a species named in his honor, *Varanus auffenbergi*, which, notably, is not one of the large species.

A contemporary of Auffenberg was **Eric Pianka** (1939–2022), whose work began by studying the effects of wildfires on local ecology, and focused on both desert lizard ecology generally (Pianka 1972, 1986) and varanid ecology (King and Pianka 2007, Pianka 1968, 1969, 1970a, b, 1971, 1982, 1994a, b, Thompson and Pianka 2001, and others) in particular. His work was a significant shift from his contemporaries in that he focused on *Odatria*, Australia's dwarf monitor species, including *Varanus brevicauda* and *Varanus eremius*. **Dennis King** (1942–2002) was a frequent collaborator with Pianka (King and Pianka 2007, Pianka and King 2004) who contributed to the area of varanid ecology (King and King 1979, 1999, King and Pianka 2007) and evolution, particularly of *V. rosenbergi* and its allies (Green and King 1993, King and Green 1999), and providing the first paper on monitor chromosomes and aspects of their molecular biology in relations to phylogeny (King and King 1975, King et al. 1982, 1991).

The 1970s marked the starting point of serious attempts to keep, breed, and study varanids in captivity, projects in which **Hans-Georg Horn** (1935–2019) was a leading contributor. Horn was the first to breed many varanid species in captivity (Engelmann et al. 2003), though credit was usually given to the first zoos to do so. An engineer by profession, Horn made numerous contributions to herpetology in general, with long-term partnerships in the private sector, zoos, museums, and universities. Among the fields to which he contributed were monitor behavior (Horn 1985, 1994, Horn et al. 1994, Mendyk and Horn 2011), physiology (Horn 2007), systematics (Böhme et al. 1994), reproduction, and captive care (Horn 1977, Horn and Visser 1989, 1997). His influence extended to helping establish and co-edit the varanid-focused International Herpetological Symposia held in Bonn, and the special publication of the proceedings in *Mertensiella*, produced by the Deutsche Gesellschaft fur Herpetologie und Terrarienkunde (DGHT). He and his family often hosted foreign varanophiles and shared his phenomenal knowledge gleaned from personal experience and an international web of herpetologists.

I have written the accounts above with some sadness because some of the people were both colleagues and friends who are still missed. They have left behind a formidable contingent of impressive herpetologists who have made important contributions to the understanding of monitor biology and evolution and, in turn, have encouraged the generation beyond them. Their influence upon

FIGURE 2.6 The dry preserved skin of *Varanus olivaceus* in the National Museum of Natural History (NMNH 25760) that allowed Walter Auffenberg (1976) to discover the long-unknown location from which the species comes. Stuffed skins such as this were the common method for preserving large lizards before the early 20th century; see Fig. 2.7F and G.

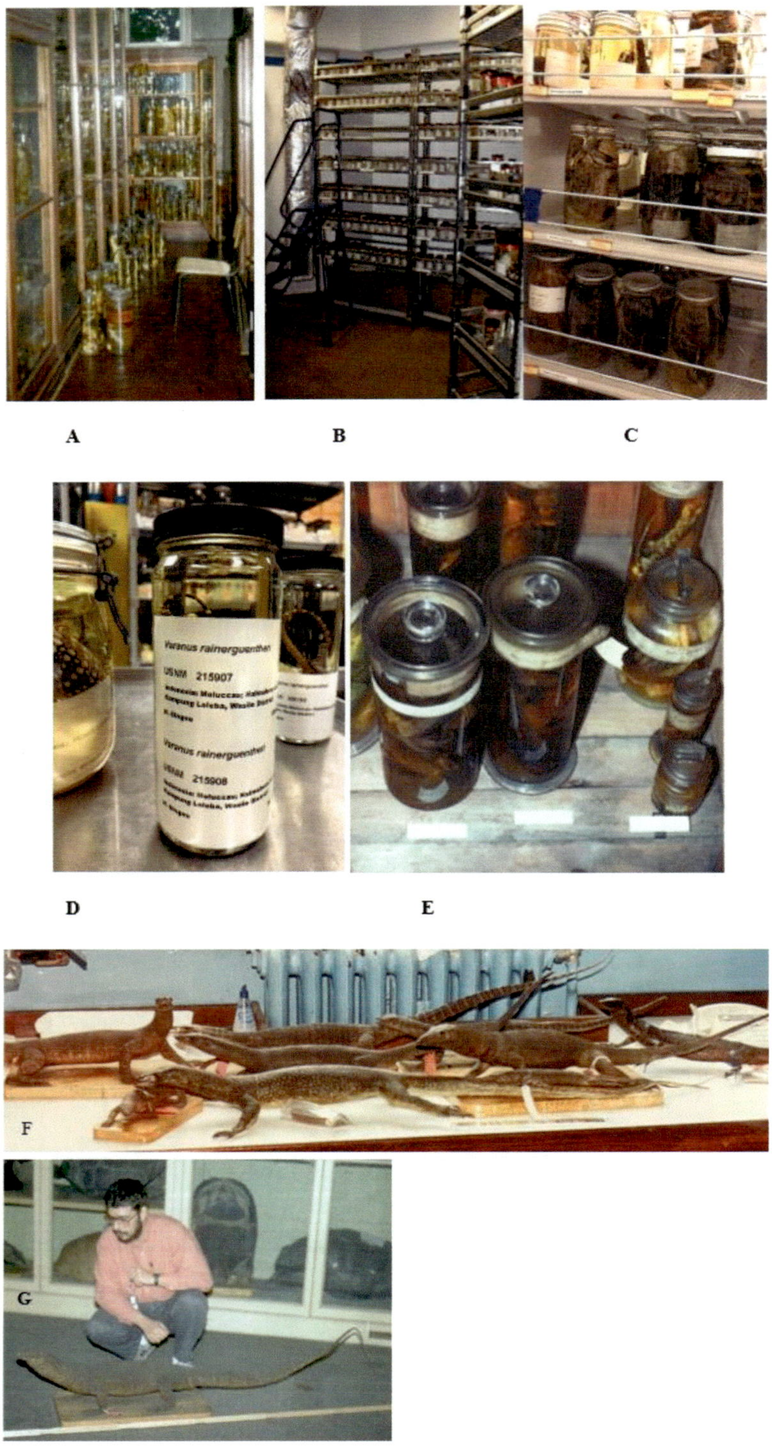

FIGURE 2.7 Examples of contemporary collections of preserved herpetological specimens, specifically of *Varanus*. (A) Natural History Museum, Berlin; (B) University of Kansas Museum of Natural History, Lawrence; (C & D) National Museum of Natural History, Washington, DC; (E to G) The Natural History Museum (= British Museum of Natural History), London, showing (E) wet specimens and (F) dry mounted skins. (G) The author with the holotype for *Varanus giganteus*, a stuffed and mounted skin.

the volume of varanid publications continues, dealing both fossil and living lizards. Collectors from the 17th through early 21st centuries have filled museum herpetology departments with specimens that sometimes still yield new species (Fig. 2.7), and plenty for continued morphological and biogeographical work. The 21st century has also brought a renewal of exploration into areas of the world where reptiles have very rarely been investigated. While it is possible that earlier explorers may have filled in many more gaps in our knowledge of the insects, shells, and birds of those places, herpetologists still have a long way to go to catch up, and discovery of new varanids is inevitable.

Part II

Morphology

The large size of many monitor lizard species has made them common subjects of morphological research. The species studied, however, are overwhelmingly six of the largest, particularly the Komodo Dragon and Bengal (Indian) Monitor, and show relatively close relationships and similarities. The small- and medium-sized monitors have received far less scrutiny until the 21st century and have shown distinctness and variation among and between each other. So far, few species have been studied in large enough numbers to describe the full range of ontological or intraspecific variation. This section provides a broad examination of a variety of varanids of all sizes and age classes and shows areas where variability may be of systematic importance for comparative studies and recognizing cryptic/sibling species.

DOI: 10.1201/9781003463528-4

3 Characters and Character States of the Varanidae

3.1 INTRODUCTION

The details and theory behind phylogenetics and monophyly are central to systematics and need an introduction prior to discussing broad aspects of morphology and relationships. As amniotes, varanids possess many features that, if described here, would be redundant when including the anatomies of turtles, birds, or other lizards (e.g., symplesiomorphies including the presence of four pentadactyl limbs, paired eyes, possession of an amnion-surrounded embryo, a closed circulatory system). The anatomical features of importance for distinguishing squamates from other amniotes, and in particular varanids from other lizards, are those used in making an evolutionary assessment, and subsequent classification. For example, the relative length of monitor tails carries little information about them as a group, but the length (as a percentage of total length), scalation, dexterity, and shape of tails are often species-group specific. Biological realities such as sexual dimorphism, convergence, and ontogeny have led to misidentifications. For example, the sexes of anglerfishes, suborder Ceratioidei, had been considered not only two distinct species, but members of different orders, though the only error at that scale in the Squamata was the initial failure to recognize *Sphenodon* as a rhynchocephalian. Taxonomy prior to about 1975 was vulnerable to errors based on misinterpretation of such features because the methods were based on building lineage models on shared common characters, or degree of morphological similarity (Mayr and Ashlock 1991). Though those methods became known as evolutionary taxonomy, it produced many phylogenies that defied the evolutionary histories of distinct groups or created models that disagreed in final form. How did taxonomists get past the problem of subjectivity in establishing groups such as genera? Are current methods of taxonomy more objective than earlier versions, and are computer-generated phylogenies accurate enough to supplant older approaches to systematics? The evolutionary taxonomists certainly did their best to be objective and were as committed to building the best phylogenetic models that they could. The advent of new methods increased the likelihood of having more accurate models, with the caveat that analyses are only as good as the data employed and the reliability of the analytic methods used. The long-used and often subjective methods of earlier taxonomists (Mayr 1963, Mayr and Ashlock 1991) were replaced by a more objective system to evaluate natural relationships, or evolutionary pathways, developed by Willi Hennig (Hennig 1966, 1969) and subsequently modified (Ax 1987, Gauthier et al. 1988, Wiley 1980, Wiley and Lieberman 2011). The methodology is known as phylogenetics or cladistics (Farris 1986, Hennig 1950, 1966, Wiley 1980). Hennig built his system around apomorphies, characters that appeared among different species because they were inherited from a common ancestor and were not the product of convergence. Characters shared by one group but not another are synapomorphies and the ordering of taxa by the evolutionary route that requires the fewest steps is deemed to be more likely to reflect reality. Models based on the fewest likely steps to achieve a particular tree topography (the evolutionary steps hypothesis) use the principle of parsimony, another core feature of phylogenetics. Characters can come in either/or discrete states (e.g., spotted or uniform), or multiple (e.g., spotted, uniform, strongly banded) states, that are coded for processing by computers. Because the number of character states and the number of trees generated with an equally small number of steps is usually immense, this method was not practicable prior to the use of computers except when using small data sets. This chapter looks at the characters important in establishing an evolutionary history and defining taxa within the Varanidae; Chapters 10 and 11 look closer at interrelationships within the family and with its closest allies.

3.2 PHYLOGENETICS AND CLADISTICS

Systematics and taxonomy have always had their own jargon, almost all of which is still in use. The central unit in taxonomy is the taxon (pl. taxa), representing any rank within the Linnaean system, including species, genus, family, class, and onward. It simply refers to the entirety of a named group. Hennig (1950, 1966) gave his phylogenetic system several new terms. In phylogenetic practice, ancestral features are known as plesiomorphies, while more recently acquired or unique states are apomorphies, terms that replace "primitive" and "advanced," and are interchangeable with "ancestral" and "derived." The ancestral features shared by species or outgroups are symplesiomorphies and have little information content for any but the largest scale phylogenetic analyses, at the phylum, class, or ordinal levels. However, unique features (apomorphies), or features shared only with other species evolved from a common ancestor (synapomorphies), are much more useful to zoologists, because they represent a trail of inherited characters from ancestor to descendant. The core of this system eschews the traditional ranks and morphology-based taxon diagnosis and replaces it with a statement of relationships. The three definitions that are used to describe particular types of phylogeny are node-, branch-, and

DOI: 10.1201/9781003463528-5

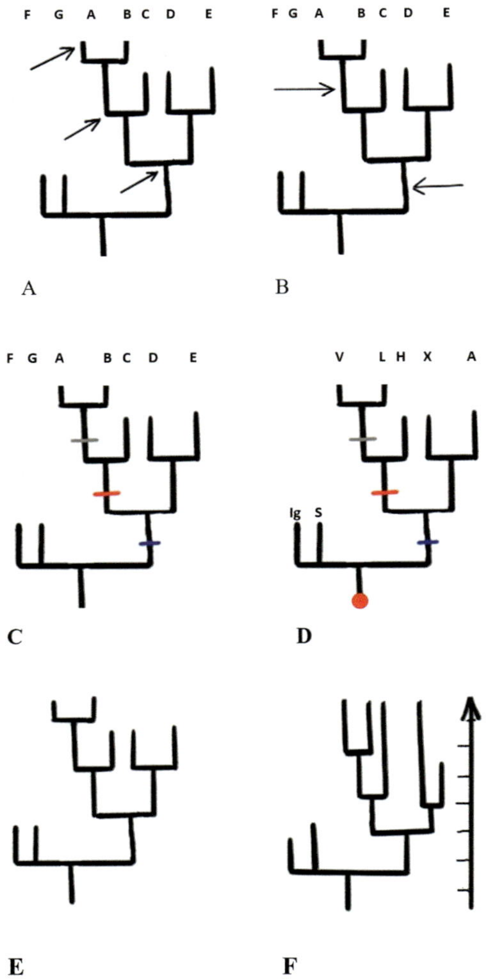

FIGURE 3.1 A – node-based, B – branch-based, and C – apomorphy-based types of cladograms used in phylogenetic systematics. A – nodes (arrows) represent taxa and root branching points. B – branches (arrows) represent characters and the lineages that carry specific character states. C – apomorphies for a given lineage are marked by bars and may indicate evolution or loss of a particular character state. D – an example using the Anguimorpha where the gray line represents synapomorphies uniting V and L but not H, the red line unites V, L, and H, and the blue line unites V, L, H, X, and A. In each case Ig and S are excluded outgroups. The red dot is the root for Squamata, blue line represents Anguimorpha, red line represents Platynota, and gray line represents Varanoidea. Abbreviations: A – Anguidae, H – Helodermatidae, Ig – Iguania, L – Lanthanotidae, S – Scincomorpha, V – Varanidae, X – Xenosauridae.

apomorphy-based. Node-based definitions are stated "the last common ancestor of A and B and all of the descendants of that ancestor" (Fig. 3.1A), where A and B represent specific taxa such as species, genera, or families. A branch-based definition would read "the Platynota encompasses the most recent common ancestor of *Varanus* (A) and *Heloderma* (B) and all of that ancestor's descendants" (Fig. 3.1B). Finally, an apomorphy-based definition (Fig. 3.1C) would read "the first ancestor of the Varanidae to possess apomorphy X and all of its descendants." Phylogenetics clearly distinguishes between statements of relationship based upon apomorphies and synapomorphies (Fig. 3.1D), and descriptions of the physical characteristics of taxa.

Of course, such terms are relative (Forey et al. 1992, Funk 1991). The various "-omorphy" terms refer to the evolutionary roots of a suite of characters. The long, forked tongue of a monitor is unique and derived when compared with other lizards (thus, an apomorphy) and found elsewhere only among the distantly related teiids and snakes (apparent synapomorphy, likely convergence). But the tongues of the three groups are morphologically different and arose convergently from unrelated ancestors (plesiomorphies), making the forked tongue a feature that arose independently at least three times. However, within varanids the particular features of the trait are universal and the structure unique, so it is considered a symplesiomorphy when studying intrafamilial relationships.

Special terms are used when describing traits. For features relating to physical or physiological aspects of an organism, the words "traits", "features," and "characters," the last being the preferred term, are synonymous. Characters specify a gene, which means the class of features, such as eye color, number of midbody scales, tail scalation, and condition of nasal bones. In contrast, "character state" corresponds to alleles, describing the specific expression of those genes as blue or brown, 124 or 190, keeled or spinose, and fused or paired, respectively. In many cases, polyallelic genes may be present, producing several distinct phenotypes (e.g., spotted, banded, and unicolored individuals from a single clutch as seen in *Varanus varius*), while other character states are the result of multiple alleles, codominant or incompletely dominant genes, numerous other gene interactions (deletions, insertions, inversions, jumping genes, etc.) and ecophysiological factors (Malhotra and Thorpe 1993, Thorpe and Brown 1989). Because morphologists can usually only examine phenotypes, this parallel between genes/characters and alleles/character states is both practical and useful in anatomical studies. In coding character states for computer analysis, each state is evaluated against one or more outgroups, taxa not part of the direct recent lineage of the group under study. This allows determination of character polarity, the assignment of ancestral (0) and derived (1 or higher number) values that are entered as the code. Often, additional derived states must be included, and they are given sequentially increasing numerical values, e.g., (2) and (3). Funk (1991), Wiley (1980), and Wiley (1991) termed a transformation series. For example, a varanid species may (0), or may not (1) have a distinct dark temporal stripe, or have one with (2), without (3), or with two (4) light bordering stripes. Ideally, a transformation series depicts how one character state led to the next. The level of complexity that can now be pried from molecules, particularly DNA, has added a huge suite of characters that must be analyzed by complex machines and makes tens of thousands (Estes et al. 1988, Gauthier et al. 1988, 2012, Pyron 2017) or more characters available. The vast number of data points in such analyses can partially compensate for a few characters of questionable value or that have been incorrectly assessed. Gauthier et al. (2012) have produced an invaluable illustrated list of 610 squamate characters and have coded that suite for fossil and extant squamates, including †*Saniwa* and three species of *Varanus*. Not all of their characters are applicable to all squamates.

3.3 CHARACTERS OF TAXONOMIC SIGNIFICANCE

All characters are informative, but not all are informative at every level of investigation, e.g., ordinal, familial, and species levels. A useful phylogenetic analysis uses only characters that can be used to distinguish one taxon from another. For example, an analysis of relationships among monitors will not benefit from inclusion of the character "limbs" and its states absent (0) and present (1), because they all have limbs, though the character could be informative if comparing varanids to other genera or higher taxa. Certain derived character states found in monitors are unusual among amniotes. For example, Gaulke (1989) noted that there are no significant ontogenetic changes in body proportions between young and adult monitor lizards of the *Varanus salvator* complex (there are significant changes in SVL-tail and limb-body proportions) but this observation does not agree with data I have collected for several other species of *Varanus,* such as *V. exanthematicus* and *V. jobiensis.* The general vertebrate trend is to produce a large-headed neonate that grows to produce a small-headed adult (frogs being a conspicuous and species-rich exception). Because Gaulke's character varies within varanids, it is useful when building phylogenies.

Monitors are described as long-necked lizards, and in fact their necks are long, if measured subjectively against other squamates. However, the length of the head and the neck is nearly equal among adults of most varanid species, corresponding with similar body proportions of many anguids, iguanians, and scincomorphs. In those species where the neck is longer than the skull, there appears to be independence from other ontogenetic features, where the relative proportions of these regions remain constant throughout the individual's life. Nevertheless, the relative elongation of the neck and its musculature allows monitors to swallow with the aid of sigmoid motions on horizontal and vertical planes, an ability very limited in most other lizard taxa. This mobility is not needed for small prey items but it is useful when large items or bones are being consumed.

For a taxon to justify recognition distinct from other taxa, phylogenetics requires it to be monophyletic, that is, there must be a clear line of descent from a most recent common ancestor (MRCA) and all the species included in the group. The monophyly of the Varanidae and its fossil members has been supported by several authors (Ast 2001, Conrad et al. 2008, Dong et al. 2022, Estes 1983a, Estes et al. 1988, Gauthier et al. 2012, Lee 1997, McDowell and Bogert 1954, Mertens 1942a, Pyron 2017, Rieppel 1980, Vidal et al. 2012), with the family content sometimes expanded to include *Lanthanotus* (Estes et al. 1988, Pregill et al. 1986) by some authors. Numerous synapomorphies give support to this hypothesis in terms of cladistic theory (Ast 2001, Conrad 2008, Estes et al. 1988, Pepin 2001, Pyron 2017). *Lanthanotus*, whether given a monospecific family or placed within the Varanidae, is undoubtedly a varanoid platynotan and very closely related to *Varanus*, for which reason it is included in the character list below.

3.3.1 Characters and Character States of the Varanidae

Uniform characters possessed by the Varanidae are given in Chapter 11. Characters that demonstrate variation, that is bi- and multistate characters, are presented here. Morphological variation among *Varanus* can be subtle, blatant, or mixed, the last referring to characters that can exist in more than one state within a species. The following list features 182 variable characters that I have incorporated in my phylogenetic analyses with polarity values assigned on the basis of my earlier outgroup comparisons (Sprackland 1991a, b, 1995b). The characters demonstrate that monitors are not as morphologically uniform as often claimed. Meristic characters are marked with an asterisk (*).

3.3.2 Head

1. Nostril nearer tip of snout than anterior border of orbit (0) equidistant to both (1) or near nostril (2). In using a clear mm. rule, if any portion of the central point between rostral and anterior orbit falls within the naris, the nostril is medial (Fig. 3.2).
2. Nostril round (0) oval (1) or a long slit (2).
3. Nostril faces laterally (0) dorsally (1) or anterodorsally (2).
4. Snout distinctly depressed anterior to nostrils (Fig. 3.2), step-like in lateral appearance (0), or snout gradually tapers to depressed tip, rounded and not step-like in lateral appearance (1). The depressed tip is universal in *Varanus* either for the full distance between the rostral and a point just anterior to the nostrils, as in *Varanus prasinus* (0), or for a very short section of the extreme tip of snout that barely extends more posterior than the rostral region, as in *V. exanthematicus* (1).
5. Nostril region is not within a raised process (0) or is within a swollen raised process (1).
6. Nostril without (0) or with (1) a valve or flap that allows closure while submerged.
7. Closed upper and lower jaws not distinct (0) (Fig. 3.2) or distinct (1) (Fig. 3.3) when mouth is closed. The mandible can be countersunk so seamlessly that upper and lower jaws are difficult to discern when the mouth is closed, or the premaxillary area is just very slightly beyond the mandible and the tips of both jaws and in distinct rounded tips.

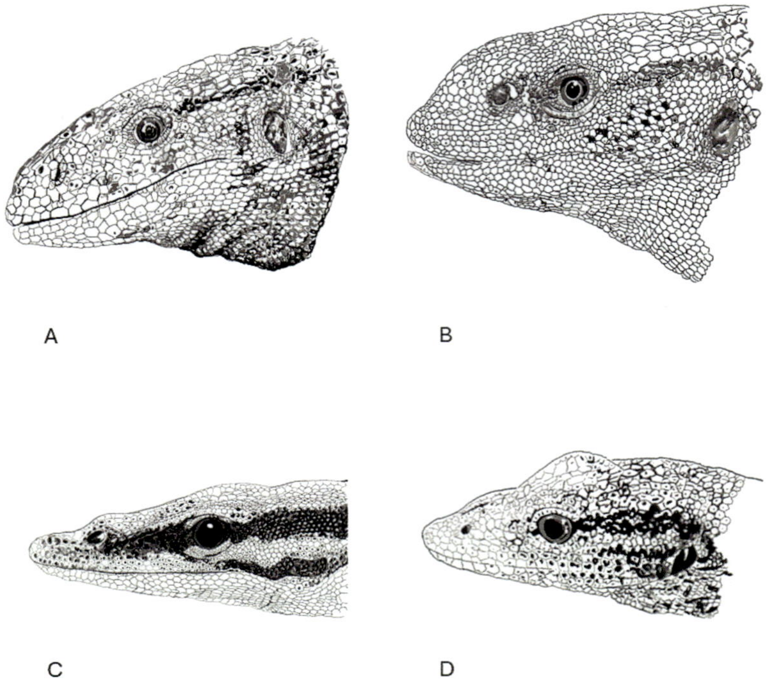

FIGURE 3.2 Character states for heads of *Varanus*. A – *V. flavescens* with anterior oval nostril and showing rounded snout without premaxillary depression. B – *V. albigularis* with extremely retracted oval nostril and showing high, domed snout with short, distinct premaxillary depression. C – *V. pilbarensis* with round nostril equidistant between rostral and anterior rim of orbit, and with pronounced anterior snout depression. D – *V. prasinus* with small, round, anterior nostril and distinct snout depression. In A and B the temporal (postocular) stripe is distinct and thin and is irregular in shape, while in D the stripe is broad, also with irregular edges. In A, B, and D there are often single light-colored flecks in the stripe. In contrast, C shows a thick, solid, and well-defined stripe that extends anteriorly to the nostril.

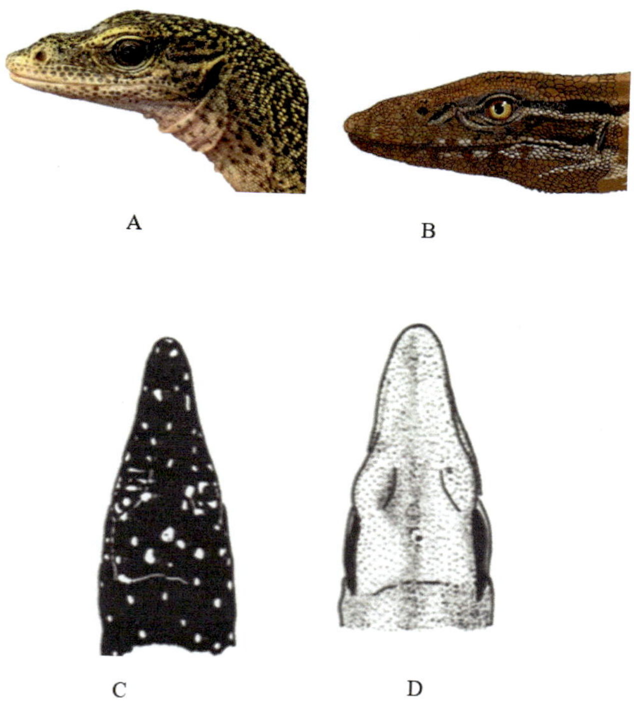

FIGURE 3.3 A and B show irregular pair (A) and solid, light-bordered postocular stripes in *Varanus* cf. *jobiensis* and *V. griseus*, respectively. C and D show extremes in head patterns, from (C) heavily marked with light scales and (D) no head markings, in *V. indicus* and *V. jobiensis*, respectively. (C and D from Sprackland (1995b).)

8. As seen from above, snout with parallel, blunt (0) (Fig. 4.4h) or pointed (1) (Fig. 4.4a, e) shape.
9. Supraocular scales subequal (0) (Fig. 4.5e) or enlarged (1). If enlarged, they form a row the length of the orbit and are distinct from other supraocular and supraciliary scales (Fig. 4.5a, c, d). Individuals may have one or more large scales longitudinally broken into two, rarely three, smaller sections (Fig. 4.5b), but these are also usually larger than other supraoculars. Such cases are rarely displayed on both left and right sides.
10. Supraoculars (Fig. 4.6) subequal to (0), or larger (1) or smaller (2) than interorbitals. Most varanids will display states (0) or (1), but among the *Odatria* group state (2) is an important diagnostic between some similar species.
11. Supraciliary scales normal, small (0), or slightly fringed (1).
12. Canthus rostralis rounded and indistinct (0), or acute, at least from anterior supraorbital region halfway to nostril (1). The canthus rostralis is defined here as the region extending longitudinally from the most anterior part of the supraocular scales to the nostril, though the acute structure does extend closer to the rostral region in some Australian species such as *Varanus tristis*.
13. In transverse section snout appears square, box-like (0) with canthus rostralis not more medial than row

of supralabials, or somewhat triangular (1) being narrower across canthus rostralis than at labials.

14. Face (rostral to orbit) dark (0), with few light markings (1), strongly mottled (2), or light and unmarked (3).
15. Dorsal surface of head without spots (0) or spotted (1).
16. Preocular stripe present (Fig. 3.2) (0) or absent (1).
17. Temporal stripe black (area with NO light speckling, Figs. 3.2C and 3.3B, D) (0), light and irregular or broken (1) (Fig. 3.2A, D and 3.3A), pale and continuous only (2), or absent (3). Nearly all *Varanus* have a postorbital stripe that runs from the posterior edge of the orbit, where it is generally widest, to the ear opening, terminating along the upper edge of the opening. Temporal stripe pattern variation can be conspicuous or subtle; in dark or melanistic specimens it is visible under good light as a region that lacks lighter coloring, such as speckles (Fig. 3.2).
18. Temporal stripe parallel border ventral only (0) (Fig. 3.3A), dorsal and ventral (1) (Figs. 3.2C and 3.3B), dorsal only (2), or absent (3). Though the feature is age-independent, it is most discernable in young specimens and young adults. Any light coloring tends to fade with age.
19. Gular region (between dentaries, not throat) pale (0), strongly barred (1), or spotted (2).
20. Throat (not gular region, but area posterior to dentaries) pale, unicolor (0), or spotted/barred with black (1).
21. Cranial scales large and polygonal (0) (Fig. 4.4a–c) or small and pebble-like (1) (Fig. 4.4d, e, g).
22. Individual cranial scales smooth or very feebly keeled, requiring magnification to ascertain (0) (Fig. 4.4a, b, f, h) or clearly keeled (1) (Fig. 4.4c, d). The overall texture of the scales may appear rough if the grooves between them are wide enough to form deep gaps (Fig. 4.4b, g; see 16).
23. Cranial scales with (0, Fig. 4.22a, b) or without (1, Fig. 4.22c, d) pronounced encircling grooves.
24. Length of head equal to (0), longer than (1), or shorter than (2) length of neck. The head length is the rostral to upper center of ear opening, corresponding to posterior edge of quadrate; neck length is from upper center of ear opening to gular fold.
25. Supralabials number 22-23 (0) or 25-28 (1).
26. Number of subnasal scales 2 (0), 3 (1), 4 (2) or 5+ (3) (Fig. 4.3b, c).

3.3.3 Skull

27. Maxillary bones, when viewed from above are acute, coming to a point (0) (Figs. 4.4a, b, e, g and 5.2b, c), or nearly parallel, producing a blunt muzzle (1) (Figs. 4.4f, h and 5.1). This character differs from number 5 because maxillae with and without overlying soft anatomy may give different degrees of acuity.
28. Dorsal surface of maxilla is a continuous near-straight slope (0) (Figs. 5.3c and 5.8b, e), has a long horizontal anterior section followed by a short, rapidly rising slope (1) (Figs. 5.3b and 5.8d, g), or has a long horizontal anterior section followed by an abruptly vertical rise and a short, tall posterior process (2) (Figs. 5.3c and 5.8a, f).
29. Anterior dorsal rise of surface of maxilla not (0) or is parallel (1) to posterior lacrimal-prefrontal border.
30. Where present, ventral edge of posterior maxillary process is subequal (0) to or broader than dorsal surface.
31. Jugal articulating surface of posterior maxilla extends very slightly (0), about one-half the length (1), or more than one-half the length (2) of posterior maxillary process.
32. Tines of jugal articulating surface small and subequal in length (0), or longer, with lower tine longer than upper (1) (Fig. 5.8a, d).
33. A line drawn from anterior point of jugal to anterior prefrontal articulation vertical or nearly so (0, Fig. 5.3b, c), or strongly angled with jugal articulation distinctly more posterior the prefrontal articulation (1, Figs. 5.2a and 5.3a).
34. Lateral surface of maxilla in adults is nearly flat (0) or slightly convex, bulging (1, Fig. 5.2d).
35. Maxillo-frontal/lacrimal recess somewhat oval (0), or flattened, more rectangular (1).
36. Anterior of maxillo-frontal/lacrimal recess closed with distinct wall formed from crista transversus (0), or opens to anterior recess (1).
37. Lower (dental) edge of maxilla nearly straight (0) (Figs. 5.2d and 5.4i) or curved (1) (Figs. 5.2a and 5.4f–h).
38. Lateral maxillary foramina (except large posterior nutrient foramen) small but distinct (0, Figs. 5.2a, d and 5.3a–c), or absent (1).
39. The premaxillary process of the maxilla and the lateral posterior process of the premaxilla meet (0, Fig. 5.7b) or do not meet (1).
40. Orbital process of maxilla single (0) or with two extensions/tines (2).
41. Maxillary facial-prefrontal crest triangular (0, Table 5.2b, e, o, p, y, z, δ), curved posteriorly, shaped like a shark's fin (1, Table 5.2a, c, j, q–t, v–x, α–δ, ζ, η, ι–π), or roughly rectangular (2, Table 5.2d, f, h, i, k–m, u, ε).
42. Seven to nine (0), or more than nine maxillary teeth (1) (Fig. 5.4h).
43. Nasal bone in adult single (0) or paired (1) (Fig. 5.5).
44. Dorsal surface of premaxilla with flat table (0, Figs. 5.2a, 5.4b, e, k and 5.7b) or sharp and blade-like (1) (Figs. 5.2d).
45. Ascendens process of premaxilla straight (0) (Fig. 5.4g, h) or convex (1) (Fig. 5.4f, i) dorsally.
46. Ascendens process of premaxilla uniform in width (0, Figs. 5.2a and 5.4b), much broader at rostrum (1, Figs. 5.2d and 5.4e), or of varying width (2).

47. Ascendens process of premaxilla without (0) or with (1) (Fig. 5.4e) a median dorsal ridge.
48. Ascendens process of premaxilla nasal articulations subequal in length (0, Fig. 5.4d), dorsal longer (1), or ventral longer (2) (Fig. 5.4a).
49. Ascendens process of premaxilla divides nasals (0) or does not (1) (Fig. 5.5).
50. Ascendens process of premaxilla extends half (0) or more than half (1) of rostrum to orbit length.
51. Ascendens process of premaxilla terminates away from (0) or near (1) anterior-most portion of nasal (Fig. 5.5).
52. Anterior rim of rostrum strongly (0), slightly (1) curved, or pointed (2) (Fig. 5.4k).
53. Anterior rim of rostrum depressed at tip (0) or not (1).
54. Ethmoid foramen large and distinct (0), or small and indistinct (1).
55. Ethmoid foramen facing anteriorly (0), dorsally (1), or laterally (2).
56. Ethmoid foramen on rostrum surface (0) or in lower portion of ascending process (1).
57. Vomerine process of premaxilla notched with two posterior prongs (0) (Fig. 5.4a, c, d), or not (1) (Fig. 5.28a).
58. Vomerine process of premaxilla equal in length (0), longer (1), or shorter (2) than rostrum.
59. Maxillary articulation of premaxilla small (as seen dorsally) (0) or well-extended (1).
60. Premaxillary teeth sockets 7 (0), 9 (1), or varies (2).
61. Premaxillary turbinal capstan-shaped (0) or U-shaped (1).
62. Median anterior notch of nasal(s) shallow (0), deep (1), or very deep, >1/2 length of nasal (2) (Fig. 5.5).
63. The bony external naris extends posteriorly with the frontal bones meeting the naris (0) or approaching very closely (1). The nasals and prefrontals have little or no contact with each other, while the nasals and maxillae are never in contact.

    In contrast to the view proposed by Estes et al. (1988), McDowell and Bogert (1954) suggested that the plesiomorphic condition is a posterior or more central naris, and evolution has secondarily elongated the varanid snout and moved the position of the naris in concert with modifications of the dorsal maxillary edge. Though there is a pattern of the nostril being more anterior in varanids with elongated snouts (e.g., *Varanus mertensi, Varanus salvadorii, and V. salvator*), the pattern is not universal (e.g., *Varanus dumerilii*, *Varanus griseus*, and *Varanus rudicollis*); shorter-snouted monitors may have the nostril proximal to the orbit (e.g., *Varanus albigularis*) or have an enlarged oval nostril midway between rostrum and orbit (e.g., *Varanus bengalensis, V. exanthematicus,* and *Varanus flavescens*). The resultant nasal foramen is the largest found in extant lizards, with the closely related *Lanthanotus* showing an enlargement about midway between that seen in *Heloderma* (average anguimorphan structure) and *Varanus*. Neonates of examined taxa *Varanus rosenbergi* (Andrews et al. 2017), *Varanus indicus* (Gregorovicova et al. 2012), *Varanus panoptes* (Werneburg et al. 2014), and *V. mertensi, Varanus macraei,* and *Varanus pilbarensis* (pers. obs.) have the nostril in an anterior position. Regardless of the direction of the ontogenetic elongation, the varanid naris coupled with the highly developed lateral excavation nasalis maxillare are unique to the Varanidae.
64. Nasals separate (0) or fused (1) in the embryos. Estes et al. (1988) do not document the source for this observation, but I have not found this to be true. In the three species for which embryological data are published (*V. rosenbergi,* Andrews et al. 2017, *V. indicus,* Gregorovicova et al. 2012, and *V. panoptes,* Werneburg et al. 2014; SEE Chapter 9), early through late stage embryos show distinct, separate nasals, as do embryos of *V. mertensi, V. macraei,* and *V. pilbarensis* (pers. obs.). It would thus be more accurate to state that "nasals become fused in late-stage embryos of some species." Approximately half the living species (primarily the Australian *Odatria*) have paired nasals even as old adults (e.g., *Varanus acanthurus*, *Varanus caudolineatus*). Mertens (1942a, b) used paired nasals as a major character state dividing *Odatria* from other varanids, but fused nasals are present in at least one odatrian (*Varanus eremius*). Among other varanids, the eastern African *Varanus ocellata* (formerly *V. exanthematicus*, in part: Bayless and Sprackland 2000a, b) and the western Pacific *V. indicus* (*sensu stricto*; see Sprackland 1995) have paired nasals throughout life. In the fossil †*Saniwa ensidens*, the nasals seem to be fused in the holotype (NMNH 2185), and at least one subsequent specimen (Rieppel and Grande 2007), but the condition of the preorbital region in the presently available fossils makes the character state ambiguous. Paired nasals are found in other lizard families and may be plesiomorphic. However, it may be argued that if African varanids are ancestral, and the Australian odatrians derived, possession of paired nasals may have secondarily evolved twice in varanids (Ast 2001, Sprackland 1991a), with heterochrony limiting the development of snout elongation and accompanying fusion of the nasals. In any event, this character is inconsistent within *Varanus* and too plastic to be accepted as a varanid synapomorphy.
65. Anterior processes of nasal fork-like (0) or irregular (1) (Fig. 5.5).
66. Lateral nasal border straight (0), curved (1), or mixed (2) (Fig. 5.5).
67. Nasal broadest posteriorly (0) or elsewhere (1) (Fig. 5.5).

68. Frontonasal suture largely straight (0), symmetrically notched (1), or highly irregular, asymmetrically notched (2) (Fig. 5.5).
69. In dorsal aspect septomaxilla is distinctly triangular (0, Fig. 5.6A, B, H, L, M, O, S), roughly rectangular (1, Fig. 5.6D, G, P), bottle-shaped (2, Fig. 5.6F, Q), or amorphous (3, Fig. 5.6E, I, J, N, R).
70. Dorsal surface of septomaxilla with distinct dome (0, Fig. 5.6A, H, P, R), slight depression (1, Fig. 5.6D, I, K), depressed with partly covered rear-facing fossa (2, Fig. 5.6F, J), or irregular or rugose (3, Fig. 5.6D, K).
71. If triangular there is no (0, Fig. 5.6A, H, L) or a distinct (1, Fig. 5.6B, D, F, I, M, Q) posterior notch on septomaxilla.
72. A thin longitudinal dorsal crest on septomaxilla is not (0) or is present (1, Fig. 5.6F, G, S).
73. Anterior projecting process on septomaxilla absent (0, Fig. 5.6A, B, E, G, H, K), single (1, Fig. 5.6R), or paired (2, Fig. 5.6R).
74. Posterior border with two (0, Fig. 5.6B, D, I, K, M), one (1, Fig. 5.6H, J), no (2), or three (3, Fig. 5.6C, F, O) projections.
75. If septomaxillary posterior projections are present, the lateral (0, Fig. 5.6B, H, I, J, N, R) or medial (1, Fig. 5.6A, C–F, H, M) is longer, or they are subequal (2, Fig. 5.6K, Q) in length.
76. In dorsal view, edges of septomaxilla have no (0, Fig. 5.6A, B, F, I, J, L, M), one (1, Fig. 5.6C, E), or two (2, Fig. 5.6D, E, G, H, P, R, S) flat lateral flanges.
77. Location of septomaxilla posterior (0, Fig. 5.6A lower), anterior to (1, Fig. 5.6A middle) intranasal crest of maxilla, or passes medially on both sides (2).
78. Lacrimal foramina double on the orbital margin (Fig. 5.10). The lacrimal foramina are contained entirely within the lacrimal bone (0), or the upper foramen is formed from lacrimal and prefrontal (1) bones. Paired foramina are unique to all living and fossil varanids plus the allied *Lanthanotus* but the morphology varies, so is a synapomorphy for *Varanus*, †*Saniwa*, and *Lanthanotus*. In the Paleogene †*Saniwa* the lacrimal foramina seem to lie entirely within the lacrimal and are of near equal diameter, though my examination of the fossils shows that the lacrimal region is not so well-preserved as to state unequivocally that the foramina are both entirely within the lacrimal. In *Varanus* the large upper foramen consists of parts of the lacrimal and prefrontal bones, and the lower foramen is wholly confined to the lacrimal. In *Lanthanotus*, both foramina are formed by the lacrimal only. The morphology of the foramina is unique to *Varanus*, and thus an apomorphy.
79. Osteoderms do not fuse to skull (0) (McDowell and Bogert 1954) or do fuse to underlying bone (1). When present on the head in *Varanus*, the osteoderms are embedded in the skin, which is attached to the dorsal cranial bones. The attachment is such that, during dissection, it can appear that the osteoderms and skull are in contact. Among living varanoids, *Heloderma* has fused cranial osteoderms; *Lanthanotus* does not. Osteoderms are not evident in †*Saniwa*.
80. Supratemporal process of parietal bone is narrow (Pregill et al. 1986) (0, Fig. 5.12f, h) or broad and dorsolaterally flat (1, Fig. 5.12d, e, j).
81. Parietal body is broad and rectangular (0) (Fig. 5.12e, j) or minimal or absent with longitudinal blade-like medial structure (1) (Fig. 5.12a, g, h). The parietal varies considerably within and among platynotans (see Chapter 5). In *Lanthanotus*, *V. dumerilii*, *V. rudicollis*, and other species, the bulk of the parietal is rectangular, longer than wide, with no or minor medial constriction, but with broadly flaring posterior processes. Among adult *Varanus*, the parietal may be so narrow medially as to produce little in the way of a dorsal table (*V. albigularis, Varanus ornatus, Varanus olivaceus*) or may be so broad as to nearly close the supratemporal openings between the postorbital arch and the parietal (*V. caudolineatus, V. dumerilii, V. exanthematicus*). Intermediate conditions are more typical, with the hour-glass shape seen in *V. salvator, V. indicus*, and *V. prasinus* being most common among the remaining species. The narrowest parietals are found in several large species, including *V. albigularis*, †*V. prisca*, *V. olivaceus*, and *V. rudicollis*. In stark contrast, juvenile varanids have broad, nearly square parietals that assume a more constricted central portion as they grow. In dwarf species (*Odatria*) the parietal remains largely squarish throughout life, resembling the condition seen in *V. exanthematicus* of all ages, supporting the hypothesis that where there is sagittal crest development it is positively correlated to relative growth (McDowell and Bogert 1954:49). The young of *V. albigularis* also possess broad, squared parietals, though as they grow, they develop into the most extremely constricted condition within living varanids, and very close to the parietal condition seen in the post-Pleistocene †*Varanus hooijeri* (Brongersma 1958), with large descending lateral surfaces and no flat dorsal parietal area. †*Saniwa* has an unusual configuration in that the posterior portion of the parietal is semi-rectangular, and the lateral edges are slightly concave. Among living *Varanus*, the condition in *V. komodoensis* is typical, though with much reduced lateral descending surfaces.
82. Parietal crests are pronounced (0) (Fig. 5.12c, d, j), extremely pronounced (= large flanges) (1) (Fig. 5.12a–c, g), or small, barely distinguishable (2) (Fig. 5.12d–f).
83. Anterior parietal does (0) (Fig. 5.12a–c, g–i) or does not (1) form distinct lateral processes (Fig. 5.12e).

84. Dorsal surface of parietal constricted at center, forming hour-glass shape (0), with extreme constriction and little or no flat dorsal surface (1), or roughly square with no prominent constriction (2).
85. Lateral flanges are with moderate (0), extreme (1), or huge (2) downward slope.
86. Lateral parietal foramen is large and distinct (0), or absent or very small (1).
87. Posterior post-parietal process is expanded into distinct pteroticum (0) (Fig. 5.12a–c, g, j) or terminates in an acute point (1) (Fig. 5.12d, f).
88. Width of supratemporal fossa is less than (0) (Fig. 5.12g–j), or greater than (1) width of parietal at mid body.
89. Supratemporal arch is thin (0) or so thick as to nearly eliminate supratemporal fossa (1) (Fig. 5.12i, j).
90. Crista prootica with single (0) (Fig. 5.14a, d), two (1) (Fig. 5.14c), or more than two (2) (Fig. 5.14b) lateral crests.
91. Prootic alar process inclined about 50^0 off horizontal (0), or 15^0–20^0 off (1) (Fig. 5.16).
92. Parasphenoid does (0) or does not (1) reach to near contact with the inferior parietal surface.
93. Parasphenoid is straight or with upward curve (0), or with downward curve (1).
94. Parasphenoid is (0) or is not (1) in contact with orbitosphenoids. If it is not in contact, it is extremely close and connected by membranous, longitudinal pilae.
95. In anterior view of the braincase the posterior vidian canal is not (0) or is (1) visible.
96. Distal process of basisphenoid broad with processes extending anteriorly and posteriorly (0, Figs.5.18e, f, h and 5.19f, g, l), having only an anterior process (1, Fig. 5.19a, b, e), or ending in a small, generally round process (2, Figs. 5.18g and 5.19a–d, h–k).
97. Lateral recessus scala tympani, fenestra ovalis, and hyomandibular foramen form a straight line (0, Fig. 5.14a) or are not in a straight line (1, Fig. 5.14c, d) under crista prootica.
98. Posterior ventral edge of basisphenoid is transversely straight (0, Fig. 5.19a, e, g, h, j, l), has a central concave notch (1, Fig. 5.19b, f, k), or has a central convex notch (2, Fig. 5.19b–d).
99. Lateral ventral basisphenoid process pronounced (0) or barely distinct (1).
100. Incisora prootica forms rounded (0, Fig.5.14a) or angular (1, Fig. 5.14b–d) surface.
101. Anterior inferior process of proötic extends about as far as superior surface (0), or not as far as superior surface (1).
102. Distal edge of crista prootica with acute, pointed process(es) (0, Fig. 5.14b, d) or smooth (1, Fig. 4.14a, c).
103. The outer conch of the quadrate is small (0) or enlarged (1, Fig. 5.23) (McDowell and Bogert 1954). Though the outer conch is thin and reduced in *Lanthanotus*, this is not the case in many *Varanus*. In *V. bengalensis*, *V. dumerilii*, *Varanus macromaculatus*, and *V. salvator*, the outer conch is robust, broader dorsally, and flaring at the articular quadrate facet (Fig. 5.25H, G, and J, respectively). Palci et al. (2020) reviewed quadrate diversity across squamates, but the only varanid in the study was *Odatria eremius*, a dwarf species that is expected to show more juvenile-type traits than seen in larger species. *O. eremius* differs from other *Odatria* in having fused nasals, so it does not necessarily represent other odatrians. In the arboreal *V. prasinus*, the conch is broad but not well-defined (Fig. 5.25C) from the narrow pterygoid flange (= tympanic conch of Palci et al. 2020), and similar but noticeably thinner in its close relative, *Varanus beccarii*. The largely aquatic *V. mertensi* has a similar morphology to *V. prasinus* except the tympanic column does not change in diameter (pers. obs.). The flange is quite robust in *Varanus niloticus* (Fig. 5.25I) and *V. macromaculatus*, species that are aquatic and terrestrial. This agrees with the observation by Palci et al. (2020) that there is little if any evidence that quadrate morphology is linked to a species being aquatic, fossorial, or terrestrial (Palci et al. 2020, fig. 2).
104. Hypoglossal foramen is small, and only one nerve emerges (0) or enlarged, containing vagus and hypoglossal nerves (1) (cranial nerves X and XII) (McDowell and Bogert 1954).
105. Coronoid and surangular processes large (0, Fig. 5.27a, b, e) or reduced (1, Fig. 5.27c, d, f, g) on posterolateral dentary. There is little overlap between the dentary and any of the post-dentary bones, excepting the ventral extension of the splenial bone along the interior dentary surface (Figs. 5.28 and 5.29).
106. Splenial and dentary do not (0) or both move (1) as a function of the intramandibular joint. The splenial in *Varanus* is much larger and more extensive than that of *Lanthanotus,* while in *Heloderma* it is nearly vestigial. Both *Varanus* and *Lanthanotus* can expand the mandibular gape laterally because of the joint, *Heloderma* cannot. The condition cannot be determined from fossils. Convergent in *Lialis*.

3.3.4 Soft Anatomy

107. Free part of tongue bifid for >50% in *Varanus* (1, Fig. 4.10A, B), 40-50% of its length (0) in *Lanthanotus*, (convergent in snakes, gymnophthalmids, teiids, and amphisbaenids). The forked tongue of varanids has often been cited as a synapomorphy with snakes, but this character has independently evolved at least four times in the squamates. In all forked-tongue species, the organ is used extensively in finding food and recognizing potential mates and competitors. In the same way that other sense organs (e.g., ears and eyes)

are paired to provide stereo input, it is conceivable that the bifid tongue serves a similar function, providing the olfactory equivalent of binocular vision.

108. Width of tongue is broad, touching buccal surfaces of dentary when retracted (0, Fig. 4.10E–H) or much thinner, not touching insides of mandibles when retracted (1, Fig. 4.10C, D). Broad tongues are possessed by *Heloderma* and *Lanthanotus*, while *Varanus* universally have narrow tongues.
109. Constrictor colli muscle does not (0) or does (1) extend anteriorly to cover first ceratobranchials (Rieppel 1980, pers. obs.).
110. The color of the tongue (in life) is pink (0), purple (1), blue (2), or yellow (3) (Figs. 4.11 and 6.2 and Table 3.1). This character warrants further study. Tongue color differences distinguish several similar sympatric species, e.g., *V. niloticus/V. ornatus, V. albigularis/V. exanthematicus, and Varanus chlorostigma/Varanus doreanus*. Given that tongue color has proven to be species specific, it is possible that the Australian *V. varius* consists of two or three species, given that purple-, pink-, and yellow-tongued lizards exist in allopatric populations in New South Wales, Victoria, and Queensland (Smissen et al. 2013, Sprackland 2009). A similar situation may exist in *V. salvadorii*, which has yellow- and blue-tongued populations (pers. obs.). Because tongue color quickly fades and becomes some shade of gray in preservative, it is a useful character only in live and freshly killed specimens.
111. Supralabial scales number 22-23 (0), or 25-28 (1).
112. Angle of snout is drawn from eye-snout tip lines from both sides of skull, in DEGREES.*
113. NUMBER of rictal scales (counted from last supralabial across skull to opposite last supralabial) (Fig. 4.3b).*
114. NUMBER of internasals across snout.*
115. NUMBER of supraoculars (Figs. 4.3d and 4.7).*
116. *Head*: SVL proportion (Fig. 4.4).*
117. *Head*: neck lengths, percent (Table 3.2). Within Varanus head and necks are very nearly equal in length (Fig. 3.4), but within the larger Anguimorpha, and possibly Platynota, the proportion varies.
118. *SVL*: tail proportion (Fig. 4.4 and Table 3.3).*

TABLE 3.1
Distribution of Tongue Colors Among 30 *Varanus* Species

Dark Blue/Purple	Yellow/Cream	Pink
albigularis	*doreanus*	*acanthurus*
exanthematicus	*komodoensis*	*beccarii*
giganteus	*salvadorii**	*bogerti*
*indicus**	*varius**	*caudolineatus*
*niloticus**		*dumerilii*
rainerguentheri		*finschi*
spenceri		*glebopalma*
*varius**		*jobiensis*
yemenensis		*mertensi*
		olivaceus
		*ornatus**
		prasinus
		primordius
		rudicollis
		*salvadorii**
		spinulosus
		storri
		telenesetes
		tristis

Note: An asterisk (*) indicates taxa that may contain undescribed species or are taxonomically in flux.

TABLE 3.2
Comparative Head and Neck Lengths in Several Species of *Varanus* (SVL/NECK) Given as Percentages

Species	SVL/NECK
acanthurus	0.97
bengalensis	0.98
caudolineatus	0.89
exanthematicus	0.98
flavescens	0.96
gouldii	0.90
indicus	0.98
prasinus	1.0

TABLE 3.3
Range of Variation of Ratios for Three Basic Morphometric Characters, Snout-Vent (SVL) and Tail (TL) Lengths, and Lengths of Heads (Head) for the African Savanna Monitor, *Varanus Exanthematicus*. SVL in Mm; Tail and Head Given as Percent of SVL, Rounded to Nearest Whole Number

SVL	TL	Head
415	146	19
378	164	20
318	214	–
306	188	–
243	173	21
230	169	20
177	160	21
167	170	21
145	131	19
133	132	22
124	170	23
118	137	24
118	186	25
N = 13 $\bar{x}$ = 220.9	$\bar{x}$ = 164.6	$\bar{x}$ = 21.4

119. Rostral to eye (mm).*
120. Rostral to nostril distance (mm) (Fig. 4.3b).*
121. Nostril to orbit distance (mm) (Fig. 4.3a).*
122. Length of eye.*
123. Posterior edge of orbit to top of ear (mm) (Fig. 4.3a, b).*
124. Head width across orbits (mm) (Fig. 4.3b, d).*
125. Length of rictus (postorbital temple) (Fig. 4.3b).*
126. Height of head at quadrate (Fig. 4.3a).*
127. Length, femur (Fig. 4.4).*
128. Length, tibia (Fig. 4.4).*
129. Length of leg, groin to end of longest digit (Fig. 4.4).*
130. Humerus, length (Fig. 4.4).*
131. Forearm, length (Fig. 4.4).*
132. Arm length, axilla to end of 3rd digit (Fig. 4.4).*
133. SVL (Fig. 4.4).*
134. Tail length (Fig. 4.4).*
135. Total length (Fig. 4.4).*

3.3.5 Body

136. Nuchal scales smooth (0) or distinctly keeled on anterior portion of neck (1) (Figs. 4.19–4.22).
137. Nuchals oval, nearly round (0), or elongate (1) (Fig. 4.22).
138. Anterior nuchals subequal to (0), or smaller than (1) occipital scales (Fig. 4.22).
139. Nuchals subequal to (0) or larger than (1) dorsals (Fig. 4.21a).
140. Nuchals close, touching or almost touching (0), or distinctly separated, by a distance equal to one half their width or more (1) (Fig. 4.22).
141. Nuchal region banded (0) or not banded (1).
142. Turquoise coloring absent on tail, dorsum, and limbs (0), or present (1).
143. Dorsum is covered in single-scale light spots (0), or solid spots of 1–3 scales (1), very large (10 plus scales) solid spots (2), hollow rosettes (3), Bismarck pattern (4), or with solid hexagonal spots (5) (Fig. 4.32).
144. Dorsal pattern is symmetrical, forming bands (0), or random (1) (Figs. 4.32 and 4.33).
145. Dorsal pattern is the same (0, Fig. 4.33a–e, g, I, j, l) or different (1, Fig. 4.33f, h, k) on front and rear halves of body.
146. Light colored dorsals equal to (0), or larger than other scales (1).
147. Dorsal scales are smooth (0), keeled (seen under magnification) (1), or strongly keeled (magnification not required for identification) (2).
148. Ventnor light and unmarked (0), lightly marked (1), strongly mottled (2), or uniformly dark (3) (Fig. 4.16).
149. Ventral scales are smooth (0) or keeled (seen under magnification) (1).
150. All ventrals are rectangular, subequal (0) oblong (1) or anterior ventrals smaller, square (2).
151. Ventrals lack (0) or have (1) slight posterior point.
152. Ontogenetic changes in body proportions present (0) or absent (1).
153. Pre-anal scales absent (0) or present (1) (Fig. 4.17).
154. Three pairs of ribs attached to sternum (0) as in *Varanus*, or two (1) as in *Lanthanotus* (convergent in iguanids, agamids, gekkonids, lacertids, cordylids, and anguids). I question why Pregill et al. (1986) used two different character states as a synapomorphy for *Lanthanotus* and *Varanus*.
155. Number of midbody scale rows counted at the point equidistant between axilla and groin.*
156. Number of ventral scale rows counted from gular fold to cloaca (Fig. 4.16).*

3.3.6 Limbs

157. Longest toe shorter (0) or longer (1) than head.
158. Second digit of pes is more than 2× length of shortest (0) or no longer than 1.5× length of shortest (1).
159. Digits are with enlarged dorsal terminal scale (0, Fig. 4.18a, d, g, i, k, m) or not (1, Fig. 4.18b, c, e, f, h, j, l, n, o).
160. Plantar surfaces are without (0) or with (2) dark nodules.
161. Enlarged subdigital lamellae are absent (0) or present (1) on at least one digit of rear leg.
162. Arms are without (0) or with (1) enlarged greave scales (Fig. 4.27).
163. Adpressed limbs do (0) or do not (1) overlap.
164. Claws are distinctly curved (0, Fig. 4.18b, c, g, k, l, m, n), long with curve (1, Fig. 4.18a, f, h, j) or long and barely curved (2, Fig. 4.18d, e, i, o).
165. Claws are mostly conical (0) or laterally compressed (1).
166. Proportional length of forelimb to hindlimb (F/H), in mm.*
167. Number of subdigital lamellae, third digit, manus.*

3.3.7 Tail

168. Tail distinctly compressed along posterior two-thirds (approximately twice as tall as wide) (0, Fig. 4.26a, g, h), intermediate between fully compressed and fully rounded (only slightly taller than wide) (1, Fig. 4.26d, e) or cylindrical (2, Fig. 4.26b, c, f) its whole length.
169. Posterior third of tail with broad (6 or more scales wide) (0), or narrow (1–3 scales wide) (1, Fig. 4.34 0, 1) Light, distinct bands; or bands made of irregular rosettes or spot clusters (2, Fig. 4.34 1) or spots present but not forming distinct bands (3, Fig. 4.34 3).

170. Tail without (0, Fig. 4.26b, c, f) or with (1, Fig. 4.26g, h) double dorsal keel.
171. Caudals keeled, either feebly (0) or strongly keeled, mucronate (1).
172. Caudals do (0) or do not (1) form distinct annuli.
173. Tail pattern is same (0) or different (1) along its length.
174. Ventral caudal scales are larger than (0, Fig. 4.24h), equal to (1, Fig. 4.26f), or smaller than (2) lateral scales.
175. Circumcaudal scales are posterior to cloaca NUMBER.*

3.3.8 Miscellaneous

176. Sex; F=0, M=1.
177. Cloacal tufts (in males) absent (0) or present (1, Fig. 4.25).
178. Paired horns in the apical regions of the hemipenes (Branch 1982).
179. Hemipenial group A (0), B (1), C (2), D (3), E (4), F (5), G (6), or H (7) (Branch, 1982).
180. Karyotype group A (0), B (1), C (2), D (3), E (4), or F (5) (Holmes et al. 1975, Sprackland 1991b).
181. Immunology antiserum group A (0), B (1), C (2), D (3), E (4), F (5), or G (6).

4 Monitor Anatomy I
External Morphology

4.1 INTRODUCTION

The most conspicuous and noted variations in monitors are their comparative sizes (Figs. 4.1 and 4.2), a range unmatched in any other vertebrate genus. At the small end of the scale are *Varanus sparnus* and *Varanus brevicauda* that grow to about 230–245 mm total length and the slightly larger *Varanus storri* and *Varanus primordius* (Fig. 4.1E) that reach 400–450 mm, to the giant living species *Varanus komodoensis, Varanus niloticus* (Fig. 4.1C), *Varanus salvadorii* (Figs. 4.1D and 4.2), and *Varanus salvator* that may reach or exceed 2540 mm, and becoming the largest with †*Varanus prisca* that exceeded 4000 mm. The maximum sizes of many of the better-known varanids are also imprecise, as in the Mangrove Monitor (*Varanus chlorostigma*) known to reach 1435 mm (4.7 ft.) in the wild, but some captive individuals have exceeded 1580 m (5.2 ft.; Fig. 4.1B). Private keeper Katy Bruce has sent me documentation of her Black Dragon (*Varanus komaini*) measuring nearly 3050 mm (10 ft.) total length, much larger than other available records, which are usually close to 2100 mm (6.8 ft.)

FIGURE 4.1 Variation in varanid body sizes and shapes. (A) *Varanus indicus* is typical of the elongated, long-limbed species in which the snout is long, relatively thin, and pointed. (B) Though *V. indicus* commonly exceeds 1 meter in total length, some specimens, such as this alleged 1.4 m specimen, have been unofficially reported. In contrast, (C) *Varanus niloticus* generally grows to a total length of 2.1 m; this specimen was reported to be 2.74 m long and possibly a record. The photo was provided without further data, courtesy of Mark Bayless. (D) Swiss reptile zoo owner and photographer Fred Kundert in 1991 showing author the size and speed of his 2.4-m tree crocodile monitor, *Varanus salvadorii*. The lizard consumed the adult rat in a single quick swallow. (E) *Varanus primordius* is one of the Australian dwarf goannas in the *odatria* group. Several of the dwarf species have caudal scalation ranging from strongly keeled, as here, to spinose. The species may grow to a total of 450 mm.

DOI: 10.1201/9781003463528-6

FIGURE 4.2 New Guinea's tree crocodile monitor, *Varanus salvadorii*, is probably the longest of the varanids that has been estimated by some biologists to grow to 4.8 m (16 ft.) But its rarity had made confirming the reports impossible. The species has become common in captivity and is frequently bred, making accurate measurements of numerous large adults possible. The specimen shown in the upper photo is posing next to its keeper, who, with the thick-soled shoes, measures 1.90 m (6.25 ft.), just a few inches longer than the monitor's tail. The SVL of the Lizard is about 0.5 m (20 in.) for a total length of 2.4 m (7.9 ft.). Both komodo dragons and water monitors have been measured slightly in excess of 3.0 m (10 ft.), A record not yet verified for the tree crocodile monitor. However, specimens with an SVL over 0.6 m have been measured, so it remains a possibility that it may match, if not exceed, its related species in length. In mass, however, the tree crocodile monitor is considerably exceeded by its relatives even at shorter lengths, a likely reflection of the New Guinean species' nearly completely arboreal habits. Photo courtesy Katy Bruce. The lower picture shows a tree crocodile monitor with Stacey Crutchfield giving an idea of the scale of the large lizard and the length of its partially extended Tongue. Photo courtesy of Tom Crutchfield.

for the species. It is possible that the diet and predator-free environments (along with veterinary care when needed) factor into the development of extra-large captives, but we must also consider that, aside from a few large, endangered, and commercially important species, there have been very few long-term field studies or significant collections of age and sex classes of monitors. For all we know, larger sizes may be common in some of the species.

There is considerable phenotypic distinction among taxa, including various head, body, limb, digit, and tail proportions, differences in scalation, color, and configuration (notably atop the head), nostril and tail shapes, and preferred habitats, and yet there is little chance of mistaking a monitor lizard for anything else. In fairness, certain African and European lacertids (*Eremias* and *Timon*) and American teiids (*Aspidoscelis* and *Tupinambis*) are remarkably similar in appearance, diet and habitat preferences, and behavior to *Varanus*, but even cursory comparison shows the great morphological distinctions among the different genera. Though goannas lack any form of elaborate crests, beards, dewlaps, or frills that some lizards have, they can alter their appearance and apparent size during displays to intimidate rivals and potential predators or attract mates.

4.2 GENERAL FEATURES OF THE BODY

Externally, monitors are unremarkable lizards in terms of special features. The body is elongated with consequent lengthening of snout, neck, limbs, body, and tail (Figs. 4.1–4.4). This elongation, in turn, allows monitors to raise the head and neck in a periscope-like fashion to a degree unavailable to most other lizards, but convergent to postures often assumed by snakes (e.g., *Crotalus*, *Thamnophis*). The extent of elongation is quite variable, ranging from the streamlined, "typical" body plan, as in the Mangrove Monitor, *Varanus indicus* (Fig. 4.1A, B), to the high-headed, short-limbed, spiny-tailed Blunt-spined Monitor, *V. primordius* (Fig. 4.1E), and from Africa's stout-bodied, relatively short-tailed Nile Monitor, *V. niloticus* (Fig. 4.1C), to New Guinea's bulbous-snouted, gangly bodied, long-tailed *V. salvadorii* (Fig. 4.1D). The range of other external morphological variations is the subject of this chapter.

4.3 THE HEAD

Monitors are frequently described as lizards in possession of long heads and necks, but the description has limited meaning (Figs. 4.3–4.7). In fact, varanids are proportionally very much in line with many other lizards, including lacertids, pygopodids, scincids, and teiids in that the length of the head (as measured from rostral to dorsal edge of ear opening; Fig. 4.3) is often precisely equal to that of the neck (posterior ear opening to gular fold, Fig. 4.3A, B). Similarly, the head-neck to body ratio falls within the norm for that seen in other lizards. The locations for standard measurements are given in Fig. 4.4. Because varanid snout height ranges from quite flat (Fig. 4.8A–B) to broad and blunt-snouted (Fig. 4.8C, D) it is easy to assume that head-neck length varies toward the longer extremes, while

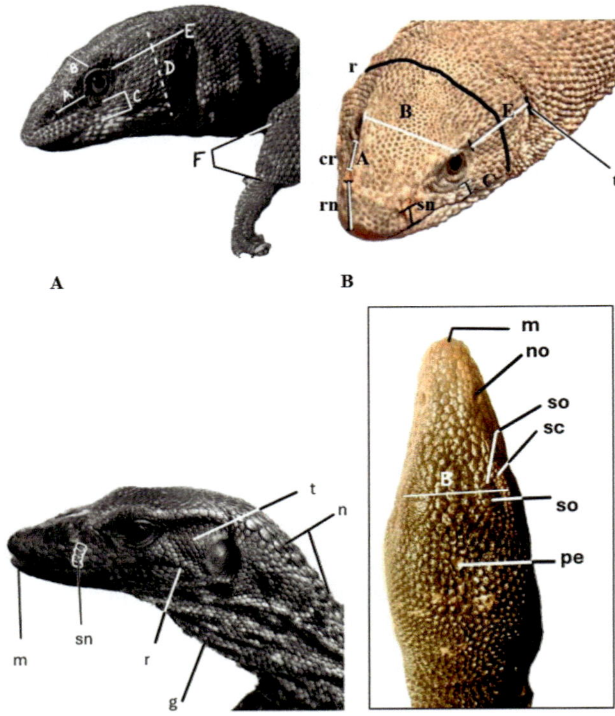

FIGURE 4.3 Landmarks of the head and their use in taking meristic data. Species shown are (A, B) *Varanus primordius*, (C) *Varanus dumerilii*, (D) *Varanus bogerti*. Abbreviations: A – nostril to anterior orbit; B – width of head across orbits; C – suborbitals; cr – canthus rostralis; D – height of head at temples; E – posterior orbit to ear length; F – antebrachial scales; g – gular region; m – mental scale; n – nuchal scales; ne – nostril to eye length; no – nostril; pe – parietal eye; r – angle of the mouth, the line for counting rictals from angle of mouth on each side; rn – rostral to nostril length; sc – supraciliaries; sn – subnasals; so – supraoculars; t – temporal scales; ty – ear opening.

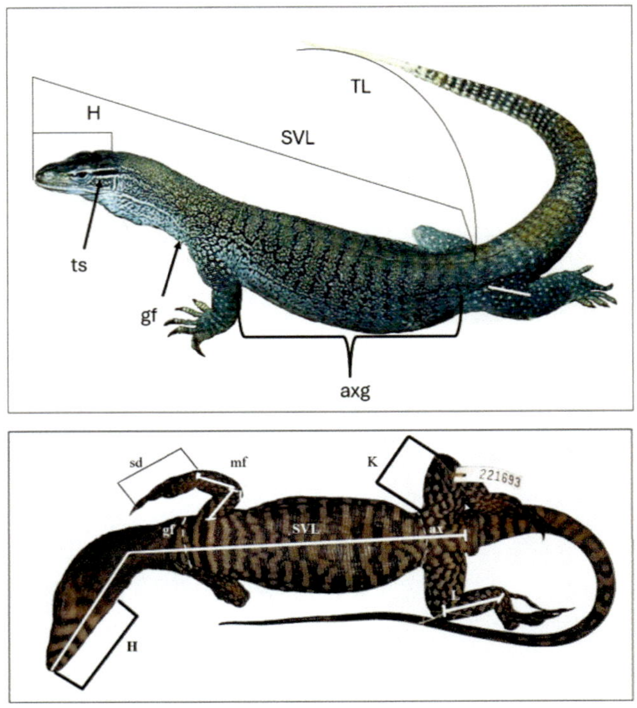

FIGURE 4.4 Landmarks and measures of the body of a monitor lizard. A – snout-vent length (SVL). B – tail length (TL). C – ventral scale rows, counted from gular fold to line across groin. D – partially extruded hemipenis (male). E – mental; scale. F – mental groove. G – gular fold. H – points to measure forearm. I – points to measure the humerus. J – subdigital lamellae of the longest toe. K – measure for femur. L – measure for crus. M – gular region (throat). N – axilla (armpit). Upper, *Varanus flavirufus*, *Varanus rudicollis* (USNM 221693).

in fact the head length is very close to the same percentage of snout-vent length in each species.

Snout configuration within the non-*Varanus* platynotans *Heloderma* and *Lanthanotus* is broad, dorsoventrally depressed, with a round nostril nearer the tip of the snout than the orbit, and the skull expands laterally just anterior to the orbit. The broadest snouts belong to adults of *V. salvadorii* and *V. komodoensis* (Figs. 4.5H and 4.9A), in which the cheeks (= maxillae) are nearly parallel, the latter also with the most anterior-facing nostrils in the family.

The head is covered in small, flat to slightly convex scales (Figs. 4.3 and 4.5), lacking clear divisions into loreals, internasals, parietals, and other scales that are diagnostic in other families. Nevertheless, they are species-specific in numbers and placement (e.g., number of subnasal, temporal, and supralabial rows, Fig. 4.3). The largest scales tend to encircle the frontoparietal region (Figs. 4.3D, 4.5, and 4.22), with the single scale surrounding the parietal organ typically largest of these. Most parietal scales are asymmetrical and not situated in equivalent left and right sides of the head (Fig. 4.4). However, the scales that encircle the parietal organ consist of one, rarely two, symmetrically placed rings. In many species (*V. salvator, Varanus prasinus, Varanus rudicollis*, etc.) there is a series of transversely enlarged supraocular scales, numbering three to seven, each broader than long (Figs. 4.5 and 4.6). There are three variations regarding proportions of the supraoculars and nearby scales: the supraoculars may be larger than, smaller than, or subequal to the interocular scales (Fig. 4.5 and 4.6), and each condition has taxonomic significance. In the *Odatria*-group and some other species, the dorsal cranial scales have a pebble-like structure, in which there are noticeable sutures between individual scales (Fig. 4.3B, D). In many other varanids, the scales may be smooth and glossy (Fig. 4.5A, F, H), or slightly keeled and dull (Fig. 4.5D). Scales may be either symmetrical (rare) or asymmetrical.

The number of postocular scale rows (Fig. 4.3A, B) ranges from a low of 12 in *Varanus timorensis* to a high of 20 in many other species. Most monitors have three or four subnasal scale rows (Fig. 4.3A, B), but *Varanus griseus* stands out with eight rows. There are three to five subocular scale rows (Fig. 4.3A–C), with three being the most common.

The canthus rostralis (Fig. 4.3A, B), the region on a line between the front of the eye and the area above the nostril,

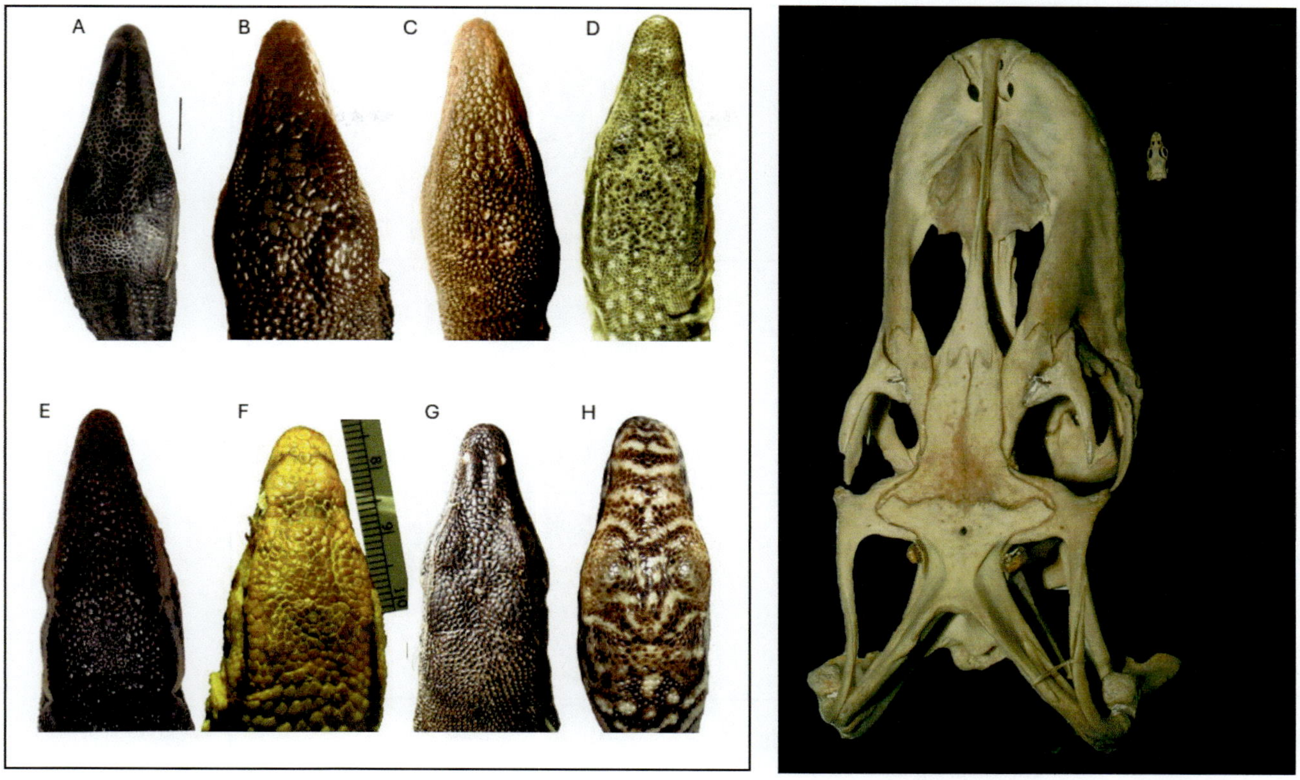

FIGURE 4.5 The range of variation in the shape of the head of *Varanus* runs from narrow and attenuated to having nearly parallel checks and a blunt snout. Shown are (A) *Varanus rudicollis*, (B) *Varanus beccarii*, (C) *Varanus bogerti*, (D) *Varanus pilbarensis* (hatchling), (E) *Varanus rainerguentheri*, (F) *Varanus dumerilii*, (G) *Varanus mertensi* [courtesy Angalino Stefani], and (H) *Varanus komodoensis* (juv.).

may be one of three forms: (1) acute with a shape-edged keel, giving the head a square aspect in sagittal section, (2) rounded and indistinct, yielding a roughly triangular sagittal aspect, or (3) intermediate. In juvenile monitors, there is an almost universal acuity of this region, formed by the comparatively large anterior supraorbital osteology. As the cranium undergoes ontogenetic growth, the snout "migrates" anteriorly, usually altering the configuration of the canthus rostralis. Among the closest allies of *Varanus*, *Lanthanotus* lacks a distinct canthal region (Fig. 4.10D) as both a juvenile and adult, as does *Heloderma* (a sister group to monitors, Fig. 4.10E, F). Adult *V. komodoensis* display a particularly weak canthus rostralis (Fig. 4.9A) in stark contrast to the structure in *Varanus mitchelli* (Fig. 4.9B) and

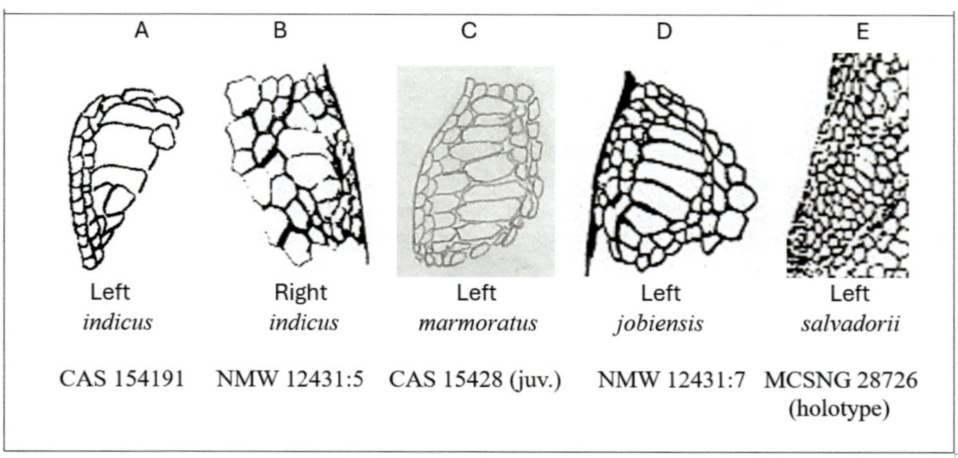

FIGURE 4.6 Variation in supraocular scale size, number, and pattern. Enlarged scales are usually distinct from those surrounding them (A, D), but in some cases they may be transversely divided (B, C) with parts subequal to the other supraoculars. In relatively few large varanids, the supraoculars may be indistinguishable or smaller than neighboring scales, a character state also useful in diagnosing several species of *Odatria*. A–D from Sprackland 1995b. E from Peters and Doria 1878.

Varanus prasinus SJSU uncat.	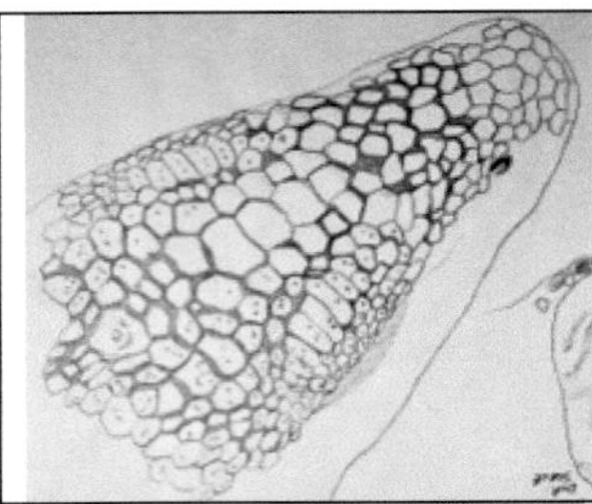
Varanus "timorensis" CAS 126913 Papua New Guinea	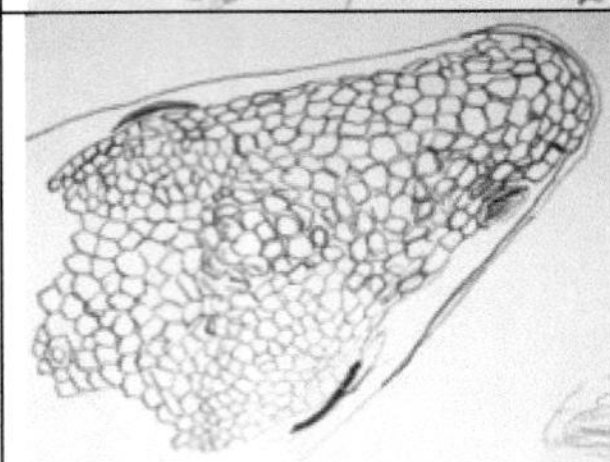
Varanus indicus	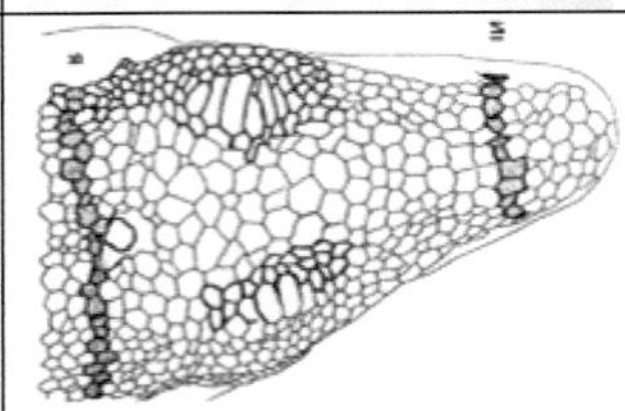
"scalaris" BMNH 1926.2.25.58 Australia: Northern Territory: Crocodile Island	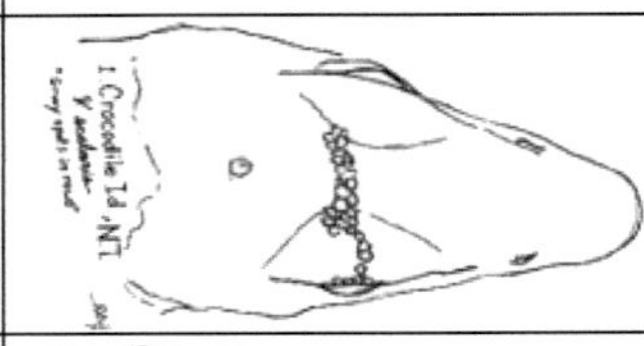
similis BMNH 82.3.27.12 Australia: Queensland: Prince of Wales Island: Torres Strait	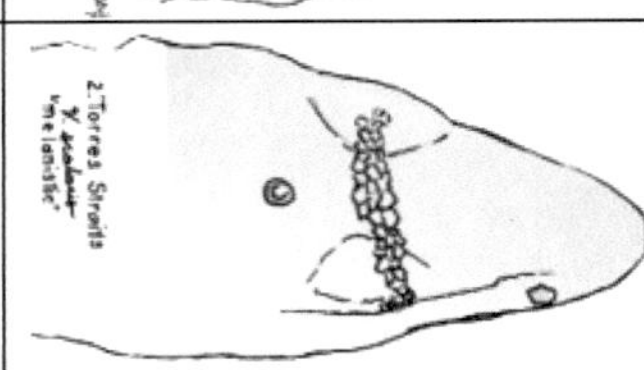
"scalaris"	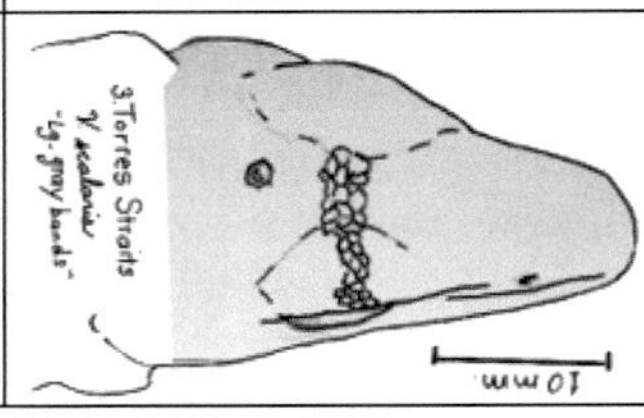

FIGURE 4.7 Variation in arrangement of supraocular and interocular scales in *Varanus*. In *Varanus prasinus*, cranial scales vary conspicuously in size, the interocular scales being the largest. The supraoculars are transversely enlarged but smaller than most other head scales. The supraoculars are slightly larger than interoculars in *Varanus timorensis* and not in a symmetrical arrangement. Like *Varanus prasinus, Varanus indicus* has transversely expanded supraoculars, but subequal to interoculars. The darkened anterior and trans-parietal scales depict where counts were made for internasal and rictal scales, respectively. The *V. "scalaris"* and *Varanus similis* specimens belong to an unresolved species group (see Chapter 11); This series probably all *V. similis*. Note that the interocular and supraoculars are slightly unequal except in the "*scalaris*" specimen, which lacks data, in which the former are larger.

Varanus bengalensis (Fig. 4.9C). Thus, an indistinct canthus and broad anterior snout are apparently plesiomorphic conditions.

The eyes of monitors have round pupils even in the sometimes crepuscular *Varanus glebopalma (*but see Sweet 1999), and well-developed, opaque eyelids. All species are diurnal and possess three types of optical photoreceptors: parietopsin, parapinopsin, and the newly described lepidopsin (Romero and de Souza 2025), giving monitors acute color vision. Squamates that possess a parietal eye, including varanids, lacertids, anolids, and *Sphenodon,* tend to have these three opsins and cone types, and all but *Varanus* also secrete pinopsin, while taxa lacking the organ produce only one (gekkonids) or two (chamaeleonids) opsins. The relative area of the eye that is exposed varies little among varanids, with the largest in the semi-arboreal *Varanus jobiensis*. The upper lid is extremely reduced, and the larger and mobile lower eyelid covers the eye completely when

FIGURE 4.8 Anterior dorsoventral depression in varanids differs from that of (A) *Varanus dumerilii* with a steep anterior drop followed by distinct, tall rise at the nostril. (B) *Varanus keithhorni* with depressed snout tip that is more level than in A; nostril mound distinct but not as much as in A. (C) *Varanus salvadorii* with no depression of anterior of snout, but nostrils in conspicuous raised prominences. (D) *Varanus exanthematicus* with the merest trace of depression of the rostral region, followed by a very tall snout and nostrils well below prominences.

closed. It is pale in color, opaque, and has a smooth texture. The inner nictitating membrane is pale bluish-white, semi-transparent, and completely covers the eyeball when closed (see Chapter 6). The upper eyelid is rudimentary but mobile, with two or more rows of supraciliary scales. The functionally blind and semiaquatic *Lanthanotus* has slightly translucent eyelids, allowing some light to enter the eye while the lizard is submerged (Fig. 4.10D). The species' reaction to light is minor, perhaps reflected in having the smallest eyes among platynotans. Largely aquatic monitors may have semi-transparent lower eyelids.

A universal feature of varanids is the depression of the rostral region at some point in their lives. The most anterior portion of the snout is slightly (Fig. 4.8A, B) or noticeably compressed relative to the nasal region (Fig. 4.7C, D) and is clearly seen in juveniles. In adults of some species, including *Varanus exanthematicus, V. bengalensis, V. niloticus, V. primordius*, and most notably in *V. komodoensis*, this feature is obscured, being small or absent relative to the massive snout. In most species, however, the depressed rostral lip region looks as if the lips were pressed together in clay. The size of the depression is the longest and most conspicuous in species that frequently burrow (e.g., *Varanus dumerilii* [Fig. 4.8A], *Varanus giganteus, Varanus gouldii, V. griseus, V. indicus, Varanus nesterovi, V. niloticus, Varanus panoptes*), but also present in species less likely to excavate (e.g., *Varanus beccarii, Varanus glauerti, Varanus keithhorni* [Fig. 4.8B], *V. mitchelli* [Fig. 4.9B], *V. prasinus*).

Two of the most plastic characters in varanids are the shape and position of the nostrils (Figs. 4.8 and 4.9). It has been proposed (McDowell and Bogert 1954; Sprackland 1991a, b) that a posterior nostril position, close to the orbit, is a plesiomorphic condition (e.g., *V. griseus, Varanus albigularis, V. bengalensis* [Fig. 4.9C], *V. exanthematicus* [Fig. 4.8D], and *V. dumerilii* [Fig. 4.8A]). In such lizards, the nostril is an oval slit, larger ventrally. At the other extreme is

FIGURE 4.9 Range of variation in the position of the nostrils (arrows) from (A) anterior (*Varanus komodoensis*), (B) equidistant between canthus of orbit and rostral (*Varanus mitchelli*), and (C) posterior (*Vbengalensis bengalensis*) position on the snout. Anterior nostril position is the least common. (D) *Varanus mertensi* showing the white tissue that passively closes the nostril with buccal air pressure while submerged.

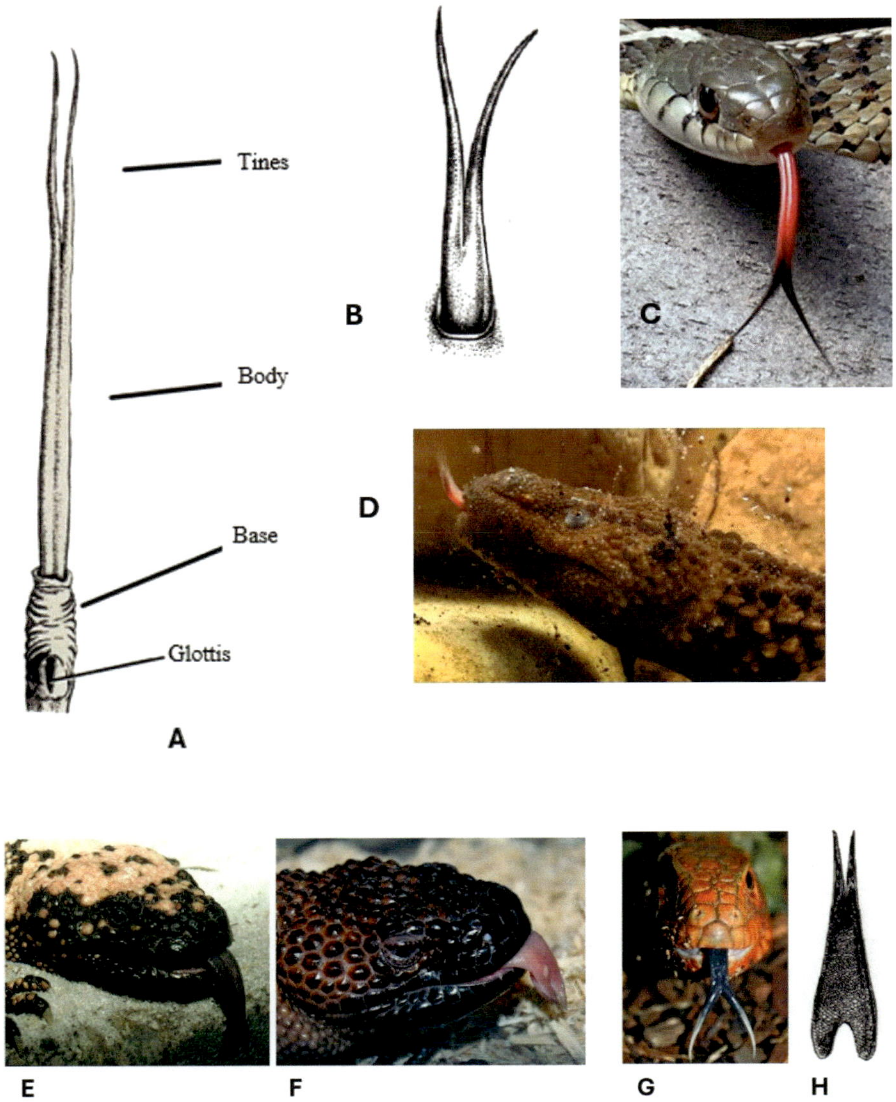

FIGURE 4.10 Bifid tongues of squamates, similar in being extremely protrusible but differing in morphology. A long, thin tongue of a *Varanus* extended (A) and in its glottal sheath (B). (C) Only snakes such as this (*Thamnophis sirtalis*) have equally thin, or thinner, tongues, but are less well-sheathed, and almost all snakes possess a notch on the ventral part of the rostral scale that allows protrusion of the Tongue without the need to open the jaws; no lizard has this trait. (D) The earless monitor, *Lanthanotus borneensis*, has a tongue very similar to that in *Varanus*, but much wider and not quite as protrusible, and more closely resembling the organ in *Heloderma*. Note the species-specific tongue coloring in (E) *Heloderma suspectum* and (f) *Heloderma horridum*, a variable character found widely within *Varanus*. The distantly related Teiidae has independently evolved protrusible bifid tongues that, though considerably different, more closely resemble those of *Lanthanotus* and *Heloderma,* as seen in the South American *Dracaena guianensis* (G) and *Aspidoscelis* (H). (B and H from Schwenk, 1995; Drawing by M. J. Spring.)

a round nostril situated near the tip of the snout (*V. komodoensis* [Fig. 4.9A], *V. salvator, V. gouldii, Varanus acanthurus*), with *V. salvadorii* [Fig. 4.8C] and *V. komodoensis* [Fig. 4.9A] having the proportionally largest round nostrils in the genus. In juveniles of *V. niloticus* and *V. dumerilii*, the nostril is round, or nearly so, becoming longer and oval as adults. Several species have medially placed nostrils (e.g., *Varanus similis, V. timorensis, V. indicus, V. prasinus, V. glauerti, Varanus tristis*) while one species has very long nostrils that span the distance of both extremes, forming a huge, narrow slit (*V. rudicollis,* Fig. 4.21A). All monitors have lateral nostrils, though *V. komodoensis* (Figs. 4.5H and 4.9A), *Varanus mertensi* (Fig. 4.5G and 4.9D), and *V. mitchelli* (Fig. 4.9B) have dorsolaterally placed nostrils, most pronounced in *V. mertensi*. The highly aquatic *V. dumerilii* has the most divergent situation for an aquatic monitor, having laterally placed nostrils that lie midway between the rostral and anterior border of the eye (Fig. 4.8A). The species *V. dumerilii, V. mertensi, V. mitchelli*, and the *V. salvator* group are the most aquatic varanids, with all but the latter having features that close the nostrils while submerged. *V. dumerilii* has a scale flap attached to an alary muscle, so that the nostril can be closed by a tiny "door." Australia's *V. mertensi* has a passive valve. A layer of white epithelium is forced upward by air pressure from the buccal cavity as the lizard submerges (Fig. 4.9D). Oddly, this ability to seal off the nostril has not, to my knowledge, been reported prior to my account (Sprackland 1989a) except by Brehm (1883), who noted,

> Two large cavities in the interior of their upper snout enable them to remain in the water for longer periods. These cavities are connected to the nostrils, can be filled with air, and can be closed off by movable edges of the nostrils.

The cavities to which he refers are the large nasal chambers, and it is passive air pressure when the monitor is submerged that pushes the "soft edges" into positions that shut the nostril. Thus, though natural selection has favored a nasal flap for each of these semiaquatic species, each has independently evolved a unique structure for the purpose. The flap mechanism, if present, is not yet known for *V. mitchelli*, also of Australia. In both Australian species, the nostril sits in a small prominence and faces dorsally (Figs. 4.5G and 4.9B). Other species that spend considerable time in water (*V. komodoensis*, the *V. indicus* and *V. salvator* groups, and *V. niloticus*) show no special adaptations for sealing off the nasal region. The first three of these taxa have frequently been seen at sea, the only varanids known to commonly frequent salt water.

Nostril placement is variable (Fig. 4.7–4.9). For many taxa, nostrils are simple holes in an unspecialized region of scales, the situation is universal in species with lateral nostrils. A large percentage of species have a small hemispherical mound, the summit bearing the nasal opening (Figs. 4.8A, B and 4.9B, C). The diameter and height of the nasal mound are typically smaller and lower in primarily terrestrial forms, and broader and higher in those that spend considerable time in the water, including the *V. mertensi*, the *V. salvator* and *V. indicus* complexes, and *V. komodoensis*. In those taxa, the nostril is in the anterior of the muzzle. The *V. prasinus*-complex, highly arboreal lizards, have the mound midway between the orbit and the snout tip. Two New Guinea species, *Varanus doreanus* and *V. salvadorii* (Fig. 4.8C), both primarily arboreal, have similar enlarged nasal capsules that create a tall and rounded snout profile (a lacertilian "Roman nose"), a situation like the even more enlarged nasal region of *V. rudicollis* (Fig. 4.21A). All three are large (>1.2 m total length), highly arboreal insectivores/carnivores, though none has been well-studied in the wild. In *Varanus olivaceus*, there is a much smaller nasal region, but the nasal capsule is sufficiently large to shape the nasal mound into a short ovoid region, a simpler version of that seen in the three carnivorous species. *V. olivaceus* is also a large arboreal varanid (total length >1.8 m), its diet consisting of fruits, nuts, and large snails.

Structurally, the nasal passages of varanids are similar in general morphology and have been thoroughly described for one species by Bellairs (1949). The naris is the distal point of an elongated nasal chamber that may extend posteriorly to near the anterior orbital rim. The chamber lies medial to the salt glands, if present (*q.v.*), ending posteriorly at a ventral choana that opens into the mouth. The nasal chamber is made of soft epithelial tissue, sometimes with a columnar cellular lining, and may be partially enclosed by thin cartilage. The whole chamber rests on the septomaxilla, beneath which the vomeronasal (Jacobson's) organ lies. There is no direct connection between the vomeronasal organ and the nasal chambers; the salt gland opens at the very edge of the nostril or posteriorly at the lachrymal ducts. There is a correlation with the size and length of the nasal chamber and canthus rostralis; a short or flat nasal chamber is associated with an acute canthus rostralis, while a large chamber accompanies a rounded or indistinct canthal ridge. This subject is covered further in Chapter 6.

In the same way that bilateral seeing and hearing organs improve hunting and defensive capabilities, the bifid tongue may function to provide a "stereo" taste and smell function. For active foragers, such as monitors, such an instrument could improve tracking by providing inputs for strong/weak signal directions and facilitating homing. Varanid tongues are used to help find food, but only rarely to manipulate food.

There is notable variation in tongue color, the organ being uni-, bi-, or tri-colored. The hue and pattern are consistent within species (Sprackland 1999b, c), but no particular pattern seems unique to any species. Variable tongue color (Fig. 4.11) is therefore a feature that has taxonomic importance at the species level (table 1) and may represent a good example of neutral mutation and selection – that is, tongue color has neither a positive nor detrimental effect on the lizards (Sprackland 1995b) – or selection for intraspecific recognition (Sprackland 1999b, c). Most commonly, the tongue is pale pink or coral (Fig. 4.10A), or a dark blue-black (Fig. 4.10C). The least common is a yellow or pale cream-colored organ (Fig. 4.11D) that is characteristic among some members of the *V. indicus* complex,

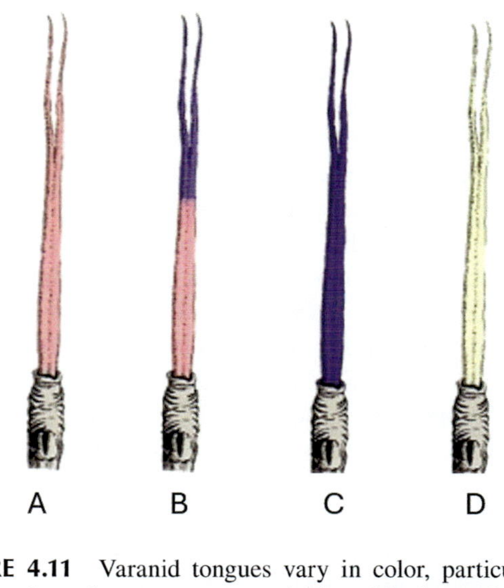

FIGURE 4.11 Varanid tongues vary in color, particularly in regions where they occur sympatrically with close relatives. In the related *Heloderma*, this also holds true (see Fig. 4.9E, F). (1) Pink (may be darker). (2) Bicolor blue/purple and pink. (3) Blue or purple. (4) Yellow.

FIGURE 4.12 Variation in tongue color in possibly allopatric populations of the Australian Lace Monitor, *Varanus varius*. Three colors are known for this taxon: (A) Pink (Eden, NSW, southeastern population), (B) purple (Yathong Nature Preserve, central NSW, west of the Great Dividing Range), and (C) yellow (roughly from Brisbane North). Map file is licensed under the Creative Commons Attribution-Share Alike 2.5 generic license. Noted localities based on live and preserved specimens examined by the author.

some populations of *V. varius* (Fig. 4.12), and the Komodo Monitor, and in the *V. indicus* group has been linked to both specific taxa and localities (Bohme et al. 1994, Sprackland 1995b, 1999b, c). In *V. varius*, where three tongue colors have been noted (Sprackland 2009; Fig. 4.12), it was suggested that cryptic species may yet be named (see Smissen et al. 2013). There are also cases where hatchling and young Lace Goannas have a pale-yellow tongue, but this changes to pink within the first year or so post-hatching. A few species have a clearly distinct bluish-purple set of tines but a pink posterior portion of the tongue (Fig. 4.11B).

The tongue is uniform in structure among varanids, being thin, deeply bifid, and much longer than in any squamate excepting the Chamaeleonidae. Unlike the latter, the monitor's tongue is in no way involved with the physical capture of food and only rarely is used to manipulate food in the mouth. In most monitors, the full length of the tongue equals the length from rostral to shoulders, though only about half of the organ is actually extended. Varanids have considerable control over the tongue, and they can limit how far it extends and how wide the anterior tines may spread. Though the tongue brings particles into the mouth and to the nasal chamber and Jacobson's organ, it lacks taste buds (Schwenk 1984, 1985). Varanids have a unique sheath for the tongue, a fleshy cylinder that lies inferior to the esophagus and extends a short distance anterior into the buccal cavity. These five characters – tongue length, tongue width, lack of taste organs, smooth surface, and a protective sheath – readily distinguish monitors from other forked-tongued lizards.

All *Varanus* have a large, external ear opening (external auditory meatus), with the tympanum usually near the surface (Fig. 4.3C). The anterior rim of the tympanum attaches to the posterior excavation of the quadrate bone. A cartilaginous extracolumella is clearly visible, its distal end a small tubercle in the middle of the tympanum. The related *Lanthanotus* lacks a tympanum and external ear opening, and the quadrate, extracolumella, and stapes closely resemble the condition seen in *Varanus*.

4.4 THE BODY

4.4.1 General Morphology

The neck may be covered dorsally in tiny, fine scales, or alternately by large, shield-like structures (Figs. 4.13–4.15). In the African savannah monitor, *V. exanthematicus*, the

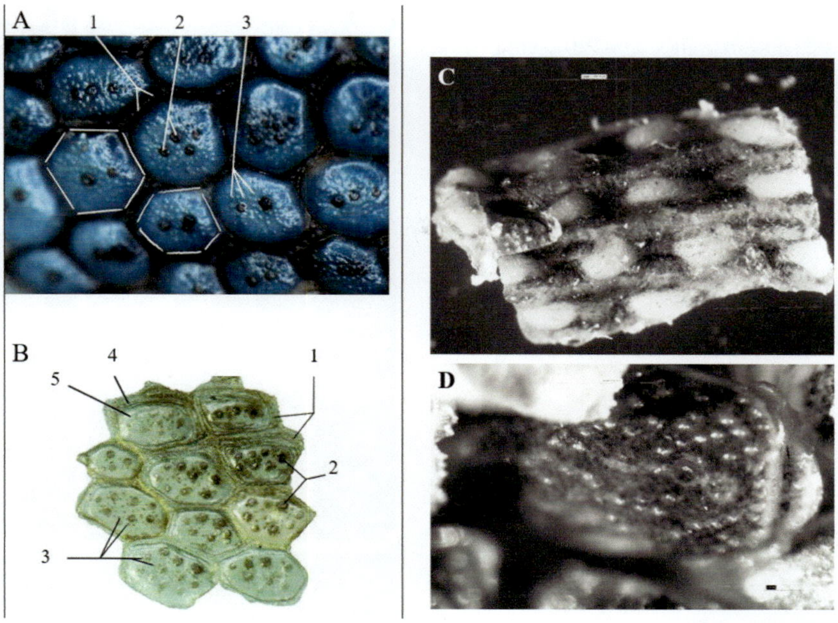

FIGURE 4.13 Structures and shapes of nuchal scales of *Varanus prasinus*. (A) External view of anterior scale group, showing smooth (granular) surface and black cutaneous sensory organs (sensilla). The granular interstitial spaces are the areas that provide skin flexibility. (B) External view of a group of shed scales from adjacent to those in (A). (C) Underside (medial) of scales in (A) above, their borders slightly obscured by the sheet of connective tissue that bonds them to the cranial bones. (D) Dorsal view of a single nuchal, greatly enlarged, showing the multitude of small hemispherical structures (3 in photos A and B). Abbreviations: 1 – tertiary granular scale rings between individual scales; 2 – large discoid sensilla, typically in groups >2 per disc, unlike the larger single structure of elongated and keeled scales (see Fig. 4.14A); 3 – small raised structures typical of all epidermal scales; 4 – secondary scale rings; 5 – body of scale, forming a discoid and almost universally hexagonal structure (white dashed lines). The hexagons may be nearly round, as in these nuchals, or elongate, as in ventrals and caudals.

round nuchals are twice as large as the dorsals and are unique among monitors in being both enlarged and smooth. In contrast, the nuchals of *V. dumerilii*, though similar, are slightly keeled and about 0.5 times larger than the dorsals. *V. rudicollis* and *Varanus bogerti* have pyramidal, keeled nuchal scales, while *Varanus spinulosus* has small, conical, spine-like neck scales, though these are barely noticeable without magnification.

The dorsum of monitors is usually covered with uniform scales that are small, large, or intermediate when compared with different species, and reflected in mid-body scale counts, with small-scaled species having higher counts than larger-scaled species. An exception is *V. dumerilii,* which somewhat resembles *Lanthanotus borneensis,* having heterogeneous scalation (Sprackland 1989a, 1992; Fig. 4.14A). Species with “large” scales tend to have lower midbody scale counts (e.g., 96) than those with “small” scales (e.g., 150). Dorsal scales are complex structures, composed of a polished round or oval plate on top of a raised tubercle (Fig. 4.19). The plate may contain one or several tiny scale-organs probably related to the sense of touch (Figs. 4.13 and 4.14). Scale organs are found on all body scales, including cranial and ventral scales, being fewer in number on the ventrals. The structure and taxonomic significance of these scales have come under some study (i.e., Bauer et al. 2023, Bauer and Russell 1988, Busack unpubl., Düring and Miller 1979) but are still subjects requiring research. Sprackland (1989d) postulated that labial scales are used to help identify food from background objects. They are discussed further in Chapter 6.

Varanids with compressed, or partly compressed, tails sport a pair of “crests” that are formed by caudal scale rows along the sharp curve on the dorsal surface of the organ. These crest scales are never larger than other caudal scales and, in fact, are the smallest scales on the tail. Re-creation illustrations of prehistoric varanids with dorsal crests (i.e., Norell *et al.*, 1992) are without foundation. The enlarged superior vertebral spines in the Pliocene †*V. prisca* were sometimes thought to be the foundation for a dorsal crest. However, these structures are well within the scaled size expected in such a large varanid (estimated at 4–6 m) and are not likely indicative of a sagittal crest.

Ventrally, the scales are square (anterior) or rectangular and arranged in a series of columns and rows, as occurs in the Lacertidae. Teiidae, and Xantusiidae (Figs. 4.4 and 4.16). Specific shape varies among taxa. They may be twice as long as broad and have rounded posterior edges, as in *V. indicus* and *V. timorensis*, or rectangular with straight borders and slightly rounded posterior edges (as in *V. exanthematicus*). In the tree-dwelling members of the *V. prasinus*-complex, each ventral is an elongate, slightly keeled scale that terminates with a short median spine, structures that may facilitate their lives in the trees. In all cases each ventral scale has,

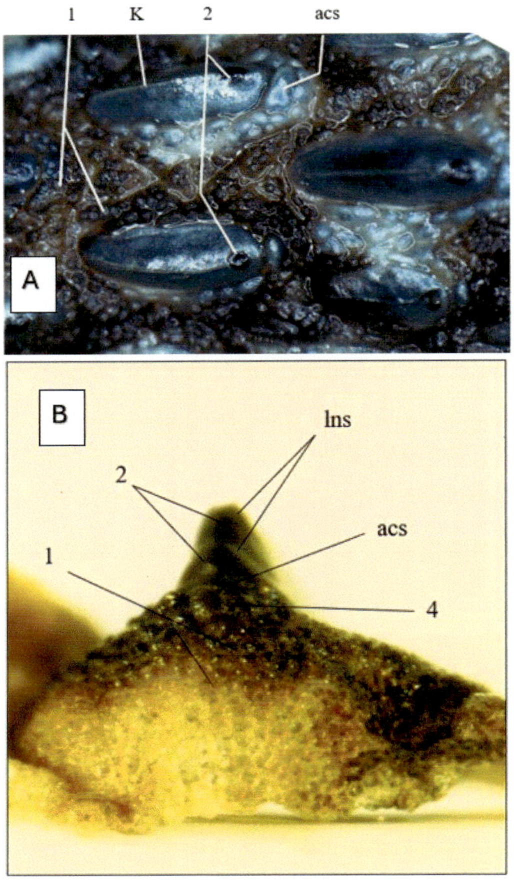

FIGURE 4.14 The nuchal scales of *Varanus rudicollis* are the most strongly keeled of any known *Varanus*. (A) Series of nuchals just left of the mid-sagittal line. (B) Transverse posterior view of a single nuchal scale; note how the "single" scale is composed of numerous much smaller scales. Abbreviations as in Fig. 4.12, plus: 1ns, primary distal scale element; asc, posterior secondary accessory scale; K, distinct sagittal keel.

along its posterior edge, a series of one (very rarely two) row of minute, granular scales.

Ventral scales may be smooth or keeled, but keeling is usually visible on scales further from the centerline of the body, or under magnification (Fig. 4.16). Not surprisingly, some authors have scored specimens as having both keeled and smooth ventrals because the condition is sometimes difficult to ascertain. Juveniles typically have small, square anterior ventrals and rectangular posterior (post forelimb) scales, while adults usually have only rectangular scales. In monitors, there is a distinct and non-expansible gular fold, just anterior to the shoulders, but this structure is absent in the allied *Lanthanotus* and *Heloderma*. The number of ventral scale rows (as well as midbody scale rows) is somewhat diagnostic for species. Authors have used a variety of methods for counting ventral scale rows, resulting in some published discrepancies in diagnoses. For monitors, I count from the midpoint of the gular fold, continuing to the cloaca (Fig. 4.4).

Monitors, like all squamates, have a common transverse external opening for the urogenital tracts called a cloaca (Figs. 4.16 and 4.17B–D). Immediately anterior to

FIGURE 4.15 Scales of *Varanus*. (A) *Varanus dumerilii*, dorsals, between shoulders RGS. (B) *Varanus "similis,"* ventrals, midbody USNM 128490. (C) *Varanus indicus*, dorsals, mid-dorsum USNM. (D) Same, Ventrals. (E) *Varanus pilbarensis*, juv., dorsals anterior to shoulders RGS. (F) Same, ventrals, midbody. (G) *Varanus rudicollis*, anterior (left) and more posterior nuchals RGS. (H) *Varanus flavescens*, nuchals USNM 267002.

the cloaca, some species (e.g., *V. salvator* group, *Varanus togianus, V. niloticus, V. dumerilii*) possess a pair of pre-anal pores (Fig. 4.17). Their function is uncertain, but it is likely linked to scent marking, as in iguanines. Other species, representing the odatrian monitors from Australia, have a patch of enlarged, spine-like scales along the lower lateral surface of the tail, just anterior to the cloaca (Fig. 4.26). In both cases, males have larger, distinct pores or tufts (never both), while in females, these structures are reduced or absent. In addition to hemipenial differences, the Indonesian *Odatria timorensis* differs from other odatrians (to which it has long been associated) by the absence of these pre-cloacal tufts, though some Australian *Odatria* similarly lack them (*Odatria semiremex, Odatria scalaris*).

4.4.1.1 Scales

The skin is the largest organ of most vertebrates and is more complex than often acknowledged. Skin performs many functions beyond holding the body together and protecting it from the external environment, in which it has several roles. Reptile scales are epidermal in origin, thus not homologous with the dermally derived scales of fishes, and do not separate individually from the body. Instead, they are of two types: ventrals and sometimes caudals that overlap in the manner of roof tiles, and those that lie distinct,

FIGURE 4.16 Ventral scale and marking variation in *Varanus*. (A) *Varanus* sp. n., related to *Varanus jobiensis*, with a pale, unmarked ventral surface. (B) *Varanus rainerguentheri* (USNM 579388) with a mottled ventral pattern. (C) *Varanus rudicollis* (USNM 259133) with distinct ventral banding that extends onto the tail. Note that none of these species has preanal pores (cf. with Fig. 4.17).

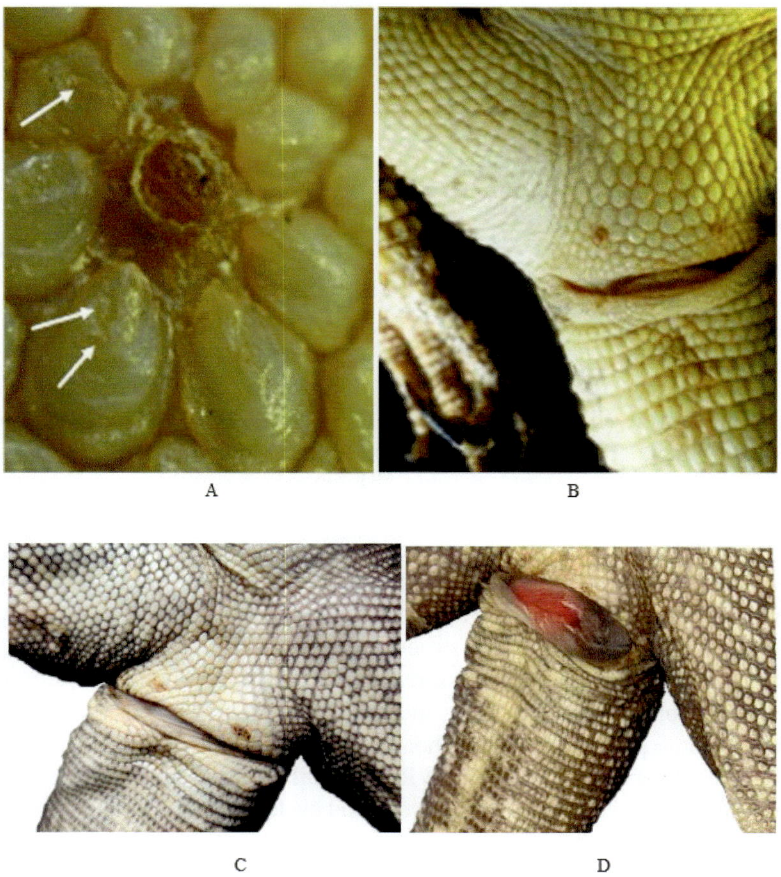

FIGURE 4.17 Preanal pores of (A) *Varanus dumerilii*, one of the few varanids to have such organs. (A) The right preanal pore, enlarged. The surrounding scales have one or more apical nodes (arrows). Note how in both scales the specialized scale is in the midst of a well-defined circle of other scales. (B) Peri-cloacal region showing the preanal pores, smooth ventrals, cloaca, and feebly keeled subcaudals. (C) In this male *Varanus togianus*, preanal pores are present and though pigmented, are less distinct and not surrounded by a circle of scales. (D) A female *V. togianus* lacks preanal pores but does have a few scales with tiny dark spots. (C and D) courtesy Michael Stefani.

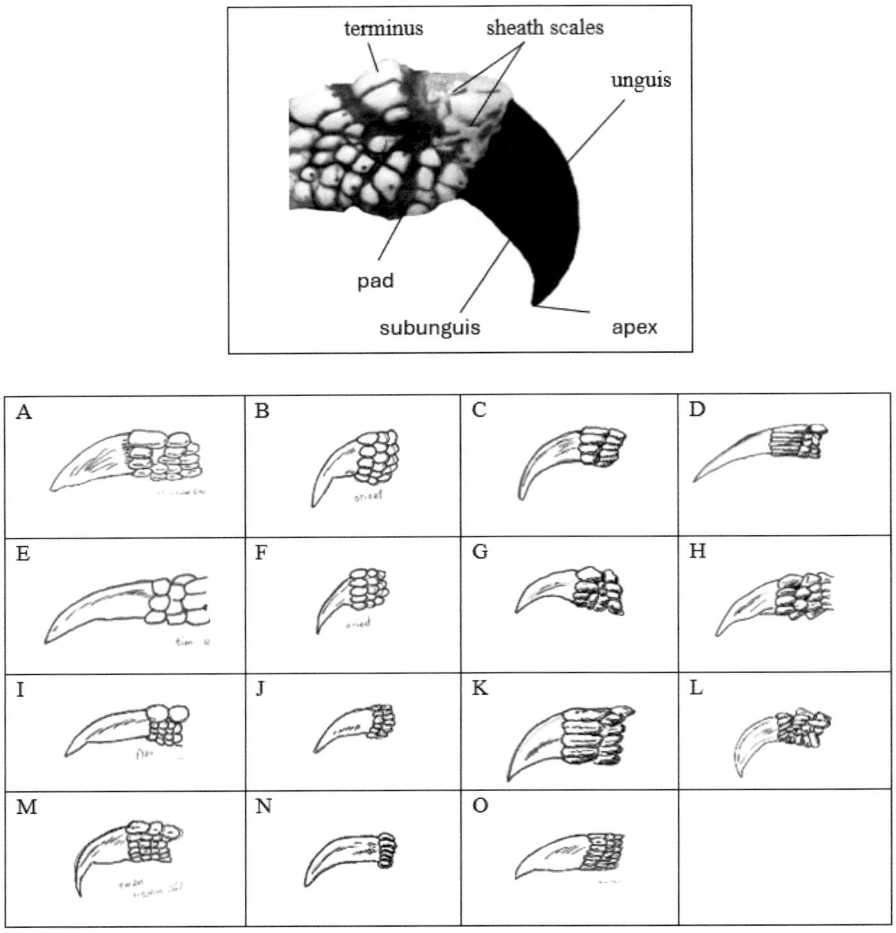

FIGURE 4.18 Variation of claw morphologies among species of *Varanus*. Upper: Diagram of major components of a claw of *Varanus prasinus*. (A) *Varanus gouldii*, USNM 64083. (B) *Varanus orientalis*, no data. (C) *Varanus indicus*, FMNH 41315. (D) *Varanus griseus*, no data. (E) *Varanus timorensis*, USNM 204026. (F) *Varanus orientalis* no data. (G) *Varanus prasinus*, RGS. (H) *Varanus bengalensis*. (I) *Varanus flavescens*, USNM 267002. (J) *Varanus "similis,"* USNM 128485. (K) *Varanus salvator*, RGS. (L) *Varanus jobiensis*, no data. (M) *Varanus exanthematicus*, USNM 161744. (N) *Varanus rudicollis*, RGS. (O) *Varanus rosenbergi*, USNM 63120.

adjacent to each other. Varanid scales vary both among and within species, with deceptively thin epidermal layers that are quite strong and pliable (Figs. 4.19 and 4.23). The composition that makes lizard skin flexible, durable, and impenetrable – the characteristics that define its value as leather –also makes it the outermost barrier of defense against threats, from microbes to claws and teeth. Skin restricts the penetration of ultraviolet radiation (UVR) and the evaporative loss of body water. Skin also hosts structures that are sensitive to tactile stimuli, including texture, density, and pressure, and to some degree, to heat and taste. Indeed, the chromatophores present and under hormonal control are involved in adjusting the entry of UVR to the sub-epidermal body. Skin is flexible where expansive movement is needed, including neck, limbs, and torso, and completely immobile elsewhere, including skull, digits, and tail. Squamate integument has been well studied (Düring and Miller 1979, Fuchs and Fuchs 2003)Guibé 1970).

Lizard skin follows the general plan of other vertebrates, having functionally and physically distinct layers broadly divided into the inner dermis and thinner outer epidermis (Fig. 4.19). The epidermis consists of an outer β keratin layer and an underlying α layer (Fig. 4.18). The β layer, also known as the *Oberhautchen* or stratum corneum, is a major component of the skin, found only in squamates, and is most conspicuous as a layer during ecdysis (shedding), covering the entire exterior except the tympanum and eye. The keratins differ structurally, the α forming helices and β pleated sheets (Lillywhite 2008). The β keratins are stiffer than other dermal components, and form the hard parts of scales, while the more flexible α keratin lies inferior to the β keratin layer and is most prevalent in soft, mobile areas and permits flexibility of the skin, such as at the hinges between scales and eyelids (Figs. 4.13 and 4.14A). This layer also individually encloses the small, round or oval osteoderms. Though often considered to be dead skin, the epidermis contains a considerable number of active cells and is rich in collagen fibers and chromatophores, but relatively poor in blood vessels. Varanids shed their *Oberhautchen* in patches, sometimes quite large. The outer surface of the *Oberhautchen* may have microstructures that reflect the structure of the immediately lower α and β layers, and includes pits, nodules, hair-like sensory structures, and keels and spines. It is the condition of the *Oberhautchen* that determines whether a scale is smooth or keeled (Figs. 4.14 and 4.15).

Basal to and physiologically more active than the epidermis lies the dermis, comprising a deep, collagen-rich stratum compactum, which binds the dermis to underlying muscle or bone and protects the upper stratum laxum (Fig. 4.19). In the former case, the skin is immovable (and most difficult to remove in dissection). The stratum laxum contains a large variety of capillaries, chromatophores, collagen fibers, osteoderms, smooth muscles, and nerves. The number of chromatophores, particularly melanophores, exceeds that of the epidermis. The collagen fibers provide flexibility to the skin and make up much of the tissue inside large, keeled scales of species such as *V. rudicollis* (Figs. 4.14G, 4.19, and 4.21), *V. exanthematicus*, and *Varanus nuchalis*. Varanids have very few cutaneous glands, but those that are present are also rooted in the stratum latum. Damage to the dermis, more than the epidermis, may lead to scarification.

Varanids possess enlarged crestal scales which vary considerably among taxa (Figs. 4.19G, 4.20B, G, and 4.22). Each scale is composed of several units (Fig. 4.22). Basal is a square, polygonal, or hexagonal region surrounded by

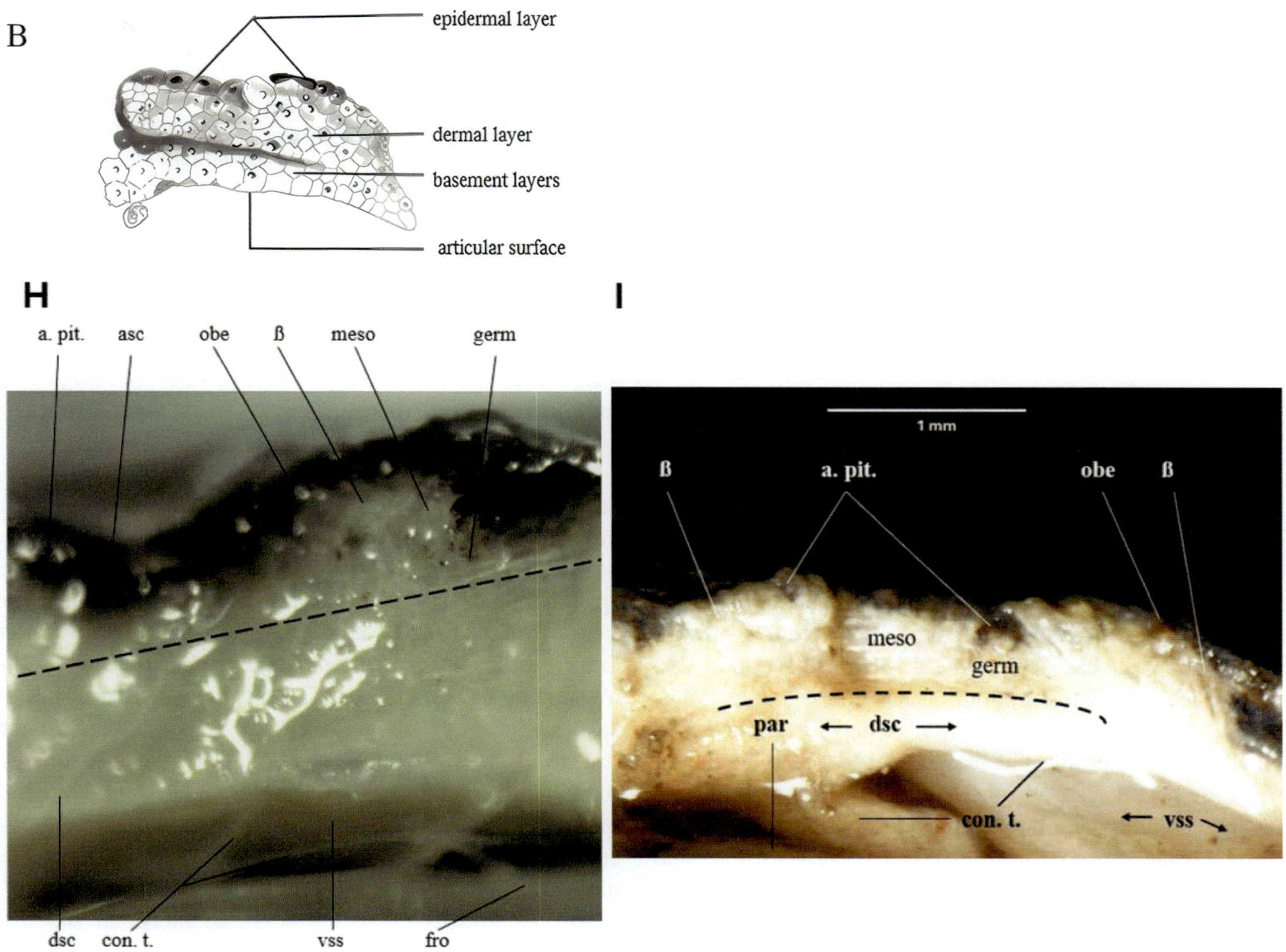

FIGURE 4.19 Head and nuchal scales of *Varanus prasinus* (A–D, I) and *Varanus rudicollis* (E–H). (A) Diagram of a single cranial scale just anterior to keeled nuchals in a slight ventrolateral aspect, partially dissected to reveal the spongy layer of dermis. (B) A photograph of the same nuchal. (C) Diagram of a scale from the left frontal region, and (D) photograph of the same. (E) Photograph of cross-section of two nuchal scales from mid-nape and (F) diagram of the same. All magnifications are 25×. (G) Partially dissected large nuchal scale, posterolateral view, with left lateral epidermis excised. Note that although its proportions are exaggerated, it follows the same anatomical configuration as flatter scales, but with a thicker epidermal layer around and anterior to the apical "pit." Magnification 35×. (H) Greatly magnified (35×) sagittal section of a single cranial scale, in situ, from the frontal region of *Varanus prasinus*. The dotted line divides the upper germinativum layer from the lower dermis. (I) Same scale as seen in (C and D) with layers demarcated and labeled. Abbreviations: 1 = tertiary scale rings between complete scale units; 4 = secondary scale row; a. pit. = secondary sensillum; apical "pit" = main body of sensillum; asc = posterior secondary accessory scale forming posterior-most portion of scale; α = epidermal alpha layer; β = epidermal outer beta layer; con. t. = connective tissue between scale and bone; dsc = stratum compactum of dermis; dsl = stratum laxum of dermis; fro = frontal bone; germ – stratum germinativum; mel = melanophore; meso = mesos layer; obe = *Oberhautchen* layer of a secondary cutaneous sensillum; par = parietal bone. Bars equal 1 mm.

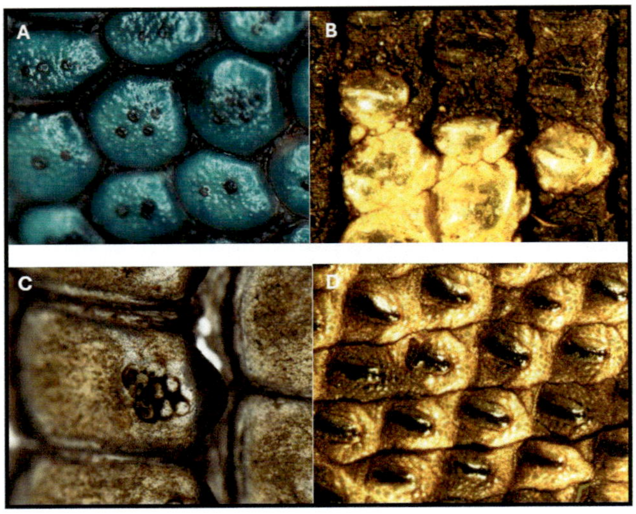

FIGURE 4.20 Scalation variety in *Varanus*. (A) *Varanus prasinus* posterior nuchal scales with two or more sensilla (black) and granular surfaces. (B) *Varanus salvadorii* with keeled and characteristic irregular dorsal scales, differing in size and shape. (C) Enlarged view of dorsal scale of *Varanus primordius* with unusual arrangement of nine sensilla. Most of the scales in this species have at least four sensilla. (D) *Varanus mertensi*, showing large hexagonal scale bases and relatively small cap scales. Sensilla are clearly visible.

equal-sized units. These are separated by an interstitial gap of α- and very little β -keratin that forms the hinges, giving the integument flexibility to bend. Scale units align in well-defined columns and rows such that bending of the body is done along the hinge row, not the scales. The largest structure of a scale above its base is a "crestal" (McDowell and Bogert 1954), which varies from oval to oblong in shape, usually slightly broader posteriorly, and extending most of the length of the base (Figs. 4.13, 4.15G–H, 4.20, and 4.22). Atop each crestal is a smooth, usually glossy, oblong apical cap that covers about a quarter of the crestal, including much of the posterior as well as the dorsal and upper lateral regions (Fig. 4.22). Caps are usually dark, even when seen on light colored scales, but there are exceptions (Fig. 4.23A, B). A distinct posterodorsal sensillum (Figs. 4.13, 4.22, and 4.23; see below) is found in almost all caps and ventrals, but may be less distinct among the gulars, ventrals, and caudals. Anterior to the crestal sensillum, there may be a low sagittal ridge of very small hemispheric cells; the ridge is larger and more distinct anterior to the cap (Figs. 4.13A and 4.14), and the larger the cells, the more keeled is the appearance of the scale. The large crestal is encircled by two or more rings of much smaller scales. A single large posterior secondary accessory scale (Figs. 4.13, 4.14, 4.19, and 4.21) is present in many species (e.g., *V. dumerilii*, *V. prasinus, V. rudicollis*) or is entirely absent (e.g., *V. mertensi, V. varius*). Though the base layer is square or nearly so (rectangular in ventrals and many caudals), the crestal and its rings commonly form a hexagonal outline (Figs. 4.13A, B and 4.20).

Crestal and other scales bear smaller mechanoreceptors (= sensilla; Bauer and Russell 1988, Düring and Miller 1979, Guibé 1970; Figs. 4.23 and 4.24), which have also been recorded in geckos (Bauer et al. 2023), Lacertids (Allam et al. 2019), and iguanians (Ananjeva et al. 1991, Preiss 1922), including *Calotes* (Paranjape 1974) and *Agama* (Harris 1963). Though similar organs are seen in many squamates, monitor sensilla tend to be symmetrical and number two or more per scale (Fig. 4.20A and B). In many lizard taxa, the sensilla are concentrated or limited to labial (Ananjeva et al. 1991) and caudal scales (Bauer et al. 2023), and one hypothesized function of caudal sensilla is to signal precisely where on the tail autotomy should occur when the appendage is seized (Russell et al. 2014); if so, the paucity of caudal sensilla in the autotomy-lacking *Varanus* supports that idea. This may indicate a link between non-autotomous lizards and the much longer retractor penis muscle, compared with lizards that have autotomy (Rachel Keefe, pers. comm.). The sensilla of *Varanus* are characteristic of body scales and are most numerous on cephalic, nuchal, and limb scales. Ventral scales have sensilla, but they are reduced in some species and many juveniles. Sensilla are universal if caps are present, except on the tail, where they may be absent ventrally or from some lateral scales. Universally, the sensilla or cap scales are larger than those on the rest of the scale and are situated posterodorsally (Figs. 4.21 and 4.23). In any single specimen, there is almost always one such large sensillum per cap (Fig. 4.14), but there are exceptions where two or three may be present (Fig. 4.20A, C).

The sensory organs come in different shape among taxa, but the most common are shaped is a form resembling an inverted pie pan, being convex or flat in live and freshly killed lizards (Fig. 4.24). Some authors have called these "pit organs," but a sunken pit is present only in dehydrated or soon-to-be-shed scales, and most noticeable in the sensilla found on the cap scales (Fig. 4.22). Each sensillum is surrounded by a contiguous ring of much smaller hemispherical structures, which delineate the disk's circumference (Figs. 4.21A, B, 4.23, and 4.24). Unlike the morphology seen in some gekkonids (Bauer et al. 2023), varanids seem to lack hairlike microbristles on their sensilla. There has been speculation about the function of the disks, from tactile mechanoreceptors (Ananjeva et al. 1991, Russell et al. 2014), for which there is considerable anecdotal support. The suggestion that they may be used to detect heat is one I strongly doubt, largely because veterinarians and keepers have voluminous evidence that the lizards may suffer severe, even fatal burns from remaining too long on heat lamps (Sprackland 1992, 2009). An effective heat detector would presumably have evolved in response to such possibilities as taking up heat from a hot substrate, such as rock or sand, and the paucity of ventral sensilla supports this hypothesis. Regarding the sensilla on the posterior edges of the scales of *Agama*, with their hairlike extension, Harris (1963, Fig. 4.5) is probably correct in stating, "It is possible that these are tactile sense organs" (p. 17). Sensory papillae are also known in crocodilians, where they detect changes

FIGURE 4.21 The strongly keeled head and nuchal scales of (A–C) *Varanus rudicollis* and (D) *Varanus nesterovi*. (A) Live rough-necked monitor in which the large, keeled nuchals and flat, polished cranial scales are clearly visible. Notice that the large scales are restricted to the nape, and while the body has similar scales, they are smaller and barely descend ventrally. Lateral and ventral scales also differ in size and appearance. (B) Diagram and (C) photograph of a single nuchal scale in left lateral view. The lower rows form rings around the enlarged uppermost crest. All scales contain discoid surface structures, but only the most distal scale has one, rarely more, large apical cap, or "pit," a sensory structure. (D) *V. nesterovi* has the most acutely pointed nuchal scales known in *Varanus* (photo courtesy Enrico Lo).

in water flow, such as occurs when an animal steps or drops into the water. Discovered by Soares (2002) and named dome pressure receptors (DPR) (also integumentary sensory organs, or ISO by Leitch and Catania 2012), they are found only around the mouth in *Alligator*, especially the lower jaw, and body-wide in the more fully aquatic crocodiles (Leitch and Catania 2012, Soares 2002). I have looked for a similar arrangement in varanids, and though species differ on the placement of labial sensilla, there was no discernible pattern, e.g., there were not more sensilla on largely aquatic species (*V. dumerilii, V. mertensi, V. mitchelli*) compared to those that are more terrestrial or arboreal, but rather fewer. Leitch and Catania demonstrated that the mechanical sensitivity of crocodilian sensilla is greater than that of primate fingertips. No equivalent study has been conducted with *Varanus.*

The number of sensilla per scale varies from one to more than the common maximum of eight (Fig. 4.19); cranial and facial scales have the greatest number.

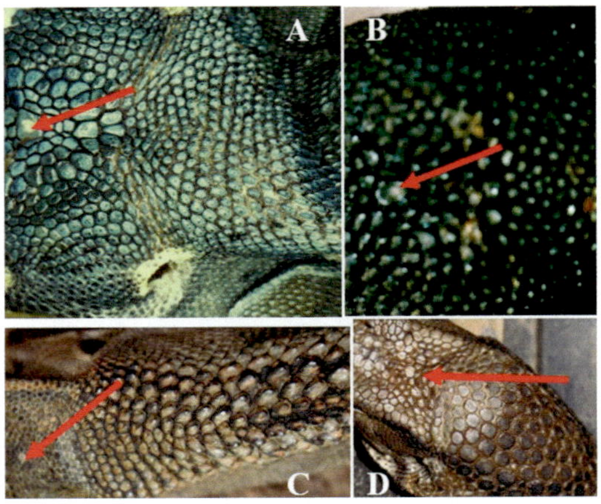

FIGURE 4.22 Cranial and nuchal scales compared in four species of *Varanus*. (A) Cranial scales of *Varanus prasinus* are nearly smooth and flat, followed by grain-shaped keeled nuchals. (B) In the closely related *Varanus bogerti* both cranials and nuchals are keeled, and in contrast to *V. prasinus*, the cranials are cone-shaped. Both species are from the New Guinea region. (C) The small, smooth cranials of *Varanus rudicollis* are in stark contrast to the very large, keeled, triangular nuchals. (D) The raised and slightly keeled cranials, and the flat, round, and smooth (or very feebly keeled) nuchals of *Varanus dumerilii* clearly distinguish the species from *V. rudicollis*. Both species are from Southeast Asia. The arrows point to the parietal organ, and the posterior curvature of the parietal bone can be seen at a short-distance caudad (to the right). In (A and D), the left tympanic opening is clearly visible.

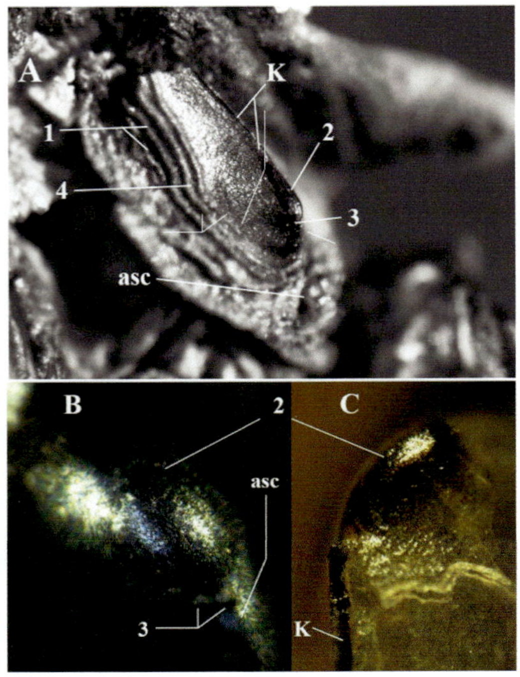

FIGURE 4.23 The enlarged nuchal scales of *Varanus rudicollis*. (A) Dorsolateral view of a single scale as seen in Fig. 4.6 in situ, showing an isolated scale and its placement between other scales. The secondary and tertiary scale rows and the posterior secondary accessory scale are all clearly visible. Note the shape and height of the scale in the right background. This species has the largest keeled scales within the Anguimorpha. (B and C) Greatly enlarged images of the apical pit of the nuchal scale in (B) left posterolateral, anterior at upper left, and (C) left lateral (anterior at bottom) views. 1 – tertiary scale rings between complete scale units; 2 – large, usually single, posterior apical "pit," the cutaneous sense organ; 3 – small raised structures typical of all epidermal scales; 4 – secondary scale row; asc – posterior secondary accessory scale; K – distinct sagittal keel.

4.4.1.2 Head Scalation

The cranial integument consists of hexagonal (though usually non-symmetrical) scales that do not fall into the clearly defined prefrontal, frontal, nasal, parietal, and other specific types common among iguanines, skinks, and xantusiids. The scale types that monitors share with the other lizards are (a) the universally present single parietal scale that houses the parietal eye (Figs. 4.3 and 4.22), and (b) the series, when present, of enlarged supraoculars, though these, too, are not always symmetrical on left and right sides (Figs. 4.5 and 4.6). I distinguish varanid cephalic scales into four types: dorsal, labial, temporal, and gular, each with distinctive traits. Dorsal cranials are the largest scales of the head, most conspicuously so from the interorbital to mid-parietal sections (Fig. 4.22). The most prevalent forms of dorsals are raised, non-overlapping scales separated from their neighbors by a recess encircling their circumference. The height of a scale and the width of the recess determine whether the scales, collectively, present a smooth, rough, or pebbled appearance. Smooth scales lack a distinct keel (e.g., *V. prasinus, V. dumerilii*) and appear glossy (Fig. 4.24A, D). Rough scales have a slight keel (Fig. 4.22B, C) that gives the surface a somewhat wrinkled appearance like crepe paper (e.g., *V. exanthematicus, Varanus rudicollis*), while pebbled scales are comparatively tall and rounded (e.g., the *Odatria* group including *V. acanthurus, V. primordius, V. storri*). Neither rough nor pebbled cranials appear glossy.

Dorsal cranials are also distinct in that they tightly attach, via connective tissue, directly to the bones of the skull, allowing no independent movement. This is particularly notable when conducting a dissection, when the skin seems to be part of the underlying bone. Nevertheless, there are numerous neuronal and circulatory branches, and sensilla per scale. Labial scales are those of the anterior lateral region, from below the nostril to the jugal, just posterior to the orbit. Labials are more symmetrically hexagonal than dorsals and are either elongated laterally or are nearly rectangular and taller than wide. The labial skin is less firmly attached to the underlying muscle than are facial scales and is capable of movement independent of the bones, e.g., the lips can be raised for examination of teeth. The rostral falls between labial and dorsal in nature, with a firmly attached dorsal and somewhat looser lower section. *Varanus* labialis tend to have fewer sensilla than dorsals, and some scales lack them entirely.

The third group consists of temporal scales (Figs. 4.3 and 4.25), characterized by (a) forming neat linear rows and columns, (b) being nearly uniform in size and shape,

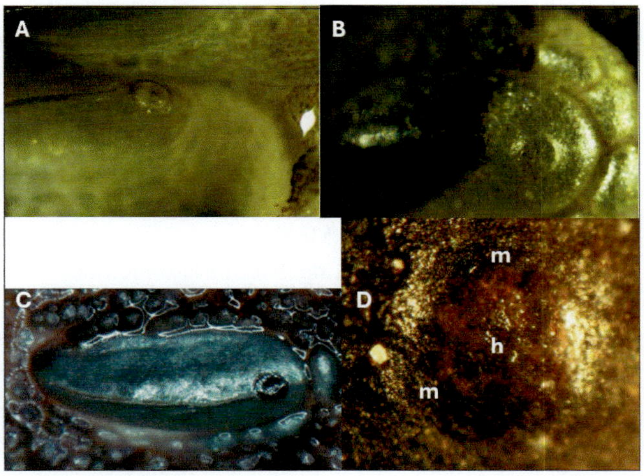

FIGURE 4.24 Single nuchal scales with mechanoreceptory sensilla of (A) *Varanus rudicollis*, with atypical pale sensillum, (B) *Varanus varius*, with pale sensillum of bicolored scale, (C) a strongly keeled scale from *Varanus prasinus* with typical sensillum color and position; note the small ring of hemispheres surrounding the sensillum; (D) extreme closeup of single sensillum from *Varanus togianus*, clearly visible are the pie-pan shape, encircling moat (m), and numerous tiny hexagonal units (h) that form the organ.

(c) having fewer sensilla than other scales, and (d) being freely mobile, attached to the underlying muscle and not bone, all of which are characteristics of ventrals. Temporal scales are rounded hexagons and are very often the most clearly separated of all scales. Gular scales may resemble temporals, being closely packed and either hemispherical or round, and rarely conical. They are often smooth or nearly so and have few sensilla.

4.4.1.3 Dorsal, Nuchal, and Ventral Scalation of the Body

Dorsal scales may be very similar to each other, with variation among species. Careful inspection shows the non-overlapping scales to be elongated hexagons with a more intricate morphology. Each scale is a complex of several smaller scales and has a square base covered in minuscule granular scales (Fig. 4.24; see above).

Dorsal trunk scales in many species are homogeneous in size and shape, and like the nuchals, each is a complex of hemispherical papillae that differ considerably in size and are clearly demarcated from the ventrals. There are exceptions where the dorsals differ in size from each other to some degree, most noticeably in *V. dumerilii* and *V. rudicollis*. Like the nuchals, dorsal scales are composed of different sizes of smaller scales and crestals. The mechanism allowing varanid skin such flexibility and strength can be seen by examining the crease of a bend in the body. The individual scales form a rigid pair of lines on each side of the crease, resembling a zipper row.

The nuchal region is the most likely to be covered with scales of visibly different sizes (Figs. 4.20 and 4.22). The pattern is small, often smooth, rounded scales just posterior to the foramen magnum, followed by the largest scales of the body, which gradually become smaller laterally. As the nuchals meet the shoulders, they grade into the dorsals of

FIGURE 4.25 Males of several Australian *odatria* species possess a ring of enlarged scales at the base of the tail immediately behind the legs, as seen in this male *Varanus tristis*. Scale bar with drawing = 10 mm.

the trunk. Nuchals may be smooth and round (*V. dumerilii, V. prasinus*), pyramidal and strongly keeled (*V. bogerti, V. flavescens* [Fig. 4.15H], *Varanus nesterovi* [Fig. 4.21D], *V. nuchalis, V. rudicollis*, or somewhere along a wide spectrum of intermediate combinations. In *V. dumerilii*, the anterior nuchals are smooth and glossy (Fig. 4.22D), while the central, largest scales have low but distinct keels, one of the few cases where keeled scales retain their glossy surfaces.

Ventrals are imbricate elongated rectangles with either a rounded posterior edge or a small apical tip (Fig. 4.15B, D, F). In all species I have examined, the most central ventrals are keeled, though it may require a lens to confirm. All nuchal and body scales are arranged in distinct diagonal bands and rows, while ventrals are aligned in columns and rows, the number of each being diagnostic for many species. Caudal scales are arranged in straight, versus diagonal, bands, but show modifications not seen in body scales (see below).

Excepting some head scales, varanid body scales tend to be longer than broad, and have three externally visible layers (Fig. 4.20). Most basal in dorsal scales is the wrinkled and usually rectangular epidermal layer, which is often black or dark brown and may be concealed by the upper layers unless the scales are spread apart. Above them is a ring of very small round or oblong scales that form one to four concentric rings around the outermost scale. The most posterior of the small scales tends to form a single ring border into which the other scales merge. The anterior scales may be obscured by the overlapping preceding scale, but all are visible with minimal magnification. Atop the two lower layers is a single large symmetrical scale that may be round, oblong, oval, or rectangular with a convex dorsal surface. This structure is rarely, if ever, concealed by adjacent scales. Ventral scales are similar in structure but are always overlapping, with a rectangular morphology. *Heloderma* has similar dorsal scales, though there is less distinction between the two lower layers, and scales are round and extremely distinct on the head and face. In contrast, the cranial scales show much less distinction of layers in *Varanus*. In *Lanthanotus,* the scalation more closely approaches the varanid condition but differs in having similar dorsal and fascial scales that show three distinct layers; possessing a non-overlapping vertebral row of scales that are smaller than the other dorsals; and having heterogeneous lateral and dorsolateral scales, including two or three longitudinal rows of much larger scales. The elongate ventral scales of *Lanthanotus* and *Varanus* are identical; in *Heloderma*, the ventrals are square, a morphology rarely seen in monitors.

Varanid scales may be acutely and obviously keeled (see above), but they are at least very slightly keeled across the family. Even "smooth-scaled" species often show some degree of a longitudinal keel under magnification (Fig. 4.20A, C).

Scale counts are important and basic meristics used in herpetological taxonomy, including varanid studies. A significant drawback to scale counts is getting uniformity among researchers. For example, ventral scales are often counted from axilla to groin (Fig. 4.4), or axilla to the last clearly defined transverse row, this being as many as 15 scales away from the cloaca, neither of which is a precisely fixed location. I count from the nuchal fold to a point at the anterior of the cloaca, and favor counting five scale rows to either side of the center, and often make three or more counts to ensure accuracy. Most taxonomists clearly state their method for scale counting, but this is sometimes omitted, especially in older literature.

Not all scales fall within the descriptions above. Scales of the cloacal region can be quite unusual and, where distinct, are sexually dimorphic characters. A few of the large species possess a pair of preanal pores (Fig. 4.17), which may be more pronounced in males, and in rare cases, only one pore is present. Males of several species of *Odatria* have, lateral and slightly posterior to the cloaca, a ring of enlarged, pointed scales (Fig. 4.26). These are reduced or absent in females. Males of *V. glauerti* have such enlarged spinose scales, most black with white tips, and *V. glebopalma* has several large, flat scales much larger than those surrounding (Sweet 1999).

4.5 THE TAIL

The monitor's tail morphology is as plastic as the nostril position (Fig. 4.27). Almost all prior descriptions of the transverse shape of the organ described it as either round or laterally compressed. In reality, it is more variable. The ancestral tail was putatively round in cross-section (Sprackland 1991a, b), as it is in most living lizards (notable exceptions, such as the agamid *Hydrosaurus* or iguanid *Basiliscus* being highly derived, and unrelated, species), including the allied *Heloderma* and *Lanthanotus*, and most of the remaining Anguimorpha excepting *Shinisaurus*. The *V. griseus* complex and *V. salvadorii* have round tails, as do all odatrians except the highly aquatic *V. mitchelli*, in which lateral compression must be viewed as a non-homologous derived character. Several species (e.g., *V. albigularis, V. niloticus, V. spinulosus, V. panoptes, V. spenceri*) have an intermediate caudal anatomy, the anterior third (or more) being rounded, the latter two-thirds compressed. The tree monitors (e.g., *V. beccarii, V. bogerti, V. prasinus, V. telenesetes, V. keithhorni*) have rounded tails that are also very elongated (>1.25 SVL) and fully prehensile, acting as a fifth grasping limb (Greene, 1986; Sprackland, 1991b, 1992). The Tree Crocodile Monitor, *V. salvadorii*, has a somewhat triangular tail, distinctly round at the base and for a distance of about one-tenth its length. In his description of the Nile Monitor, which he referred to as *le Triangulaire*, Lacépède wrote, "This triangular-tailed lizard lives in Egypt; what distinguishes it from others is the three-sided pyramid shape of its long tail to its tip" (Lacépède 1839: 214; translated from the French by Teri Sprackland). By the time of Mertens (1942a), he and subsequent authors had determined that tail shape was reduced to being round or laterally compressed, an oversimplification.

FIGURE 4.26 Varanid tails differ in length relative to head and body, degree of rugosity and flexibility, and in shape transversely. A distinct full-length tail is seen in *Varanus doreanus* (A), an arboreal species that forages along rivers and streams. Species with cylindrical tails include the terrestrial, relatively short-tailed spiny-tailed goanna, *Varanus acanthurus* (B), and the arboreal long-tailed *Varanus prasinus* (C). In *Varanus gouldii* (D) and *Varanus komodoensis* (E) the tail does not fit neatly into compressed or cylindrical classes. The base and as much as the anterior third are cylindrical, and the remainder triangular. In young of both species, the more typical compressed morphology is pronounced. *Varanus indicus* may have a flattened posterior third to half of its tail, but, as can be seen here, the base (F) and sometimes much more than half of its length may be round. As in most monitors, the caudals are equal in size and form rings of defined annuli. This 3048-mm black dragon, *Varanus komaini* (G) of the *V. salvator* clade, shows its definitively compressed tail, the upper quarter always like this except for the first tenth or so near the hips. Notice also the very distinct paired row of keeled scales that form a low crest; This, too, is always present regardless of whether the Lizard is in good health or poor (photo courtesy Kathy Black). A lateral view of the tail from hip to the halfway point of *Varanus dumerilii* (H) that displays several unusual features: (A) caudals are not uniform but highly heterogeneous in size and form, (B) ventrals longer than other scales – The scale bars are equal size, the upper spanning four scales, the lower only three, (C) this species shows marked caudal compression from The hips, and (D) because of The variation in sizes and lengths, The caudals do not form clear annuli. The object in the photo is a part of a tree branch.

In many species, notably those that have an SVL greater than about 240 mm, the subcaudals may be 1.5–2.5 times the length of the lateral caudals (Fig. 4.27H). Those species have fully or partially compressed tails, and the caudals do not form fully symmetrical annuli. The remaining species, those with round tails covered in subequal scales, have distinct whorls that symmetrically encircle the tail, and those with the least keel can coil the tail to the greatest degree (Fig. 4.27B, C, F, G).

Scales of varanid tails are almost never smooth, the keel visible without magnification in adults of all species (Fig. 4.27G, H). Caudals, like ventrals, are longitudinal rectangles at least twice as long as wide, imbricate, and bear a small posterior sensillum. The posterior edge may be rounded or pointed. As with dorsals, each caudal may have a large crest surrounded by two or three rows of very small hemispherical cells. Caudals form distinct rows that may either be equal in size and form bands (Fig. 4.27A, C–G), or they may be divided such that the ventral caudals are about twice as long as the lateral and dorsal caudals (Fig. 4.27B, H). In the latter case, only the most dorsal caudals from a pair of longitudinal rows will form a low crest. Round-tailed varanids always lack such a crest and have subequal scales that form complete bands (Fig. 4.27C, D). Among the smoothest scales are those in the tree monitor group (*V. prasinus* and kin; Figs. 4.27C); at the other end of the spectrum are the long spines found in the aptly named *V. acanthurus* (Fig. 4.27B) and its relatives. Strongly keeled scales are seen in several of the smallest monitors, including *V. primordius* (Fig. 4.27C) and *V. storri*. In all of these examples, the tails are round in section.

Alternatively, the monitor tail may be strongly compressed, particularly so in aquatic species (e.g., *V. dumerilii,* young *V. komodoensis, V. mertensi, V. mitchelli, V. niloticus,* young *V. salvator*, *V. semiremex,* and *Varanus*

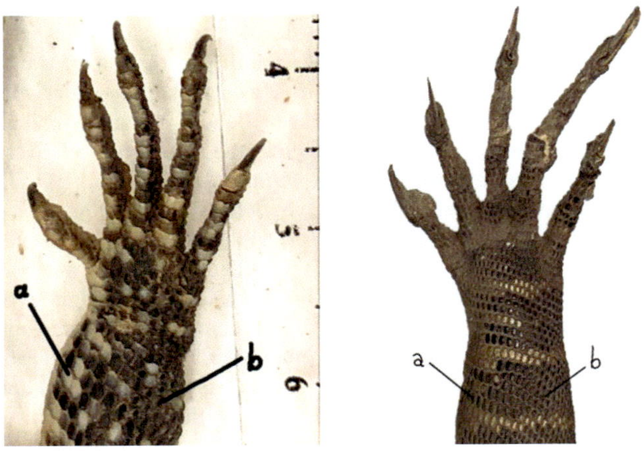

FIGURE 4.27 Two types of inner (A) and outer (B) forearm scale arrangements in (left) *Varanus rainerguentheri* with enlarged "gauntlet" (antebrachial) scales on the inner-facing surface. These are both larger and nearly smooth in contrast to the smaller keeled scales (B) around the forearm. At right, the arm of *Varanus varius* has barely distinguishable differences between inner and outer scales. Both figures show the right arms in dorsal view.

togianus). However, species that rarely or never swim (*V. gouldii, V. komodoensis, V. albigularis*) may also have at least partially compressed tails; the closely related and semiaquatic *Lanthanotus*, however, has a round, prehensile tail that aids in both climbing and moving underwater (Sprackland 1972a, 1998, 1999d). All flat-tailed species use the tail as a defensive weapon, lashing it with considerable force at potential aggressors. Round-tailed species are less likely to display this behavior. Some, such as *V. griseus* and *V. prasinus* do. Monitors have lost the power of caudal autotomy, as have *Lanthanotus* and *Heloderma*, making it a synapomorphy among Platynota (Varanoidea). The ability to regrow a lost tail is common among squamates, including *Sphenodon* and the Diploglossa, but is absent in most snakes, amphisbaenids, and many Iguanians.

4.5.1 Tail Proportions and Uses

The variation between the proportions of monitor head-body length and tail length is one of the most conspicuous characteristics of the family, along with actual size and coloration. Arboreal goannas have particularly long tails, as much as 200 percent of SVL in *Varanus salvadorii*, and well over 150 percent in *V. prasinus, V. beccarii*, and *V. keithhorni*, all of which live in and near the canopies in tropical forests and only make infrequent, brief forays on the ground. Species that spend much time in the water (*Varanus macromaculatus, V. salvator, V. mertensi*) have laterally compressed and relatively long tails, but comparatively shorter and more robust than those of tree monitors. Residents of arid regions, including savanna and deserts, have the relatively shortest tails, rarely more than 1.33 times the SVL, and include both giant (*V. komodoensis, V. albigularis*) and dwarf (*V. acanthurus, V. brevicauda, V. primordius*) species, and are found on all continents that are home to varanids. It is this link to habitat and body proportions that has, in part, made consensus difficult on just how long the extinct †*V. prisca* was in life (Fejérváry 1918, 1935, Hecht 1975, Molnar, 1990, 2004, Owen 1859). Available skeletal elements assuredly support claims that it was a true giant, its skull bones dwarfing those of adult Komodo Dragons. The bones represented in museum collections are remarkably intact (see Chapter 10), but not representative enough to make a reliable length estimate. The question of its habitat and how much or little it took to the water are important and missing variables in any equation trying to put a firm size estimate on the species. The size and structure of its bones, however, coupled with the likely paleoenvironment in which it dwelt, argue for a species larger than *V. komodoensis*, and following the SVL: tail lengths curve I have based on living varanids' proportions (using specimens of *Varanus* with adult SVL > 400 mm and excluding the outlier *V. salvadorii*) gives me the following comparison:

> *Varanus komodoensis*, adult ♂: skull 22 cm, SVL 150 cm, tail 160 cm, total length 310 cm †*Varanus prisca*, adult ♂: skull 40 cm, SVL 275 cm*, tail 260 cm*, total length 535 cm *estimates.

Thus, a lizard *V. prisca*, even if considerably shorter than 7 m, would still be a quite formidable predator at 5.35 m (17.5 ft). My estimate also favors the short tail length based on physiological considerations. A large-bodied predator has high energy demands in terms of basal, thermoregulatory, and prey-capture physiology. Though an energy-storing tail could be useful to such a lizard, its mass would also be an energy drain, making the physiological demands more difficult (see below). The monitor's physiology and needs would more closely resemble those of contemporary large carnivores, such as lions and tigers, which never reach the proportions of many of their sympatric herbivorous mammals. Neither would a varanid with the tooth structure of †*V. prisca* likely be an opportunistic omnivore, as are the bears. Based on these factors, I suggest that the species has a tail slightly shorter than SVL, and possibly of less mass than other estimates have given.

Depending on the species and circumstances, a varanid's tail may be employed in one or more distinct ways as a defensive weapon, against biotic and abiotic threats, and in either active or passive modes. Species with spinous or strongly keeled caudal scales (*V. acanthurus, V. citrinus, V. primordius, V. storri*) have been observed using the tail passively to block the entrance to a burrow or crevice in which they have taken refuge. Presumably, a potential predator small enough to enter such a retreat will be deterred when it encounters a prickly barrier. In my opinion, it is more likely that plugging a den with the tail serves to retain moisture and block heating from the outside; the stout tail with its thick scales also probably reduces water loss from the lizard. As for the rough scales of dwarf monitors having a deterrent effect, I have seen three species (*V. storri, V. acanthurus*, and *V. primordius*) move their tails in a way that scratched my hand – did not break skin but may have deterred a small animal (Sprackland 1980). The sawing-like action contrasted with what other monitors in the hand have done, specifically coiling and whipping the appendage, or lifting it and emptying their intestines. This latter response is so widespread amongst vertebrates that it must be a plesiomorphic character!

As an active form of defense, monitors use the tail to great effect as a whip. Tail-whipping behavior has been reported for at least 200 years and is a signature reaction of an anxious varanid to a potential or actual threat. Brehm (1883) said that a monitor in Manila, if caught by a person, "…defends itself bravely with tooth and claw, also delivering powerful blows with its tail." Ditmars (1956) famously noted how a water monitor, after being placed in an enclosure exposed to sunlight, altered in personality and lashed at the keeper who was bringing his meal. It is probably impossible to have worked with live monitors for any appreciable time without having witnessed similar tail-whipping behavior, and while only a large lizard could cause any pain, the speed and aim of the action would startle almost anyone. I have been unable to find any corroboration of large specimens breaking a dog's or man's leg; that story has been told on many occasions and is probably greatly exaggerated. It is very likely that any animal on the receiving end of such an assault would be startled enough to give the lizard time to flee.

Many lizards are known to store fat and water in their tails, with geckos and beaded lizards having appendages that may dilate or shrink in girth within a few days. *Varanus* that have this ability, and the rate of nutrient storage or consumption varies by species and age (pers. obs.). In species with hyperactivity, the rate of change takes the least time. Captive specimens that I offered regular daily food items were experimentally tested to determine how quickly the stored materials in their tails would be used or resupplied. In *V. spinulosus*, specimens may go from a satiated and robust tail to the indented state that displays the outlines of caudal vertebrae in as little as three days after a feeding. Adults of *V. exanthematicus* and *V. indicus* are apparently more conservative in their energy expenditure, with the change of state in the range of ten to fourteen days. I had expected small species to react rapidly, but in the three dwarf *Odatria* I tested, the quickest was *V. acanthurus* (about one week to change either way), while *V. primordius* and *V. storri* showed no sign of tail shape change after going without food for ten and 12 days, respectively. Because the tail in the *V. acanthurus* did shrink, I do not consider tail shape, i.e., round in section with heavily keeled scales, a factor for why change was unobserved in the other species.

These examples were not intended to be specific experiments but occurred under different circumstances (such as being cared for by someone besides myself while I was away), and I have only intentionally experimented with the frequency of food intake and anterior tail circumference for the *V. spinulosus* and *V. primordius*. The relationship with regards to metabolic rate is, I think, worthy of study. Because varanids may inhabit areas where prey are not always plentiful, use of the tail as a storage organ may be linked with the loss of autotomy in the Platynota (Bogert and Del Campo 1960, Price 2017).

Along with the Chamaeleonidae, and a scattering of genera in other families (*Cophotis, Corucia*) the Anguimorphan lizards have species with the most prehensile tails of non-snake lizards (terminology inspired by Caldwell 2020), most especially within *Abronia, Barisia*, and *Varanus*. Common among these genera is the inclusion of arboreal species, to which a fifth grasping organ can be useful. Though all varanids can climb (particularly as juveniles), there are degrees of arboreality, and these are reflected in their tail morphology. Greatest flexibility, and thus the ability to form the tightest coils, is found in arboreal specialists (*Varanus prasinus* group and *V. salvadorii*). This is not to say that other tree-dwelling or saxicolous species lack caudal prehensility; rather the degree of their flexibility is less than that of the specialists. Among the *Odatria* species, the long-tailed *V. glebopalma, Varanus pilbarensis*, and frequent climbers with more typical tail length, such as *V. orientalis, V. scalaris, V. similis*, and *V. tristis*, have the requisite small, subequal caudal scales in well-defined and spaced rows that allow the range of motion to form tight coils. Intermediate-sized members of the *V. indicus* clade can also coil and grasp well, though not as well as with the prior two groups. Still, the gracile *V. jobiensis* can coil to a considerable degree and uncoil the tail into a whip that can be aimed with great accuracy (pers. obs.).

4.5.2 SVL: Tail Relationships

The proportional relationships of varanid tail length to snout-vent length (SVL) are at their greatest at hatching and generally become more equal in adults. Within the species with the longest tails, the ratio changes little, and their number includes the species that are highly arboreal (*V. prasinus*-group, *V. salvadorii*) or live in tall rocky habitats (*V. glauerti, V. pilbarensis*), species where the tail meets or exceeds twice the SVL. In contrast, the proportionately smallest tails are seen among the largest species (*V. albigularis, V. exanthematicus, V. komodoensis, V. niloticus*) and the smallest (*V. brevicauda, V. bushi, V. gilleni*). Varanids have a habit of standing upright, forming a tripod of hind limbs and tail. In heavy specimens, the tail certainly becomes a counterbalance that makes the posture possible. Bipedal running has been recorded for numerous species of goanna (Schuett et al. 2009), including the massive Komodo Dragons (Suzuki et al. 1991). In all other lizards that run bipedally (*Basiliscus, Chlamydosaurus, Crotaphytus, Hydrosaurus*) loss of half or more of the tail, not surprisingly, has a serious detrimental effect on balance, posture, and likely speed. That bipedality has been the result of strong selection pressure, i.e., the tail is not just useful but also important for the species' survival, may explain why these species also lack the ability to autotomize the appendage.

In assessing the relationship between adult SVL and tail length, I noticed that, as a rule, the larger (bulkier) the species, the shorter the tail. This is conspicuous in Africa's *V. albigularis, V. exanthematicus*, and *V. yemenensis*, and less pronounced in *V. niloticus*. The characteristic also holds for Asian *V. macromaculatus* and relatives and *V. komodoensis* but not in the equally long but much less massive *V. salvadorii*. Of these taxa, only the latter is arboreal as an adult, where a grasping appendage is more useful than a counterbalance for bipedality. Proportionately, though, the shortest tail belongs to the appropriately named *V. brevicauda*, followed by most of the other spiny-tailed *Odatria* species.

4.6 LIMBS

Their limbs reflect the ecological diversity of monitors. Variable features include limb proportions in length and musculature, scalation of forelimbs, and digital length, scalation, and claw structure. As with many other anatomical features in goannas, limb structure has considerable breadth, from the massive stout limbs with moderate digits (*Varanus komodoensis*) to long thin limbs and digits (*V. beccarii*), and from the combination of limb anatomy, one can see clear links with their environments.

4.6.1 Limb Proportions and Uses

Of all external anatomical parts, the limbs, claws, and tail vary most in proportion to the head, neck, and body. Adpressed limbs may overlap, just meet, or fail to meet, and these correlate with limbs that are thin, intermediate, or stocky, respectively. Thus, *Varanus prasinus* has long, thin, overlapping limbs when adpressed, while the short, stocky legs of *Odatria brevicauda* do not even meet. There are obvious relationships between limb morphology, habits, and even habitats. Thus, we have long-limbed tree dwellers, short-limbed burrowers, and intermediate generalists that include semiaquatic and terrestrial monitors.

Two distinct scale size-related patterns are seen in the limbs of the Varanidae. First, differential proportions exist between the antebrachials (between ulna and radius) and remaining forearm scales, and second, the antefemorals (crus scales) and remaining leg scales vary among species. In the first, and more commonly encountered condition, the scales of the limb are equal in size completely around the limb, being subequal in size with the adjacent brachial scales, or there is only a very gradual and minor gradation from slightly smaller antebrachials and the remaining scales (Fig. 4.27). In the second case, the antebrachials and/or antefemorals on the anterior surface of the limbs are noticeably larger than the others (cf. *Cnemidophorus* and *Aspidoscelis*). There is variation in the phenotype for this situation, from species in which there are comparatively minor (but always distinct) differences in scale sizes, to those in which the scales are truly gauntlets and greaves, strong protective armor found primarily among digging species and those that attain a snout-vent length in excess of 230 mm. In addition to the states of having either enlarged or subequal antebrachials, there is an additional intermediate state in which the inner antebrachials are somewhat enlarged but gradually merge in size with their smaller numbers, a condition seen, for example, in the Solomon Islands *V. spinulosus*.

4.6.2 Digits and Claws

All monitors have four well-developed, pentadactyl limbs. In some monitors (the *Odatria* group, e.g., *brevicauda, primordius, storri*), the adpressed limbs do not overlap; such species are excellent burrowers and may possess undersunken mandibles (*acanthurus, primordius,* Figs. 4.2A and 4.8C). In the remaining species, the limbs broadly overlap. All species have sharp, compressed claws, but the shape of these structures varies from almost straight (*V. griseus*) to strongly recurved (*V. prasinus*) (Fig. 4.17).

All varanids have claws that are at least slightly curved (Fig. 4.18). The deserticolous *V. griseus* has the straightest and proportionally longest claws (Fig. 4.18D), in which the inferior edge is extremely acute, knife-like, and almost straight. The claw is triangular in section, the dorsal surface very slightly rounded, and there are no lateral grooves. *V. timorensis* has similar claw morphology but also has a distinct longitudinal groove (Fig. 4.18E). There is a ring of smooth scales encircling the base of each claw, the scales being twice as long as, or longer, than, those in the penultimate distal row. The uppermost scales of the second row are twice as long as the inferior rows. This condition is very closely approximated in *Varanus flavescens* (Fig. 4.18I), in which the terminus of the claw ends in a short conical point that extends below the inferior edge of the rest of the claw. The distal-most dorsal scales of the digits are about twice as large as the other scales, but unlike the case in *V. griseus*, the distal-most ring is not made of enlarged scales, but of two longitudinal dorsal rows larger than the much smaller lateral and ventral scales. The Australian *V. gouldii* (Fig. 4.18A) and *V. panoptes* have claws most similar in shape to *V. griseus*, though proportionally much shorter in length, and with digital scales as seen in *V. flavescens* with large dorsal and much smaller remaining scales, but the inferior edge is clearly divisible into three, not just two, sections: the anterior conical point, the posterior straight sharp edge, and a middle section that is also blade-sharp, but has a slight convex curve. *V. bengalensis* claw structure falls intermediate to the water and desert monitors, being long (the longest digit about 0.66 head length) and distinctly curved dorsally. The inferior surface is like that in *V. salvator*, but the apical tip is longer and more robust (Fig. 4.18H, K). In some adult specimens, the claw may be twice as broad as the claw in an equally sized water monitor. Unlike the other species mentioned, *V. bengalensis* has numerous minuscule scales (not pictured) between the larger digital scales. The digital scales, however, are mostly undifferentiated from each other. *V. bengalensis*, *V. flavescens*, *V. gouldii*, *V. panoptes*, *V. rosenbergi*, and *V. griseus* are all residents of dry desert and semi-desert areas where soil is often densely compacted and, therefore, hard. All these species also excavate their own burrows (as well as commandeer burrows of other animals), often both deep and extensive; *V. panoptes* burrows have been found that are more than two meters deep (Doody et al. 2018). Though primarily arboreal, *V. rudicollis* has scythe-shaped claws, differing from other species in having a long straight proximal section and a short distal hook very similar to those of *V. bengalensis*, though the proximal portion of the claw may be proportionately longer.

The collar of enlarged scales is also found in the *V. salvator*-complex, where they are elongated, keeled, and diminish gradually in size from the largest dorsal to the smallest inferior. The next row contains subequal dorsal and lateral scales. Comparatively, the claws in *V. salvator* are half the length of those of *V. griseus* (Fig. 4.18D), about 25 percent taller, and possess a distinct lateral groove midway up each surface (Fig. 4.18K). The lower edge of the claw is sharp, and the tip is conical in section and acute. The claw is distinctly curved. In both *V. griseus* and *V. salvator*, the largest toe is about 75 percent of head length.

The arboreal tree monitors (*Varanus prasinus*-complex and *V. indicus*-complex) may have the sharpest claws among the varanids, the tips coming to extremely fine pin-like points. Both upper and inferior claw surfaces are quite curved, the latter particularly so, and the claw is highly laterally compressed except along the most dorsal edge. Scales of the digit form rings of subequal size, excepting the dorsal pair, which are about twice the size of the others. There are very small scales between the larger ones, but these are fewer in number and not as evenly distributed as those in the Bengal monitor. In tree monitors, the longest toe equals head length.

Despite their presence in such diverse habitats, the palmar surfaces are surprisingly uniform across species of monitors. Inferior digital scales are small and granular; those of the palmar regions are similar, though often very slightly conical, the apical tip is most ventral. Palm coloration tends to be light, though in some tree and rock-dwelling species, including *V. prasinus*, *V. glebopalma*, and *V. spinulosus*, the soles are black. In *V. glebopalma*, the soles have hemispherical nodules, and *V. spinulosus* has a row of comb-like scales under the two longest toes.

4.7 COLORATION AND PATTERN

Monitor lizards range in color from melanistic to most colors of the spectrum, with species that are blue, green, yellow, orange, red, brown, black, and gray. Only purple and violet seem to have eluded the varanid palette.

4.7.1 Head and Neck

In almost every species, the color and pattern of the head and body are similar to each other. Many taxa have alternating light and dark vertical labial bars, the specific pattern of each possibly as distinct as fingerprints. In some specimens, the bars seem continuous from supra- to infralabial scales, but in others, they may miss alignment to varying degrees. Mandibular scales tend toward horizontal elongation and stay largely consistent in forming nearly symmetrical hexagons.

The most conspicuous pattern feature of the head is the temporal (=post-ocular) stripe, a horizontal black bar that runs from the posterior of the orbit to the upper edge of the ear opening. Because the pattern within a species is consistent, it is a useful diagnostic character. This bar occurs in four forms (Figs. 4.28 and 4.31):

1. Most common is a single solid black stripe (Figs. 4.28A, B, F, H and 4.29F–H), which, in many cases, particularly in the *V. indicus* complex, has a thinner or dashed parallel inferior stripe (Fig. 4.28A);
2. The bar may be broken into a thick dashed line (Fig. 4.28A, C);
3. The bar may be cryptic, i.e., black on a dark background, and thus discernible only under certain lighting conditions, such as in faded preserved or very young lizards (Figs. 4.28I, E and 4.29F, H), or by absence of any light spots, and;
4. The bar is absent, the least common situation (Fig. 4.29A–E).

Gulars are usually the smallest head scales and are either semi-hemispherical or conical (Fig. 4.30). Ventral patterns vary from monochromatic and lighter than dorsal and lateral head scales, or may be mottled, spotted, or have diagonal bands that form V-shapes that point posteriorly. The gular pouch is highly distensible both when the lizard is in threat mode or when consuming large food items. The gular pouch is underlain by three roughly equidistant folds, the first two loose and able to expand considerably. The posterior fold, which spans the distance between the anterior insertions of the forelimbs, is immobile and cannot expand. Rather, it seems to act as a hinge between the esophagus and the stomach, regulating gular expansion and the maximum size of an object that can be swallowed.

Nuchal markings in the forms of solid or spotted transverse bands, when present, are more defined and narrower than those on the body, the latter usually broader markings. Among certain species, the most anterior band is an extension of the temporal streaks, starting at the posterior orbital rictus, extending back to the upper part of the ear, then arcing medially to meet on the nape a short distance posterior to the rear of the skull, and forming a caudad-pointing V (Fig. 4.34). Conspicuous examples are *V. albigularis*, *V. caspius*, *V. dumerilii*, and *V. griseus*, and though the markings remain conspicuous in adults, they gradually lighten as the lizards age. These species retain throughout life the anterior V as the most distinct of the 3 to 5 more

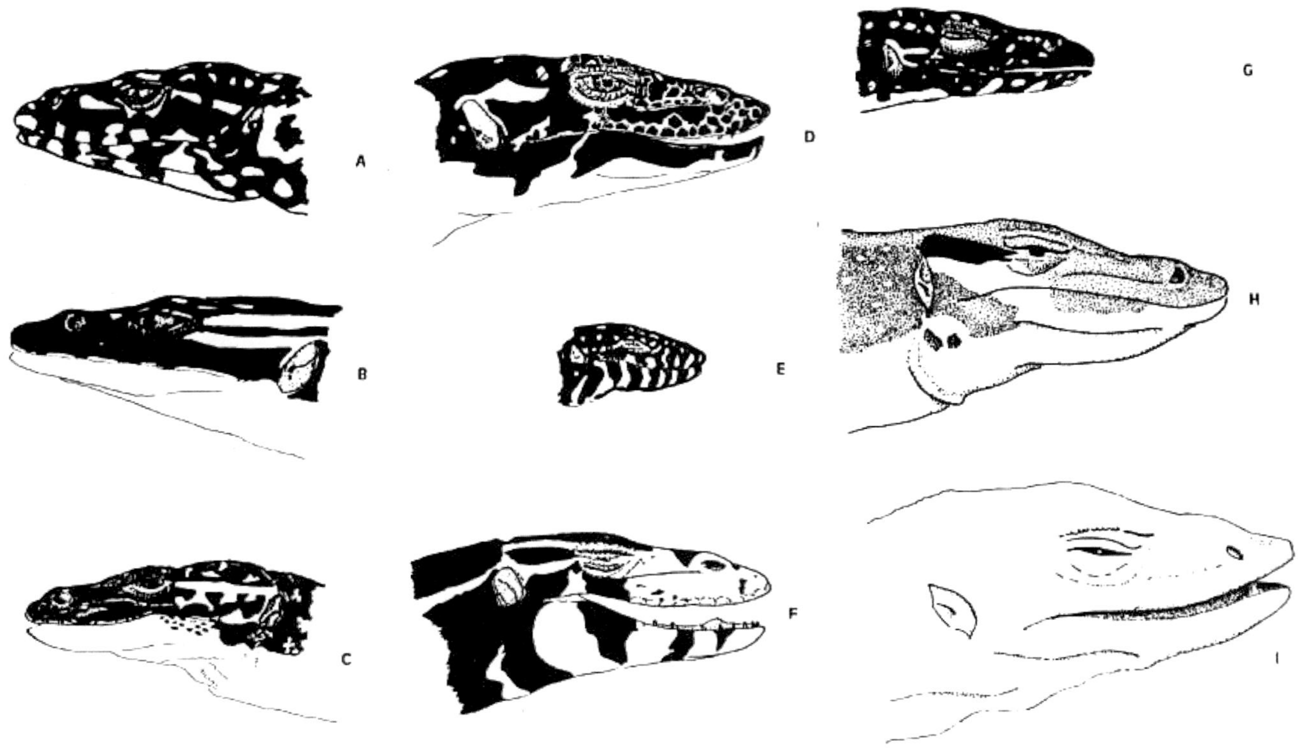

FIGURE 4.28 Facial patterns seen in *Varanus* showing the wide variety of postocular (temporal) streak. Shown are diagrammatic representations of the *salvator* (F), *spinulosus* (I), and *indicus* (the rest) groups that illustrate the major temporal streak and facial patterns seen throughout *Varanus*. (A) *Varanus doreanus*, ZSM 11/1970, Bulolo, Papua New Guinea. (B) *Varanus bennetti*, yng. CAS 111549, Palau. (C) *Varanus indicus*, ZSM 129/1975, seram. (D) *Varanus indicus*, RMNH 3150, amboina. (E) *V. indicus*, yng. SMF 11585, Aru Islands, paratype for *V. indicus rouxi*. (F) *Varanus togianus*, ZMA 15460, Togian Islands, Sulawesi. (G) *Varanus finschi*, ZFKM 26347, New Britain, holotype. (H) *Varanus jobiensis*, RGS 188, West Irian. (I) *Varanus spinulosus*, NMW 23387, Ysabel, Solomon Islands, holotype. (From Sprackland [1995b] and notebooks.)

posterior and increasingly larger bands of the neck. The ensuing dorsal bands may remain intact and distinct or break up as the monitor grows. In the variegated pattern *V. varius*, the five nuchal bands remain distinct, and the most anterior closely aligns with the rear of the skull as a thick hood-like marking; in the Bell's color phase, there is a single thick band across most of the anterior nape, followed by alternating broad black and orange-tan bands on the torso (Fig. 4.34). However, they may start as hatchlings and young monitors; at least half of the species lose distinct nuchal bands roughly coincident with attaining reproductive age. Speckles or irregular short dark lines may remain, but they no longer represent what we can define as transverse bands. Examples include monitors of all sizes, including the large *V. bengalensis, V. komodoensis*, and *V. exanthematicus*, and, interestingly, nearly all the *Odatria* (Fig. 4.31).

4.7.2 Body Color and Patterns

Marking shapes

There are eleven major dorsal patterns borne by varanids, consisting of some combination of a limited number of marking types, which I divide into spots, lines, bands, or rosettes. The primary marking shapes, though they vary in pattern formation across the genus, fall into just a few distinct configurations:

1. The largest spots are round **and** solid or mostly so, as seen in *V. salvadorii* and *V. spinulosus*.
2. Lines are narrow, composed of a transverse, linear set of light-colored scales, usually no more than two scales broad, three or more in length.
3. Bands are distinct transverse markings, largely a single predominant color (e.g., *V. caspius, V. dumerilii*) or dark with flecks of lighter hues (*V. griseus*) or light with dark flecks (*V. prasinus*).
4. Rosettes are flower-shaped, either round or oval, and have either a black center or the center is the same as the predominant dorsal color.

Spots are found in all species of *Varanus*, ranging from tiny clusters of three or four scales to large concentric rings of two or three different colors. Small unicolor spots are seen, for example, in *V. bengalensis, V. bitatawa, V. chlorostigma, V. keithhorni*, and *V. indicis*, while large concentric spots are common to *V. albigularis, V. caudolineatus, V. exanthematicus*, and *V. flavescens*. In *V. exanthematicus*, for example, the spots may consist of a solid pale tan or yellow center surrounded by a narrow brown ring, and sitting in a gray torso. Large solid spots are seen in *V. brevicauda, V. giganteus, V. salvadorii, V. semiremex, V. spinulosus*, and many *V. albigularis*. Rosettes are probably unique to *Varanus*; they are rings, flower-shaped, round, always larger than any adjacent solid spots, and most characteristically with a dark center. When present, rosettes rarely appear randomly dispersed but tend to form regular transverse bands. Rosette patterns are most developed among the water monitors (*Varanus salvator*) of SE Asia and most *Odatria* monitors (*acanthurus, scalaris, tristis*) of Australia, and widely in species from south central Asia east to Australia and the Solomon Islands (e.g., *V. gouldii, V. flavirufus, V. rosenbergi, V. timorensis, V. spinulosus*). Note that a species possessing one type of marking shape does not exclude it from having others. *Varanus salvadorii* and *V. spinulosus* may have all three, and some species,

FIGURE 4.29 Facial patterns across 15 species of the genus *Varanus*. (A–C) *Varanus jobiensis* in common, melanistic, and intense pink varieties, respectively. (D) *Varanus melinus*. (E) *Varanus doreanus*. (F) *Varanus indicus*. (G) *Varanus yuwonoi*. (H) *Varanus timorensis*. (I) *Varanus flavescens*. (J) *Varanus rudicollis*. (K) *V. bengalensis*. (L) *Varanus dumerilii*. (M) *Varanus keithhorni*. (N) *Varanus beccarii*. (O) *Varanus mertensi*. (P) *Varanus flavirufus*. (Q) *Varanus salvator*. (R) *Varanus salvadorii*. Of these, only E fully lacks a temporal streak, while N has one concealed by the total melanism of the body. In H, I, and Q there is the banded snout pattern. The snout morphologies of J and K are quite similar in several characters, but the scutellation and proportions differ. Additionally, J has a gracile body and long tail, while K is more robust and has a comparatively shorter tail. *(Continued)*

FIGURE 4.29 *(Continued)*

including *V. rudicollis* and *V. yuwonoi*, may have all three but limited to different parts of the body. In *V. rudicollis*, all but large adults may have broad bands composed of small spots near the shoulders, and narrower bands made of large solid spots from axilla to hips, while in *V. yuwonoi* the posterior and anterior parts of the body are clearly differentiated mid-torso into two very differently colored and patterned halves. Rosettes, being exclusive to *Varanus*, are a synapomorphy.

In adults, the patterns possible include (Figs. 4.32 and 4.33):

1. Dark body with well-defined rows of cross-bands composed of rosettes, as seen in *Varanus salvator*, *V. timorensis*, and *V. prasinus*.
2. Dark body with thinner, well-defined rows of cross-bands composed of rosettes that are interspersed by *irregular* spots between and sometimes within bands (*V. cumingi, V. giganteus*).
3. Body with broad transverse bands of light spots interspersed by solid dark bands (*V. griseus*)
4. Dark body with numerous to extremely numerous light (usually yellow) spots that form no distinct pattern (*V. indicus, V. nebulosus*).
5. Dark body with extremely numerous spots, each no larger than a single scale, that form no distinct pattern, *combined* with a denser cluster of similar colored scales that produce distinct cross bands (*V. jobiensis*) (Fig. 4.33).
6. Dark body with well-defined, solid circular spots that form distinct but thin bands (*Varanus salvadorii* and *V. spinulosus*).
7. Dark body that, in adults, lacks any pattern (*V. beccarii, V. komodoensis,* and *V. komaini*), which are *patternless* melanistic forms. Juveniles of such taxa, though, often sport spots or rosettes that fade with age.

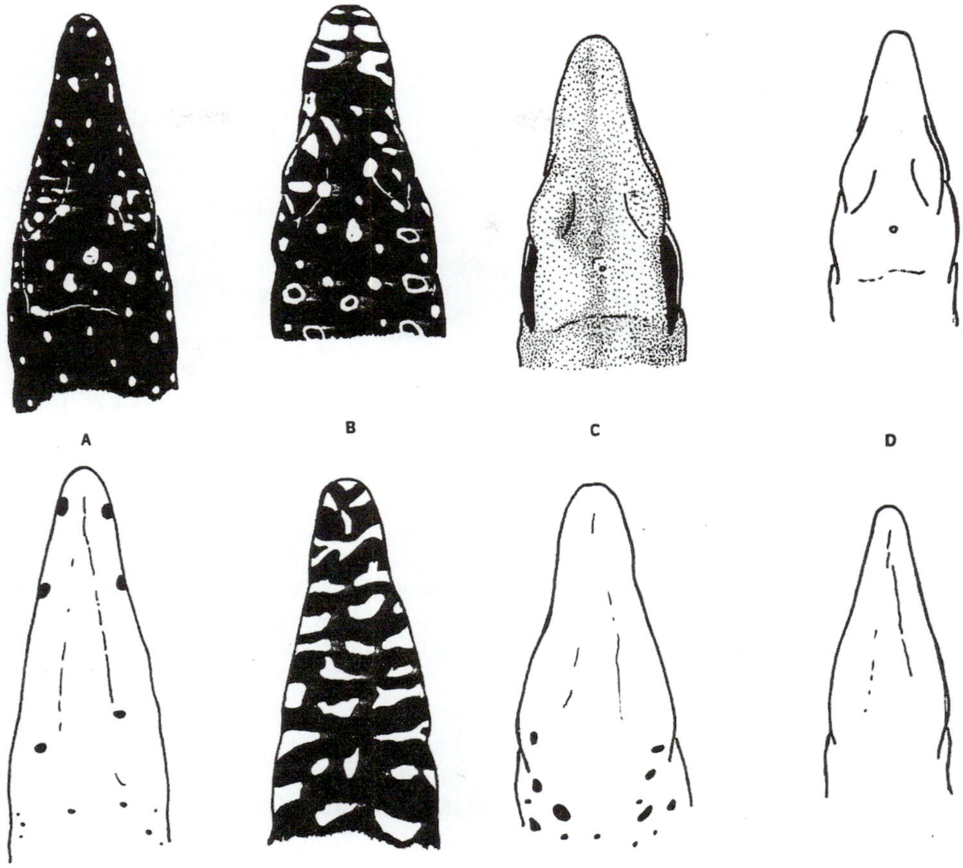

FIGURE 4.30 Pattern variety of the dorsal and gular surfaces of the head in adults *Varanus*. (A) *Varanus indicus*; (B) *Varanus doreanus*; (C) *Varanus jobiensis*; (D) *Varanus spinulosus*. See also Fig. 4.5 from Sprackland (1995b).

FIGURE 4.31 Color and pattern variation in Australia's Lace Monitor, *Varanus varius*. (A) The "Bell's phase," a normal variant that is often produced from egg clutches with variations of the "normal" or dark phases. (B) One of the three main dark phase color and pattern varieties, in this case with clear dark and light transverse dorsal bands and an unpatterned face. Both this specimen and the one depicted in a were observed within a few hundred meters of each other in the Outback of central New South Wales. (C) The widespread dark color morph from coastal New South Wales, in which dorsal bands are less distinct, facial bars are present, and yellow coloring is ventrally widespread and bright. The monitor in (D), like that in (C), has a pattern like (C); there are fewer and less defined mandibular bars, and much less widespread or intense yellowish markings. This specimen was found near Brisbane.

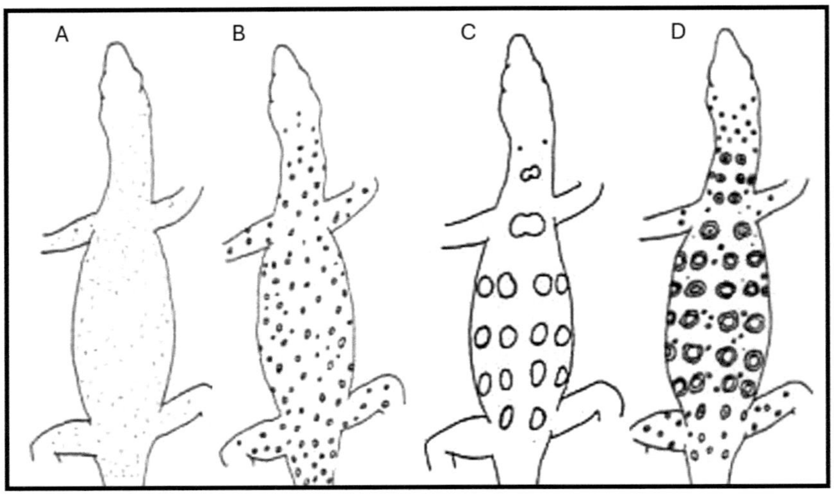

FIGURE 4.32 Examples of the range of combinations of dorsal markings that can form the eleven basic patterns in *Varanus*. (A) Randomly spaced tiny flecks that do not form a pattern or join to make spots or ocelli. (B) Background covered in small dots that can be formed by 2–5 light colored scales, but do not form a distinct pattern of transverse bands. (C) Spots present, sometimes restricted to the nape, sometimes just to the body, but always larger than flecks, always form circles, and almost always align into clear dorsal bands. (D) Pattern can vary considerably but is consistent in having distinct, if irregular, transverse bands, and ocelli distinct, hollow, and usually round or slightly hexagonal.

8. Body equally dark brown or black, and gray, both forming bands that may meet ventrally (*V. rudicollis*).
9. Light body with dark cross bands (*V. dumerilii*).
10. Light or dark body with spots forming dorsolateral cross-bands (*Odatria gilleni V. caspius, V. griseus,* and *V. albigularis*).
11. Body dark with transverse thin light bands of rosettes, many being more square or hexagonal than round, *separated* anteriorly and posteriorly by regular lines of light (*V. finschi*). (I have designated this the Bismarck pattern, unique to the species from the Bismarck Archipelago.)

Though color and pattern were long considered to be of minor taxonomic significance, they are useful not only for species recognition, but as field marks as well. The existence of hybrids developed in captivity and intergrades, which are naturally occurring hybrids, has both confounded innumerable attempts to draw clear lines to demarcate species (Burbrink et al. 2022, De Queiroz 2007); this is covered in Chapters 10 and 11. Sometimes specimens will display modified ("aberrant") coloration, such as the rare blue form of the Green Tree Monitor (*V. prasinus*) where the yellow pigment chromatophores fail to develop. Albinos are extremely rare in the field, and both their locality data and other morphological characters should help solve identification problems. On the other hand, melanistic specimens are somewhat more common than albinos. They tend to occur within a particular range, within largely insular populations. Species such as *V. beccarii* and *V. bogerti*, close relatives of *V. prasinus*, are melanistic and larger than *V. prasinus*. A particularly interesting trimorphism exists in the large Australian Lace Goanna, *V. varius*, which occurs in three known color and pattern sets (Fig. 4.31). They are (a) the dark-bodied form with numerous small random light spots, (b) one with distinct bands, and (c) the black and tan banded variation. So different from previously known *V. varius* were the first banded specimens brought to Europe that they were treated as a distinct species, *Varanus belli*. There are several reasons why *belli* was subsumed into *varius*, most telling being the fact that dark, spotted goannas and *belli* varieties would emerge from the same clutch of eggs. Shortly after this was learned, it was hypothesized that the banded specimens were all female, but detailed examination of lizards showed no sexual link. Today, the banded specimens are still informally referred to as the *belli* or Belli (not italicized) phase by amateurs and professionals both.

4.7.3 The Tail

Color and pattern in varanid tails are not always consistent within a species, and in fact often change during the growth of the lizard. As with the dorsum, the tail may contain spots, solid bars, or bands of rosettes, but if bars are present, they may be relatively broad and solid, not necessarily comprising spots. They may occur as broken lines around the tail. In some species, notably among Mangrove Monitors (*V. indicus*-complex) there may be a drastic change in tail patterns; juveniles may have broad light and dark bands that, in adults, become a series of large spots or random smaller spots (Sprackland, 1995b Fig. 4.37). In many species, the tail is monochromatic (e.g., *beccarii, prasinus, primordius*). The young of many members of the Mangrove Monitors (*V. indicus* group) may sport bright or pale blue tail markings as young that fade to light gray or white in the adults. In eastern New Guinea, many young Mangrove Monitors have broad white caudal bands that

FIGURE 4.33 Naturally occurring color and pattern types in *Varanus*. (A–C) show patterns composed of well-defined rosettes or ocelli that form transverse dorsal bands. (D) Random light flecks on dark dorsum. (E) Numerous tiny flecks of light color that form distinct transverse bands interspersed by thinner dark bands. (F and H) Only a few varanids have split patterns where the anterior and posterior parts of the dorsum have distinct, non-mixing patterns, in this case with anterior markings similar to those seen in E and the posterior with bands of large spots (very rarely rosettes). (G) Relatively unmarked dorsum in adults, (H) Body devoid of true rosettes but has numerous light markings that tend to resemble spots "melted" into each other and not forming a clear pattern. Dorsal bands are extremely broad, dark, and with sparsely distributed light markings. (I) Body marked with distinct transverse bands, the dark predominantly solid, the broader light with numerous dark flecks; No rosettes. (J) Dorsum with large solid light spots, often with a thin dark border. Bands are distinct but not as symmetrically as in other banded species. (K) Body dark with numerous light specks in which bands with 2–5 light markings form small ocelli or somewhat circular rings, and the dark gaps between such bands with additional bands made of small straight groups of 2–4 light scales that form dotted lines. (A) *Varanus salvator*. (B) *Varanus macromaculatus*. (C) *Varanus giganteus*. (D) *Varanus indicus*. (E) *Varanus jobiensis*. (F) *Varanus spinulosus*. (G) *Varanus komodoensis*. (H) *Varanus rudicollis*. (I) *Varanus dumerilii*. (J) *Varanus caspius* (courtesy Sherman Minton). (K) *Varanus albigularis*. (L) *Varanus finschi*.

diminish into thin rows of gray spots as they grow. The incredible variation in juvenile patterns in this group played a significant role in leading herpetologists to name so many taxa, both species and subspecies, that have subsequently been relegated to junior synonymies. Most of the species in this group retain the basic pattern of coloration as adults, even though the colors and size of light markings may be drastically reduced (Fig. 4.34).

4.8 ONTOGENETIC VARIATION AND SEXUAL PATTERNS

Very little has been studied or published on the ontogenetic changes that occur in varanids as they age. The most pronounced change in color and pattern was seen in *V. beccarii*, *V. dumerilii*, and *V. rosenbergi*, with the discoveries most well documented from captive-bred specimens. The teeth – particularly the posterior-most – in some large species change from the common compressed recurved form to a blunted, molar-like morphology. Brief descriptions of changes in the intensity of coloration have also been noted elsewhere, but to date, there has been no overview of ontogenetic changes in monitors. Here I present information on four areas of ontogenetic change: head shape, SVL: tail proportions, color and pattern, and dentition.

For the sake of continuity, I divide the age classes as follows: neonate = hatchling to three months; young = three months to one-half adult SVL; subadult = one-half to three-quarters typical adult SVL; adult = at or near (over or below) typical SVL; old adult = onset of post-adult ontogenetic change(s).

4.8.1 Head Shape

Across species boundaries, neonate monitors will more closely resemble each other than they will in their adult forms, and to such a degree that keying them to species can be difficult. This applies especially to members of the *Odatria tristis* group (see Chapter 11) and several of the *V. indicus* group. Regardless of the adult form, young monitors always show an acute, depressed head configuration, with a small, round nostril nearer to the snout tip than to the orbit. This includes such blunt-snouted species as *V.*

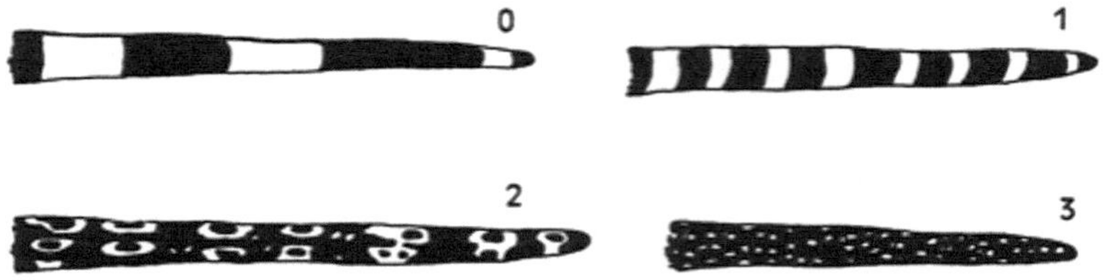

FIGURE 4.34 Diagrammatic representations of the tail patterns seen in *Varanus*. Numbers refer to their character state values as used in my phylogenetic analyses (Sprackland 1991a-b, 1995b). The value 0 = tail with distinct broad light bands. The value 1 = tail with thin light bands, sometimes terminating for the last fifth of the tail. The value 2 = banding pattern made of ocelli or fragmented versions. The value 3 = tail with extremely small, irregular light spots with no trace of transverse bands or any distinct pattern. The pattern seen in juveniles may change considerably in adults.

bengalensis, *V. flavescens*, *V. nebulosus*, and *V. niloticus*. Each of these species undergoes an elongation and curling of the nostril, and its position is distinctly mid-snout during the young and early sub-adult phases; the final posterior position is coincident with the rise of the canthus rostralis region, which begins in the sub-adult phase. The final relative snout height, which gives the blunt-muzzle appearance, is greatest in *V. bengalensis*, least in *V. nebulosus*. Exceptions to this rule are the Afro-Arabian *V. albigularis*, *V. exanthematicus*, and *V. yemenensis*, which all start life as they will end it, with box-like heads. In these taxa also, there is very little change in shape or relative position of the nostril.

Relative head width is a secondary sex trait. There is little or no difference seen between the sexes in head width through the sub-adult stage. At that stage, however, and in healthy lizards, muscles of the postorbital region become enlarged in males. The difference may be slight, and it may be impossible to determine without comparison among several adult animals. When measuring the width of the temporal area of lizards of equal SVL length (i.e., 350 mm), the male may be larger by only one or two millimeters, or as much as five millimeters. More reliable for sexing, though, is to evert the hemipenis.

4.8.2 SVL: Tail Proportions

The proportion of head-body to tail length is generally related to adult size, with comparatively short tails going with large bodies, and long tails with short bodies. In neonates and young lizards, the general appearance is always one of a streamlined body and long, thin tail (primary exception: *V. exanthematicus*). As the lizard grows, its tail and body grow at different rates. Species such as *acanthurus* initially appear to have short tails because, with growth the tail becomes broader, but when measured, it shows itself to be more than 1.5 times SVL.

There are, naturally, exceptions. *Varanus salvadorii*, which is unofficially (e.g., specimens in excess of 3500 mm have been measured, but neither published nor saved with a voucher specimen) the longest living lizard, has the longest SVL to tail ratios in the varanids, and in fact approaches that of the long-tailed lacertids, *Takydromus sexlineatus*. With *V. salvadorii*'s tail about 300–340 percent of SVL, no other varanid's proportions come close. The medium-sized tree monitor group has tails longer than more terrestrial monitors of similar SVL. In both cases, though, the species are highly specialized for arboreal life, and the round prehensile tail is an adaptation for that life. Furthermore, despite their differences in size, the ratios of tail to SVL are close. Other monitors with tails exceeding 190 percent SVL include *beccarii* (1: 2), *kingorum* (1: 2.3), *glauerti* (1: 3), and *glebopalma* (1: 1.8). At the other extreme are the smallest monitors, *brevicauda* and *bushii*, which have tails about equal to their SVL. For most of these species, tail length lies at the upper proportion, though in the tree monitors, the tail proportion may increase with age.

4.8.3 Color and Pattern

Hatchling monitors overwhelmingly resemble their parents in pattern and color, though juvenile monitors, especially hatchlings, are often much more pronounced. There are, however, some exceptional cases where juvenile coloring is very different from that of adults. In neonates and young, rosettes may appear as spots, the dark center not developing until the subadult stage. In *V. dumerilii* of Borneo, western Indonesia, and Malaya, we see the unequivocally most dramatic change in color and pattern from hatchling to adult monitor. Neonates (Fig. 4.35) are black dorsally with five thin cream or pale-yellow bands spaced across the dorsum, the most anterior being thick and solid, with each succeeding band thinner and less solid. The three anterior bands are composed of small spots. The head, face, nape, and first post-nuchal band are bright orange. There are three to four black horizontal labial bands and a thick, solid black temporal streak extending from the eyes to the mid-nape, where the left and right ends meet. Ventral coloring is light orange or ochre with seven or more thin bands that meet the dorsal pattern laterally. The tail is black with broad pale-yellow bands, broadest ventrally. I have hypothesized (Sprackland 1992, 2009) that this neonate pattern mimics two highly venomous snakes that are sympatric with the monitor, the Blue Coral Snake (*Calliophis bivirgatus*), and the Red-headed Krait (*Bungarus flaviceps*) (Sprackland, 1992, 2010). The pattern begins to change at six to eight weeks of age, with the orange and light bands becoming

FIGURE 4.35 Extreme change in coloring and pattern seen in a neonate (5 weeks), young (90 days), and adult *Varanus dumerilii*. Note that the bright neonate coloration has already begun to dull by the third month, sometimes sooner. The adult retains the pattern in shades of brown and tan, with no orange, red, or black remaining. Dated photos by Aiden Bradley; added by the author.

a pale tan, the black bands turning brown, and the lizard now resembling young King Cobras (*Ophiophagus* spp.) more than most kraits. By the ninth month, the lizard has grown considerably and taken on the dark brown and tan adult coloration (Fig. 4.35).

Coloring and pattern in young Komodo Dragons (Fig. 4.36) bear a remarkable resemblance to Australian Lace Monitors (*Varanus varius*). The snout has broad black bands, and there are four broad pale cream or yellowish nuchal chevrons, their points facing caudad. There are small light spots between each chevron and on the dorsum. The top of the head is mottled cream and brown. The temporal streak common to most varanids is unique, the posterior ends meeting near the occiput to form a triangular or T-shaped light mark. The black temporal streak running superior to the light streak terminates slightly anterior to and above the ear opening; this configuration is also unique. Temporal stripes terminate at the ear opening in all other varanids. The humerus has four distinct bands of closely set light spots, and the forearm has five; the bands become fainter distally. Only very poorly defined bands of well-spaced spots encircle the hind limbs. The tail has dark bands that do not meet subcaudally, and those of the proximal half are wide with three interior bands of light spots; the central row is the largest. Distally, the caudal bands have but one row of small light spots. The complexity of the juvenile pattern, along with the variety of colors, completely fades with age. Animals from Komodo Island eventually become entirely gray, with very faint or no caudal bands. There is no remaining pattern of spots or chevrons anywhere on the body. Lizards from Flores Island, east of Komodo, also become a patternless gray, but often differ in possessing pale yellow eyelids and supraciliary scales, and small traces of pale green and pink scattered on the head and face.

Neonates of the Black Tree Monitor, *V. beccarii*, hatch with a predominantly black body but are covered in very small and sparsely distributed light green spots, each only one or two scales in size. While the green coloring reflects the affinity of *V. beccarii* with *V. prasinus*, the body becomes completely black within the first five months. Juvenile *V. beccarii* closely resemble the Australian Canopy Monitor, *V. keithhorni*, of all ages. Australia's Pilbara Rock (*V. pilbarensis*, Fig. 4.37) and Heath Goannas (*V. rosenbergi*) have bright pink or orange heads at hatching, and the body bears a much more intense version of adult coloring.

If we account for minor changes in accepted prey size (e.g., the shift from eating small insects to eating large insects), the ecological differences between juvenile – young goannas differ enough from those of subadult – adult lizards that they could, in practice, be studied as separate species.

4.9 CONCLUSIONS

The external morphology of *Varanus* has evolved a remarkable variety in spite of having made rather few modifications in form. The head alone varies in 11 conspicuous features: acuity, height, breadth, nostril shape and position, tongue color, anterior snout form, canthus rostralis morphology, cranial scalation, sensilla numbers and distribution, and color pattern. The ontological differences between

FIGURE 4.36 Extreme change in coloring and pattern seen in a boldly marked neonate (upper) and drab adult Komodo Dragon (*Varanus komodoensis*).

young and adult reflect adaptations to different lifestyles of the age classes that may be as ecologically distinct as separate species.

Scutellation is remarkably diverse in at least four ways. First, three-dimensional scale structure ranges from round and hemispherical to elongated with a strong medial ridge, making scales smooth, rough, or keeled. Though the head, torso, and limbs never bear spiny scales, the tails of several dwarf species do. Tail dimensions and caudal texture can be diagnostic in establishing how it is used and in which habitat(s) a species may dwell. Second, monitor scales have a layered anatomy with a squarish basal layer, two to four rings of tiny, round, hemispherical microscales in distinct rows, and a large cap scale that is the most obvious structure. Within an individual, the scale anatomy is consistent, except for the ventrals, but may differ in length relative to each other. Ventral scales may be nearly square in neonates and young monitors but are rectangular in subadults and adults. They have the same structural anatomy as the dorsal scales but differ in that the majority (between axilla and groin) are imbricate. Dorsal and limb scales are very rarely overlapping; caudals frequently are. Third, extremely small, embedded organs are present over much of the body, and almost universally on the torso, nape, and tail. Sensillum position, when single, is caudad in a posterolateral position atop a cap scale. When in clusters of three to eight, they are on the flat surface of the scale and localized to head scales and extreme anterior nuchals. These sensilla are known amongst most squamates but have not been well-studied. Their precise function(s) remain uncertain, though the prevailing hypothesis suggests a tactile function. Fourth, scale coloring falls into two patterns, either unicolor or bicolor, and while only certain species have bicolor and unicolor scales, others have only unicolor. Whatever the color of a scale, sensilla tend to be black except if they are on the light surface of a bicolored scale or are on a completely light

FIGURE 4.37 Extreme ontological change in coloring and pattern seen in an *Odatria* species, *V. pilbarensis*. (A) Hatchling emerging from egg, and (B) A clutch of neonates having fully emerged. (C) Young Lizard, about three months old. (D) adult male showing the distinct coloration, and (E) adult female with a similar pattern but subdued colors. A courtesy Ralf Vitt; B courtesy Robert Grabowski.

scale on a lizard that has bicolored scales. In species with monochrome scales, the sensilla are always black.

I have discerned four types of individual markings that form eleven distinct marking patterns across *Varanus* that produce many less distinct variations. The patterns and colors displayed in neonates are largely consistent with those in adults, but some degree of ontological modification is nearly universal. That is, fairly small solid spots in hatchlings may expand into hollow (dark-centered) rings or rosettes as the monitor grows. Varanid coloration and pattern contribute to crypsis in all adults, though a few species have hatchlings that have non-cryptic colors that may be part of mimicry of toxic sympatric species, and this is a subject worth serious study.

Certain parts of a varanid's body clearly show adaptations firmly linked to natural history and environment. For example, monitors that dig considerably have longer and thicker claws and digits than others, and the anterior forearm often has enlarged antebrachial scales. Long thin limbs and digits with strongly arched claws are found in primary arboreal species. A laterally compressed tail is a characteristic weapon among many large species and is further adapted to be thinner in largely aquatic goannas. It is also evident that monitors are not restricted to having either cylindrical (= round) or flat tails, for several species have distinct triangular appendages. Tail length runs from about 90 percent of snout vent length in some dwarf species to over 250 percent in arboreal and saxicolous monitors. The tail in *Varanus salvadorii* probably gives the species the right to the title of longest living lizard (Fig. 4.2), though its mass is far inferior to *V. komodoensis,* as verified by photographic data and reliable personal communications from people who have worked with the species for many years. Tail morphology seems closely related to habits but may also be important in nutrient storage. Except for a small number of outliers, tail length is inversely proportional to snout-vent length in adults. In juveniles the tail is always at least slightly longer than the SVL.

Collectively, varanids are not among the most colorful of lizards, but their exceptions can be extraordinary. Species that are bright green, yellow, or blue are certainly something to behold, even though species in other families can be far more variously and intensely colored; consider the South American *Cnemidophorus* and *Ameiva*, Caribbean *Anolis*, Central American *Abronia*, African *Agama*, European lacertids, and of course the chameleons, all of which do far more with the spectrum than monitors. Recent work on squamate opsins shows that varanids possess cones that produce three of the four opsin types known in lizards, which is likely the result of an evolutionary reversal that lost the fourth type. How that affects varanid vision is unknown.

5 Monitor Anatomy II
Internal – the Skeleton

5.1 INTRODUCTION TO VARANID SKELETONS

"The" Varanid Skull does not exist. Many published accounts of varanid skulls are based on one or a few specimens. Estes (1983) stated that the study of single examples of a species does not truly represent the conditions in that species, and that no one specimen is truly "typical" of an entire taxon at any level. While his statement highlights the obvious, the reality is that there are not nearly as many skeletal materials as wet materials in museum collections. Many taxa are represented by few or no specimens, and those in condition that spans excellent to horrid. Collections that do have large skeletal holdings, more often than not, have numerous specimens but few species, and these tend to be of the larger taxa. The skeletal collections studied by Robert Mertens (1942a–c) in Frankfurt and Berlin span nearly every taxon known at the time, from his smallest species, *Varanus caudolineatus*, to the obligatory *Varanus komodoensis* (Fig. 5.1).

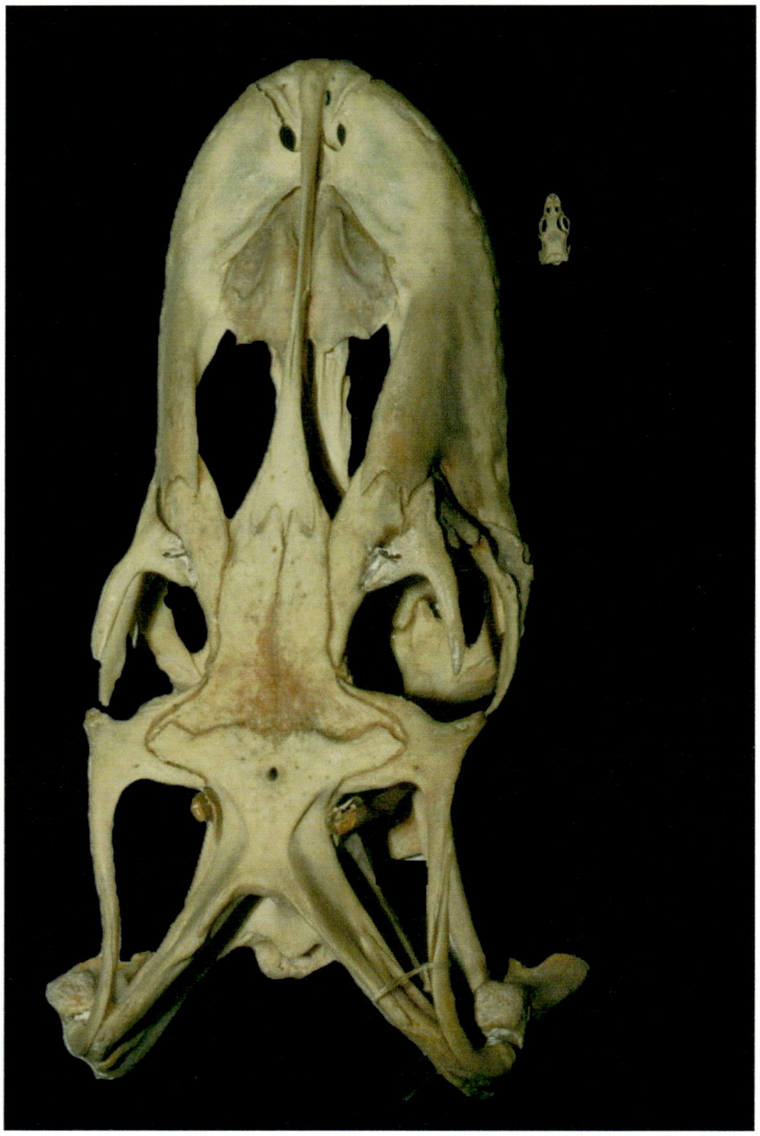

FIGURE 5.1 Robert Mertens presented a photograph of the skulls of the largest (*Varanus komodoensis*) and smallest (*Varanus caudolineatus*) skulls in his study (Mertens 1942a: tafel 18). In 1991 I reshot the pair during my own research at the Senckenberg Museum.

DOI: 10.1201/9781003463528-7

Those collections are also unusual because the majority of specimens have good collection data; it is extremely common to see "no data" on labels for skeletal material, and the number of named varanids has doubled since 1942 and continues to increase.

Varanid skulls do share numerous similarities, e.g., all monitors have an elongated snout that is longer than the postorbital cranium, but the proportions vary by species, ontogeny, and, to a lesser degree, sex of the animal. Baron Géza Fejérváry used the Komodo Dragon as an extant animal to clarify the taxonomic positions of squamate fossils, first with *Varanus* and †*Megalania* (now *Varanus*) (Fejérváry 1918). His posthumous second and much more comprehensive 1935 work revisited his study of fossil lizards and included considerable comparisons to other living *Varanus*. Despite flaws in his papers because of limited or misinterpreted material, both of Fejérváry's works remain important contributions to understanding varanid biology. Karm Narayan Bahl (1937) began his work "to identify the large number of foramina in the skull," which expanded to become the most extensive review of the complete skull of a varanid (*Varanus bengalensis*), closely resembling the work of Oelrich (1956) for *Ctenosaura*, de Quieroz for iguanines (1987), Surahya for *V. komodoensis* (1989), and of Ledesma and Scarpetta (2018) for *Elgaria*. The scope of Bahl's paper makes it an essential reference to varanid researchers over 80 years after its publication.

Robert Mertens produced a stream of mostly taxonomic papers on varanids from the 1920s through the 1970s, naming many new species and subspecies, and describing intraspecific variation. His greatest product was the three-volume *Die Familie der Warane* (Mertens 1942a–c). The Abhandlungen (Mertens 1942a) provides a general overview of the features of the cranial bones, then gives a predominantly morphometric comparison of each bone in his species accounts, accompanied by 159 plates providing dorsal, ventral, and lateral views of all the taxa known at the time, and the 11 figures provide excellent references to the general morphotopography of varanid skulls. Mertens's Der Schädel (Mertens 1942b) provides more detailed descriptions of the (mostly) articulated skulls of almost every taxon known at the time, and his illustrations include dorsal and lateral views of the specimens. The photographs in the volumes are not of high enough quality to be used for detailed study of most features, nor was it possible to improve them for the rereleased edition (Mertens 1997). However, though he discusses some variation among species, he does not provide any illustrations or descriptions of the individual bones that cannot be examined from an articulated skull, nor was that his intent:

> The following craniological treatment is not intended for the anatomist, but mainly for the taxonomic zoologist. It does not contain a detailed description of the various skull elements that have been disarticulated….
>
> (Mertens 1942b:120, RGS translation)

After World War II, Mertens visited major natural history collections in the Allied nations, notably Australia and the United States. From these collections, he described several new taxa and published several important papers (Mertens 1951, 1953, 1962a, b, 1971) that are essentially addenda to *Die Familie der Warane*, and include descriptions of the skulls of the new lizards (the skull of adult *Varanus salvadorii* in 1950, and several Australian species in 1959).

The next important review was by Angus Bellairs (1949), providing a detailed account of the snout of "*V. monitor*" (either *bengalensis* or *nebulosus*) and a comparison of that species with other lizards and snakes. *Atlas Komodo* was produced by Soeparmi Surahya (1989) and expands considerably on Bahl's paper (1937) to cover the entire anatomy in a highly detailed account of a single species, *V. komodoensis*.

Though the skulls of some varanids have been very well studied, the actual number of taxa so examined is small, usually of a single species, and represents species rather closely related to each other: the east Asian *V. bengalensis* (often as "*Varanus monitor*," Bahl 1937, Bellairs 1949), *Varanus salvator* (Bellairs 1949, Säve-Söderbergh 1946, 1947, Watkinson 1906), and *V. komodoensis* (Auffenberg 1980; see especially Surahya 1989, De Jong and Brongersma 1927, Fejérváry 1918, 1935, Mertens 1942b); the African *Varanus niloticus* (Hoffmann 1890), and *Varanus exanthematicus,* and the Australian extinct *Varanus prisca* (De Jong and Brongersma 1927, Fejérváry 1918, 1935, Head et al. 2009, Hecht 1975, Lee 1996), and *Varanus varius* (McDowell and Bogert 1954), where the latter focused on comparing anguimorphs with putative ancestry to snakes. *Varanus griseus* was long depicted as representative of a "typical" varanid skull (Gadow 1901, Hoffmann 1890, Smith 1935), but was not the object of in-depth analysis. *V. komodoensis* has been the subject of many studies by various authors, probably because of interest in this largest of extant lizards and the ease of working with such a large skull. In contrast, the skulls of the smallest species (*Odatria*) have barely been examined, and the differences between large and small taxa involve far more than the orders of magnitude sizes of the species. Excellent overviews of varanid skulls have been presented by Borsuk-Białynicka (1984) (mostly on fossil taxa), Evans (2008), Conrad (2015), Bochaton et al. (2019), Dong et al. (2021), Gauthier et al. (2012), Head et al. (2009), Kuhl (1820), Mertens (1959), and Rieppel and Grande (2007).

In very few studies are varanid skulls illustrated in exploded views, and in fewer are the anatomies compared with other members of the genus. The skull has been studied as an integral unit to find correlations between its structure and ecological diversity, focusing largely on the diverse Australian taxa (McCurry et al. 2015, Openshaw et al. 2016). A detailed introduction to the subject was published online (Bhavya, N.D.), and despite its unverifiable source and author, it is a very useful atlas for general varanid skeletal morphology. Like other studies, it is based on a single species. No single taxon is representative of the entire, quite diverse group.

The premaxilla is the most anterior bone of the skull, forming a single arrow-shaped structure extending from the rostrum to at least half the length of the bony naris (Figs. 5.3 and 5.4F–I). There is a long dorsal extension of the premaxilla, the processus premaxilla nasalis (Fejérváry 1935, Mertens 1942b), or dorsal ascending process (used here), that meets the anterior edge of the nasal bone or divides the nasals when the latter are paired, as in *Odatria, Varanus ornatus, and some populations of indicus* and *olivaceus*. Generally, the dorsal surface is a nearly straight line (e.g., many *Odatria, V. salvator, Varanus dumerilii*; in the latter, the dorsal surface becomes concave in adults), but in a few taxa (e.g., *V. exanthematicus, V. salvadorii, V. bengalensis, Varanus olivaceus*), it is convex, particularly so in older lizards. Of greater diagnostic value is the breadth of the dorsal surface of the dorsal ascending process, which may either be very thin and blade-like or somewhat expanded such that the process resembles a "T" in cross-section. This is one character that helps distinguish among three morphologically similar African taxa; in *Varanus albigularis* and *Varanus ocellatus*, the dorsal surface is blade-like, while in *V. exanthematicus* it is quite broad (Bayless and Sprackland 2000a, b). The similar Asian species, *V. bengalensis* (broad) and *Varanus nebulosus* (narrow), are also distinct. There is a unique compressed low crest near the anterior tip of the premaxilla in *V. bengalensis* (Fig. 5.3C). The ascending process is comparatively broadest in *V. dumerilii*. This dorsal surface may be of uniform width or taper at either end, but usually flares widely as it meets the anterior premaxillary process; in some taxa, it broadens near the nasals. In the juveniles examined, the process is unbowed and nearly level.

Bochaton et al. (2019) give an extremely useful presentation of major variations among specific bones of the skull in extant varanid species and focus on how such variations can be successfully used to identify fossil and subfossil monitors. Their illustrations make the paper indispensable for any varanid anatomist. A putative stem varanid from Eocene China (Dong et al. 2021) compares diagnostic features of varanids, sensu lato, with very well-preserved and nearly complete fossil materials, and offers considerable description and comparison of variations in individual bones among several species. In these and several other papers there is the fusion of neontology and palaeontology in the manner of Fejérváry (1935), and some of these are quite extensive in the review of fossil bones with extant counterparts (Borsuk-Białynicka 1984, Caldwell 2003, Conrad et al. 2008, Estes 1983, Gao and Norell 2000, Georgalis et al. 2020, Lee 1996, 1997, Norell and Gao 1997, Norell et al. 1992, Rieppel and Zaher 2000). Most often, *V. bengalensis* (as *Varanus monitor* in older literature; see Sprackland 1982a) is used as the contemporary representative varanid for comparison with fossil varanid material (Brongersma 1958, De Jong and Brongersma 1927, Fejérváry 1918, 1935, Georgalis et al. 2017, Ivanov et al. 2017).

Comprehensive reviews of cranial structure in other lizard groups have been well represented. Oelrich's (1956) paper on the anatomy of the head of the iguanid *Ctenosaura pectinata* has long served as the starting reference for other lizard anatomists and includes details of both hard and soft structures and is accompanied by useful color illustrations. Jollie's (1960) study was based largely on dissections of *Tupinambis nigropunctatus* (= *Tupinambis teguixin*) and published illustrations of other species. Harris' treatment of *Agama agama* (1963) is not as detailed but covers the anatomy of the entire body. Those papers are rare examples of complete anatomical reports, including muscles, blood vessels, nerves, and bones. Several genera of iguanine lizards are osteologically compared in considerable detail in the paper by de Queiroz (1987). Klembara et al. (2017) present an exhaustive review of the cranium of the anguid *Pseudopus apodus*, but do not cover mandibular osteology. Their coverage of an anguid represents a species closer to the varanid lineage than either Harris's or Oelrich's iguanian-based papers, and extends the material to include the ontogeny of the cranium. For most lizard families, let alone extant genera, few studies match the exhaustiveness and usefulness of that done for the American anguid, *Elgaria panamintina* (Ledesma and Scarpetta 2018), which, though based on a CT scan study of just two specimens of the target species, compared them with "several traditionally prepared articulated and disarticulated specimens of the closely related species *Elgaria multicarinata*." The purely osteological study includes the cranium, mandible, and hyoid structures, and the scans allow image examination of individual bones in disarticulated views. Anguimorph skull osteology is covered along with most other squamates in McDowell and Bogert (1954), and varanid osteology and myology are presented by Rieppel (1980). In a similar but far less voluminous review, Conrad (2015) provides osteological descriptions of two *Varanus palawanensis* and compares them with several closely related species of the water monitor group (= subgenus *Soterosaurus*), showing features that readily distinguish among several similar taxa. The authors of most of these papers note that there is a dearth of specimens, notably skeletal materials, for many species, and thus comparative observations are often not possible. Taxonomy is also sometimes an impediment because old specimens may be reassigned to new species – precisely the case with *V. palawanensis*, which was named in 2010, but Conrad's specimens were collected and identified in 1962 and 1987, and without additional data, it is often impossible to correctly assign skeletal material to a species (see Bochaton et al. 2019). This is particularly true for taxonomically "active" groups that include the *indicus, salvator*, and *tristis* complexes, plus many of the subgenus *Odatria* taxa.

Fossil varanids rarely contain complete skulls, and cranial materials almost always lack (or lack sufficiently intact) diagnostically informative osteology, such as premaxillae and septomaxillae. Perhaps the most complete reviews of varanid braincases are Norell and Gao (1997), and Rieppel and Zaher (2000) who compare excellent fossil mosasaurid materials with extant *Varanus*, and I use their terminology and Bahl's (1937) for braincase structures. The braincase is also explored in some detail by Borsuk-Białynicka (1984), who provides terms for the structural types of the basisphenoid-basioccipital complex as seen in ventral view. Fossil taxa represented by cranial material rarely have fully intact bones. Consequently, the systematics of those species have been inconsistent even at the higher taxonomic categories. This review is limited to taxa assigned to the genus *Varanus* and those that are unambiguously closely related to *Varanus*. †*V. prisca* of the Australian Pleistocene is known from numerous fossils that include several well-preserved maxillae, frontals, dentaries, and vertebrae, and Brongersma (1958) described the Holocene †*Varanus hooijeri* from Timor from subfossil vertebrae plus several good cranial elements that include parietal, maxillae, dentaries, and proötic. A review of the relationships between extant and fossil varanids will be presented in Chapter 11.

There is no one-source text that includes the details that a professional anatomist will need, and though this contribution is far from exhaustive or complete, it is nevertheless an attempt to bridge that gap between researcher and widely scattered literature. These are my primary goals for this chapter:

a. Coverage is comprehensive. I describe entire bones, each examined in isolation from all sides and fully labelled. My focus is on comparing the range of variation in *Varanus* osteology.
b. The bones described are all based on post-hatchling lizards. I do not discuss embryological developmental aspects of varanid osteology, a field seriously in need of study, and refer readers interested in ontogeny from the embryo stage to consult Andrews et al. (2017). Muscles are referenced when relevant to osteological features, and I have tried to include the major cranial fossae, fenestrae, and foramina, and, for the latter, what passes through them.
c. Varanid skulls show a great deal of variation among age classes as well as species, and I have shown where common features are systematically important or are matters of non-lineage convergence.
d. Because of the variety of abbreviations used in the literature, I have attempted to provide a standard for terms (see appendix) and have relied upon three main sources for my terminology and labels. For general osteology, I follow McDowell and Bogert (1954). For additional terms, including foramina, I follow Bahl (1937) and Evans (2008).

In many cases, I have illustrated a structure more than once, generally from different perspectives to avoid a common textbook error, e.g., mentioning an important structure in the text and then not depicting it in accompanying illustrations. Table 5.1 contains synonyms and dated terms for several bones.

5.2 SHARED CHARACTERS OF THE SKULLS OF *VARANUS*

Varanus has 57 or 58 bones in the skull, including cranium and lower jaw (= mandible), depending upon whether the nasals are single or paired, the skull with 43 or 44 bones, the mandible with 14 (Figs. 5.2 and 5.3 and Table 5.1).

Cranial variations are primarily of size and development or are closely linked to habits and diet. Cranial kinesis in varanids has been studied, with Frazzetta (1962) leading a long series of papers that do not always share similar results (Condon 1987, 1989, Hofer 1960, Iordansky 1968, McCurry et al. 2015, Rieppel 1980, Smith and Hylander 1985, Wilken et al. 2019). Much, but not all, of what appears to be osteokinesis is demonstrably muscle mass movement, especially in the temporal and supratemporal areas, but four mobile sutures have been well confirmed (Bahl 1937, Frazzetta 1962). During at least the early life of a monitor, there is considerable frontoparietal (mesokinesis of Frazzetta 1962) and atlanto-occipital kinesis. The small degree of vertical movement is probably useful mainly for maxillary tooth row motion, facilitating the movement of prey caudad. The postdentary (intramandibular) joint demonstrates the greatest mobility in the skull, except for that of the quadrate. The sum of the kinesis of that joint combined with the minor frontoparietal and atlanto-occipital movement provides an ability to "walk" the teeth to help move prey caudad, like what is seen in most snakes. Unlike most snakes, varanid mandibles lack bilateral independence. Thus, varanid cranial kinesis is closely linked to prey manipulation and less so to increased prey size. Frazzetta (1962) demonstrated the kinetics of both cranial and mandibular joints in the seizing and swallowing of food. McCurry et al. (2015) provided a detailed comparison of the cranial kinesis in 12 species of *Varanus*, a study that thoroughly demonstrates the wide interspecific variation of monitor skulls, and explicitly shows a positive correlation between skull morphology and the lizards' ability to exploit a variety of food sources.

Smith and Hylander (1985) did not find cranial kinesis in *V. exanthematicus* to any great extent, and Auffenberg suggested that snail crushing by *V. olivaceus* may have led to reduced kinesis in that species. Rieppel and Zaher (2000) suggested that kinesis is age-related, with calcification of joints as the lizards age, in agreement with the hypothesis of McCurry et al. (2015).

The maxilla extends posteriorly about half the length of the orbit but is excluded from the orbital border by the prefrontal, lacrimal, ectopterygoid, and jugal. The inferior posterior process of the maxilla ends in a pair of vertically oriented tines, their size varying among species. The

TABLE 5.1
The Acuity of a Lizard's Snout Can Be Measured Based on a Pair of Lines Drawn from the Orbits to Snout Tip and Representing the Angle in Degrees

Inst.	Cat. No.	Taxon	Age	Angle	SVL
NMNH	v2185 **H**	*†Saniwa ensidens*		20	
CAS	104407	*Acanthurus*		28	108
BMNH	1905.11.29.4	*beccarii*		30	
CAS	16883	*bengalensis*		32	199
CAS	uncat.	*bengalensis*		33	
BMNH	76.7.6.2	*bogerti*		38	
BMNH	64.7.22.3 **H**	*caudolineatus*		42	198 (total)
NMW	12465	*dumerilii*	juv.	35	103
NMNH	30974	*dumerilii*		33	251
NMNH	30975	*dumerilii*		34	310
NMNH	166730	*dumerilii*	juv.	31	83
BMNH	97.10.27.59	*eremius*		38	
NMNH	161744	*exanthematicus*		48	839 (total)
NMNH	267002	*flavescens*		43	
CAS	135137	*griseus*		37	135
NMNH	196455	*griseus*		30	
CAS	uncat.	*indicus* "Australia"		24	
FMNH	41313	*indicus (*Solomons)		31	
CAS	Uncat.	*indicus* (Australia, ex. AM)		24	
CAS	111549	*indicus* (Palau)		31	118
FMNH	41313	*indicus* (Solomon Islands)		31	
SMF	231891	*komodoensis*	ad.	16	
NMNH	519762	*komodoensis*	yng.	27	160
BMNH	96.7.8.1	*kordensis*		43	129
BMNH	57.10.24.34	*mitchelli*		38	88
NMNH	128387	*orientalis*		31	182
NMNH	128489	*orientalis*		30	183
NMNH	128491	*orientalis*		31	
NMNH	128492	*orientalis*		42	
NMNH	215908	*rainerguentheri*		28	108
NMNH	306185	*rainerguentheri*		32	101
NMNH	215907	*rainerguentheri*		30	100
NMNH	61320	*rosenbergi*		34	327
NMNH	63119	*rosenbergi*		33	299
NMNH	259133	*rudicollis*		23	560
CAS	uncat.	*salvator*	ad.	28	
NMNH	128490	*similis*		30	
NMNH	128491	*similis*		31	
NMNH	128493	*similis*		31	
NMNH	128482	*similis*		30	192
NMNH	128490	*similis*		30	
NMNH	128492	*similis*		42	
SMF	59011 **H**	*storri*		34	114
NMNH	379388	*timorensis*		40	172
BMNH	1.3a **H** for *Odatria punctata*	*tristis*		32	205
		Average		32.04	
		Range		20–48°	

Note: This table gives comparative data for numerous species of Varanus of different sizes. Note that snout acuity changes considerably between hatchling and adult *Varanus komodoensis*, but not in *Varanus rudicollis*. **H** = holotype.

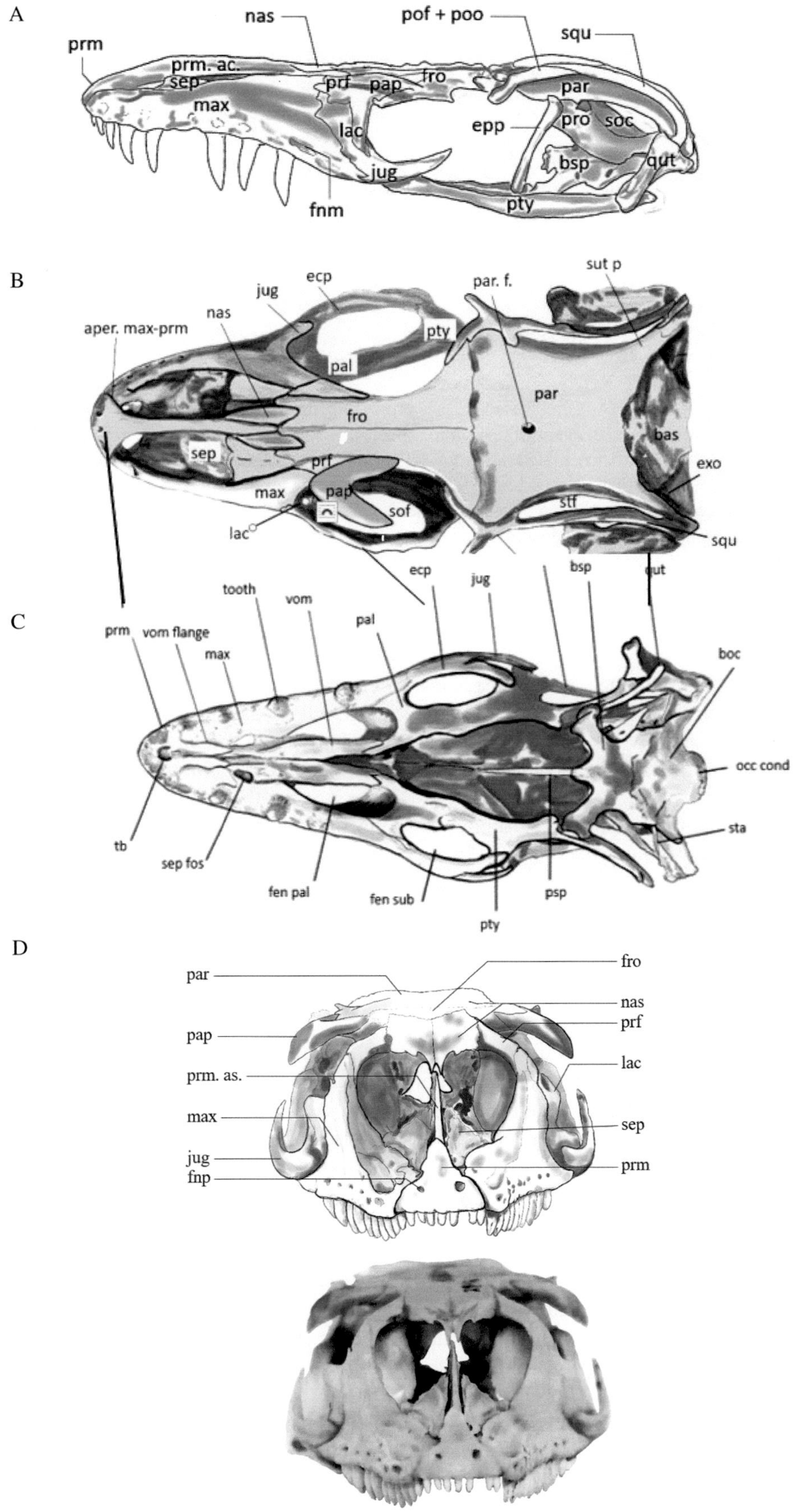

FIGURE 5.2 Cranial bones of four species of *Varanus*. (A) Left lateral view of *salvadorii* (RGS 410); (B) dorsal view of *caudolineatus* (SMF 40086); (C) ventral view of *niloticus* (USNM 220294); (D) anterior view of *nebulosus* (USNM 220280). Not to scale.

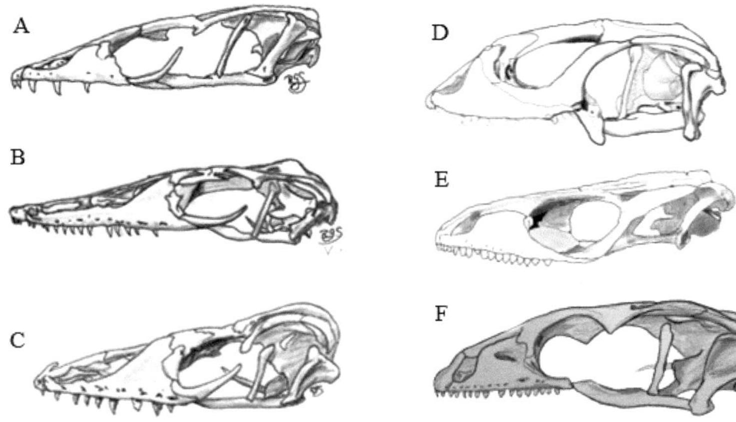

FIGURE 5.3 Skulls of three varanids (A–C) contrasted with skulls of other lacertilians. (A) *Varanus caudolineatus* SMF 40086), representative of the *Odatria* group of small Australian monitor species. (B) *Varanus rudicollis* (ZSM 18/1919b) is an atypical monitor with an extremely elongated snout, a rather flat head, and numerous small teeth. (C) *Varanus bengalensis* (CAS uncat.), representative of many of the largest monitors, with a distinctly awl-shaped head, a high braincase region, and a heavily ossified snout. No varanid has a high-Domed skull or the greatly enlarged pterygoids as commonly present in (D) *Crotaphytus collaris* (USNM 29151), an iguanian. Lizards with dorsoventrally compressed skulls generally have somewhat blunt, rounded snout tips, as in *Crotaphytus* and (E) *Xantusia henshawi* (USNM 229870). Many lizards have a complete postocular arch (D, E); in monitors, the arch may be completed by a short length of cartilage, but not by bone, while in many other lizards the arch (and jugal) may be entirely lacking, as in (F) *Pygopus lepidopodus* (USNM 56223). Not to scale.

morphology of the maxillae is extremely varied (see below) and is a useful diagnostic character. Paired lacrimal foramina are universal, the lateral entirely within the lacrimal and the larger medial bordered medially by both lacrimal and the prefrontal.

The skull anterior to the orbits (snout) is usually subtriangular, seen in dorsal or ventral aspect, though in species with a pronounced canthus rostralis, the shape may be square. The ratio of snout to postorbital skull lengths is smallest in hatchling varanids, e.g., it lengthens with age, and greatest in the extremely long-snouted *Varanus rudicollis*. Maxillary and dental tooth rows consistently terminate pre-orbitally, and teeth in extant taxa are restricted to those bones and the premaxilla. The premaxilla divides the left and right sides of the upper skull with a long, attenuated dorsal ascending process, which may deeply penetrate or fully divide the ypsiloid nasal(s), thus extending caudad about 75 percent or more of the length of the snout. The large fenestra nasalis is broad, nearly as long as the ascending process of the premaxilla, and is largely oriented horizontally. The septomaxillae are large and have arguably the most variable morphology of any cranial element among varanids, with at least 31 distinct morphologies among adults (see below). They encase most of the Jacobson's organ ventrally and abut medially with the nasal capsule. The upper temporal arches are universally present, and the jugal is never articulated posteriorly, except sometimes by thin cartilage or connective tissue. The supratemporal fenestrae are large except in *V. exanthematicus* and *V. dumerilii*, in which the elements roof over most of the region; the lateral temporal fenestrae are universal and large. Palatal and post-palatal fenestrae are very large, gradually widening posteriorly. Skull bones overall are gracile, certainly in young and small lizards, but can become quite robust in the adults of large (>1.7 m SVL) taxa whose diets evolve to include tougher prey materials such as snails and large bones (e.g., *albigularis, bengalensis, komodoensis, mertensi, olivaceus*). This robustness is not seen in other large taxa that consume hard-bodied items such as clams, crayfish, crabs, and moderate-sized vertebrates (e.g., *dumerilii, giganteus, indicus, salvadorii*). The braincase, made of several occipital elements, encloses a nearly cylindrical tube about one-third the length of the postorbital skull, and, more anteriorly, by the descending and nearly touching inferior fontanelles (cristae cranii) of the long, narrow frontal bones. A parietal foramen is universally present.

The mandibles have a fused anterior symphysis, and the dentaries extend to, or very slightly posterior to, antorbital rim. A large posterior adductor fenestra fully pierces the bone near the Meckelian joint, the smaller medial opening is slightly anterior to the lateral. There is a distinct intramandibular joint formed by the posterior portion of the dentary that allows considerable lateral kinesis; the medially positioned splenial inhibits inward flexion (Frazzetta 1962). A deep and well-defined Meckelian canal from the anterior symphysis to the anterior of the coronoid is universal, as is a deep oval Meckelian fossa formed by the medial surfaces of the surangular and prearticular. Teeth are present on the premaxilla, maxillae, and dentaries, arranged medially between a dental crest and basal dental shelf. Premaxillary tooth positions number seven or nine, the most anterior lying medially anterior to the others. All young varanids have comparatively long, pointed teeth that become slightly curved as the lizards grow. All teeth except the premaxillaries are notably laterally compressed, possess a sharp apex, fluted (plicidentine) sides, and a feeble posterior serrated ridge. Each tooth has a central inferior lingual nutrient foramen. Larger species may also have a minor anterior serration. Tooth morphology in species that regularly take hard-bodied prey as adults (*albigularis, bengalensis, exanthematicus, niloticus*, and the frugivorous, snail-eating *olivaceus*) develops conical molariform teeth, always in the posterior half, sometimes along the entire tooth row (SEE D'Amore 2015 for *niloticus*). This change in tooth morphology is absent in the hard-bodied prey predators *dumerilii, giganteus, indicus,* and *salvadorii*, all of which retain acutely pointed, sharp teeth throughout life. Premaxillary teeth are subequal in size and are smaller than other teeth. Central maxillary teeth are usually the largest, and dentary teeth generally alternate in size between large and medium or are nearly uniform with each other. Dentition is pleurodont, teeth are replaced posterolingually and continuously into old age (but perhaps not into *very* old age: see Mertens 1942b and below). The presence of palatal teeth is unknown in extant and subfossil varanids, but they are present in †*Saniwa*, a fossil genus of varanid.

Posterior to the intramandibular hinge are the remaining mandibular bones. The coronoid is convoluted with descending lateral and medial processes. There is a pair of small, extremely thin anterior tines that insert into an upper posterior notch in the dentary. The coronoid crest is large, nearly vertical, and sits in the middle of the dorsal surface of the bone. The surangular is the predominant lateral post-dentary bone and bears a convex surface with a feeble lateral crest and two distinct small foramina. There is a large medial adductor fossa for jaw muscles, and an ovoid posterior articular surface that meets the articular bone. Beneath the surangular are the angular, medial surface of the post-dentary mandible. As monitors grow, the lateral suture between the surangular and prearticular may become indistinct. Most posterior is the articular, which meets the surangular along a distinct, often broad suture. The anterior surface bears a crest with a concave posterior surface, forming the quadrate process, the unarticulated ventral joint that allows quadrate kinesis. The more posterior retroarticular process is a lateral ramus that extends beyond the joint.

5.3 THE CRANIUM

5.3.1 Premaxilla

The premaxilla is the most anterior bone of the skull, forming a single arrow-shaped structure extending from the rostrum to at least half the length of the bony naris (Figs. 5.2, 5.4, 5.5, and 5.7). There is a long dorsal extension of the premaxilla, the processus premaxilla nasalis (Fejérváry 1935, Mertens 1942b), or dorsal ascending process (used here), that meets the anterior edge of the nasal bone or divides the nasals when the latter are paired, as in several *Odatria, V. ornatus, and some populations of Varanus indicus* and *V. olivaceus*. Generally, the dorsal surface is a nearly straight line (e.g., many *Odatria, V. salvator, V. dumerilii*; in the

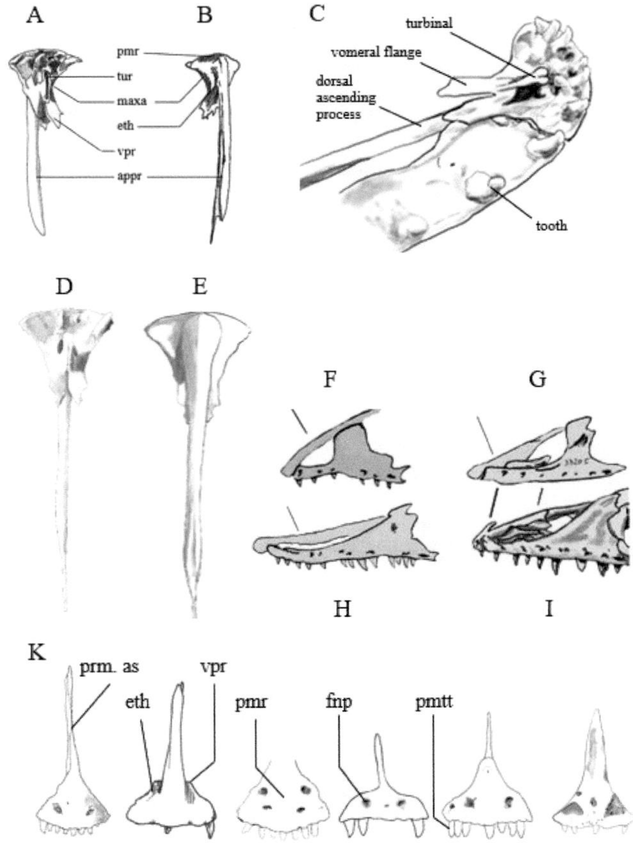

FIGURE 5.4 Premaxillae of *Varanus* showing the general morphology and the range of variation in *Varanus salvadorii*, showing (A) right ventrolateral and (B) left dorsolateral views. (C) Partial ventral view of the palate showing the relationship of the premaxilla to the adjacent bones. (D) Ventral and (E, F) dorsal views of *V. salvadorii* premaxilla. (G, H) Variations in premaxillary curve marked in (G) *Varanus storri*, (H) *Varanus griseus*, (I) *Varanus rudicollis*, and (J) *Varanus bengalensis*; the left marker in (J) shows the characteristic "hump" of the snout of this species. (K) Anterior views showing foramina variation in premaxillae of (L to R): *Varanus niloticus*; *Varanus salvator*; *Varanus exanthematicus*; *V. bengalensis*; *Varanus nebulosus*; *Varanus dumerilii*.

latter the dorsal surface becomes concave in adults), but in a few taxa (e.g., *V. exanthematicus, V. salvadorii, V. bengalensis, V. olivaceus*), it is convex, particularly so in older lizards. Of greater diagnostic value is the breadth of the dorsal surface of the dorsal ascending process, which may either be very thin and blade-like or somewhat expanded such that the process resembles a "T" in cross-section. This is one character that helps distinguish among three morphologically similar African taxa; in *V. albigularis* and *V. ocellatus* the dorsal surface is blade-like, while in *V. exanthematicus* it is quite broad (Bayless and Sprackland 2000a, b; Fig. 5.2D). The similar Asian species *V. bengalensis* (broad) and *V. nebulosus* (narrow) are also usually distinct. There is a unique compressed low crest near the anterior tip of the premaxilla in *V. bengalensis* (Fig. 5.4I). The ascending process is comparatively broadest in *V. dumerilii*, where the dorsal surface may be of uniform width or taper at

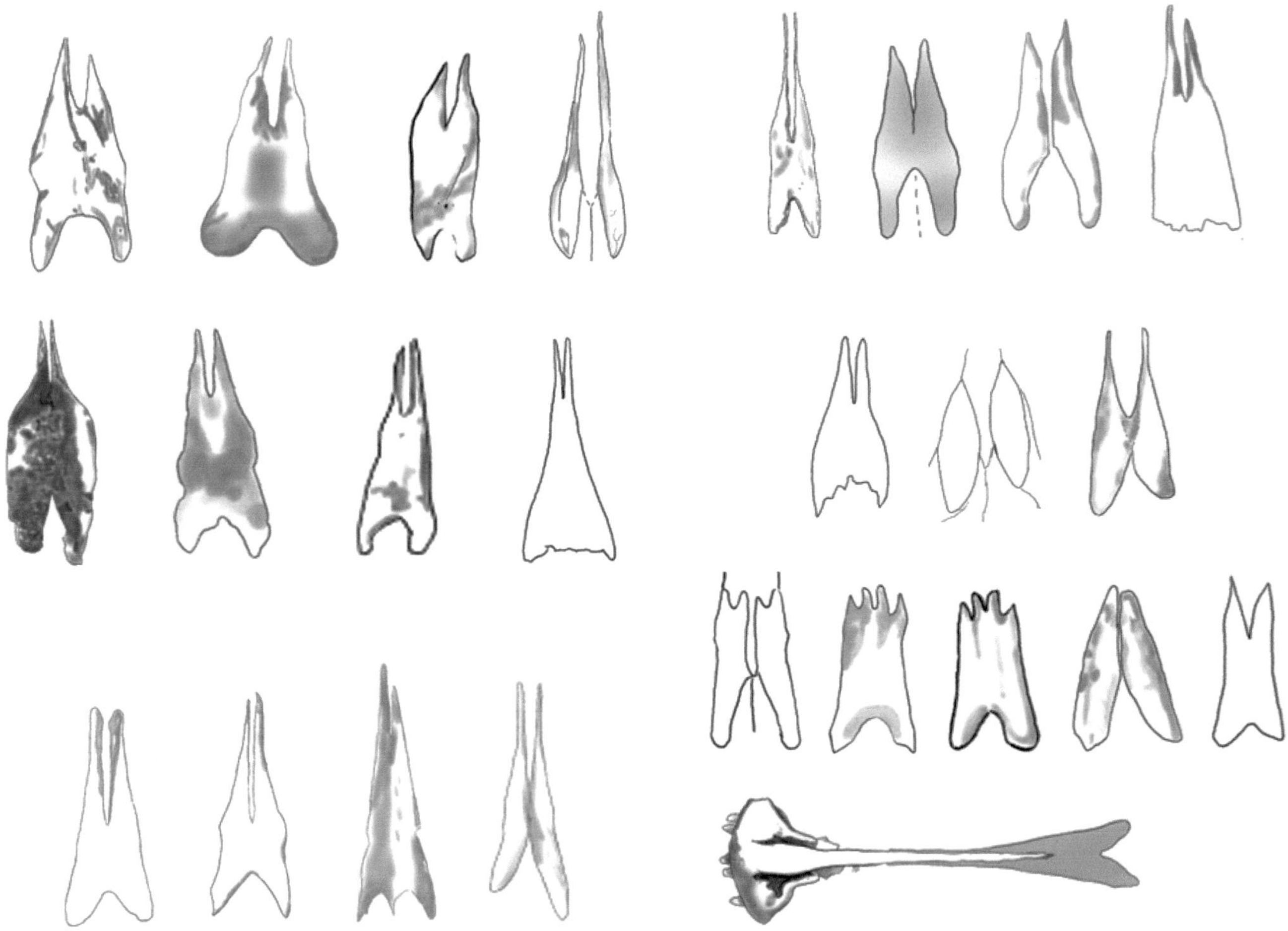

FIGURE 5.5 Diagrammatic representation of nasal bone morphology in *Varanus*. Dorsal outlines of (left to right) top row: *niloticus* (MVZ 4786); *niloticus* (USNM 220294); *albigularis* (USNM 72027); *albigularis* (USNM 216260); *exanthematicus* (FMNH 51638). Second row: *griseus* (USNM 220282); *bengalensis* (USNM 14753); *bengalensis* (MVZ 77048); *dumerilii* (RGS 203); *rudicollis* (ZSM 18/199). Third row: *komodoensis* (USNM 101444); *marmoratus* (USNM 287413); *salvator* (USNM 163810 Vietnam); *olivaceus* (FMNH 223181); *olivaceus* (USNM 222400). Fourth row: *gouldii* (USNM 25760); *giganteus* (MVZ 80070); *caudolineatus* (SMF 40086); *timorensis* (RGS 189). Fifth row: range of nasal variation seen in the *indicus* group, from left to right: Caroline Islands (SMF 32808); Guam (USNM 122469); Guam (USNM 323717); Saipan, Marianas (MVZ 46017); Aru Islands (SMF 11635). Anterior at top, not to scale.

either end, but usually flares widely as it meets the anterior premaxillary process; in some taxa, it broadens near the nasals. In juveniles examined, the process lacks curvature, though in small species, it can be quite rounded. Premaxillary curvature, then, is not directly correlated with absolute skull size.

Most varanids have unpaired nasals (Fig. 5.5), and in these species the dorsal ascending process comes to an acute point that penetrates at least a quarter of the length of the nasal, and the most posterior position of which may be diagnostically useful, at least in the water monitor subgenus *Soterosaurus* (Conrad 2015). In monitors with unpaired nasals, the premaxilla does not contact the frontals; in species with paired nasals, the premaxilla completely separates the nasals and terminates in contact with the frontals. The posterior ventral part of the dorsal ascending process is vertically forked and underlies the anterior nasal(s), and the dorsal nasal process overlies the nasal, making the premaxillary-nasal bones interdigitate along their suture. The ventral process is equal in width to the dorsal in blade-shaped premaxillae, but narrower in taxa with the broad-dorsum premaxilla.

At the point where the anterior dorsal process and posterior edge of the anterior process meet, there is a distinct round or oval posterior dorsal premaxillary foramen (= ethmoid foramen or foramina dorsalis of Mertens 1942b:126; Fig. 5.4K). It may lie facing dorsally, as in *V. exanthematicus*, or have a more vertical position at the extreme anterior side of the dorsal process, as in *Varanus marmoratus*. The presence of this foramen is universal in varanids, but its size is highly variable.

Some species also possess a second, larger pair of anterior dorsal premaxillary foramina. They face forward and dorsolaterally, and are clearly visible in anterior, dorsal, and lateral views of the premaxilla. They are seen, for example, in *exanthematicus*, *indicus*, and *marmoratus*. In *dumerilii,* the foramen is extremely large, triangular, and occupies most of the anterolateral portion of the anterior premaxilla (Fig. 5.4) and is considerably more complex than described by Bochaton et al. (2019:222). The posterolateral wall of each foramen extends from the dorsal process to the apex of the lateral edge, and behind this is the laterally oriented ethmoidal foramen. In species that lack the anterior dorsal premaxillary foramina, including *albigularis*, *niloticus*, *salvadorii*, and *indicus* from Halmahera, the labial edge of the bone is smooth, only rarely showing any indentations. In *Varanus macromaculatus* from southern Thailand, there are two or three very shallow anterior fossae, while most specimens show no anterior fossae or foramina. In *salvadorii* the premaxilla is unusual in that the long dorsal process originates far anterior, just above the rostrum, with a minor ridge along its ventrolateral edge that gives it a rod-like shape (Fig. 5.4A–E). The dorsal surface is smooth and convex and remains so in contact with the nasal. The part corresponding to the anterior half of the anterior premaxillary process (= anterior body) bears a low, distinct medial ridge (Fig. 5.4B, E). There are a pair of posterolaterally-facing, small ethmoid foramina, and the anterior foramina are lacking.

The anterior body of the premaxilla is squarish to nearly triangular (Fig. 5.4K), the square angled such that the rostrum and most lateral portions are corners. The edge between the lateral points and the junction with the dorsal process is the maxillary articulation, a thin ridge that lies under the anterior-most part of the maxilla. The rostrum most typically comes to at least some semblance of a point, but in taxa including *V. komodoensis*, it is almost completely rounded. The degree to which it is rounded or pointed may be partially diagnostic for distinguishing among some of the largest species (Bochaton et al. 2019). Laterally, the premaxilla-maxilla suture is variable, ranging from nearly no gap (*V. salvator*, *Varanus timorensis*) to a distinct separation between the bones laterally (*V. bengalensis*, *V. komodoensis*, and all remaining *Odatria*).

The ventral surface of the anterior premaxilla is more variable than the dorsal ascending process. On one extreme, the lateral portions of the maxillary articulations may be completely smooth, lacking foramina, fossae, or any complex topography. In *V. salvadorii*, for example, the surface is smooth and rounded, the posterior structures (vomeral flanges) sloping gradually downward (Fig. 5.4A, D). In *V. salvator* and *Varanus bivitattus*, however, the surface is slightly rugose, and a pair of pronounced rectangular depressions lies adjacent to the central dorsal process. These depressions form short channels, with the ethmoid foramina piercing the innermost vertical wall of the structure (Fig. 5.4B). In many taxa, the depression is shallower and may bear a ventral, dorsal-facing foramen. In those species with a smooth anterior surface, the dorsal process is also smooth, while in those with rugose premaxillae, the dorsal process, too, is rugose.

The premaxillary ventral vomerine flange is universally flat and bifid (Fig. 5.4A, C, D). The general *Bauplan* of this process is a laterally flattened table divided by a medial septum, the anterior part of which forms an enlarged, rounded process. The lateral shelf has paired posterior processes that extend ventrally to the ventral process, the distance being species-specific. The posterior processes vary from short and somewhat rounded to much longer and pointed and underlie the anterior vomers. The anterior process forms the medial surface of the ethmoidal foramen, and in structure may be nearly non-existent or an elongated table with a deep central depression. In some taxa, the turbinal is spool-shaped, and in others it is reduced so that it resembles a pair of vomerine teeth. The range of morphological variations of the premaxilla observed in *Varanus* is given in Table 5.3.

5.3.2 Naris and Nasal

The most variable part of a monitor's anatomy is likely the internal naris structure, including the nasal (Figs. 5.2 and 5.5). While the external nostril shows variation in shape (round, oval, slit, or tear-like), size (large, intermediate, small), and position (anterior, median, or near orbit), the internal chambers (= turbinals of some authors) can be simple invaginations, a recurved tube, or a labyrinthine tunnel. Aquatic and semi-aquatic taxa have more complicated nasal chambers, perhaps acting as a pressure-based mechanism that inhibits the entrance of water while the lizard is submerged, but only two such species are known to have valves, or flaps, that retard water entry (*dumerilii* and *mertensi*). The variation of the external nostrils is discussed in another section. Of the cranial bones, the nasals are among the simplest in structure.

Covering only the rearmost section of the nasal chamber is the nasal bone, either paired or a fused central element. The nasals are paired in all neonate skulls I examined but may not be the case in *rosenbergi* (Andrews et al. 2007). Presence of paired nasals is seen in many of the subgenus *Odatria* (e.g., *acanthurus*, *caudolineatus*) but not others (*eremius*), and rarely in other taxa. In the polymorphic *V. indicus* group are some populations in which the nasals are consistently paired, while across most of the range, they are fused. This requires re-examination because members of the *indicus* group have undergone taxonomic revision recently, and certain character states have already been linked to some of the new (or resurrected) species. The nasals are not as morphologically variable as the septomaxilla but do have a range of shapes. Adult *V. komodoensis* have a somewhat triangular main element that bears an anterior pair of tines, and two thick posterior corner processes. In *Varanus giganteus*, the main element is more hemispherical in shape, also bears paired tines, and lacks distinct posterior structures. Instead, the lateral edges of the bone swing back beyond the central body, forming wing-like processes, and the frontonasal suture is highly asymmetrical. *V. niloticus* has a lozenge-shaped central body that terminates in paired broad posterior processes and has an elongated rectangular anterior process that ends in very short anterior tines. The anterior half of the nasal consists of two prolonged fork-like tines in *V. bengalensis*, while the anterior half becomes a very broad triangular structure. The variety of nasal morphologies falls into five broad categories (Fig. 5.5):

1. Broad, triangular body with a distinct pair of acute anterior tines that embrace the posterior ascending dorsal process of the premaxilla; e.g., *albigularis, griseus, komodoensis, olivaceus*.
2. Narrow, triangular body with a distinct pair of acute anterior tines that embrace the posterior ascending dorsal process of the premaxilla; e.g., *bengalensis, dumerilii, giganteus, gouldii*.
3. Anterior and posterior sections are considerably different in width, demarcated by a sudden change in appearance; e.g., *marmoratus, niloticus* (part), *salvadorii, salvator*.
4. Nasals, whether separated or united by a small lateral process, oval or kidney-shaped, with feeble or no anterior tines; e.g., *caudolineatus, exanthematicus, indicus* (Marianas and Caroline Islands), *olivaceus, timorensis*.
5. Somewhat rectangular nasals with four distinct, subequal anterior processes; *indicus* (*bennetti*?) (Guam).

Among *V. ocellatus*, many of the subgenus *Odatria*, and in some *V. olivaceus* and *V. indicus*, the nasal bones are unambiguously paired (Fig. 5.5), and in adults they do not meet even at their most posterior points. They originate along the last quarter of the dorsal premaxillary process and terminate well posterior to the anterior-most parts of the frontals. However, in *V. olivaceus*, both paired and fused nasal conditions occur (Fig. 5.5). A similar situation is present among different populations of the large *V. indicus* complex, probably representing characteristics of different species. This group, once seen as comprising a few very similar species, has been shown to represent several distinct taxa. There are geographically linked similarities in nasal morphology; for example, I have seen only one form in lizards from Guam, in which there are four distinct anterior processes. Specimens from the Caroline Islands have the same gross morphology but differ in having the nasals fully divided, while Marianas specimens have divided nasals but lack any trace of anterior projections. In contrast, *indicus* and *chlorostigma* from New Guinea (including peripheral islands such as the Arus) and Australia have nasals quite unlike their more northern relatives, and more like *timorensis* and *giganteus*. It will not be possible to clearly determine if the condition of the nasals is linked to specific species until more specimens are collected with locality data that reflect current taxonomy. As for the genus as a whole, it seems that (1) nasal structure varies with age, (2) has a consistent general body plan within a species, and (3) there are numerous minor variations – such as length of tines or relative sizes of each half of the bone(s) – present within each taxon.

5.3.3 Septomaxilla

The most obviously variable morphological elements, even between closely related species, are the septomaxillary bones (Figs. 5.2, 5.6, and 5.7). Surprisingly, Mertens (1942b:130–131) had very little to say about the septomaxillae, confining his remarks to a single paragraph and describing very superficial features. Considering the bone's relationship to both the nostrils and vomeronasal receptors, it is odd that this highly variable bit of osteology has been neglected. Hillenius (2020) published a detailed review of

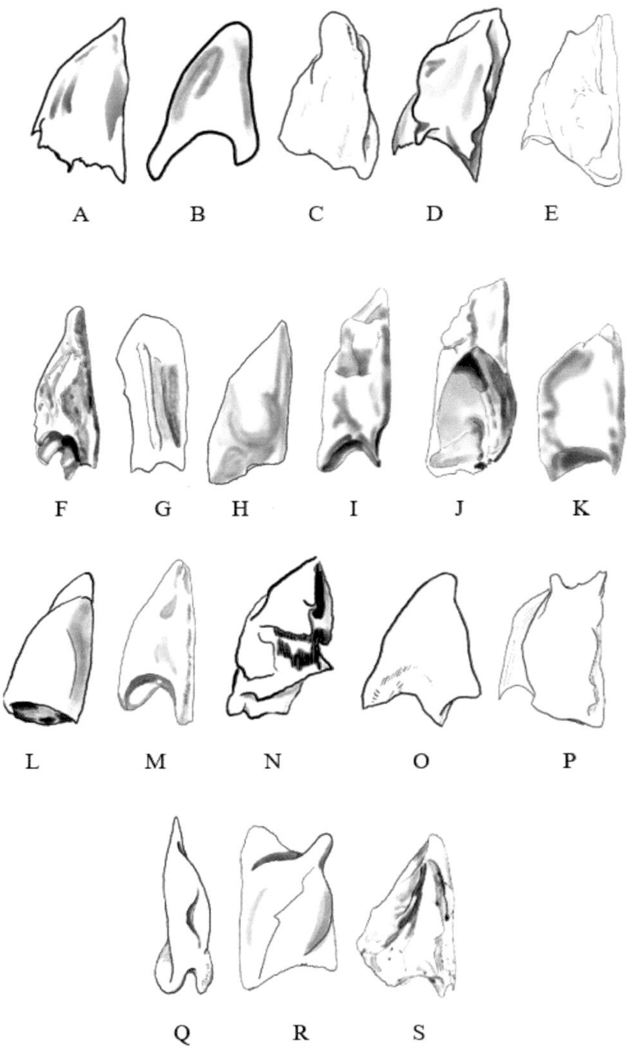

FIGURE 5.6 Variation in left septomaxilla shape, dorsal view, in *Varanus*. (A) *dumerilii* (adult-SMF 1156), (B) *dumerilii* (juv.-SMF 11557), (C) *flavescens* (SMF 11546), (D) *giganteus* (SMF 53263), (E) *gouldii* (USNM 67734), (F) *griseus* (N. Africa-USNM 99240), (G) *griseus* (Sudan-USNM 220282), (H) *indicus* (USNM 122469), (I) *marmoratus* (USNM 498916), (J) *niloticus* (USNM 220294), (K) *olivaceus* (USNM 222400), (L) *prasinus* (RGS 102), (M) *bengalensis* (USNM 14753), (N) *timorensis* (RGS 189), (O) *tristis* (SMF 11630), (P) *varius* (SMF 11617), (Q) *rudicollis* (ZSM 18/1919), (R) *salvadorii* (RMNH 4620), (S) *komodoensis* (USNM 101444). Drawn in alignment with the dorsal ascending process of the premaxilla, anterior at the top.

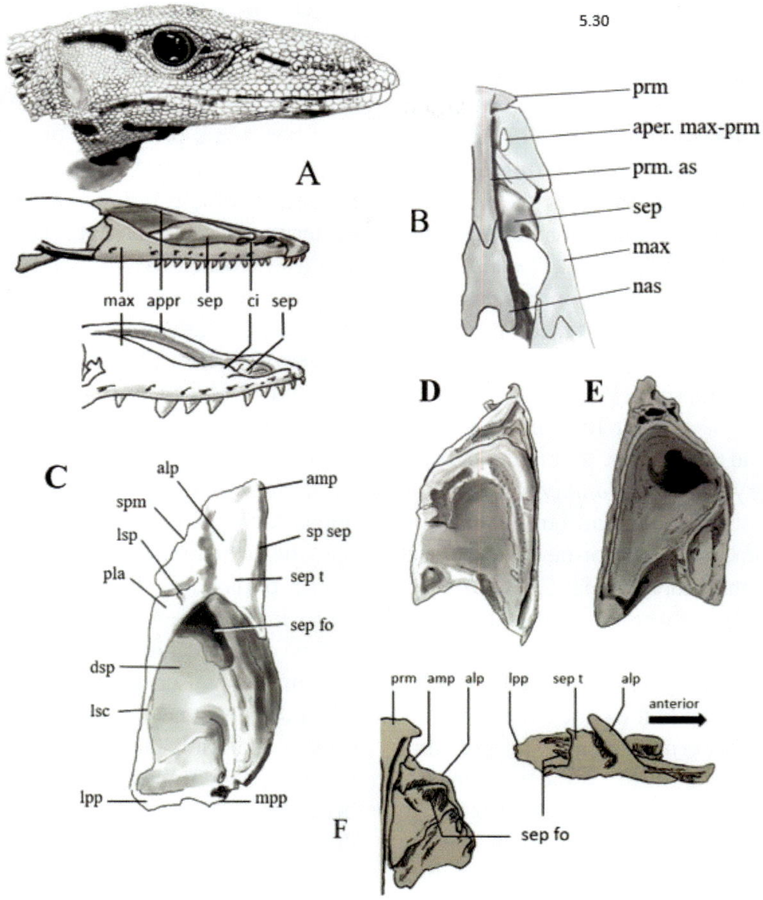

FIGURE 5.7 Septomaxilla morphology and position in *Varanus*. (A) Profile of *rudicollis* from life (upper) and anterior half of skull (lower, approximately to scale), showing adult and juvenile morphologies. Note the length of the septomaxilla. (B) Dorsal view of the right half of the snout of young *niloticus* showing septomaxilla and associated bones in situ. (C) Morphology of the dorsal surface of the septomaxilla of *Varanus niloticus*. Dorsal (D) and ventral (E) views of the septomaxilla of *Varanus olivaceus* (USNM 222400); the ventral view shows a recess for Jacobson's organ. (F) Dorsal (left) and lateral views of the right septomaxilla of an adult *niloticus* (BMNH 1935.8.3.1), showing the highly irregular septomaxillary topography of the species. Note also the well-developed covered fossa, a possibly unique feature in *Varanus niloticus*.

synapsid septomaxillae, focusing on soft-tissue anatomy, but no similar work exists for Squamata. From my observations, the septomaxillae represent a wide diversity of some 31-plus morphologies (Fig. 5.6) that range from the sub-triangular, open at the rear structure of *V. niloticus* (Fig. 5.6J), to the bladder-like structures, adorned with anterior spiny projections, seen in *flavescens*, *salvadorii, salvator*, *and varius* (Fig. 5.6C, R, P, respectively). The paired septomaxillae lie deep to the elongate premaxilla-nasal complex. The variation in septomaxillary structure among taxa makes this bone a useful diagnostic at the species level.

The septomaxilla is the floor of the anterior nasal canal (Figs. 5.2A and 5.7A, B), the tube that begins at the external nostril and terminates anterior to the junction with the duct from the nasal gland. It lies dorsal and horizontal to the vomer and is bordered anteriorly by the fused premaxilla, medially by the dorsal ascending process, and laterally and posteriorly by the maxilla. The dorsal surface and posterior border are free of contact with other bones. This region is hidden in other lizards by the nasals and anterior portion of the maxilla (Fig. 5.3), but the "curiously incomplete condition of the dermal skeleton of the varanid snout" (Bellairs 1949) leaves the nasal capsule exposed, including most of the dorsal surface of the septomaxilla. Jacobson's organ lies in a hemispherical fossa ventral to the septomaxilla. The septomaxillary region of many monitor species that live in or near deserts or salt water also contains a small salt gland (Dunson 1974) and includes *acanthurus, flavescens, gouldii, giganteus, griseus*, *mitchelli*, *rosenbergi*, *salvator*, and *semiremex* (Hazard 2004, Minnich 1979). The shape of the bone is highly variable. In varanids with a three-sided septomaxilla (e.g., *bengalensis, dumerilii, flavescens, indicus, prasinus*), the apex of the triangle is anterior (Figs. 5.6 and 5.7B–E). The posterior edge is bi-lobed to some degree, typically with the medial lobe more elongate and acute. In *flavescens, giganteus, gouldii, varius*, and others, the lateral lobe is longer (Fig. 5.6C, D, E, P), while in *Odatria* the posterior processes are nearly equal in length and rather short (Fig. 5.6N, O). The dorsal surface is generally convex, often nearly flat, and may be smooth (e.g., *dumerilii*, *indicus*, Fig. 5.6B, H) or quite rugose (e.g., *giganteus*, *griseus*, *tristis*, Fig. 5.6D, F, G, O), and in many species forms a distinct capsule under the septomaxillary table, ranging from a minor depression (e.g., *griseus*, *marmoratus*, Fig. 5.6F, I) to a pronounced deep recess (e.g., *niloticus*, Fig. 5.6J). Unfortunately, this bone is also quite thin and fragile, and skeletal preparation often damages the structure. In its simplest form, the septomaxilla is a simple, slightly convex, equilateral triangle (ex. *beccarii*, *indicus*, *prasinus*, *Odatria*). In *beccarii* there are no or very minor processes. *Varanus prasinus* (Fig. 5.6L) has a tiny median point on the posterior edge, while in *V. indicus* the small process lies medially along the premaxilla (Fig. 5.6H). In *giganteus* (Fig. 5.6D), the bone is similar to that of *beccarii*, but the medial border folds inferiorly, making a partially closed sub-septomaxillary canal. In *dumerilii*, the anterior half of its septomaxilla is triangular, but in subadults, there is an elongation into two posterior-facing processes, the more distal being slightly longer (Fig. 5.6A, B). As in *giganteus*, an inferior folding forms a canal. This is also the case in many *niloticus*, though the outer process is as long as the anterior half of the bone, the medial process is very short, and the ventral wrap of the bone extends to form a wide table between the external process and premaxillary. In *niloticus* the posterior third of the septomaxilla is extremely unusual in that the septomaxillary table is a transverse flange forming a strongly concave cup, the central portion extending ventrally, and the medial and distal edges of the bone rising sharply to form upright borders (Figs. 5.6J and 5.7B, C, F). The tube so formed is apparently unique among varanoids, not seen in the other African monitors. This flange also extends as much as 40 percent of the total length of the bone. In most specimens, the two posterior processes extend back as long, narrow structures with acute termini. The recess so formed provides support for a spherical portion of the soft nasal tube that extends slightly anteriorly before folding medially and extending rearwards. *Varanus flavescens* has a more elongate septomaxilla (Fig. 5.6C) than African species. It is almost rectangular and has a pronounced, broad, finger-like anterodorsal process that arcs slightly distally. A similar condition is seen in *varius*, though in *varius* the septomaxilla is longer (Fig. 5.6P). The most complex of the triangular septomaxillae is the form with a high crest along the dorsal surface, seen only in *komodoensis* (Fig. 5.6S). This crest gives the septomaxilla a triangular shape when viewed in sagittal section.

The other prevalent shape of the septomaxilla is a variant of rectangular or square. In *salvadorii* the bone has a predominantly square body, with a hemispherical dorsal extension from which projects an anterior, distally facing sharp horn (Fig. 5.6R). Each corner of the square is also a pointed process, the posterior medial corner being the largest. Most of the dorsal surface is a smooth convex hemisphere. In *rudicollis* the septomaxilla has a distinctly wine-bottle shape (Figs. 5.6Q and 5.7A), e.g., a long, blunted anterior "neck" and an elongated body with nearly parallel sides. There are two posterior processes, the medial slightly longer, with a long posterior-directed horn above. No other varanid has a septomaxilla like this, possibly correlated with the unusual extension of the snout in this species.

In width, the septomaxilla may be as wide as the concave horizontal process of the maxilla (fenestra exonarina anterior), as seen in *V. indicus* (Fig. 5.6H), or no more than half that width, as in *prasinus* (Fig. 5.6L). Additionally, the posterior extent of the septomaxilla may vary even between species in which the bone is roughly the same shape. For example, in *V. indicus*, the septomaxilla covers the vomer such that the latter cannot be seen in dorsal view, while in *timorensis*, a considerable portion of the vomer is clearly exposed (Fig. 5.6N); in *prasinus*, there is an intermediate condition in which a small portion of the vomer is exposed.

An anterior septomaxillary foramen lies along the dorsal surface of the fenestra exonarina anterior, the smooth dorsal-facing anterior-most section of the maxilla. The opening is extremely small in *rudicollis*, being little more than a pinhole, moderately larger in *dumerilii*, and much larger in *indicus* and *olivaceus*.

The considerable variation seen in septomaxillary structure makes it useful as a taxonomic character, especially when there is a need for a positive identification from only skeletal material. The water monitor group (*V. salvator* and relatives) has a tremendous geographical range and has long been divided into several subspecies, shown to represent distinct species (Koch et al. 2007, Welton et al. 2013). There appears to be a discrete variation in septomaxillary structure among the water monitors. For example, *V. bivitattus* from eastern Indonesia possesses unique paired, thin anterior septomaxillary horns, while the Philippine *V. marmoratus* has smooth, undecorated septomaxillae. Specimens from parts of Sumatra (*V. macromaculatus*) have, instead, a flattened square anterior process, and lizards from Thailand and Vietnam have tiny single horns, but the maxillae are somewhat longer and distinctly smoother than in lizards from other regions. Without detailed locality and identification data, it is difficult to authoritatively assign the variations to each regional taxon, but variation seems to agree with recent taxonomic changes within the group (Conrad 2015, Koch et al. 2007, Welton et al. 2013).

One drawback to using the septomaxilla as a key character is that it occurs with frustrating uniformity in the subgenus *Odatria*. In these dwarf varanids, restricted to Australia and southern New Guinea, the septomaxilla is consistently seen as a broad triangular bone with a tiny, or no, anterior horn, with variation among species essentially nil. A rear-facing dorsal fossa is seen in some taxa (*acanthurus*) but not in others (*caudolineatus*), and I have not seen adequate numbers of skulls of the group to determine the range of variation among or within *Odatria* species.

5.3.4 Maxilla

The maxillae (Fig. 5.8) and premaxilla are the tooth-bearing bones of the upper skull in all *Varanus* (Figs. 5.2–5.4, 5.7, 5.8, 5.10, and 5.14 and Table 5.2). The maxilla is the longest

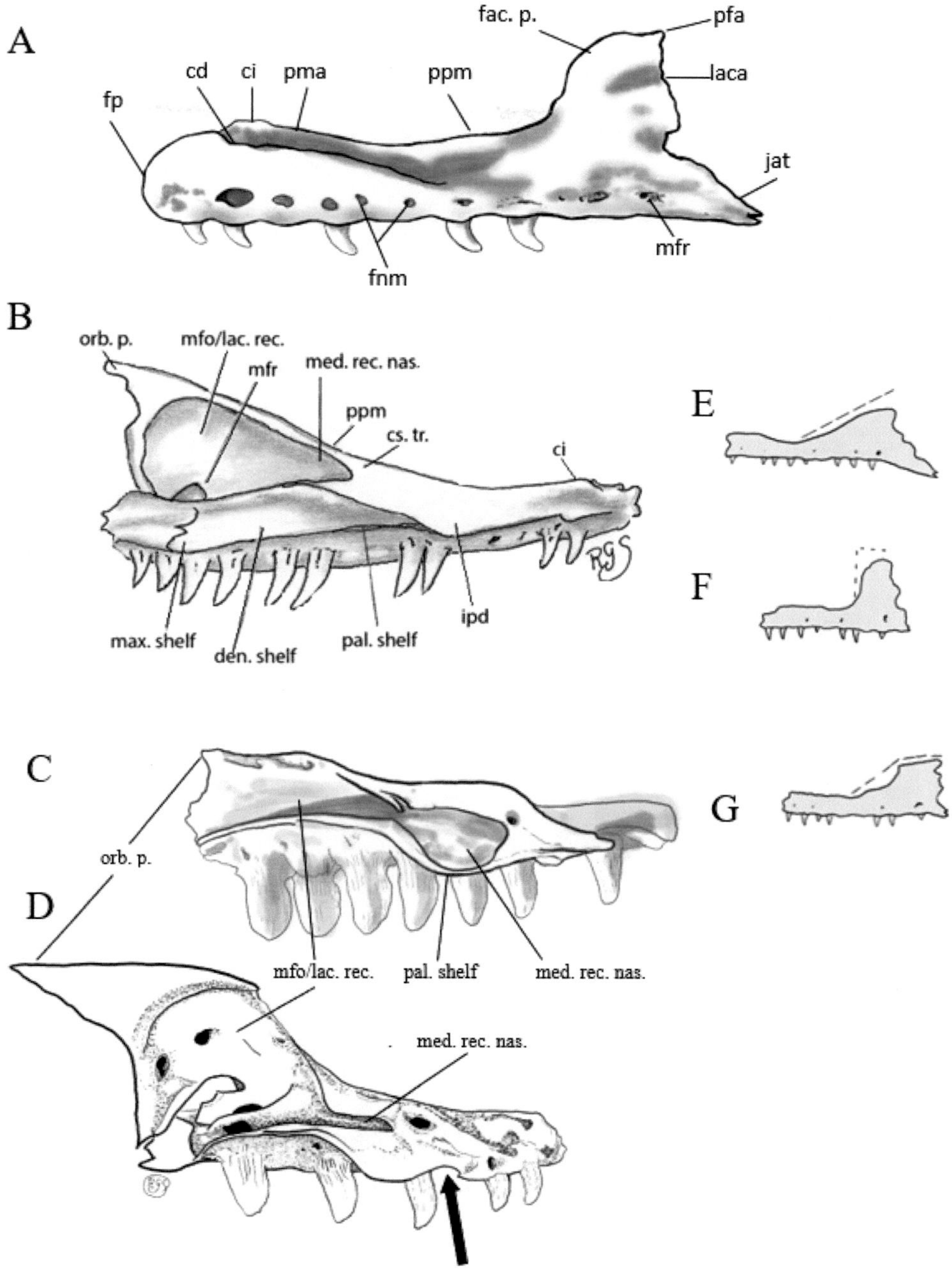

FIGURE 5.8 Morphology of a varanid maxillae. (A) Left lateral view of *Varanus salvator*; (B) same, medial view. Drawn from a skeleton in the author's collection. (C) Medial view of maxilla of *Varanus albigularis* (USNM 216261); note the constriction separating the maxilla-frontal/lacrimal recess and the medial recess for the nasal sac, and the anterior shelf produced by the medial intranasal crest. (D) Posteromedial view of left maxilla of adult *Varanus exanthematicus* showing reduced medial recessus nasalis and large prelacrimal recess. (E–G)Three types of idealized varanid maxilla based on Fejérváry (1935). (E) Hypsiprosopic, in which the level and inclined portions of the dorsal maxillary crest are subequal; (F) platyprosopic morphology, with a flat anterior and abrupt, nearly vertical posterior section; (G) mesoprosopic condition, where the dorsal maxillary crest follows a smooth curve. There are, in fact, numerous intermediate morphologies, making the classification of limited usefulness.

and largest of the lateral bones of the skull, extending from immediately posterior to the premaxilla's maxillary articulation (Fig. 5.2A) and extending to a point mid-orbit. It does not form part of the bony border of the orbital fenestra, being supplanted by the lacrimal, palpebral, and prefrontal along the anterior edge, and the jugal, which rests superior to the maxilla beneath the orbit (Figs. 5.2A, D, 5.3A–C, 5.10, 5.13, and 5.14). Universally, the maxilla is the predominant facial bone, has a slightly convex lateral surface (the degree ranging from conspicuous to nearly flat), laterally encloses most of the snout, and has inner medial shelves that both protect the Jacobson's organs and support the nasal capsules. There is a posteroventral notch in the maxilla, which acts as part of the joint with the muscles

TABLE 5.2
The Range of Variation of the Left Maxilla Across the Varanidae

A 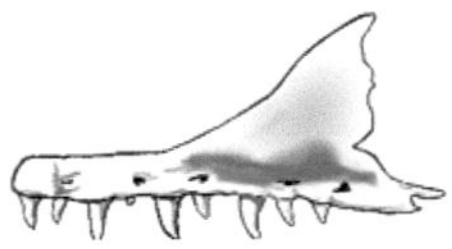 *acanthurus* SMF 11639	B  *albigularis* SMF 11343	C 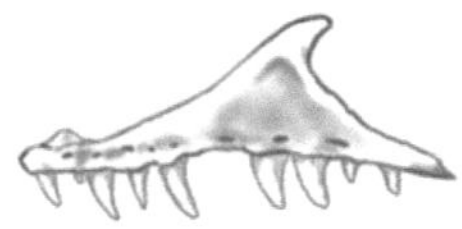 *beccarii* RGS201	D  *bengalensis* (yng.) SMF 40160
E  *bengalensis* (subad.) ZFMK 32956	F  *bengalensis* SMF 60428	G 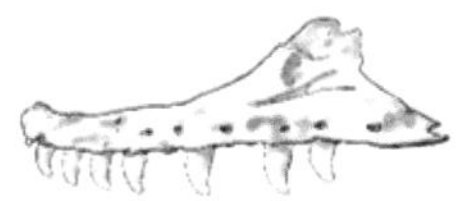 *bengalensis* (Iran) ZFMK uncat.	H  *caspius* (Iran) SMF 60428
I 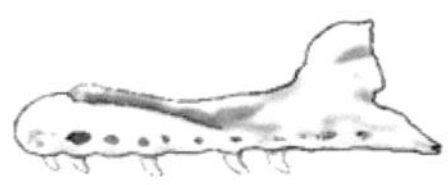 *dumerilii* RGS 203	J  *dumerilii* (subad.) ZFMK 11556	K 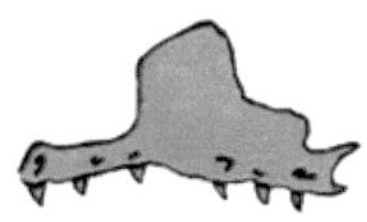 *eremius* SMF 11648	L 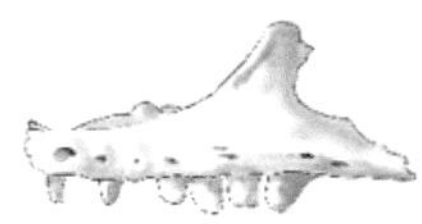 *exanthematicus* RGS 202
M  flavescens SMF 11546	N  *giganteus* (yng.) ZFMK uncat.	O  *giganteus* ZFMK 14882	P 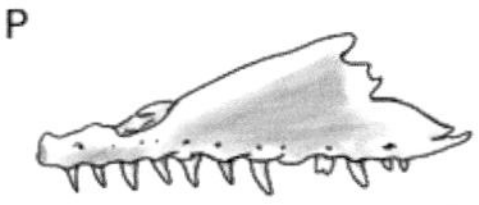 *giganteus* SMF 53263
Q  *gilleni* SMF 11627	R  *glauerti*	S  *gouldii* ZFMK 14885	T  *griseus (*Morocco) USNM 220282
U 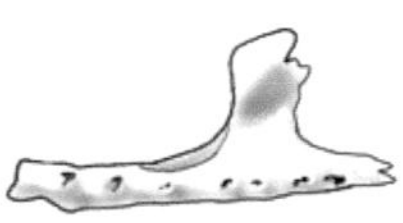 *griseus* SMF 33205	V 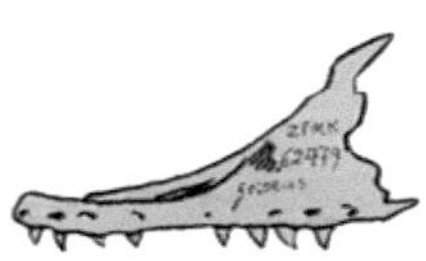 *griseus* ZFMK 62479	W  *indicus* (Guam) USNM 122469	X 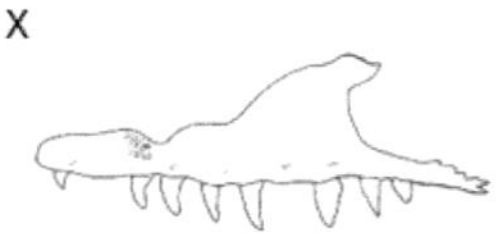 *indicus* (Aru Is.)
Y  *indicus* (Aru Ids.) USNM 33290	Z 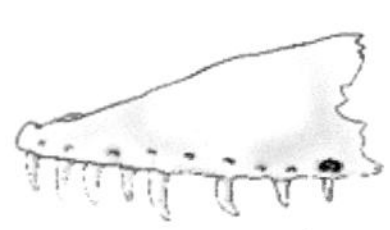 *komodoensis* SMF 23189	α 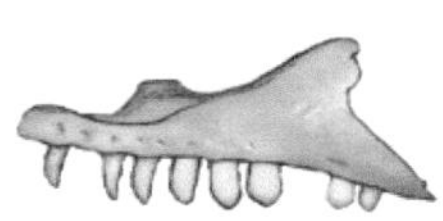 *niloticus* USNM 42260	β 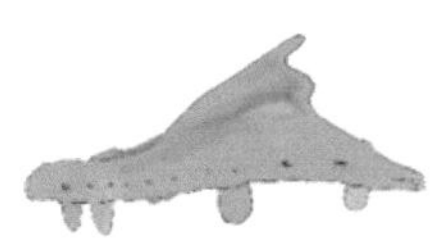 *olivaceus* USNM 222400
γ  *orientalis* ZMB 10557	δ  *prasinus* ZFMK 69454	ε  *rudicollis* (yng.) ZFMK 54253	ζ 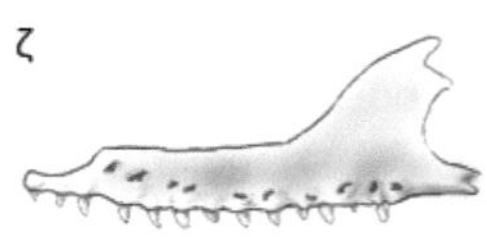 *rudicollis* SMF 40207

(Continued)

TABLE 5.2 ***(Continued)***
The Range of Variation of the Left Maxilla Across the Varanidae

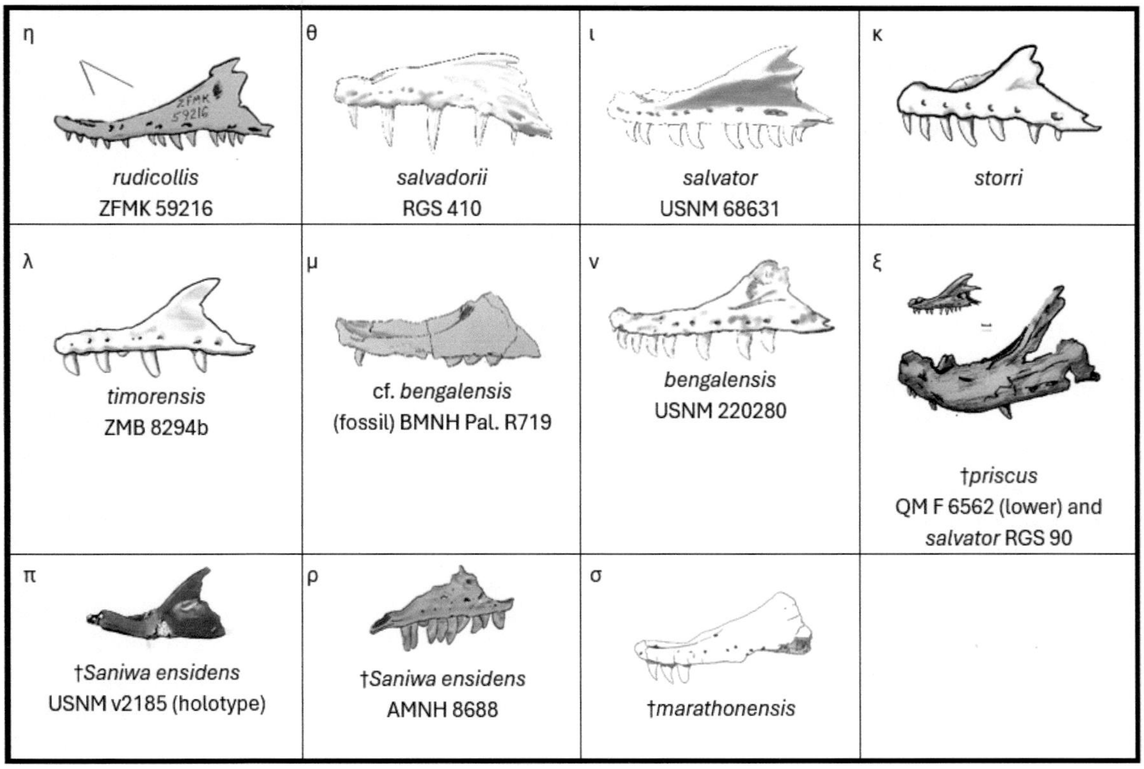

Note: Dorsal facial shape and height are the most obvious character states of the bone. In adult monitors, the strongest bites are associated with long, gradually sloping dorsal processes, though no species has a comparatively weak bite. The morphology is so distinct that species-level identifications are possible among most *Varanus* if the maxilla is intact. The variation among the five *bengalensis* (D–G, μ *ν*) reflects age differences, but the adult forms differ enough that I suspect at least one to be *nebulosus*. Most aberrant in morphologies are *eremius* with its huge ascending dorsal process, proportionally larger than in any other varanid; the sharply curved "shark fin" shape of *beccarii, gilleni,* some *indicus, orientalis, timorensis*, and the extinct *†Saniwa ensidens*; and the similarity in the otherwise quite different straight-edged dorsal surfaces and triangular shapes in *giganteus, komodoensis*, and *salvadorii*. Box ξ compares the maxilla of the largest known varanid with that of a 2.2 m extant animal.

of the anterodorsal surface of the coronoid. The distance from the tip of the premaxilla to the maxilla-coronoid joint represents the extent of the possible gape. In *Odatria* and *V. exanthematicus*, the maxilla represents only about one-third the skull length, the only adult varanids with such notably short maxillae. Among the remaining monitors, the maxilla typically nears half of skull length. Both the dorsal ascending process of the premaxilla and the lateral surfaces of the nasals lie medial to the maxillae, and between these elements lies the nasal fenestra. Immediately posteroventrally, often following a short constriction, lies the choanal fenestra (= fenestra exochoanalis), which is extraordinarily long, more than 50 percent of the length of the maxilla. In most varanids, the smooth surface of the fenestra that supports the nasal capsule ends at or shortly posterior to where the posterior process of the maxilla starts to rise, but in *V. dumerilii*, the smooth surface extends the length of the maxilla, becoming partially vertical posteriorly.

The smooth, horizontal medial shelf of the medial maxilla is the external nasal shelf (= excavatio nasalis maxillare). In structure, it is relatively thin, widest anteriorly, and has two slight depressions which are concave when viewed dorsally. In almost all adult varanids, the anterior depression is both smaller than the posterior and is largely or completely covered by the septomaxilla. In *V. dumerilii*, which undergoes a variety of notable ontogenetic changes, the triangular septomaxilla tends to entirely cover the anterior excavation throughout life. In nearly all Australian taxa – *Odatria* plus *V. giganteus*, *V. varius*, *Varanus panoptes,* and *Varanus gouldii* – the furthest anterior portion of the excavation remains uncovered and clearly visible in dorsal view. The length and breadth of the shelf are taxon dependent. In *V. salvator*, for example, the shelf is both long and broad (Fig. 5.7B); in *V. albigularis* and *V. exanthematicus* it is of medium length and quite narrow and thin (Fig. 5.7C, D). In *Odatria* the shelf tends to be short and broad. However, the septomaxilla is not necessarily underlain by the shelf; that element may be narrow enough to sit almost entirely medial to the shelf. The nasal capsule, which does rest on the shelf, is very closely linked in length with the length of the second (posterior) shelf depression.

Maxillary shape falls within a broad range of configurations (Fig. 5.8E–G and Table 5.2). Fejérváry (1935) described three distinct maxillary types: a long, gradual slope from premaxilla to frontals (hypsiprosopic, Fig. 5.8B, E); a mostly horizontal upper maxillary surface with a distinct rectangular dorsal process midway between premaxilla and orbit (mesoprosopic, Fig. 5.8D, F); and a mostly horizontal upper maxillary surface with a distinct rectangular dorsal process that lies near the orbit, along the frontal-lacrimal border (platyprosopic, Fig. 5.8D, G). Fejérváry (1935) and Mertens (1942b:129) further distinguished maxillary structures as follows:

1. Hypsiprosopic (Figs. 5.2A, 5.4H, I, 5.7B, and 5.8B, E) = where the posterior portion of the maxillary border of the nasal fenestra (excavatio nasalis maxillaris) does not extend beyond the posterior margin of the septomaxilla; the dorsal edge of the maxilla posterior to the nasal fenestra rises caudad gradually in a long straight or very slightly curved line; and possesses a long, narrow narial fenestra. Included taxa have small, round, terminal nostrils. Examples: *Varanus beccarii, V. bengalensis, Varanus doreanus, V. indicus, V. komodoensis, V. nebulosus, V. prasinus, V. salvadorii, V. salvator,* and *V. varius* plus the Paleocene †*Saniwa*.
2. Platyprosopic (Fig. 5.8D, G) = the posterior portion of the maxillary border of the nasal fenestra ascends steeply, almost 90° from the horizontal, into a short process; the process originates posterior to the lateral edge of the maxillary crest; the superior maxillary border forms a convex snout; and the narial fenestrae are broad and short. The large species have large, posterior, slit-like nostrils; the odatrians have round. Examples: *Varanus caspius, dumerilii, V. exanthematicus, V. flavescens, V. griseus, V. ornatus, V. rudicollis* (juv.), and odatrians such as *V. caudolineatus, Varanus primordius*, and *Varanus storri*.
3. Mesoprosopic (Fig. 5.8D, G) = where the posterior portion of the maxillary border of the nasal fenestra extends beyond the posterior margin of the septomaxilla and rises in a diagonal line that is often broken (i.e., upper edge has lines at two distinct angles) at the midpoint. Nostrils may be small and round, or large and oval, or slit-like, and are terminal, central, or posterior on the snout. Examples: *V. albigularis, V. rudicollis* (adult).

Though the descriptive terms for maxillary shape seem well-defined, the reality is they vary too much ontogenetically as well as between closely related taxa. Though Fejérváry's terminology is useful in describing the placement of the dorsal line of the maxilla, it lacks precision (Table 5.2). For example, both *V. exanthematicus* and *V. flavescens* are classed as platyprosopic by Mertens (1942b:129) because of the position of the maxillary rise. However, in shape they are quite distinct, *V. exanthematicus* having a near-rectangular structure with parallel vertical sides and being wider on the dorsal edge than at the base (Fig. 5.8D, F), while *V. flavescens* has a more inclined anterior maxillary edge and a notched posterior surface (Fig. 5.8G). The maxilla of *V. dumerilii* perhaps comes closest in form to that of *V. exanthematicus*, but in the former, the dorsal portion of the vertical process is not wider than the base of the process, and there is a large sub-lacrimal extension that is absent in *V. exanthematicus*. Furthermore, the maxilla anterior to the vertical process is much longer and has nearly parallel dorsal and inferior edges. The maxilla of *V. rudicollis* is considered hypsiprosopic by Fejérváry because of the relationship between the posterior septomaxilla and maxillary rise. However, the shape of the rise coupled with its extreme posterior position also qualifies juvenile *rudicollis* for inclusion with platyprosopic species (Table 5.2ε), but mesoprosopic in adults (Table 5.2ζ), introducing a rare case of significant ontogenetic change of a large bone. Despite its placement with the hypsiprosopic taxa, the large *V. salvadorii* has a unique maxillary configuration in which the ascending dorsal process follows a straight line, with no distinct posterior rise. The closest I have seen to this configuration is *V. marmoratus*. This morphology, plus other data derived largely from molecular research, has not only provided cause for raising *V. marmoratus* to species level, but also for doing so for all other former subspecies of *V. salvator*. In †*V. prisca* the maxillary ascending process is possibly unique, in being hypsiprosopic, but having a concave lateral surface that provides highly modified attachment surfaces from unusually large jaw muscles. However, as the fossils lack complete maxillae, this character cannot be ascertained. In *V. komodoensis*, the maxilla is very slightly concave, and the lateral jugal is deeply so. Assuming that such jaw modifications are indeed size-related, we could expect the jugal of †*V. prisca* to also have a deep lateral fossa.

Thus, even within one of Fejérváry's maxillary morphotypes, there is variation. The mesoprosopic "type" is a "none of the above" category that includes several combinations of characters from the platyprosopic and hypsiprosopic lizards (Fig. 5.8), and as noted by Mertens (1942b:129) the distinction between hypsi- and platyprosopic forms is obfuscated by the mesoprosopic, which often is not an intermediate condition. Consequently, Mertens abandoned use of the terms because of their imprecision.

Fejérváry's terminology for maxillary shape is useful, however, in describing the bones in the same terms, such as smooth, rugose, and convex. The lack of intraspecific consistency makes them of very limited phylogenetic use for at least three reasons. First, the definition of maxillary type can be done based on where the posterior dorsal maxillary crest rises by referencing the septomaxilla or point of anterior abruptness of the vertical surface, but not usually both. Based on the former criterion, *rudicollis* may be considered hypsiprosopic, but based on the latter criterion, it is distinctly platyprosopic. Second, the shape of the maxilla may change considerably ontogenetically, with, for example, young *rudicollis, acanthurus,* and *ornatus* shifting from hypsi- to platyprosopic as they grow. In *bengalensis* the shift is from hypsi- to mesoprosopic. Adults of *bengalensis, dumerilii,* and *rudicollis* share very similar maxillary profiles, the most conspicuous difference being the relative elongation of that bone. In *bengalensis* the maxilla is intermediate between the extremely long anterior process and minuscule premaxillary process (appearing flat-tipped) of *rudicollis,* and the extremely abrupt, nearly 90° vertical rise of the posterior maxillary edge in *dumerilii.* The shape of the dorsal maxilla of *giganteus* is most like †*Saniwa ensidens* in its long, smooth, and gradual ascent caudad, which is similar to *salvadorii,* except the latter has an irregular dorsal edge. Several of the *Odatria* group (e.g., *acanthurus, caudolineatus, eremius, glauerti*) have, if we follow Fejérváry's morphological taxonomy, a chimaera of features: an anterior maxilla low and narrow, followed by a mid-maxillary crest that is nearly 90° vertical, square, and followed by a much lower, irregular posterior process, in which each section represents about one-third the length of the bone. The condition defies the classificatory scheme and is unique among the Varanidae. Third, the distinction between conditions, like species themselves, is clearly demarcated when a limited number of specimens is compared, but those distinctions blend into a near continuum as more taxa are included in the study. Thus, the hyspsi-, meso-, and platyprosopioc terms represent relative descriptors rather than solid class-group characters.

Is there a correlation between maxillary configuration and the shape, size, and position of the external nostril? Fejérváry (1935) links internal and external naris morphology, but the links are limited and, as Mertens (1942b:129) states, of minor importance. A posterior slit-shaped nostril is present in *bengalensis, dumerilii, griseus,* and *rudicollis,* but size and shape are quite different despite osteological similarities. Furthermore, Fejérváry's system would put most of the *Odatria* group into the platyprosopic category, but the numerous species in that group have small, round, anterior nostrils.

The jugal articular process is universally more posterior than any other feature of the maxilla, and a line between that and the prefrontal articulation may be nearly vertical (*dumerilii, griseus, rudicollis*) or an obtuse slope in which the jugal articulation is far posterior to the prefrontal articulation (*giganteus, prasinus, salvator, tristis*). The forked process is nearly universal but varies slightly among species. In many, the jugal articulation process is either straight or arches slightly downwards (*bengalensis, giganteus, rudicollis, tristis,* most *Odatria*). The process in *salvadorii* is a short, single, upward-curved prong, while in †*S. ensidens* the process is simple, short, ending in a single downward-arched tip.

The fossil †*S. ensidens* resembles *V. doreanus* and *V. olivaceus* in having the large post-narial structure rise from about mid-length of the maxilla, such that the tooth-bearing section below, the rising dorsal process, and the angled posterior edge form a nearly equilateral triangle in lateral view. In †*Saniwa* and adult *V. olivaceus,* that triangular region is conspicuously convex and rounded, and in the former may have a distinct bilobed upper edge. In nearly all other varanids, the lateral maxillary surface is flat, though in *V. komodoensis* and adult *V. salvator,* the surface is concave. This latter condition of maxillary concavity extends along the first half of the jugal.

Internally, in monitors with a long, rising maxilla, the bone is strongly concave dorsolaterally, with the lacrimal and frontal forming the rear wall, and a supradental shelf forming a thick lower rim (Figs. 5.8B–D and 5.10). Posteriorly, the paired lacrimal foramina open into the maxillary cavity, a triangular space that terminates anteriorly with a sharply downturned process that is either below or just posterior to the septomaxilla. In monitors with very short maxillae, there are two basic medial configurations. In those, notably *V. exanthematicus,* with a nearly rectangular, and thus comparatively more substantial posterior lateral maxillary process, the interior chamber is much reduced, and the anterior wall is straight or very slightly concave. The posterior vertical edge has a thin inward-directed wall that abuts the lacrimal and prefrontal. The posterior process exceeds 50 percent of the length of the maxilla. The alternate condition, seen in *V. dumerilii,* consists of a very short, triangular vertical process that lacks both anterior and posterior walls. The process is extremely flat and laterally compressed; the posterior process is less than 25 percent of the length of the maxilla. In both groups, however, there is still a prominent supradental shelf with a downward-sloping anterior flange positioned above the third maxillary tooth (or socket).

The posterior vertical edge of the maxilla is highly variable. In *exanthematicus* and *dumerilii,* the rear surface is concave and relatively smooth, and the most dorsal and inferior ends are pointed, extending further caudad than any other part of the bone (Table 5.2I, L). In *V. salvadorii* (Table 5.2A) the rear edge is extremely ragged, the maxilla in lateral view resembling a glove with slightly spread fingers. These extending processes interdigitate with the prefrontal and lacrimal.

The tooth-bearing surface in most varanids is roughly straight, but in adults of durophagous species, including *niloticus, albigularis, exanthematicus, bengalensis,* and *olivaceus,* the line is somewhat to considerably curved, especially in older individuals, a situation paralleled in the teiid *Dracaena,* the curvature very likely providing additional strength to withstand the pressure of crushing hard prey (Condon 1987, Dalrymple 1979, Frazzetta 1962, Iordansky 1968, Losos and Greene 1988, McCurry et al. 2015, Wilken et al. 2017, 2019). The maxilla (and dentary) in *Dracaena* contains several large foramina, while varanids possess fewer and much smaller structures.

As the predominant facial bones, the maxillae make the major contribution to the transverse sectional shape of the snout, e.g., as seen in the anterior aspect, the preorbital region may be nearly square (*albigularis, exanthematicus, indicus*), rectangular (*griseus*), triangular (*komodoensis*, *nebulosus*, *timorensis*), or have a rounded domed appearance (*olivaceus*). This feature depends upon the degree to which the maxillae are parallel to each other but is augmented by the transverse width of the mid-skull (across the eyes and frontals) compared to that for the mouth immediately below. The broader the mid-skull, the more four-sided the appearance, while a narrower mid-skull width results in the triangular to rounded shapes. There is also a correlation of skull transverse shape with posterior premaxillary process width. Species with blade-like premaxillae have more triangular shapes, and those with broad, table-like surfaces tend to be square.

5.3.5 Teeth

Maxillary and dentary teeth are, in adults and juveniles, the largest teeth of the jaw (Fig. 5.9). In structure they have a strongly laterally depressed, sub-conical base that becomes flatter at the tip. Typically, anterior and posterior surfaces are sharp (Fig. 5.9E), blade-like in form, the latter with a serrated edge that is more pronounced in some taxa, including *V. komodoensis*, *V. indicus*, and *V. salvator*, than in others. It is most conspicuous in the extinct †*V. prisca*, which also possesses the most posterior-directed angle of the teeth. Serration is lacking in *V. beccarii*, *V. dumerilii*, and adult *V. exanthematicus* and *V. olivaceus*. Numerous fine vertical grooves line the tooth bases; these may be absent apically and are less pronounced in larger taxa. In section each tooth has a labyrinthine infolding, a condition termed plicidentine (Bullet 1942). The teeth are pleurodont, with replacement teeth aligned along the internal edge of the bone. Typically, each functioning tooth is separated from the next by an unfilled socket, with replacement teeth arising from the posterior socket.

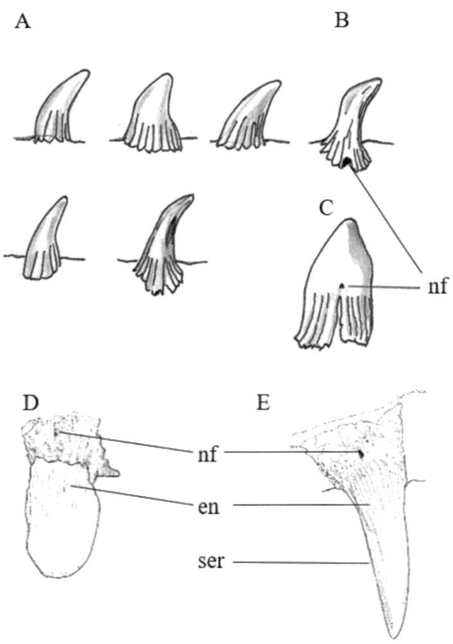

FIGURE 5.9 Teeth of varanids. (A) All varanids possess at least some slightly curved teeth, but the comparative thickness of the base and lateral compression varies widely among species and age. Few species have species-specific tooth morphologies, but include many adult *Varanus dumerilii* (B), which have a possibly unique set of posterior teeth with a distinct secondary hook. Several large species that feed on hard prey items as adults Develop Conical, mostly straight teeth (C, D). (D) Medial view of the posterior left maxillary tooth of adult *niloticus* (USNM 199568). (E) The teeth of *salvadorii* (RGS 410) are uniquely species-specific, being more laterally compressed, straighter, and considerably longer than in any other varanoid. The infolded enamel (en) at the base of each tooth gives the striated appearance (plicidentine condition). All varanids possess saw-like serrations along the posterior borders of non-molariform teeth, and usually have anterior serrations as well.

Tooth structure is variable among taxa (Fig. 5.9) and, in many cases, ontologically (D'Amore 2015). All neonate and juvenile varanids possess sharp, slightly curved teeth that correlate with the need to seize and hold small live prey. For most monitors, the teeth retain that shape, but become larger as the lizard grows. In some species, in which the diet shifts as lizards mature, the maxillary and dentary (but not premaxillary) teeth alter from the sharp form to a more barrel-shaped, blunt-tipped structure (Fig. 5.9C, D). In some cases, the development of blunt teeth is limited to the posterior half of the maxilla, while the anterior teeth become somewhat more cylindrical and retain sharp tips. Such species include *albigularis*, *bengalensis*, *exanthematicus*, *niloticus*, and *olivaceus*, monitors that consume crabs, snails, and other hard-shelled prey. However, the inclusion of hard prey does not necessarily indicate that teeth become modified. For example, the semi-aquatic cancrivorous *dumerilii* does not experience a change in dentition; the anterior teeth are nearly conical, curved, and have pronounced serrations on the anterior and posterior edges. The more posterior teeth are possibly unique within Varanidae in having a base and most of the tooth nearly straight and slightly compressed, the distal third often having a distinct rearward angle (Fig. 5.9B). Like *dumerilii*, *mertensi* also consumes crayfish, and like the former, has typical curved and pointed teeth.

The neotropical teiid genus *Dracaena* is, like the varanids noted above, a large durophagous lizard. It is interesting to compare the seemingly convergent evolution between the genera. Like *V. albigularis*, *V. bengalensis*, *V. exanthematicus*, *V. niloticus*, and *V. olivaceus*, *Dracaena* possesses greatly enlarged molariform teeth (Dalrymple 1979). Unlike the varanids, however, *Dracaena* has such teeth throughout life, while the *Varanus* all spend their pre-adult and early adult years with typical pointed dentition. In each case the number of teeth is lower in the molariform-bearing stage than in related taxa. Like *V. niloticus*, *Dracaena* are semi-aquatic and consume hard-bodied prey throughout life (Avila-Perez 1995, Dalrymple 1979, Sprackland 2009). Other varanids are typically found in drier habitats and eat a variety of soft-bodied prey until they become adults. Exceptions are the arboreal forest-dwelling *V. olivaceus* and its close Philippine relatives, which take large, hard snails from early in life (Auffenberg 1988).

The teeth of *V. salvadorii* are unique amongst lizards, being long, nearly triangular, weakly or not curved, and more compressed than the teeth of other varanids (Fig. 5.9A). Posterior serration is minimal, but the anterior and posterior surfaces are about equally sharp. The teeth interdigitate when the mouth is closed, thus acting to shear flesh during a bite. *V. salvadorii* has the longest teeth of any varanid and can be 1.5–1.8 times longer than the teeth from an equal-sized *komodoensis*. In contrast, the large *V. rudicollis* has the shortest teeth, relative to body size, of all varanids, as well as the greatest number of maxillary and dental teeth.

Gilmore (1922), in his description of †*S. ensidens* stated that both the vomer and pterygoid bore a row of small teeth. He maintained this view in his later monograph (1943:59), but later authors (Estes 1986, Rieppel and Grande 2008) accepted Gilmore's comments without seeing the feature for themselves. I have re-examined the holotype of †*S. ensidens* microscopically, and found an ambiguous small irregularity on one pterygoid, and a single, 0.8 mm conical structure at the center of the right vomer but cannot confirm that it is either a tooth or an artefact of fossilization or preparation. The only definitive varanid (for I am still inclined to see *Lanthanotus* as belonging to a separate, though varanoid, family) with pterygoid teeth is †*V. prisca*.

5.3.6 Lacrimal

The lacrimal separates the posterolateral maxilla from entering the orbit, and externally is somewhat triangular (Figs. 5.2A, B and 5.10). Internally, there is a short, stout body that articulates with the frontal, prefrontal, and maxilla. There are two lacrimal canals on each side, a condition unique to *Lanthanotus* and the Varanidae. In *Varanus* the superior foramen is always the larger of the pair (Fig. 5.10A, B). In *Lanthanotus* the canals are subequal in diameter and lie entirely within the lacrimal bone. In *Varanus*, the lower canal is also wholly in the lacrimal, but the much wider dorsal canal is bordered medially by the lateral vertical process of the prefrontal. The lacrimal has the shape of a six or nine, the lower circle enclosing the inferior lacrimal canal, and the open upper portion forming the lateral (distal) dorsal canal. The upper foramen opens anteriorly into the large maxillo-frontal/lacrimal recess, and the inferior angles downward toward the anterior suborbital fenestra (Fig. 5.10C). In the fossil †*S. ensidens* in which the canal region is visible, they are paired, the smaller confined to the lacrimal bone, the larger uncertain because of the poor resolution of its medial suture (Fig. 5.10D). A lacrimal associated with well-preserved †*Varanus marathonensis* has one distinct foramen, but the bone is incomplete (Villa et al. 2018).

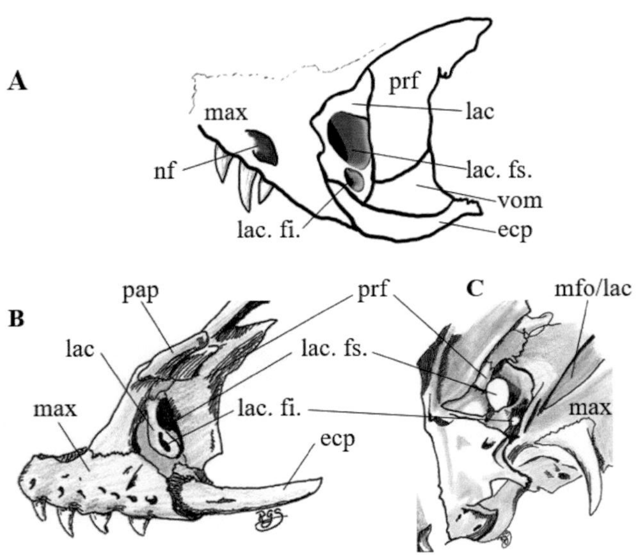

FIGURE 5.10 Lacrimal of varanids. (A) Diagrammatic illustration of the lacrimal of *Varanus salvator* in posterolateral view, showing the relative positions of the bones surrounding the paired lacrimal foramina. Note the superior foramen is bordered medially by the prefrontal, while the inferior is completely within the lacrimal bone. (B) Lacrimal region of *Varanus exanthematicus* in left posterolateral and (C) left medial anterolateral view of *Varanus exanthematicus*.

5.3.7 Frontals

The frontal bones in varanids are universally paired and situated above and for the length of the orbits (Figs. 5.2 and 5.11). The anterior section meets the nasal(s) medially and is far narrower than the straight posterior border with the parietal. The most variable feature of the frontals is the shape of the anterior processes. In all cases the anterior suture contains four facets that interdigitate with the nasal(s) and may be one of three configurations: lateral pair slightly longer than medial pair (Fig. 5.11A), medial pair longer (Fig. 5.11B), or lateral pair much longer than medial (Fig. 5.11C). The inferior facets – nasal, prefrontal, and anterior – extend further anterior than the superior surface. In degree facets range from a short process (e.g., *acanthurus, salvadorii*) reaching only slightly anterior to the upper lappet and measuring far less than one-quarter the length of the frontal, to the greatly protruding facet seen in many large species (*niloticus, salvator*), to the enormous processes seen in *komodoensis* and †*prisca* (Fig. 5.11F), where the inferior shelf may be nearly one-quarter the length of the frontal. The concave shape of the lateral edge conforms to the eyeball. The bone is somewhat hourglass-shaped (Fig. 5.11) or has straight lateral edges (Fig. 5.11B), but the anterior edge is always narrower than the posterior. The degree of constriction, when present, varies but is always narrowest at mid-orbit. The posterior portion forms part of the upper temporal arch, broadly abutting laterally with the fused postfrontal-postorbital bone. Dorsally, the frontals are flat and have a slightly to considerably rough texture. In many species, pitting occurs where elements of the skin or osteoderms adhere to the skull (Fig. 5.11D–G).

Though varanids lack the types of cranial ornamentation seen among the iguanines, some species develop a low medial frontal crest formed from upward turns of the

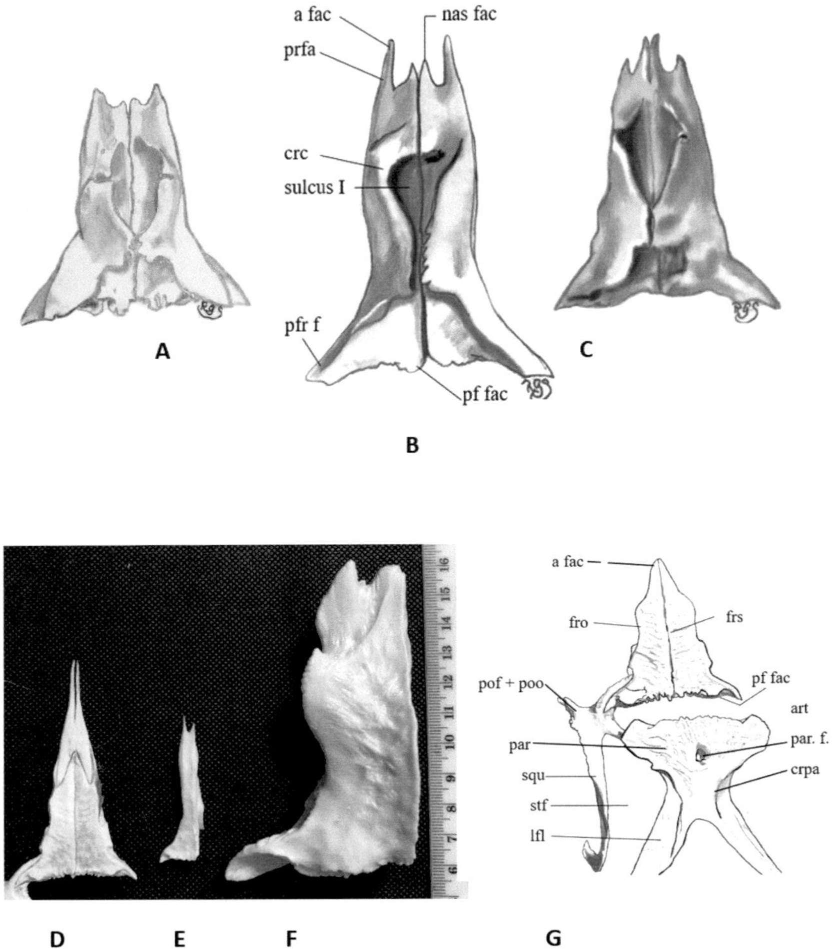

FIGURE 5.11 (A–C). Frontal bones of three morphologically different species of *Varanus* in ventral view. (A) *dumerilii*, (B) *exanthematicus*, and (C) *salvator.* Note that in *exanthematicus* the frontals form a rough rectangle, and the cristae cranii are long. In *exanthematicus* and *dumerilii*, the medial nasal facets are much larger than the anterior facets. In *salvator* and *dumerilii* the frontals form a slightly triangular shape. (D–F). Comparisons of frontals from adult examples of (D) *salvator* (1981 mm TL), (E) *salvadorii* (1889.7 mm TL), and (F) a presumably adult †*prisca* (unknown TL) to give scale among the three largest known *Varanus.* (G) Skull roof of *salvator* showing articulation of frontals and parietals.

bones along the medial suture. I strongly suspect that frontal crests are characteristics of male lizards, as they are not consistently present in all specimens for any species. I have not had access to enough osteological specimens of known sex in any single taxon to confidently support that hypothesis. It seems more common in species that may exceed a total length of 1250 mm (*bengalensis, indicus, komodoensis, niloticus, salvadorii, varius,* and †*prisca* (Fig. 5.11F), but I have also found frontal crests in smaller species (*acanthurus, tristis, beccarii*).

The inferior surfaces form elongated downward extensions (= cristae cranii) that curve medially, forming a bony sulcus chamber that partially surrounds the olfactory nerves. These downward processes nearly meet (Fig. 5.11B), but they do not fuse and retain a narrow gap (Fig. 5.11C); both halves of the frontal remain distinct throughout life. The length of the near-meeting processes of the cristae cranii varies from having a very short length (Fig. 5.11A) and thus having CN I broadly exposed ventrally, to longer and with anteriorly extended lateral ridges that provide more protection for the nerve (Fig. 5.11B). In the allied *Lanthanotus*, the descending processes also form a chamber, but the distal edges do not meet and, when compared with *Varanus*, are much more separated. In this regard both taxa closely approach but do not attain the complete ventral closure of the CN I sulcus developed in snakes.

5.3.8 Parietal

The varanid parietal is the largest and strongest bone of the braincase, providing protection to the dorsal surface in conjunction with the supraoccipital (Figs. 5.2A, B, 5.12, and 5.13). It is generally an hourglass-shaped single bone that most broadly meets the more anterior frontals, and laterally articulates with the postfrontal + postocular. The anterior third of the parietal and its posterior lateral processes (= processus paroticus) show little variation; however, the central body of the bone shows considerable variability

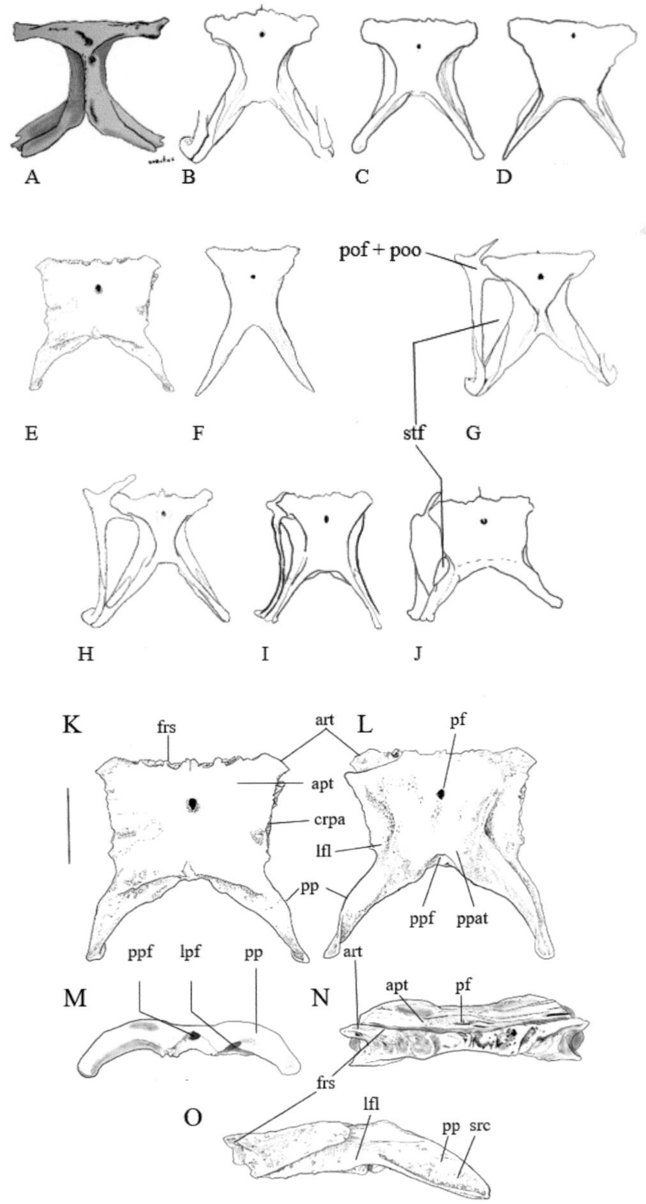

FIGURE 5.12 Parietals in dorsal aspect, showing diversity of posterior parietal table morphology and comparative width of supratemporal fossa. The range of shape goes from the bladelike table and large descending flanges in (A) *ornatus* (BMNH uncat.), (B) narrow table and large flanges in *niloticus* (USNM 220294), (C) a "typical" table with moderate flanges in *bengalensis* (USNM 14753), (D) broad table and small flanges in *exanthematicus* (RGS 202), and (E) a nearly square table and vestigial flanges in *dumerilii* (RGS 203). (F) is an outlier morphology of an elongated dorsal surface that merges without a distinct border into moderate flanges in *salvadorii* (RGS 410). Bottom row shows range of variation in thickness of supratemporal arch and size of supratemporal fossa, ranging from large and open in (G) *olivaceus* (USNM 222400), (H) *komodoensis* (USNM 101444), (I) *timorensis* (RGS 189), (J) *exanthematicus* (RGS), to small and nearly closed in (K) *dumerilii* (FMNH). The majority of *Odatria* species have parietals similar to *timorensis*, though generally slightly less attenuated. (L) morphology of the parietal of *bengalensis*. Bottom row – the frontal-parietal region of (M) *salvator* (RGS 90) disarticulated and (N) *komodoensis* articulated, showing the relationships among bones and their sutures. Figures K-O: Parietal bone of *V. dumerilii* (RGS 203) in (K) dorsal, (L) ventral, (M) posterior, (N) anterior, and (O) Left Lateral views. In O the supratemporal bone is Omitted to clearly show the shape-conforming recess of the posterior lateral flange. Note also the lack of distinct anterior and posterior parietal tables with no "waist," lateral flanges not visible from above. This species, along with the taxa with no lateral posterior parietal table (Fig. 5.12A, G), represents extreme ends of the spectrum of *Varanus* parietal morphology. Scale bar = 10 mm.

(Fig. 5.12). In its simplest form, the mid parietal is broad and flat, forming a mesa (e.g., *dumerilii, rudicollis, caudolineatus*, Fig. 5.12E, J). At the other end of the spectrum, the midbody becomes very thin, with almost no dorsal table, instead having large, broad downward processes that angle inferiorly away from the medial plane (e.g., *bengalensis, niloticus, olivaceus,* Fig. 5.12A, B, G, H). Among adults of the African varanids, there exists an excellent example of the range of parietal shapes. *V. exanthematicus* (Fig. 5.12D) has a nearly square parietal table so broad it comes closer to the supratemporal arch than in any other monitor except *dumerilii* (Fig. 5.12E), effectively closing the opening. Both *komodoensis* and *olivaceus* have a more hour-glass shape (Fig. 5.12G, H), but the rear surface is still more narrow than in most other varanids. Both *niloticus* and *ornatus* have such a narrow posterior parietal region as to effectively substitute a long blade for a flat dorsal table (Fig. 5.12A), the anterior parietal a transverse, near rectangular bar. As juveniles, *niloticus* have a nearly typical parietal, though the dorsal surface is still quite narrow compared to the lateral flanges.

There are at least three direct positive correlations between configuration of dorsal table of the parietal (= anterior + posterior parietal tables) and the (a) size and angles of the lateral processes, (b) absolute size of the animal (e.g., ontogeny), and (c) diet, which is, among larger (SVL> 900 mm) monitors, ontogenetic. All hatchling and neonate varanids have a broad rectangular dorsal parietal surface, and both the processus externalis and supratemporal-post-parietal processes are reduced or absent. This morphology holds true even for the juveniles of the largest species (e.g., *niloticus, salvator,* and *komodoensis*). Among the smaller species (e.g., *brevicauda, caudolineatus, storri, timorensis, tristis,* Fig. 5.12I), the entire dorsal surface retains the broad rectangular shape throughout life, though in *timorensis* it becomes slightly more elongated longitudinally (Fig. 5.12I). This juvenile-type parietal also lacks enlarged lateral flanges (decensus parietalis), the processes being narrow and with a nearly straight lower edge. In adults of the small species, the depth of the flanges may increase slightly, but never to the comparative extent seen in larger varanids. The distinct edge between the horizontal dorsal surface and the variably vertical flanges is the crista parietalis lateralis, and, however short, is universally distinct. In the narrow parietal table species,

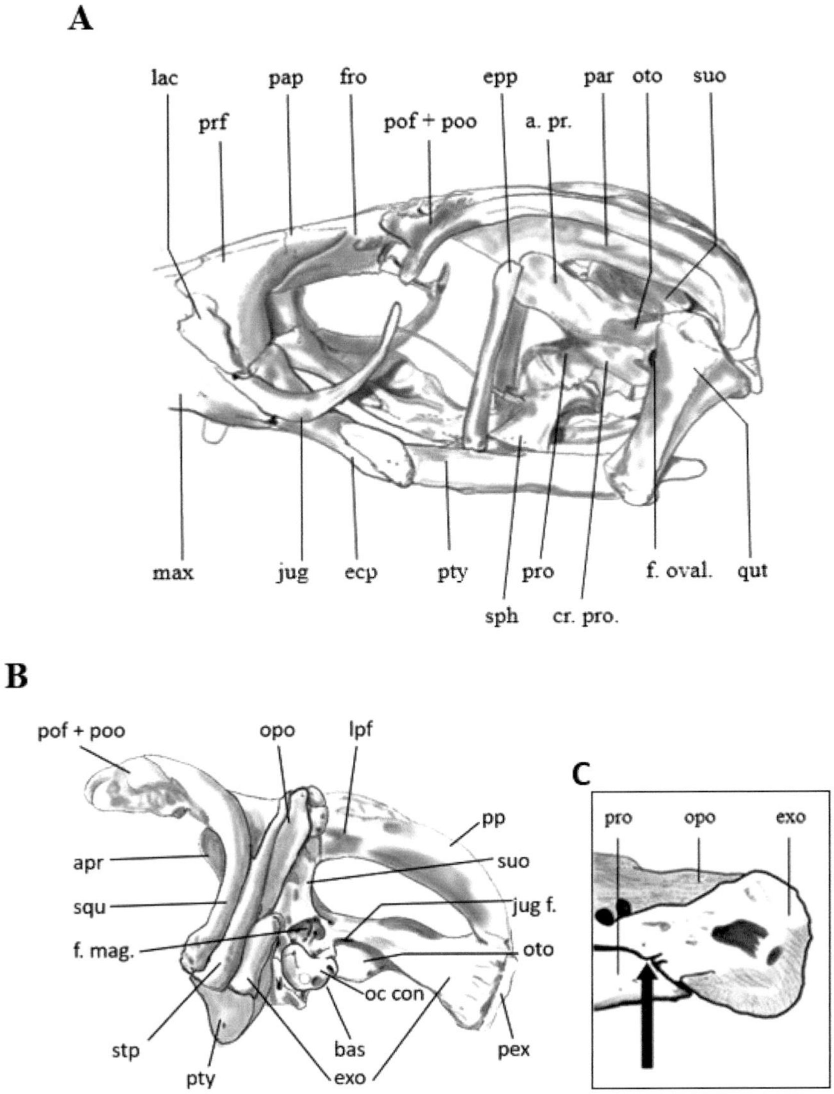

FIGURE 5.13 (A) The temporal arch and position of left lateral braincase elements in *Varanus prasinus* (RGS 102) showing locations of major bony elements of the braincase. The upper epipterygoids loosely contact the anterior proötics; the lateral sphenoid processes loosely contact the medial pterygoids in life. The lower quadrate would lie along the dorsal surface of the pterygoids. The distal stapes is hidden by the posterior quadrate, and the proximal ends would attach to the fenestra ovalis, partially seen near the upper anterior border of the quadrate. (B) Posterior braincase elements of *Varanus olivaceus* (USNM 222400) in posteroventral aspect with quadrate removed to show convergence of exoccipital, otoccipital, squamosal, and supratemporal elements. (C) Dorsal view of the posterior of the braincase showing where exoccipital, otoccipital, and proötic meet. Suture lines may be invisible without magnification or may fuse in adults. The arrow shows a large dorsolateral suture between the proötic and the other bones.

including *bengalensis*, *niloticus*, and *olivaceus*, the lateral flanges are broad, deep, have a larger surface area, and incline more sharply ventrally than in the smaller monitor species. But neither table morphology nor flange structure is universally tied to adult size. In adolescent and young adult *komodoensis,* the parietal is very similar to that of adult *indicus, salvator*, and *varius*, with the processus externalis extremely broad and the posterior table narrow, together forming a triangular dorsal surface. The posterior processes also arch distally and become quite long, often as long as the dorsal parietal table or more. Coupled with the change of shape to the triangle, the lateral flanges become greatly enlarged flaring processes that, in turn, support massive muscles of the jaw and pharynx. By far the broadest flanges are those in adult *komodoensis*, creating a surface area subequal to the entire remainder of the parietal. There is a general pattern of the degree of dorsal triangle morphology and an increase in size, depth, and surface area of the lateral flanges. In the African *ornatus*, the flanges become enormous, the posterior dorsal table has a thin blade (Fig. 5.12B). This condition is approached, but not equalled, in *albigularis* and very old *niloticus*, all of which are apparently phylogenetically closely related. It is, however, equaled in the frugivorous/molluscivorous Philippine monitors (*olivaceus, mabitang,* Fig. 5.12G), which, like the African taxa, may exceed 1100 mm in total length. The greatest expansion of the lateral parietal flanges is found in †*V. hooijeri*, an Indonesian subfossil species that was likely contemporaneous with *komodoensis*. Its parietal processes far exceed any seen in living varanids. For the African lizards, form follows function as the lizards change from a largely insectivorous juvenile diet to taking larger and often hard-bodied prey, including mammals, birds, eggs, crabs, and turtles (Cowles 1930, Dalhuijsen et al. 2015) as adults. The Philippine monitors consume hard fruits, eggs, and snails (Auffenberg 1988, Gaulke et al. 2007), all requiring a strong bite that can benefit from muscle attachments to enlarged flanges.

We see from adult parietals strong correlations between morphology and diet: species with narrow posterior parietal tables and enlarged lateral flanges all consume, at least as an important part of the diet, prey that requires jaws that can crush effectively. There is a difference between being able to seize and retain struggling prey, which all varanids can accomplish, and holding prey while also crushing shell or thick bones. Varanids such as *indicus* and *mertensi* have parietals of intermediate shape, constricted coronally but not forming a triangle shape, with descending flanges that drop sharply from the table, and create a larger lateral osseous wall for the braincase, but do not extend deeply or become broad. Both species frequently consume hard-bodied molluscan or crustacean prey, but their prey choices generally involve animals that can be easily manipulated and swallowed intact, needing no specialized crushing or tearing jaw mechanics. They, along with *dumerilii*, consume insects, frogs, fish, small mammals, and eggs as dietary staples, none of which require massive jaw muscles or modified teeth. In *dumerilii* (SVL > 1150 mm), though, the parietal is drastically different from any of the large species described above, being of the squarish juvenile type (Fig. 5.12). Their prey is like that of *indicus*, covering a wide variety of small vertebrates and large arthropods, but they are also specialized crab-eaters (Krebs 1979). Their mode of attack is to seize the crab from the rear, hold it down by standing on the carapace, twist the limbs off one by one, and swallow them entire. Eventually, they swallow the limbless crab entire. This mode of prey handling frees the parietal from the need for surfaces of attachment for large crushing musculature, allowing large Duméril's monitors to handle and consume prey as if they were small insectivorous lizards. There is little ontogenetic alteration in the parietal structure in *dumerilii*, and in all size classes, the temporal arch is broadened to an extent that the supratemporal fossa is functionally absent (Fig. 5.12J). The anterior parietal table lacks the noticeable lateral processes common to almost all other monitors (but not *exanthematicus*, Fig. 5.12D, K–O). No other varanid shows such a morphology at any age class. In *exanthematicus*, which most closely resembles the temporal state of *dumerilii*, and some Australian dwarf species, the parietal surface remains square or nearly so, but the temporal arch (postorbital + squamosal) remains thin, nearly columnar in section, and leaves room for a distinct fossa. In other species (*salvator*, *indicus*, *prasinus*, *exanthematicus*), the lateral flanges are either much less broad or, as in *prasinus* and some *indicus*, are much shorter and not as ventrally expressed.

Jutting posterolaterally from the posterior parietal table are the long, tail-like supratemporal processes (= postparietal processes), laterally compressed with broad, smooth lateral and medial surfaces. The upper edge is strongly compressed into a blade-like edge. The processes are small in hatchlings, but in adults vary from almost non-existence (e.g., *exanthematicus, gouldii, panoptes*, Fig. 5.12D) to extremely produced and longer than the length of the parietal table (Fig. 5.12A, B, G). The relative proportions can vary among species; the 51 mm parietal of a 1980 mm *salvator* has supratemporal processes 28 mm long (55 percent of parietal length), for an 1828 mm *salvadorii* (Fig. 5.12F) a 45 mm parietal has 24 mm processes (53 percent), while a 28 mm parietal from a 1220 mm *dumerilii* (Fig. 5.12K–O) has processes only 11 mm long (39 percent).

The posterior half of the supratemporal process is also termed the paraoccipital process (Fig. 5.12K–O). The lateral face of the process contains a distinct shallow recess into which fits the supratemporal bone (Fig. 5.12O). During growth, the anterior half of the supratemporal may fuse with the parietal sufficiently to be indistinguishable, but the posterior portion remains distinct, wedged between the parietal and squamosal.

Regardless of shape, in all parietals the anterior suture meets the frontals, while the small, squared lateral processes meet the postorbital-postfrontals, which form a suture nearly half the length of the parietal (Fig. 5.12G–J). Ventrally, the epipterygoids, proötics, and supraoccipital meet or nearly meet the parietal. The lateral articular surfaces of the parietal form a slightly extended inferior lapett that is overlain by the posterior frontal. The greater width of the parietal, however, tends to overlay or interdigitate with the frontal, a situation which forms and sustains a hinge across the skull, allowing for a degree of cranial kinesis. The articulation between parietal and postorbital-postfrontals is slightly flexible in young and small varanids but becomes more rigid in adults.

In all *Varanus*, the parietal is pierced by a small, round foramen, in which sits the parietal eye. In all cases, the parietal foramen lies medially in the broadest portion of the bone. The dorsal side of the opening is larger than the inferior, and though the foramen is universal, it may become obscured in adults, particularly among the largest species. In the *Odatria* group, the foramen is often barely a pinprick but can be seen under magnification. The report of a second parietal foramen and parietal eye has been reported for some specimens of †*Saniwa* (Smith et al. 2018) and will be discussed in Chapter 11.

There is a much larger medial posterior parietal foramen between the wide-arcing lateral processes. This aligns with the anterior foramen of the supraoccipital and thus forms the anterior half of the roof of the braincase. The posterior parietal foramen is bracketed by the vertical parietal fossa. The anterior and posterior surfaces of the posterior processes also each bear a single small lateral posterior foramen.

5.3.9 Temporal Arch

Varanids have a well-developed temporal arch that begins anteriorly with the fused postorbital-postfrontal bone and entwines posteriorly with the elongated squamosal (Figs. 5.2, 5.11, and 5.13). Lying flat along the posterior lateral parietal process and interior to the squamosal is the flat, curved supratemporal, which fits into a shallow lateral groove (Fig. 5.12O) of the paraoccipital process. In almost all taxa, the broadest section of the arch is the anterior, with an inner V-shaped notch that envelopes the lateral wings of both frontal and parietal bones. Posteriorly, the squamosal abuts the basioccipital and posterior lateral process of the parietal, and the medial surface of the quadrate. In species with a long temporal region, the arch is nearly straight, as in *V. exanthematicus*, and the most extreme case seen in *V. dumerilii*. Species with shorter temples have strongly bowed squamosals, in which the posterior tip may lie slightly anterior to the posterior part of the curve of the bone. Among adults of the largest species (e.g., *bengalensis, komodoensis, nebulosus, niloticus, olivaceus, palawanensis, rudicollis, salvator,* and *varius;* Fig. 5.12G–I), the temporal arch arcs medially, forming an indented temple osteology that accommodates large lateral temporal muscles. Within the *Odatria* species, a nearly straight temporal arch is typically present, often with a slight inward curve when viewed dorsally. In *exanthematicus*, *dumerilii*, and the *Odatria* group, there is a retention of the juvenile skull arch condition, in which the arcade, in lateral aspect, is only very slightly bowed, in dorsal aspect is straight or very slightly angled medially, and lies close to a square or rectangular, broad-bodied parietal; the resultant supratemporal fenestra is consequently quite narrow when contrasted with most of *Varanus*.

V. rudicollis has the relatively shortest temporal region, deviating from the condition seen in other monitors, being <0.33 times the length of the skull. In other varanids, the arch is >0.38 skull length. The posterior lateral flanges of the parietal are broad and, combined with the expanded postorbital-postfrontal, make a roof that considerably reduces the size of the supratemporal fenestra. Examination of stomach contents of very few specimens shows *V. rudicollis* eats frogs, insects, and crabs (Losos and Greene 1988), but the natural history of this species is largely unknown, so making specific links between skull morphology and diet is premature. The supratemporal fenestra is nearly closed in all age groups of *V. dumerilii*, in which the bones of the temporal arcade are broad and dorsolaterally compressed and widest posteriorly. The morphology of the bones occludes all but a small fenestra adjacent to the paraoccipital process (Fig. 5.12G).

The squamosal, supratemporal, otoccipital, exoccipital, and posterior parietal processes collectively form a paraoccipital process, a region of the skull that largely surrounds and protects the ear and associated structures.

5.3.10 Occipital Bones: The Braincase

Reptiles possess brains with a simpler morphological complexity than that of mammals, with three large lobes arranged linearly (Figs. 5.13–5.20). From anterior to rear: cerebrum, tectum, and cerebellum. In lizards, including varanids, the brain is not completely encased by bone or cartilage, the anterior portion largely within a series of tough, semitransparent connective tissues (= pilae). Extending rostrally from the equator of the cerebrum is the olfactory nerve tract (CN I), a pair of thick, semi-cylindrical fibers that extend between tough sheets of tissue that are separated medially by the elongate anterior parasphenoid process of the basisphenoid bone. The anterior braincase (Figs. 5.13, 5.14, and 20A, B) is a large opening with a broad bony rim consisting of the dorsal supraoccipital, lateral proötics, and ventral basisphenoid. The lateral braincase is comprised of the proötic, otoccipital, and basioccipital (Figs. 5.13 and 5.14). The posterior area includes the supraoccipital, exoccipitals, and basioccipital, which form a ring around the exit for the brainstem, the foramen magnum (Fig. 20C–E). One-third to half of the cerebrum, including the olfactory and optic tracts, lacks any bony encasement. Connective tissue surrounds and attaches the olfactory nerves to the inferior dorsal portion of the frontal bones through a bony shaft formed by downward, infolding fontanelles of the frontals. This largest nerve bundle leaving the brain has no bony foramen through which to pass, but instead transits through the connective membranous pila antotica, pila metoptica, and pila medialis.

The braincases of varanids have received considerable attention (Head et al. 2009, Norell and Gao 1997, Rieppel and Zaher 2000, Säve-Söderbergh 1947). Of the several osseous elements of the braincase, only a few articulate with other bony elements. The anterior portions of the proötics rise at a steep angle becoming the alar processes (Figs. 5.13A, B, 5.14A–D, 5.15, 5.16, and 5.20A, B) that broadly and loosely articulate with the lateral flanges (decensus parietalis) of the parietal, forming, with the central supraoccipital, a significant portion of the bony roof for the brain. Posteromedially to the alar processes is the curved dome of the supraoccipital (Figs. 5.13B, 5.14, and 5.18), which forms a short transverse articulation below the parietal's lateral posterior foramina. Posteriorly, the supraoccipital forms the upper border of the foramen magnum, through which the medulla oblongata leaves the skull. The exoccipitals and basioccipital form the lateral and ventral borders, respectively, of the foramen magnum (Fig. 5.20).

In anterior view (Fig. 5.20A, B), the braincase looks as follows: the anterior of the basisphenoid varies considerably in morphology among taxa and age classes, but certain features are consistent. The ascending parasphenoid process (Figs. 5.2A, 5.13B, 5.14B, 5.18H, and 5.19A, F, G, I, K) is always medial and extends upwards almost to the level of the fronto-parietal suture, though osseous articulation

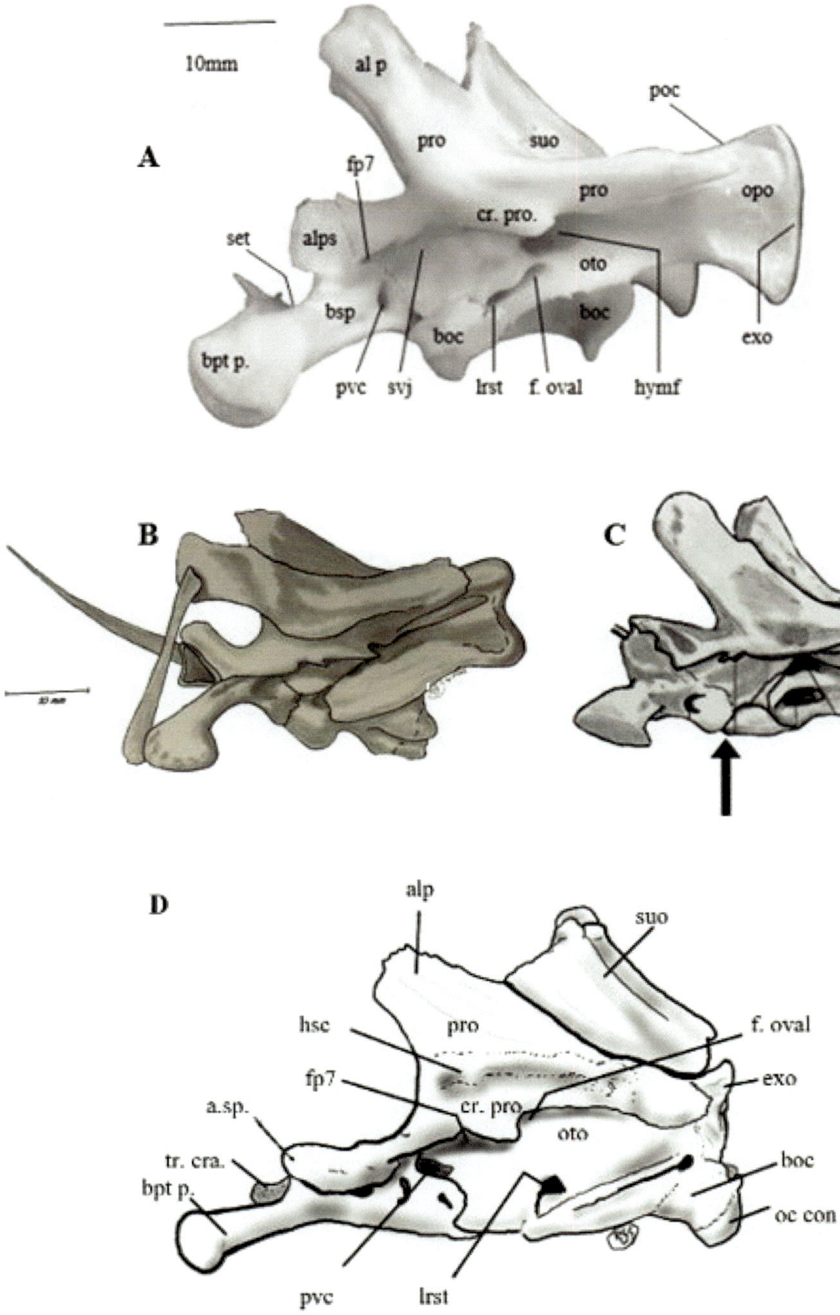

FIGURE 5.14 Bones forming the braincases of varanids, left lateral view. (A) Braincase of *Varanus salvadorii* (RGS 410). Note the anterior-pointing process on the upper basipterygoid process, a probably unique character for the species. (B) Braincase of *Varanus salvator* (RGS 90). (C) Braincase of *Varanus dumerilii* (RGS 203) with sphenoid slightly separated from other occipital bones (arrow), parasphenoid missing. (D) Braincase of *beccarii* (RGS 201) showing a morphology quite different from that of other *Varanus*.

with the roof bones never occurs. The basal width varies slightly among species, but the process is thin and bladelike dorsally and for most of its length. In juveniles and small species, the process may be translucent, but it is always a bony element. The parasphenoid process terminates posteriorly at the sella turcica (Fig. 5.14A) above and with the lateral basipterygoids that also extend somewhat anteriorly. Distally, the parasphenoid meets the small U-shaped orbitosphenoids. On either side of the sella turcica is a large anterior foramen for the vidian canal (Figs. 5.14, 5.15, and 5.20A, B); the canal extends through the basisphenoid posterolaterally and exits in a similarly large opening on the lateral surface of the basisphenoid above the proximal point of the basipterygoid process (Figs. 5.14, 5.15, and 5.18H). The large canal is easily identified as being the largest and most anterior of the lateral braincase foramina.

The basioccipital (Figs. 5.18 and 5.19) represents about half of the lower bony occipital-complex floor. Its lowermost posterior edge terminates in a blunt, rounded knob, the central part of the occipital condyle, joined laterally by

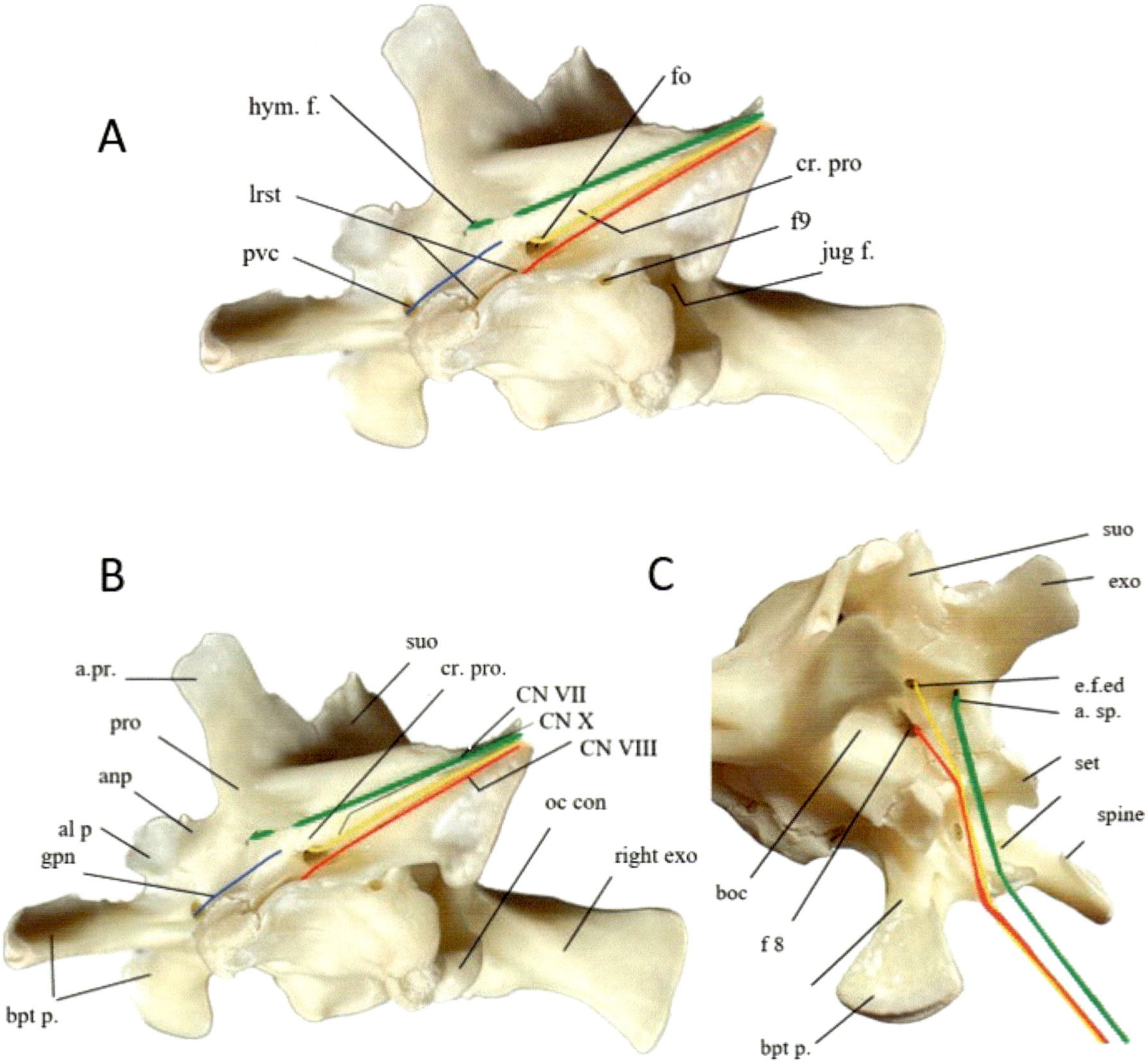

FIGURE 5.15 Foramina and associated nerves. (A) *Varanus salvadorii* (RGS 410) showing foramina and (B) the same showing nerves. (C) Right anterolateral view of foramina of basipterygoid and otic capsule of *salvator* (RGS 90). Colored threads were used to map the routes of the nerves through foramina; in fact, they all originate from the brainstem.

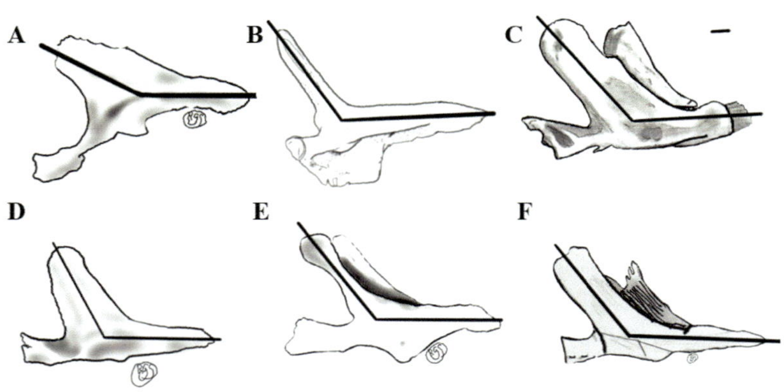

FIGURE 5.16 The variation in the angle formed by the alar process of the proötic and the lateral portion of that bone. (A) *beccarii* (RGS 201); (B) *caspius* USNM 195429). (C) *dumerilii* (RGS 203). (D) *griseus* (USNM 22065). (E) *indicus* (USNM 536545, Marianas). (F) *salvadorii* (RGS 410). The angle varies among species; in *griseus* 68°, in *caspius* (B) 58°, in *dumerilii* (C), *indicus* (E), and *salvadorii* (F) it inclines at about 50°, and in *beccarii* (A) about 28°. The supraoccipital has been removed from A, B, and D. Two of the rings of the semicircular canals lie deep to the lines.

the paired exoccipitals, and above by the supraoccipital; together they surround the foramen magnum, from which exit the spinal nerves, and which is the attachment point of the skull to the vertebral column. Medially on the basioccipital, there may be a single slightly elongated condylar process, as found in *prasinus* and *salvator*, or it may be entirely absent, as in *gilleni*, *caudolineatus*, *bengalensis*, and *rudicollis*, in which it is replaced by a deep cleft. In *indicus* there is neither process nor cleft, but a straight edge between lateral posterior projections.

Posterior to and above the sella turcica are the ascending alar processes of the proötics and the medial ridge of the supraoccipital (Figs. 5.14 and 5.15), each of which attaches to the ventral surface of the parietal. Extending laterally from the braincase are the cristae proötica (Figs. 5.15 and 5.16), horizontal rough-edged shelves of the proötics, under which lie the several foramina and fenestrae for cranial nerves, the inner ear, and blood vessels. In some species, the posterior vidian canal can be seen from an anterior view, but in most it is too lateral to be seen. In many species, the foramina for the maxillary (rostral) division of the trigeminal nerve (CN V^1) are visible dorsal to the anterior vidian (= pterygoid) canals (Figs. 5.14A, D and 5.15A, B).

Seen in posterior view (Fig. 5.20C–E), the braincase looks as follows: a superior supraoccipital that rises drastically anteriorly, forming short left and right articulations with the descending flanges of the parietal, above the point of the otic capsules (inner ear; Fig. 5.20C–E). A broad inferior basioccipital with vertical dorsal walls that meet the descending counterparts of the supraoccipital, and together form the large foramen magnum. Flaring at right angles to the foramen magnum and facing caudad are the exoccipitals, flat bones that extend posterolaterally and articulate laterally with the opisthotics (Fig. 5.13C), and lateral to the basioccipital and inferior to the exoccipitals are the small otoöccipitals (= oto-occipitals). It is within chambers formed at the junctions of the opisthotics, otoöccipitals, and basioccipital that the labyrinth system of inner ear canals forms (Figs. 5.15, 5.17, and 5.18).

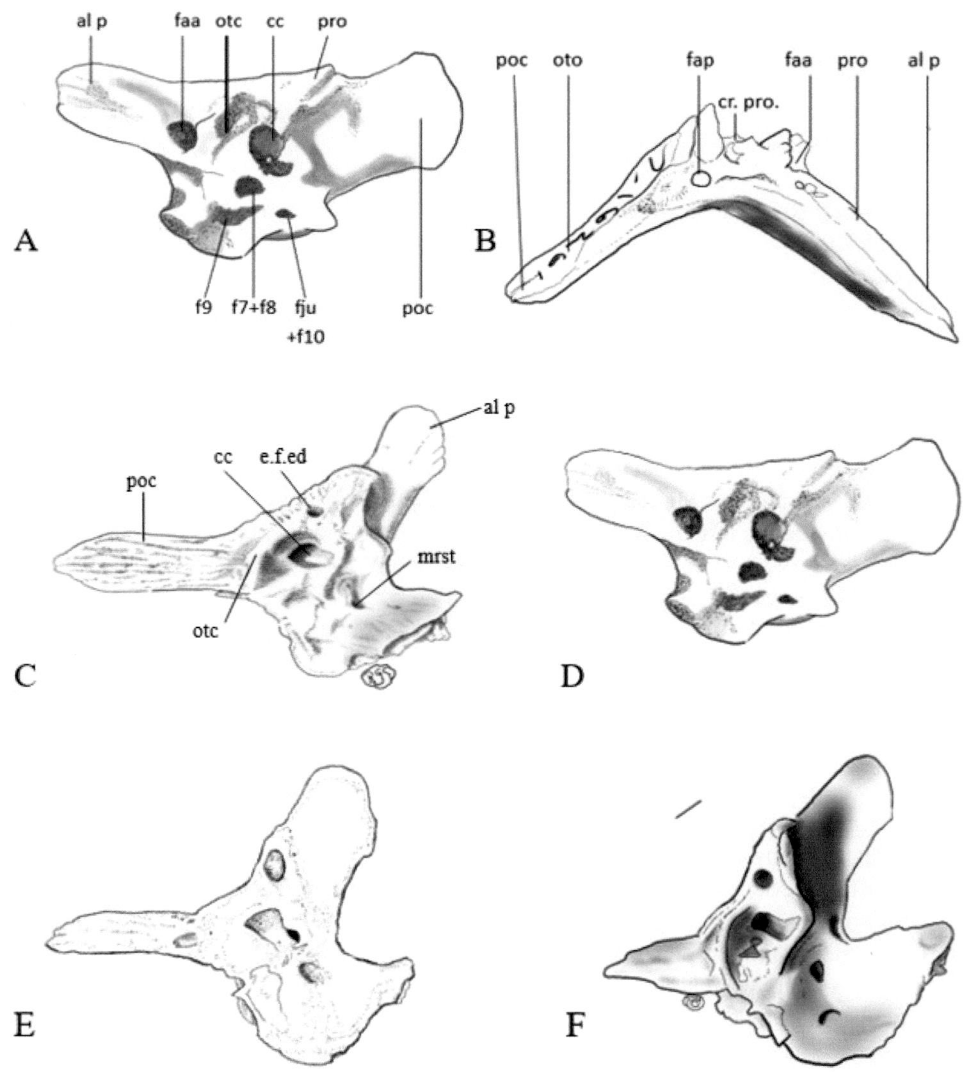

FIGURE 5.17 The otic capsule of the proötic. (A) Left medial and (B) dorsal views of *caspius* (USNM 216261); left medial views of (C) *exanthematicus* (RGS 202) and (D) *indicus* (RGS 39), (E) *niloticus* (USNM 216261) in medial and (F) posteromedial aspects.

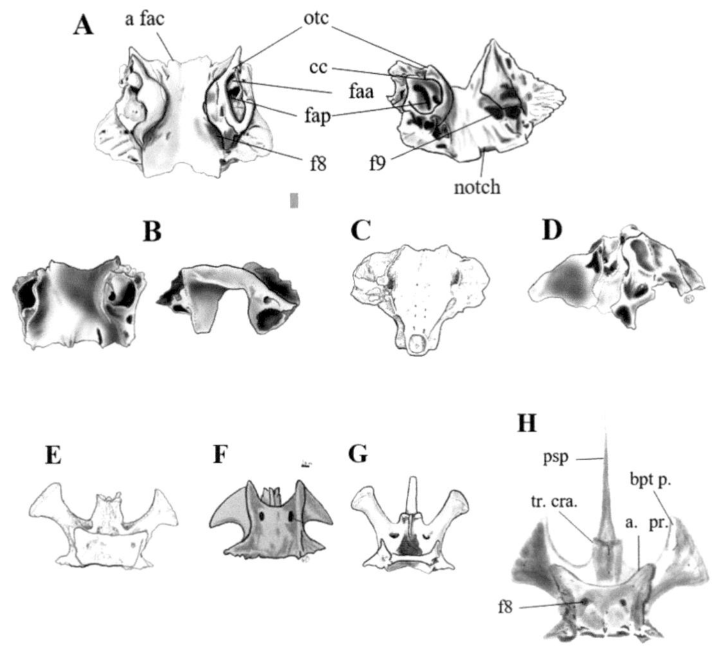

FIGURE 5.18 Elements of varanid braincases. Supraocipital. (A) The disarticulated left supraoccipital of *dumerilii* (RGS 203) showing prominent otic capsules in (left) dorsal, and (right) posterolateral views. (B) Dorsal (left) and anterior views of the disarticulated supraoccipital of *indicus* (USNM 122469); the same for (C) *exanthematicus* (RGS 202); (D) posterodorsal view for a probable *salvator* (USNM 216258) from the Philippines. Basisphenoid (sphenoid). Dorsal views of (E) *caspius* (USNM 216261), (F) *dumerilii* (RGS 203), (G) *salvator* (USNM 216258), and (H) *exanthematicus* (RGS 202).

Though the stapes attach to the fenestra ovalis of the proötics (Figs. 5.13 and 5.26), they are also attached distally to the tympanum via cartilaginous extracolumellae, and are discussed in conjunction with the quadrate, below. In living and alcoholic specimens, the parasphenoid (Figs. 5.1 and 5.14–5.17) is generally straight or may have a very minor upward arch. However, in many skull preparations, the accompanying desiccation often causes an upward arching of the bone. Functionally, the parasphenoid supports vertical sheets of connective tissue and cartilage (pila antotica and associated pilae) that divide the left and right halves of the skull, and that form the protective casing for the cerebrum and diencephalon. The tissue thus supported is a flexible mid-cranial septum, and divides the exiting CN I and CN II nerves in lieu of bony foramina. These tissues are contiguous with the more anterior bony interorbital septum. Anteriorly, the parasphenoid reaches toward the ventral surface of the middle of the parietal, where they are typically joined by connective tissues and a cartilaginous process.

The basisphenoid (Figs. 5.14, 5.15, and 5.19) occupies roughly one-half the ventral bony occipital complex, and is separated from its posterior counterpart, the basioccipital, by a transverse suture that remains distinct except in very old specimens. That suture may be one of three kinds: straight with little or no notch, straight but with posterior lateral extensions that form a notch along most of the suture, or deeply V-shaped, having long posterior lateral processes, the latter being least common, but diagnostic for *V. bengalensis*. The basisphenoid has the multiple roles of braincase floor, exit for the major anterior olfactory (CN I) and optic (CN II) nerves (Figs. 5.14 and 5.15), housing the vidian canals, buttressing the lower braincase with the pharynx via the forward pair of basipterygoid extensions, and the medial anterior process forming the slightly concave sella turcica, which supports the hypothalamus and pituitary. The basipterygoids (Figs. 5.14, 5.18E–H, 5.19, and 5.20A, B) extend as processes that arc downward to some degree in all species, especially so in older adults of the largest species (e.g., *albigularis, bengalensis, olivaceus,* and *niloticus*) and loosely articulate with the lateral posterior pterygoids. There is typically an extending, club-like posterior portion, though this is lacking in small varanids (Fig. 5.19H–K) and in *salvadorii* (Fig. 5.19G). The terminus of the basipterygoid process varies between two morphologies: broad and T-shaped, with processes extending in both anterior and posterior directions relative to the shaft (e.g., *dumerilii, exanthematicus, griseus, indicus, prasinus* [some], *salvadorii, salvator,* Fig. 19B, C, G, L), or possessing only an anterior-facing process that is often little more than a knob (e.g., *beccarii, bengalensis, caudolineatus, gilleni, niloticus, prasinus* [some], *rudicollis,* Fig. 5.19E, H–K). This variation is considerable, the broad and stout processes of *exanthematicus* compared with the small knobs of *beccarii.*

Each of these morphologies coincides with food preferences, with the slender tree-dwellers having the smallest epipterygoid processes and the species that consume hard prey with the largest, probably favoured because hard prey would necessitate a stronger, harder palate that better protects the brain from pressure during swallowing. In all

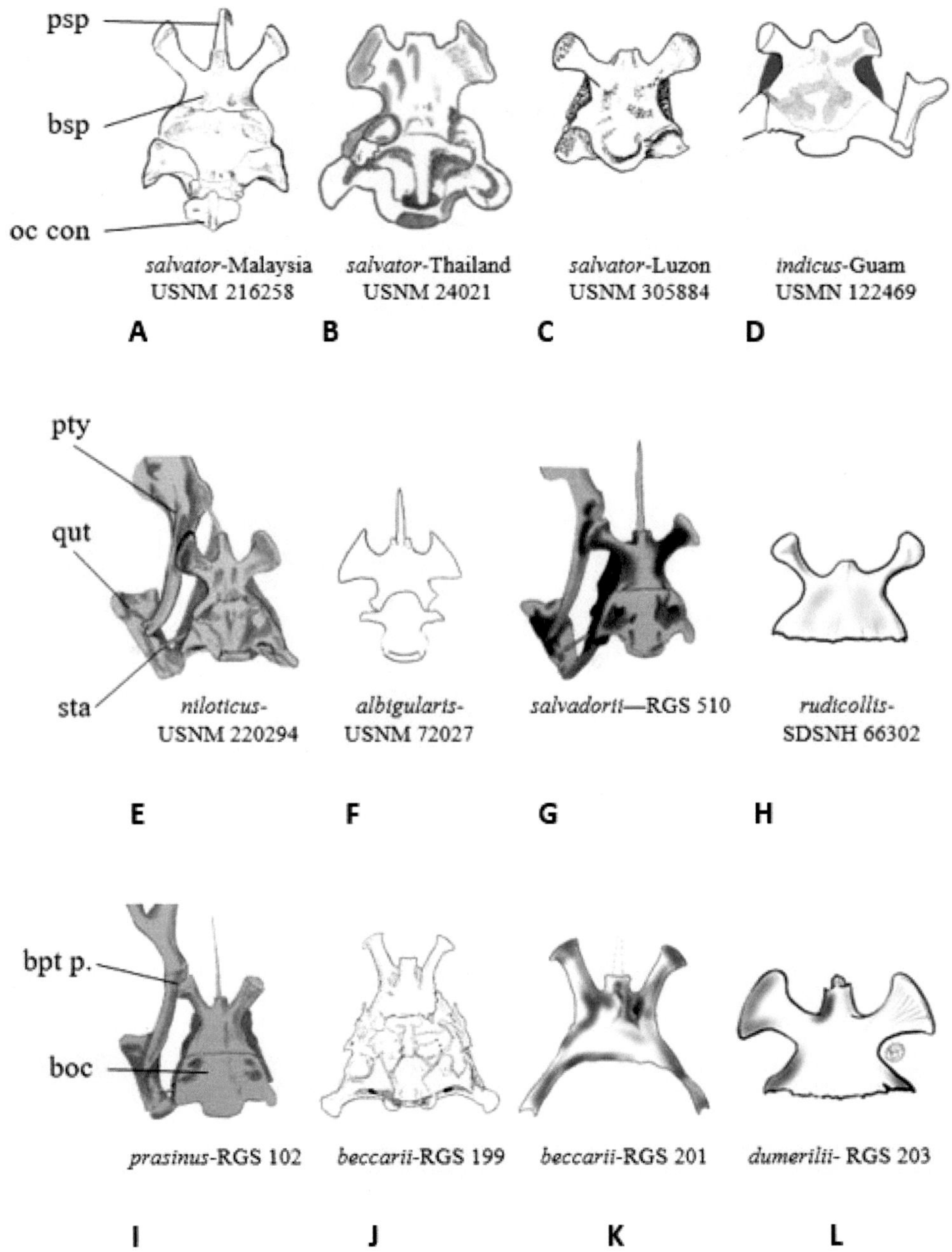

FIGURE 5.19 Comparisons of ventral elements of varanid braincases. In most specimens, the parasphenoid was damaged (A, E) or missing. Note that the basisphenoid-basioccipital suture may be well-developed (A, E, F, G, I, J), always the case in young lizards, or may fuse to near invisibility (B, C, D). The quadrate alone (D) or with other adjacent bones is depicted (E, J, I) to show the ventral structure of the quadrate Joint. Variable features include shapes of basipterygoid processes, shape of notch at basisphenoid-basioccipital suture (F, H, I), parallel or constricted midbody of basisphenoids, ratio of breadth at basipterygoid processes and posterior edge, and the presence or absence of flaring posterior basipterygoid "wings." There is near agreement in form among closely related taxa (E and F, I, J, and K, H and L). However, the variation in the *salvator* group (A–C) lends support to the reorganization of that "species" into several distinct species (Conrad 2015, Koch et al. 2007, Welton et al. 2013, 2014).

skulls of juveniles and neonates that I have examined, the pattern is consistent – that is, the epipterygoid-pterygoid articulation is either broad or narrow – but the degree varies with age; the abutment becomes relatively more massive with age, along with a decrease in the limited kinesis. As in many other features, the morphology of the epipterygoid process is unusual – and I suspect unique – in *V. salvadorii*, where there is a small, distinct dorsal spine formed from an extension of the dorsal crest of the process. In skulls I have seen, the spine may be directed either rostrad or caudad, but is symmetrical and will both point in the same direction. The central body is somewhat cube-shaped, and the posterior edge has a pair of lateral projections.

In the ventral aspect, the basisphenoid lacks the detail of its other surfaces, but its shape seems to be positively correlated with maximum adult size and prey type. In slender,

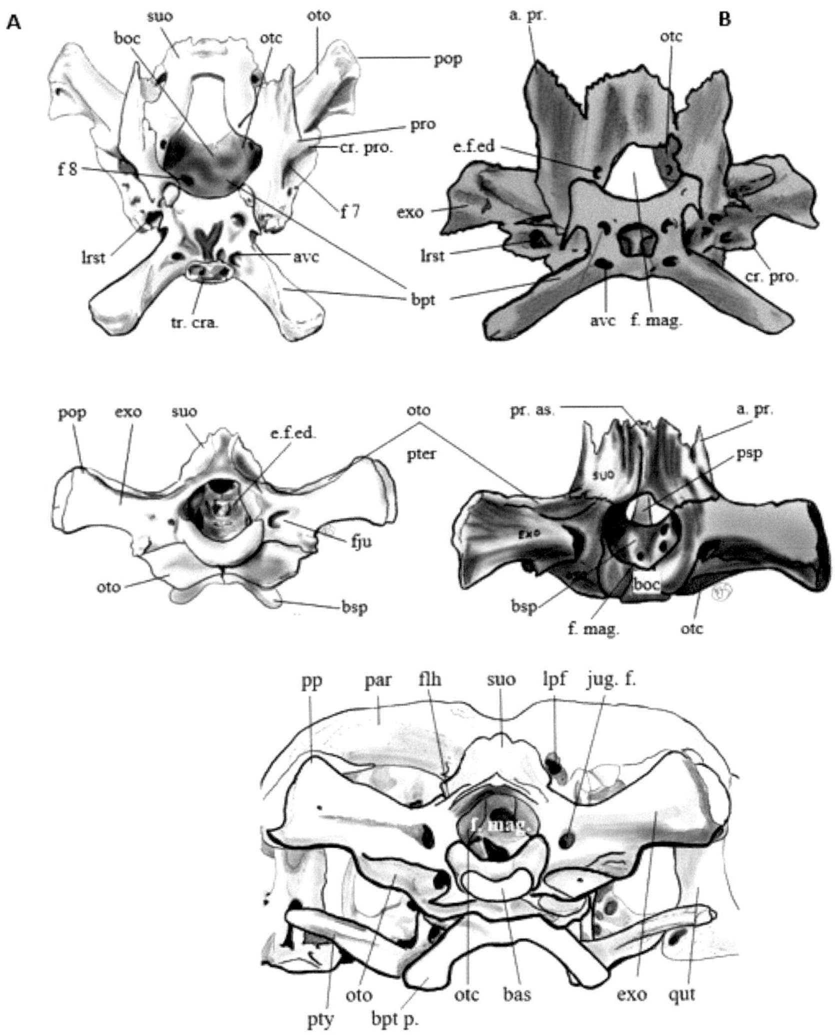

FIGURE 5.20 Elements of varanid braincases. Anterior views of braincases of (A) *Varanus dumerilii* (RGS 203), and (B) *Varanus salvator* (RGS 90), parasphenoids omitted. The most notable specific differences among the anterior elements are the shape and angle of the alar process of the proötic (see Fig. 5.18), height and shape of the alar process of the basisphenoid, and the visibility, or lack thereof, of lateral foramina. Posterior views of braincases of (C) *Varanus gouldii* (USNM 67734), and (D) *Varanus salvadorii* (RGS 410). (E) Posterior view of occiput of *Varanus olivaceus* (USNM 22400) with associated bones in place; compare with Fig. 5.13. The most notable specific differences among the posterior elements are the acuity of the process ascendens of the supraoccipital, the shape of the posterior edge of the paraoccipital process, and the size and orientation of the small foramina.

small, and arboreal species (*beccarii, caudolineatus, glauerti, prasinus*) the bone is triangular, narrow at the sella turcica, broad at the articulation with the basioccipital, and with elongate posterior sphenoid wings that closely resemble the basipterygoid processes in shape (Fig. 5.19I–K). Most species have square or slightly rectangular basisphenoids that, in older adults, may even become slightly broader anteriorly. There may be a raised medial ridge extending from the basioccipital part or the entire length of the basisphenoid. In larger species (*albigularis, exanthematicus, dumerilii, komodoensis, olivaceus, salvadorii, salvator, varius*) the ventral surface (excluding the basipterygoid extensions) is square or rectangular, equally broad along the anterior edge and the basioccipital suture (Fig. 5.19A–C, G, L). The sphenoid wings may be distinct posterolateral processes (*albigularis, exanthematicus, dumerilii,* Fig. 5.19E, F, L) or greatly reduced or absent (some *salvator,* Fig. 5.19A, C). The ventral surface lacks a distinct medial ridge, though there may be a slight convexity present.

In the anterior aspect, the basisphenoid presents several important apertures for nerves, blood vessels, and muscles (Fig. 5.20A, B). Along the mid-ventral face is the foramen for the paired trabeculae cranii, a pair of cartilaginous tubes that flank the sella turcica. The anterior-most parts of the trabeculae form the lateral walls of the parasphenoid process, and the dorsal region just posterior to the three structures is the sella turcica, the depth and extent of the fossa varying among species. In the largest varanids, the fossa is deep and forms a nearly spherical chamber for the pituitary. In other large species, however (e.g., *dumerilii*), the structure may extend nearly half the length of the basisphenoid but have a minor excavation or be flat.

The large anterior foramen, bordered above by the proötic and below by the basisphenoid, is generally oval with a central septum that usually begins interior to the opening. The anterior medial face of the basisphenoid is the lower portion of the large fenestra from which the anterior third of the brain emerges, including both olfactory tracts (CN I) and optic nerves (CN II) (Fig. 5.20A, B). Immediately dorsal to the trabeculae and much smaller in size is the internal carotid foramen; left and right foramina are widely separated, near the anterior terminus of the crest of the alar process of the sphenoid. More lateral and often barely facing forward lie the external foramina for CN VI. The interior foramina are large dorsal openings in the anterior half of the basisphenoid and are usually not visible except with a disarticulated sphenoid.

The anterior region where the proötic and basisphenoid meet forms a vertically compressed wing-like alar structures that are the most anterior bony lateral walls of the braincase. The suture ankyloses early in life, making it difficult to discern even when magnified.

Midway between the left and right epipterygoid-parietal-alar processes is the parietal foramen, the parietal eye and its nerve passing nearly straight down between cerebrum and optic (= tectic) lobes at the most posterior part of the brain, not enclosed laterally or ventrally by bony material. The posterior lateral surfaces and under the crista are foramina for the maxillary and mandibular branches of the trigeminal nerve (CN V_2 and V_3), palatal branch of the facial nerve (CN VII_{pal}), glossopharyngeal (CN IX) and hypoglossal (CN XII) nerves, and jugular vein, plus the foramen ovalis. Also present beneath the crista proötica is the lateral aperture of the recessus scala tympani, analogous, not homologous, to the mammalian foramen rotunda (= cochlear window, round window). The trochlear nerve (CN IV) exits a foramen between the alar process and the supraoccipital. The branches of CN VIII (CN VIII is variously known as the acoustic, audio-acoustic, auditory, audiovestibular, octaval, statoacoustic, vestibuloacoustic, vestibulocochlear, and probably other names. I therefore opt to call it, simply, CN VIII, perforate the medial wall of the otic capsule, but not the outer. The remaining cranial nerves exit the braincase from ventral or posterior-facing surfaces.

The opisthotic and exoccipital bones form most of the posterior part of the braincase, excepting the floor formed from the basioccipitals. Seen from behind the skull, the exoccipital forms a distinct pair of lateral wing-like extensions of the braincase, the most distal parts flaring broadly to form the paraoccipital processes. The dorsal edge of the exoccipital is a broad, flat contact zone, which is the articulation surface with the opisthotic. In the lateral view, the posterior of the proötic forms a structure similar to but smaller than the anterior alar process, and the bone beneath that is the opisthotic. The suture in adult monitors may be invisible because the bones articulate so completely. When viewed dorsally, though, there may be a trace of part of the suture along the most posterior part of the proötic (Fig. 5.13C). The bulk of the opisthotic is visible beneath the crista proötica caudad to near the end of the braincase; the very last section, visible as a thin edging, is the exoccipital. On either side of the foramen magnum, the exoccipital is pierced by its largest integral openings, the jugular foramina, through which CN X (and sometimes CN XII) exits the skull. The jugular vein does not pass through this foramen.

As for the brain itself, it is contained within bony enclosures from a point below the parietal foramen, posterior to the foramen magnum, and includes the optic tectum (second lobe of the brain), cerebellum (third lobe), and midbrain, from which most cranial nerves emerge. The cerebrum (first lobe) lies partly between the alar processes of the proötics and is bordered by the parietal anterior to the parietal foramen, the anterior-most portions of the alar processes, the epipterygoids, and pterygoids. The cerebrum is further protected by thick membranous tissue supported by the epipterygoids and parasphenoid process, and the sheets of muscle that extend downwards from the parietal and proötics. Varanid brains have not been well studied, the exception being *V. exanthematicus* (Barbas-Henry and Lohman 1984, 1986).

5.3.11 Jugal

The jugal is analogous to the mammalian zygomatic, forming the lower and posterior bony border of the orbit (Figs. 5.2, 5.21, 5.22, and 5.24). In varanids, the anterior of the jugal meets the lacrimal dorsally and abuts the maxilla anteriorly and ventrally at long sutures. Functionally, the jugal provides ventral and posterior protection for the orbit, and serves as an anchor point for small jaw muscles

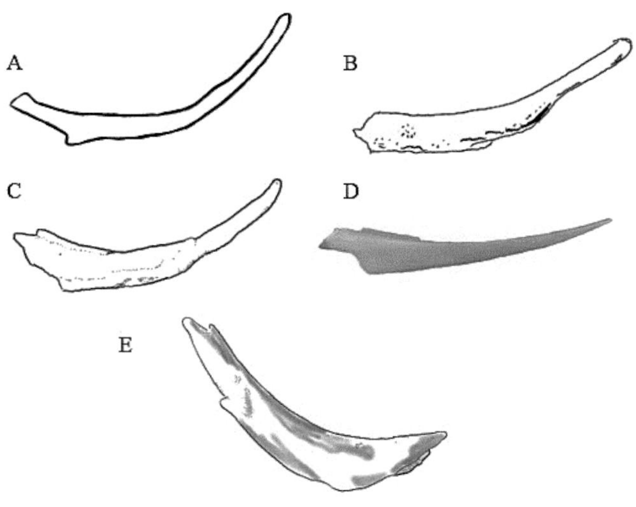

FIGURE 5.21 The variation in the jugal bone of *Varanus,* left lateral view. (A) *indicus*, Marianas (USNM 536545); (B) *timorensis* (ZMB 8294b); (C) *salvadorii* (RGS 410); (D) *niloticus* (USNM 216261). (E) *dumerilii* in left medial view (RGS 203). In A, C, and E, the posterior orbital process nearly meets the lateral postorbital; in B and D, the distance between the elements is considerable.

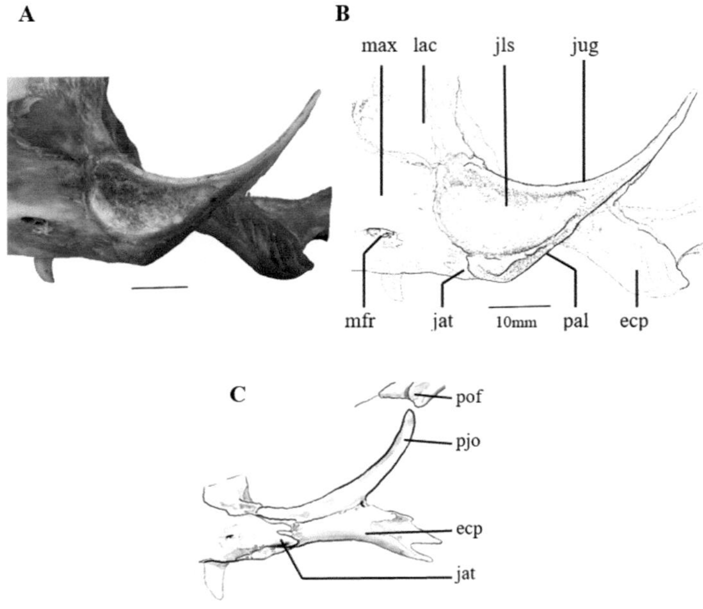

FIGURE 5.22 (A–C) The variation in the jugal bone of *Varanus,* left lateral view. Upper – photograph and illustration of the extreme concave jugal with its broad anterior section seen in *komodoensis* (USNM 101444). Other large (>2 m) species may develop a much less pronounced concavity, but none show the disparity in proportions between anterior and posterior sections of the jugal.

(see Chapter 6). The jugal separates the maxilla from the bordering the orbit. Medially, the anterior jugal meets a lateral process of the ectopterygoid and barely the palatine. The jugal is thickest at the anterior end and tapers to a fine posterior point; in no case is there a bony connection between jugal and postorbitofrontal. In species where the jugal has a pronounced curvature (Fig. 5.21B, D) cartilaginous material often links the bone with the postorbitofrontal. Where the jugal meets the lacrimal, maxilla, and ectopterygoid, it is flattened, often spatulate. There is a ventral canal that extends from that junction to about midway toward the apex.

There are two notable variations of the jugal. The anterior half usually has a smooth, rounded lateral surface, making the entire bone a curved cylinder (Fig. 5.21C), with anterior and posterior regions similar in diameter. Alternatively, the anterior half may have a pronounced lateral concavity, most developed in *komodoensis*, and seen in *albigularis*, *bengalensis*, *gouldii*, and some *olivaceus* (Fig. 5.23A, B). In these species, the compressed area makes the anterior of the jugal much wider than the posterior half. Depictions and reconstructions of †*V. prisca* show a highly compressed jugal almost identical to that of *komodoensis*, but at present, no jugal fossil is known for the species, so such representation is hypothetical.

The second notable aspect of jugal configuration is the degree to which it bows out laterally at the inflection point along its lower portion. The range of variation can be measured as an angle formed by the intersection of straight lines drawn from the orbits to the nostrils (Table 5.1), with the lower numbers aligning with broad snouts with nearly parallel maxillae, and higher numbers with tapering snouts. For skulls I've measured, the greatest difference

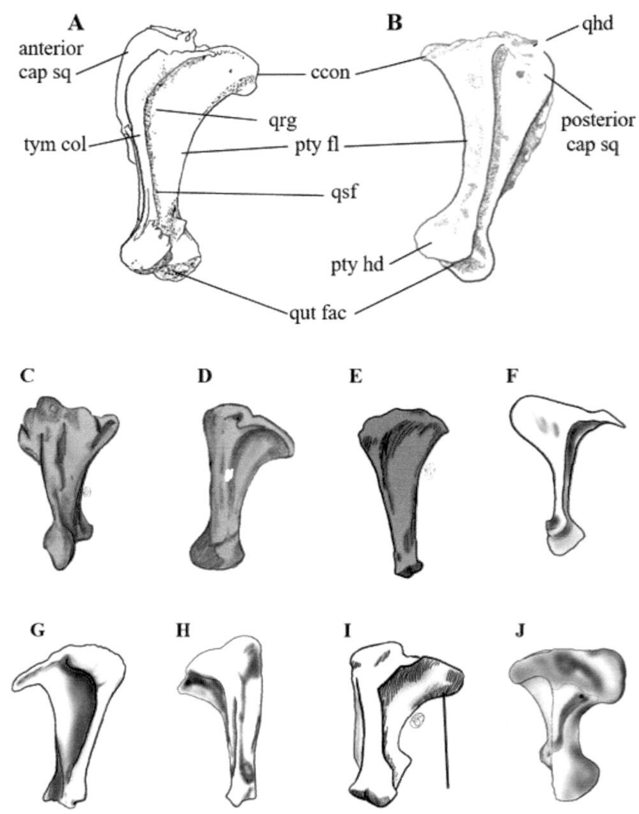

FIGURE 5.23 Quadrates of (A) *indicus* (Guam, USNM) left lateral and (B) left medial. (C) *prasinus* (RGS 102) left posterolateral, (D) *chlorostigma* (NMW) lateral, (E) *indicus* (Indonesia, RGS 39), (F) *indicus* (CAS uncat.), (G) *dumerilii* (NMW) right lateral, (H) *bengalensis* (CAS uncat.) lateral, (I) *niloticus* (NMW, 1678, bar = 10 mm) lateral, (J) dorsolateral view of quadrate of *salvator*; for posterolateral view see Fig. 5.13 lower.

in obit/nostril widths is an adult *exanthematicus* (48°, NMNH 161744), and the least difference seen is in the adult extant varanid *komodoensis* (66°, SMF 231891), and the extinct †*S. ensidens* (20°, NMNH v2185). Skull proportions are strongly linked to ontogeny; for example, a juvenile *komodoensis* (NMNH 519762, SVL 160 mm) has a snout with an angle of 27°, indicating a broad ocular region and a much narrower snout tip. In *rudicollis*, the head is narrow, but the orbital region is only slightly broader than the snout tip. This measurement, though, does not capture the full scope of the jugal's contribution to snout configuration, in that some taxa (e.g., *griseus*) have very wide orbital regions in which the jugal flares laterally and the maxillary area immediately anterior is considerably narrower. This contrasts strongly with species such as *komodoensis*, where the lateral faces of the snout are almost parallel (Fig. 5.25R).

5.3.12 Quadrate and Stapes

The most posterior lateral bone of the cranium is the quadrate (Figs. 5.23 and 5.24). Much has been written about the quadrate in terms of its kinetics, use in phylogenetic hypotheses, and its evolutionary destiny to become the malleus of the inner ear in mammals. Hendrickx et al. (2015) have established standardized terminology for quadrate morphology that I largely follow here (Fig. 5.25).

The mobile quadrate allows a greater gape than would be possible if it were static, and the hinge with the skull lies on the inner surface of an enlarged quadrate head (= superior conch), the long convex dorsal edge of the distal

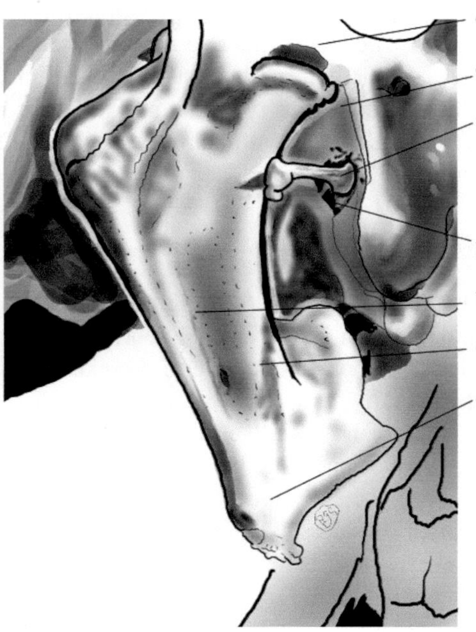

FIGURE 5.24 Left quadrate and stapes of *Varanus dumerilii* (USNM 595348) seen in posteroventral view. As depicted, the distal end of the stapes is protruding into the empty space; in life, the distal tip articulates with the cartilaginous extracolumella, which then attaches to the medial surface of the tympanum.

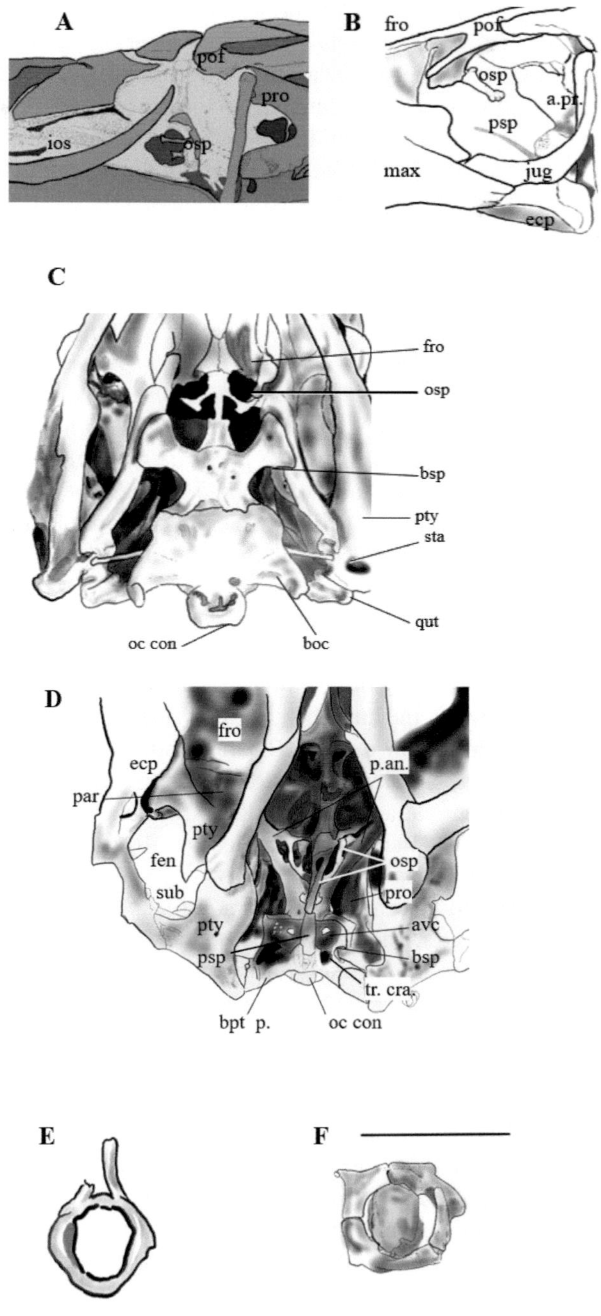

FIGURE 5.25 The orbitosphenoid in some *Varanus*. (A) *salvator* (USNM 35762, Sumatra) lateral view of postorbital fossa, showing orbitosphenoid in relation to parasphenoid (dotted lines) and medial connective tissues. (B) Left lateral view in situ in *Varanus olivaceus* (USNM 222400). (C) *dumerilii* (USNM 595348) ventral view of braincase bones in situ; orbitosphenoid (in white) and stapes clearly visible. Note the stapes directed posterolaterally, the case in all adult *Varanus*. (D) *gouldii* (USNM 67734), anteroventral view. (E) *V. olivaceus* (USNM 222400) anterior view, right dorsal process missing. (F) *indicus* (RGS), anterior view, scale = 1 mm. Abbreviations: a. pt. – anterior proötic; avc – anterior vidian canal; boc – basioccipital; bpt p. – basipterygoid lateral process; bsp – basisphenoid; ecp – ectopterygoid; fen sub – suborbital fenestra; fro – frontal; ios – interorbital space; jug – jugal; max – maxilla; oc con – occipital condyle; osp – orbitosphenoid; p. an. – pilla antotica; par – parietal; pof – postorbital-postfrontal; pro – proötic; psp – parasphenoid; pty – pterygoid; qut – quadrate; sta – stapes; tr. ca. – trabecular canal.

surface. The anterior surface of the bone is a large, smooth, rounded, or nearly straight process (= anterior pterygoid flange). The strongly concave posterior pterygoid flange extends downwards to terminate in the posterior joint of the mandible at the articular. Between the flanges, facing distally, is a lateral rod-like quadrate shaft, the anterodorsal surface of which forms the lateral process. The shaft is a nearly vertical strut which may be broad and flat or thin and nearly columnar, the former in monitors with strong bite ability, the latter in smaller insectivorous species. In *V. bengalensis, salvator*, and some *indicus, the shaft is nearly flat, being thick, broad, and with a pair of raised posterior edges. The width of the central column and flange of the quadrate correlates positively with diet; lizards that require strong grips or crushing power (e.g., bengalensis, dumerilii, olivaceus*) have very broad flanges for adductor muscle attachment, while species that consume small or easily handled prey generally have narrow columns and spindle-shaped shafts.

The most posterior point of the posterior pterygoid flange is a downward arch that terminates in an acute point that is the most superior point of attachment for the tympanic membrane, while the anterior tympanic surface is attached to the central lateral process. There is no bony structure for ventral or posterior tympanic attachment; instead, the lower tympanum attaches to the dorsal articular surface of the lower jaw, and the rear edge attaches by ligaments to the dermal border of the tympanic cavity.

Specific quadrate morphology is a taxonomically significant character. In *Odatria* the bone is a near equilateral triangle with a strongly pronounced dorsal surface, the posterior head extending back almost as far as the occipital condyle. The central lateral process is strongly curved, providing space for a comparatively large tympanum. The paired ventral articulating anterior and posterior pterygoid flanges are much narrower than the dorsal head. Among other monitors, the quadrate may be nearly straight and bar-like, as in *V. bengalensis*, or have a broad head, narrow base, and drastically constricted body, as in *V. indicus*. Most commonly, the quadrate has a broad head with a secondary, more posterior process (which may make the quadrate head some three to four times broader than the base), an elongated shaft region, and slightly expanded lower articulating flanges (e.g., *V. dumerilii, V. salvator*).

Ventrally, the quadrate ends in a pair of mandibular articulations, one medial, one lateral, that form the upper part of the joint with the articular. The articulation points may be barely developed (*V. indicus*) or be quite large and distinct (*V. salvator*). The loose upper and much looser lower articulations of the quadrate with other skull elements give the bone the ability to swing on a hinge to a considerable degree. Though the mandible cannot drop away from the quadrate joint, it can open to a degree that allows most monitors to consume very large food items when compared to what other lizards can accommodate. This highly mobile quadrate and its attendant gape are also seen in the varanoids *Heloderma* and *Lanthanotus*, but not in the putatively related *T.*

The quadrate shows almost no intraspecific variation except ontogenetically. In hatchlings and juveniles, the bone is little more than a thin, near-vertical shaft with a minor dorsal expansion, the enlarged cephalic condyle, anterior capitum squamosum, and pterygoid flange developing as the lizard grows. This is the same pattern seen in the gekkonid *Hemidactylus turcicus* (Paluh et al. 2018), in which species there is little discernible variation between adults of each sex, but juveniles have rod-like and little developed quadrates.

The needle-like stapes lies beneath the large posterior quadrate head, at a point about midway up the shaft, and attaches to the center of the tympanum via a short cartilaginous element, the extracolumella. The medial stapedial tip forms an enlarged, flattened oval footplate that meets and attaches to the oval window membrane in the basioccipital. Except in neonate and very small monitors, the oval window faces posterolaterally, and the stapes of adult monitors is aligned with its distal end further posterior than its proximal end, and the fenestra ovalis, a feature seen very clearly in CT scan sectional views. In many juveniles, the stapes is aligned in a less angled orientation. The stapes is both very thin and very small, even in large varanids, and is often lost in skeletal preparations. In almost all cases, the stapes is cylindrical and straight. Occasionally, it may be very slightly spatulate distally (e.g., adult *bengalensis* and *salvator*) or have a very minor central constriction (*mertensi, prasinus, salvadorii*). The distal portion of the extracolumella expands broadly from the stapes and attaches directly to the medial tympanic membrane by two (sometimes three) small tubercles. The larger tubercle is the tympanic base and is clearly visible externally as a small semi-circular rise near the center of the tympanum. There are often very fine striations of the tympanum radiating out from the tympanic base.

5.3.13 Vomer

The paired vomers constitute the bulk of the osseous roof of the mouth, extending anteriorly to abut the premaxilla, and posteriorly to about the middle of the external nares, then arching laterally to meet the palatines (Figs. 5.2, 5.25, and 5.26). Vomers and maxillae form a flat floor beneath the septomaxillae, supporting the Jacobson's organs. In no case do the bones actually meet medially. Vomerine configuration varies considerably among taxa. In *V. timorensis*, the vomers nearly meet behind the premaxilla but diverge immediately behind, while in *V. prasinus* the separation is more pronounced and the bones end in sharply turned posterior lateral hooks.

Among the larger (SVL > 250 mm) varanids, the vomers are wide and flat along the first two-thirds of their length; in young and small taxa, the surface is usually rounded, often sub-cylindrical.

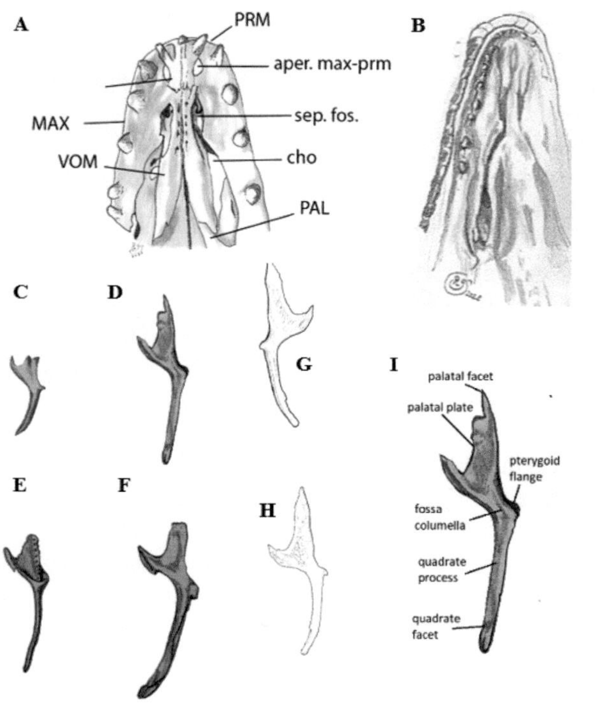

FIGURE 5.26 Palates in *Varanus*. Palatine: (A) the bony palate of an adult *Varanus indicus* (RGS 38). (B) Soft palate of *Varanus nuchalis* (USNM 305982). Pterygoid: right ventral view of (C) *Varanus exanthematicus* (RGS 202), (D) *Varanus salvadorii* (RGS 410), (E) †*Saniwa ensidens* (USNM v2185, holotype), (F) *Varanus salvator* (RGS 90), (G) dorsal and (H) ventral Views of pterygoid of *Varanus caspius* (USNM 216261). (I) Structures of pterygoid of *V. salvadorii* (RGS 410).

The anterior portion of the vomer bears a thin lateral flange, nearly hemispherical with the cup facing dorsally. This structure is the floor of the Jacobson's organ and varies considerably in size and shape among taxa. In the putatively ancestral *V. griseus* (Ast 2001, Sprackland 1989, 1991), the flange is large, extending nearly half the length of the vomer and being more than half the width of the distance between the center of the mouth and the outer surface of the maxilla. In this species, the cup is distinct, smooth, and has rounded edges. At the other end of the morphology is *V. komodoensis*, in which the flange is about one-sixth the length of the vomer, is irregular in shape, and is bi-lobed with few smooth edges.

The anterior portion of the vomer also bears a distinct dorsoventrally aligned ridge. The bone has a tube-like section running nearly half its length, the inner wall made from the ridge, and the more medial surface with a slightly ventrally turned lip. The concave central canal is lined with several foramina, the largest near the posterior extent of this canal.

5.3.14 Palatine

The position of the palatines, lying posterior to and more lateral than the vomers, marks the location of the entry to the varanid's gullet (Figs. 5.2 and 5.26). The palatine is conservative in form, a V-shaped bone, the fork forming half the anterior part of the infraorbital fenestra and much of the floor of the orbit. The long (proximal) arm of the V forms a suture with the posterior of the vomer, the more distal arm meeting the maxilla. Posteriorly, the vomer has a notch, the depth of which varies among taxa, that serves as a slot for the anterior suture with the pterygoid. In fossil varanids, such as †*Saniwa*, the V is less pronounced, with the lines almost parallel. Of all the cranial bones, the palatine seems to be the least variable.

5.3.15 Pterygoid

The pterygoid and maxilla are nearly equal in length (Figs. 5.2 and 5.26). In shape the pterygoid somewhat resembles the letter Y, the anterior medial branch joining the vomer, and the lateral part abutting the ectopterygoid. In some cases, especially in large adults, a small bit of that lateral process may be visible laterally, posterior to the ectopterygoid. The longest arm of the Y may be half to two-thirds the length of the bone and terminates medial to the lower conch of the quadrate, but does not form a suture with that bone. The inferior tip of the epipterygoid meets the longer branch, to which it attaches, and though the lateral processes of the basioccipital abut on the pterygoid, they do not form an ankylosed suture. One large dorsal foramen is located where the anterior V and posterior straight section meet. Two or three much smaller foramina lie in the center of the broad portion of the V. In *Varanus* and *Odatria*, the anterior portion of the ventral surface of the pterygoid is either flat or very slightly concave. In †*Saniwa*, though, the distal branch is strongly compressed into a lateral wall that broadly meets the ectopterygoid, such that the ventral surface is strongly concave. The medial flange also rises more than in extant varanids, so the area between the arms of the V is cup-like.

Thus, in being sutured only along its most anterior processes, the rest of the pterygoid is largely unrestricted in possible mobility. Aligned in diverging angles from each other, the pterygoids form lateral borders for a very wide and expansive throat, the lateral limit of that expansion limited primarily by the extent to which the mobile epipterygoids can move. The lower quadrate, being free of the pterygoid, can swing anteroposteriorly, thus accommodating a gape that, among lizards, is exceptionally deep and broad. Of the non-ophidian squamates, such a gape is only somewhat approached in the pygopodid genus *Lialis*.

5.3.16 Ectopterygoid

This is a sub-jugal element that forms the anterolateral part of the postorbital arch (Figs. 5.2, 5.10, 5.13, and 5.25). Anteriorly, the ectopterygoid (= transversum of earlier authors) marks the most posterior extent of the maxilla externally, meets the distal process of the palatine anterior to the orbit, and posteriorly sutures to the distal articulating surface of the pterygoid.

5.3.17 Orbitosphenoid

The orbitosphenoid (Fig. 5.25E, F) is a small, thin osseous ring that is suspended from the anteroventral surface of the parietal and acts as an enveloping support for the olfactory nerves and associated small vessels. From the orbitosphenoid, the nerves enter the crista cranii of the frontal. Because this bone is so frequently damaged or missing in skull preparations, I have omitted a comparative examination.

5.4 THE MANDIBLE

The lower jaw of varanids is composed of the dentary, splenial, coronoid, surangular, angular, prearticular, and articular bones on each side. All these bones except the articular contribute to the suture, forming a mobile intramandibular joint. The joint allows considerable lateral flexion that permits the lizards to consume relatively larger prey items than can most other lizards. The splenium is the only bone that crosses the suture (see below).

5.4.1 Dentary

The dentary of varanids is a long bone, fully half the length of the mandible, and is the only tooth-bearing bone of the lower jaw (Figs. 5.27 and 5.28). The anterior elements of each mandible fuse to form a symphysis that, depending on species and age class, has little to no elasticity, especially in juveniles. There may be one or more lateral mental fossae, and a larger posterior adductor fenestra that fully pierces the bone, the anterior opening facing anterolaterally near the Meckelian joint, the medial facing posteriorly, and generally concealed by the splenial. The dentary tooth row rarely extends posterior to the lateral

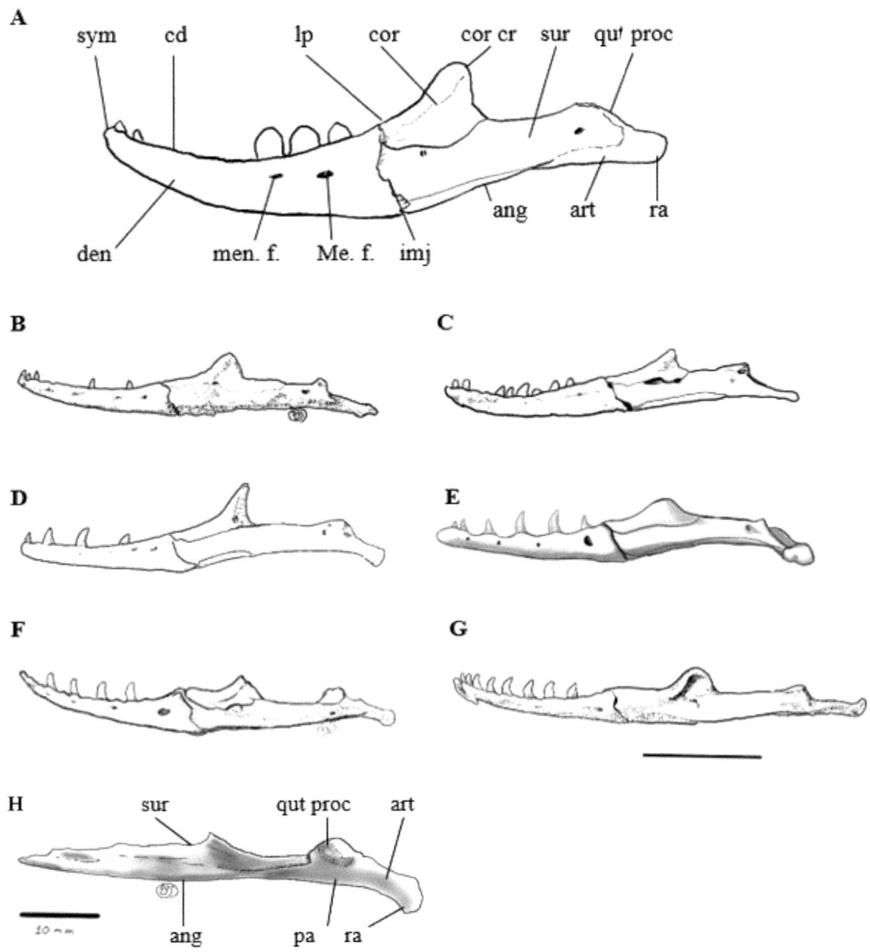

FIGURE 5.27 Left lateral views of the mandibles of seven species of *Varanus*: (A) *albigularis* (USNM 72027), (B) *acanthurus* (SMF), (C) *bengalensis* (SMF 11550), (D) *eremius* (SMF 11648), (E) *kordensis* (ZMB 8831), (F) *salvator* (USNM Sumatra), (G) *timorensis* (RGS 189). Note the angle of the retroarticular; it generally turns downward (A, B, D, F), may have A sharp turn (E), or be nearly straight (C, G). Coronoids vary from large, shark fin-shaped, and tall (A–D) or be low (F, G) or greatly reduced and rounded (E). The height of the mandibular joint is also highly variable. (H) Lower rear half of jaw of *caspius* (USNM 195429) with coronoid removed to better show surangular; scale = 10 mm.

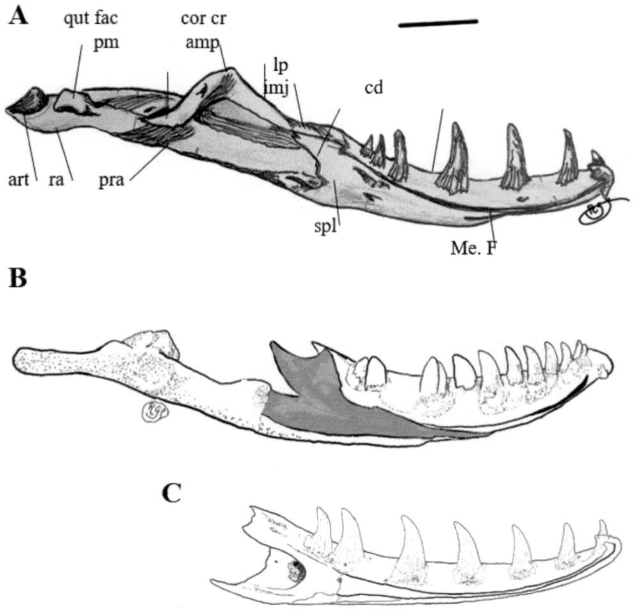

FIGURE 5.28 Left mandible in medial views of (A) *Varanus salvadorii* (RGS 410, Bar = 10 mm), (B) *Varanus dumerilii* (RGS 203) with coronoid and surangular omitted, splenial highlighted. (C) *exanthematicus* (RGS 205), lingual view of dentary with splenial omitted to reveal the extent of Meckelian fossa.

fenestra, and never beyond the medial one. The most posterior lateral portion of the dentary terminates anterior to or just below the anterior edge of the lacrimal, while the ventral element forms a broad process that may reach a point below the mid orbit. The pleurodont teeth lie on a lingually angled subdental shelf forming the upper edge of a distinct Meckelian canal that extends posteriorly to the coronoid. Tooth bases lie in shallow excavations against a dental crest, a horizontal lateral portion of the dentary along the tooth row. Anteriorly, there is a small vertical wall of the dentary that curves medially to form the symphysis. The posterior sub-Meckelian canal is deep and thickens posteriorly. The horizontal posterior border of the dentary and splenial attaches to the posterior bones (coronoid and angular) to form a mandibular hinge joint with limited but distinct lateral and vertical mobility. This joint allows varanids to engulf large food items, and also helps move food into the esophagus by providing extra dentary tooth motion that moves food posteriorly.

5.4.2 Angular

There is little variation seen in the morphology of angular bones in varanids (Fig. 5.27). The bone lies beneath both posterior dentary and anterior surangular, with a dorsal groove into which both larger bones fit. Functionally, the angular seems to limit downward hyperextension of the dentary when the jaws are open and is incapable of any movement. The bone is universally visible in lateral, lingual, and ventral views of the mandible.

5.4.3 Coronoid

This is a very irregularly shaped bone that vaguely resembles a saddle (Figs. 5.27–5.29). The extreme anterior of the coronoid is laterally compressed with a bifid anteromedial process with sharp, elongated tips that meet the dentary anterior to the intramandibular joint. The superior anterior tine is both longer and thinner than the inferior. The gap between the tines is the posterior border of the anterior inferior alveolar foramen. The superior and anterior edges are formed by the dentary, and the inferior by the splenial (Fig. 5.28). The ventral portion of this process sits shallowly between the anterior prearticular and surangular. Dorsal to the thin section is a transversely widened area of bone that has sub-equal descending processes that saddle the anterior-most dorsal part of the surangular, where it meets the dentary joint, resulting in horizontal and ovoid fenestra to either side of the surangular-coronoid suture. Most of the coronoid is composed of the predominantly dorsal, smooth saddle-like crest, which is slightly concave and covers about 30 percent of the surangular, depending on species and age class. Ventral to the "seat" is a deeply recessed coronoid fossa into which fits the anterodorsal surangular surface. The anterodorsal part of the saddle is a horizontal V-shaped process, between which the coronoid process of the dentary fits. This fork represents the dorsal limit of the intramandibular joint, which extends between the dentary anteriorly and the anterior vertical planes of the coronoid and surangular posteriorly. This joint has kinesis second only to the quadrate-retroarticular joint, and allows considerable vertical flexion that facilitates food handling. Its primary action seems to be to ratchet food rearwards into the throat, but it also allows for a broadened gape by permitting lateral expansion of the mandibles. However, the joint does not allow any flexion medially (lingually); the way the dorsal dentary lies in the anterolateral coronoid joint, coupled with the location of the splenial across the joint's medial surface, prevents any inward flexion (Frazzetta 1962).

At the posterior end of the saddle is the coronoid crest, a triangular component that is the most conspicuous coronoid feature in an articulated mandible. The crest rises vertically to meet, but not articulate with, the posteroventral process of the jugal, and serves as an attachment point for jaw abductor muscles. From the crest extends a long descending process that lies nearly vertically on the median surface of the surangular. The process abuts in such a way as to limit any inward flexion of the posterior mandible, and is likely important to provide muscle attachment and physical strength that allows the swallowing of large or hard prey items. The anterior end of the large mandibular fossa, formed by the medial surfaces of the surangular and prearticular, is obscured by the coronoid posterior descending process.

In the snail-crushing teiid *Dracaena*, there is a huge lateral ridge formed from the coronoid and surangular that serves as an attachment for very large muscles that allow the lizards to crush hard-shelled prey. There is no such

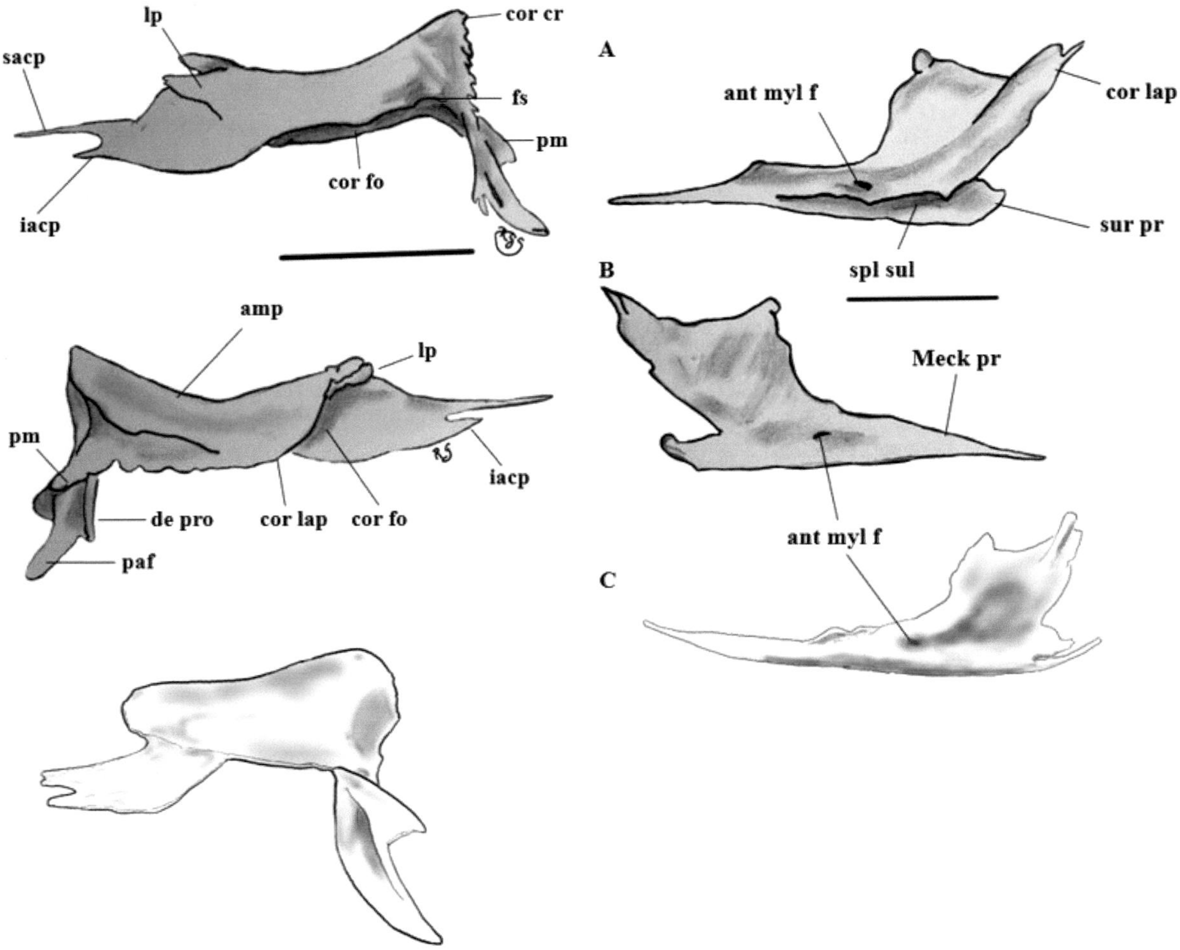

FIGURE 5.29 Left mandibular elements of left, *Varanus dumerilii* (RGS 203): upper, coronoid in lateral and, lower, medial views. Right, left coronoid of *Varanus niloticus* (USNM 216261) in lateral view. Left splenial of *Varanus dumerilii* (RGS 203) in (A) lateral and (B) medial views. The surangular foramen is mostly contained in the surangular, but the upper boarder is made by a notch in the lateral coronoid (fs). Scale = 10 mm. (C) medial view of left splenial of *Varanus albigularis* (USNM 199568) showing characteristic forked surangular process.

modification seen in any varanids, including the known snail feeders (e.g., *albigularis, bengalensis, niloticus, olivaceus, salvator*).

5.4.4 Prearticular and Articular

Varanids have a long prearticular bone and short articular running horizontally beneath the surangular and almost entirely on the median aspect of the jaw (Figs. 5.27, 5.28, and 5.30). Its anterior edge borders the intramandibular joint, the dorsal edge is concave, and the superior medial border forms the lower edge of the mandibular fossa. The anterior half is laterally very compressed and abuts laterally with the coronoid and, much more broadly, the surangular. The anterior edge is thin and has sharply pointed, thin processes that meet the lingual surface of the dentary. The posterior half is more cylindrical and encloses a long tubular canal along its ventral portion. Dorsal to that canal is a long open space, the adductor fossa, formed by the vertical central portions of the prearticular and surangular. From a lateral aspect, the prearticular is a thin ventral element visible from about mid-coronoid to the articular. Near the posterior end of the surangular, the prearticular rises along the lateral jaw and forms the broad retroarticular process; the anterior third is the socket for the lower quadrate. Behind the process is the smooth-surfaced posterior retroarticular, which contains a long, shallow fossa to which abductor muscles attach. The lateral dorsal crest of this section, posterior to the quadratic joint, is the articulating surface for the lower edge of the tympanic membrane. The anterior surface, just caudad to the surangular socket joint, is pierced by the foramen for the mandibular branch of the facial nerve (CN VII).

The articular is the heel of the mandible, representing the most posterior and, via its anterodorsal retroarticular process, forms the concave quadrate process. The anterior dorsal surface has a prominent rise of smooth bone that serves as the leading edge of the quadratoarticular joint. It is the ventral surface of the articular that gives the posterior mandible its upward arc, contrasting with the downward arc of the dentary.

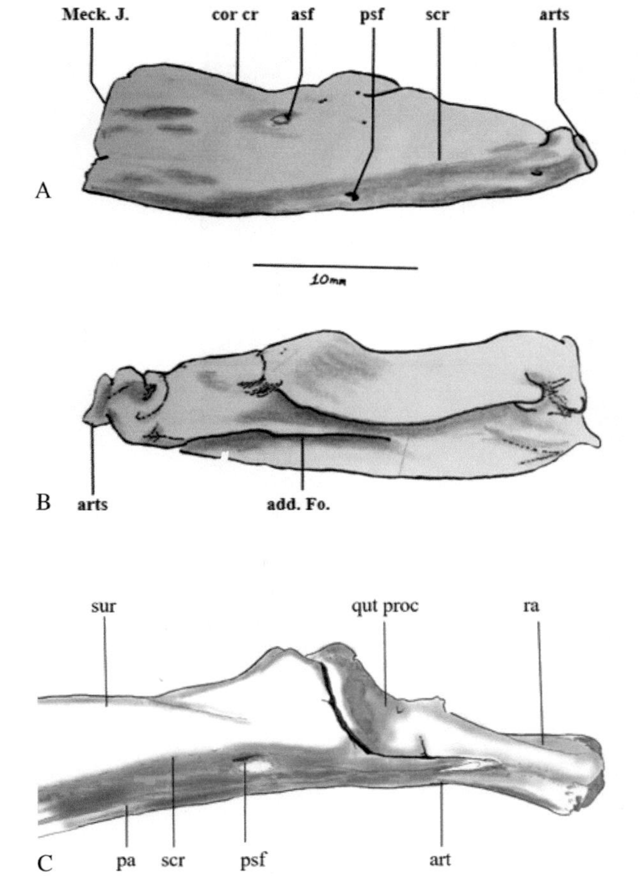

FIGURE 5.30 Left mandibular elements of *Varanus dumerilii* (RGS 203). Surangular in (A) lateral and (B) medial views. (C). Posterior mandible bones that form the hinge articulation with the cranium via the quadrate. The quadrate process is smooth and transversely saddle-like in shape. The retroarticular process and articular fuse posteriorly, eventually the suture becoming obscured as the lizard ages. Abbreviations in Appendix I.

5.4.5 Splenial

The splenial is a long, extremely thin mandibular bone that lies mostly concealed between the coronoid, dentary, and surangular in the Meckelian groove, and is primarily visible medially through the infracoronoid fenestra (Figs. 5.28 and 5.29). It is the only jawbone that traverses the intramandibular joint. The bone is axe-shaped; Mertens (1942b) describes it as "wedge shaped," "leaf-like," and "sometimes diamond-shaped." The ventral portion is an elongated, sharply pointed process that extends anteriorly, beginning a short distance posterior to the symphysis, and going posterior to the lower end of the dentary. It terminates in an acute, slightly upturned angular process that extends a short distance posterior to the vertical dorsal axis. The upper surface of the splenial is a very thin sheet of bone. It rises sharply from just posterior to the last dentary tooth and slightly below the dentary crest. This dorsal process is an irregular rectangle, its most posterior surface surrounded by the coronoid. There is one or two distinct posterodorsal coronoid processes that fit a notch beneath the coronoid's anteromedial process, and an inferior upward hooked surangular process that fits an anterior fenestra in the lower surangular. The rounded dorsal surface of the mid splenial and the ventral surface of the posterior dental shelf of the dentary combine to form the anterior inferior alveolar (nerve) foramen. There is an anteroposteriorly narrow elliptical anterior mylohyoid foramen at the anterior base of the axe-like process, above which there is a dorsally inclined longitudinal shelf with an expanded process posterior to the foramen. Most monitors possess a distinct ventral splenial sulcus. In *V. dumerilii,* there is a pronounced, slightly inwardly curved lower shelf, while in other species, such as *Varanus mertensi* and *V. prasinus*, the splenial shelf is a very minor ridge. Perforating the shelf and facing laterally or slightly dorsolaterally is the elongated oval mylohyoid foramen.

5.4.6 Surangular

The surangular is the only bone that contacts all of the other mandibular bones except the articular (Figs. 5.27 and 5.29). Extending from the dentary joint on the lateral surface of the mandible rearward to its dorsal crest, the surface of the surangular forms the socket portion of the articulation with the cranium's quadrate, thereby forming the second most mobile and flexible joint of the skull. Combined, the dentary plus surangular extend most of the length of the mandible, and the relatively short retroarticular process makes up the rest. In the lateral aspect, the surangular is unremarkable: slightly convex with a smooth vertical surface, nearly rectangular shape, a few small foramina, and a tall, nearly vertical anterior edge that forms most of the posterior border of the intramandibular joint. Except for the small anterodorsal surface where the coronoid sits, the bone is mostly visible laterally in articulated jaws. There are five conspicuous surangular foramina: a small anterior-facing foramen under an overhanging anterior shelf of the coronoid; the anterior lateral surangular foramen below the coronoid crest on the coronoid-surangular suture (anterior surangular foramen); and a posterior lateral surangular foramen that pierces the bone and opens into an anteromedially facing internal lateral surangular foramen (posterior surangular foramen). There is a concealed foramen where a medial thin flange of the surangular lies under the descending posterior coronoid process that extends deep into the surangular. Additionally, there is a small dorsal foramen just anterior to the quadratoarticular joint.

Unlike the lateral surface, the medial surangular is much less visible and much more complex in structure (Fig. 5.28A). The coronoid lappet angles steeply, with the uppermost process pointing caudad. It forms a lateral ridge that is the thickest part of the bone, and at its base curves anteriorly and horizontally to produce the deep splenial sulcus.

The basic relative positions of major cranial organs and bones are summed up in Fig. 5.31.

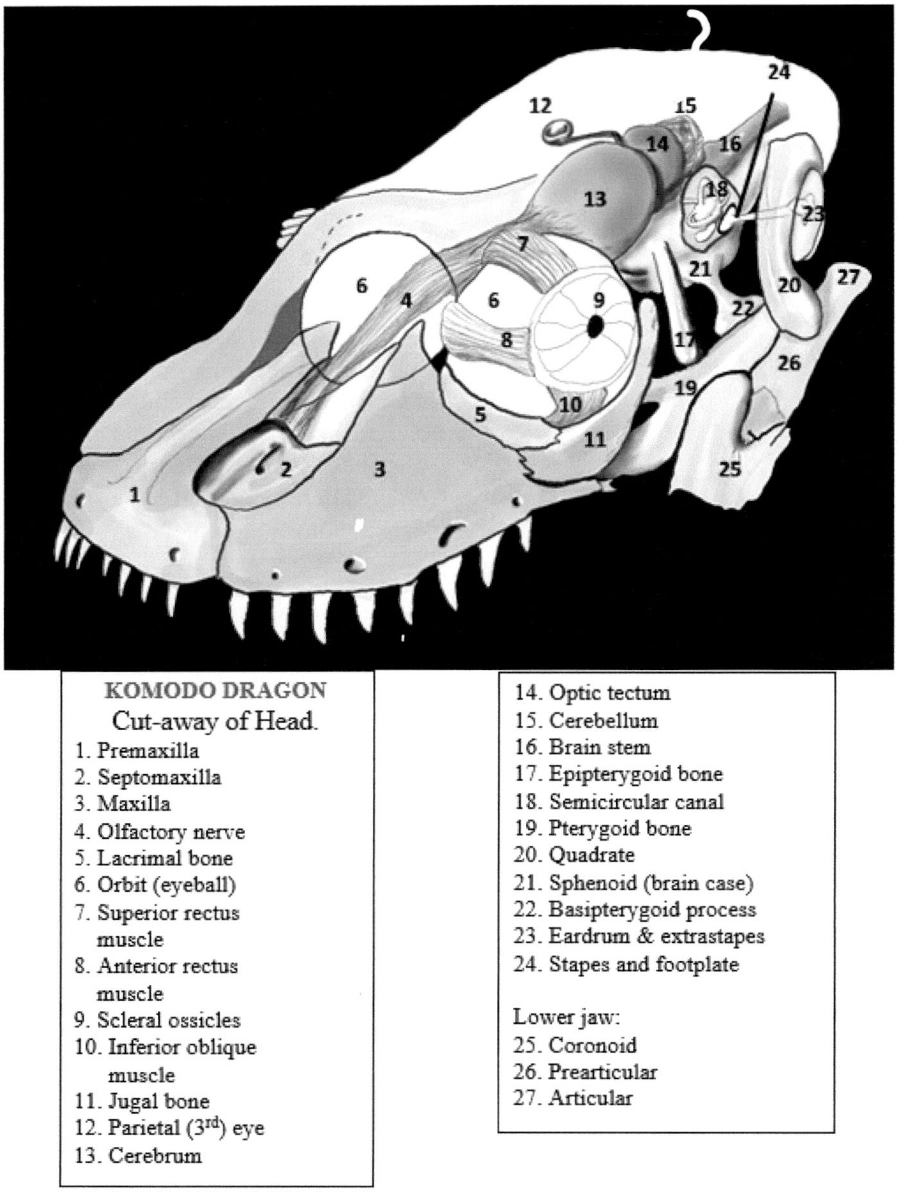

FIGURE 5.31 A transparent Komodo Dragon skull showing the relative positions of major organs and their relationship to the bones. See Chapter 6 for soft anatomy.

5.5 CRANIAL FORAMINA

Bahl (1937) produced an excellent study of the skull of *V. bengalensis* ("monitor") as an outgrowth of his initial task "to identify the large number of foramina in the skull." As documented above, varanid skulls may vary considerably among species, variation that includes foramina. Of the 30 described foramina, only a few can be said to be conservative in nature, by which I mean they are universal in presence, location, and relative size. In some cases, especially among those beneath the crista prootica, foramina may be a shared single lateral opening with a dividing septum slightly inside, or they may be lacking altogether. Much of Bahl's terminology differs from that of later authors, but I have put his terms, where they have been replaced, in parentheses. For the current usage, I follow Evans (2008). My coverage below follows a largely anterior-to-posterior sequence.

Anterior (dorsal) premaxillary foramina – These are located on the anterior surface of the premaxillary body, on either side of and slightly anterior to the ascending dorsal process, and provide access for a branch of the trigeminal nerve (CN $V_{\text{ophthalmic}}$) to innervate the anterior of the snout. The foramina are nearly universal and conspicuous (Fig. 5.4K) but are sometimes absent (some of the *salvator* group), greatly enlarged (*dumerilii*), or consist of two pairs of openings (some *exanthematicus*. The ethmoid (posterior premaxillary) foramina do seem to be universal, lying lateral to the ascending process or vertically penetrating that bone. In many skulls, the foramen lies on the flat surface of the maxillary articulation or, rarely, is in the posterior-facing curve where the ascending process and premaxillary body unite. In each case, though, the trigeminal nerve (CN $V_{ophthalmic}$) exits laterally to innervate the side of the snout just posterior to the snout tip, and the dorsal ascending process.

5.6 OSTEOLOGICAL CONVERGENCES OF THE SKULLS WITHIN THE VARANIDAE

Certain features of skull osteology and diet are intimately linked across taxa, and the relationship is duplicated in geographical areas, e.g., similar convergent morphologies among continental-level groups. This pattern is noticeable when the distribution of varanids is divided into four geographic units based on north – south alignments (Map 5.1). From west to east these are (1) Africa – Middle East, including the circum-Caspian area, (2) south-central Asia, roughly between Iran and Myanmar, (3) southeast Asia, from southern China through Indonesia to Wallace's Line, including the Philippines, and (4) Australia, which can be further divided into (a) the New Guinea province and (b) continental Australia. In each region, there exists the presence of at least two of the five basic maxillary morphologies (Table 5.2 and Map 5.2): type mA (= maxilla type A),

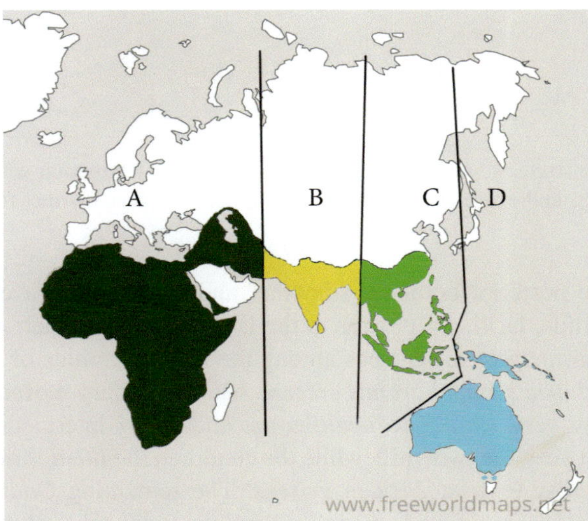

MAP 5.1 World biozones (A–D, with a and b subzones) of *Varanus* are approximate borders between large groups of related species showing similar convergent morphological divergence among continental-level groups. This pattern is distinct when the distribution of varanids is divided into four geographic units based on north – south alignments. Three patterns seem noteworthy because of their close alignment with the biozones. (1) Though some species do cross the borders, they are few in number of species and degree of cross-border dispersal, e.g., *Varanus nebulosus*, *Varanus macromaculatus*, and *Varanus caspius*. (2) Border areas (± 320 km), despite having a mix of zones, tend to be regions of low species diversity, or, as in the case between zones C and D, lower diversity to their within-zone diversity. Africa has no western continental counterpart but has had immigrants from Europe and Asia Minor. (3) The number of species within parts of each zone varies, the trend showing more species in each zone going from west to east.

in which the continuous concave processus praefrontalis that terminates in a sharp, not rounded posterodorsal processes that abut the prefrontal and lacrimal, the elongated L – shaped (mL), a continuous concave processus praefrontalis (mB), and the slightly convex processus praefrontalis (mC). Western African representatives of *V. griseus* have the distinct mA-type maxilla. The more southern *V. exanthematicus* has a similar morphology, though there is a posterior extension of the maxillary-frontal process. In both species, the anterior edge of the posterior process is nearly vertical and square. In *V. griseus* from eastern Africa, the maxilla is type mB, with or without a small posterior process. Both *V. niloticus* (adults) and *V. albigularis* have very similar morphologies to eastern *griseus*, except that in *niloticus*, there is only very rarely the extended maxillary-frontal process, and the posterior edge of the maxilla forms a steep and indented surface. Both western *griseus* and *albigularis* more closely resemble each other in maxillary shape than either does to *exanthematicus*. Though no African varanids possess the mC type, some *albigularis* come close, though the upper concave shape is extremely minor.

In South Central Asia, *V. griseus* maxillae somewhat resemble the type mL, but differ in that the vertical posterior portion of the bone lies at a distinct angle (Table 5.2). The central Asian *flavescens* is more like the west African *griseus* in having an extended posterior orbital process (less distinct in eastern *griseus*), and the vertical structure has a rearward angle. The other mainland species, including *bengalensis* and *flavescens*, both appear derived from the *griseus* model, having slightly sloping posterior processes and with two posterior tines emerging from the maxillary-frontal process and the lower maxillary-lacrimal suture. *V. bengalensis* more closely resembles *griseus* in possessing a long anterior maxillary body, much longer proportionally than in *flavescens*. While the mainland Indian taxa all have mA-type maxillae, the southernmost species, *salvator* (*kabaragoya*), displays a modified mC form. This somewhat resembles the condition seen in *albigularis*, but differs in lacking an extension of the maxillary-frontal process, thus giving the rear border of the bone a more rounded profile. In *salvator*, we have the typical mC, e.g., a long sloping processus praefrontalis that lacks a distinct posterior vertical rise. The posterior orbital process extends far more caudad than the maxillary-frontal process, and is also the morphology seen in *giganteus*, *indicus*, *komodoensis*, and *salvadorii*.

Southeast Asia presents mA and mC morphologies across most of the region, but type mP is represented only by the large Philippine taxa, e.g., *olivaceus*. The maxillary morphologies of *dumerilii* and *rudicollis* are the mL type, similar to each other, with *rudicollis* having a more continuous processus praefrontalis, with a short, depressed anterior premaxillary process, a unique condition (Fig. 5.7A). In both, the horizontal section of the premaxillary process is at least 60 percent of maxilla length. The posterior edge is irregular, with an acute, extended maxillary-frontal process, lacrimal suture, and posterior orbital processes in a nearly straight vertical line. In contrast, the similar Asian *griseus*

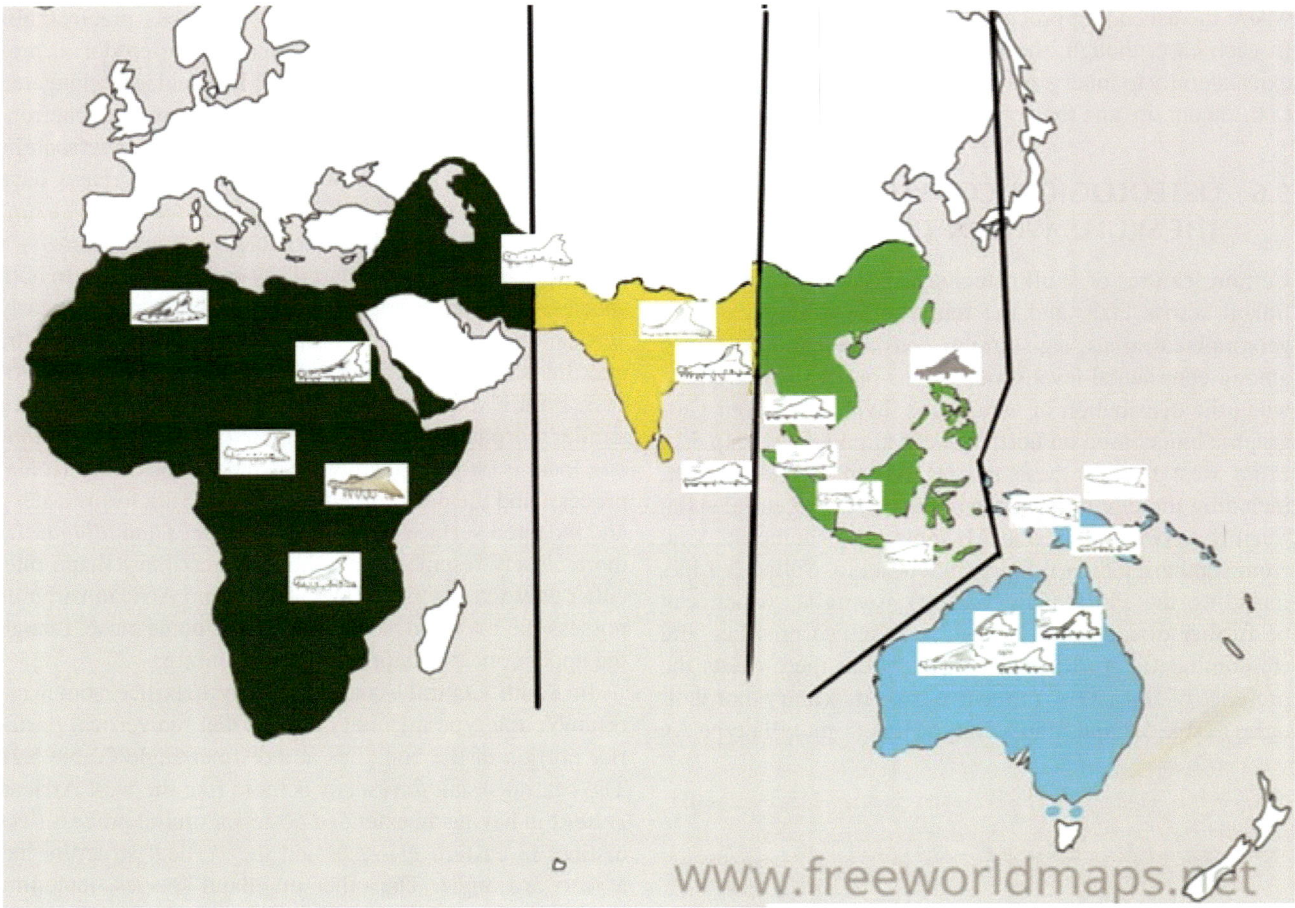

MAP 5.2 Maxillary morphology plotted within their biozones demonstrates seven concepts related to patterns of evolution within and among zones, correlations with absolute lizard size and their diets, and examples of morphological convergences distinct from evolutionary pathways. See text for details.

has a slightly angled posterior edge. Though *dumerilii* and *rudicollis* differ significantly in several other aspects of morphology, their maxillae are remarkably similar.

In *komodoensis* the maxilla has perhaps its simplest morphology, a narrow triangle with straight, subequal upper and lower edges, and a nearly horizontal rear edge. There is no trace of a depressed anterior premaxillary process except in juveniles, but it is present in most *salvator*.

The Australian regions contain all three maxillary types. *V. giganteus* and *varius* most closely resemble *komodoensis*, differing in having a pronounced posterior orbital process that reaches much further caudad than the maxillary-frontal process, and the anterior 25 percent of the snout has a very slight, often barely distinct, depressed anterior premaxillary process. The crista intranasalis is more pronounced than in *komodoensis*. In *gouldii*, the maxilla closely approaches the morphology seen in South Asian *griseus*, being of the mC type condition, though the Australian taxon has a comparatively broader posterior dorsal process, widest at the base, and the posterior edge is nearly straight, with a large posterior extension of the posterior orbital process. The semi-aquatic *mertensi* is also in this group but differs in having curved and triangular posterior dorsal processes with the apices pointing posteriorly, and the posterior border is smoother and strongly concave in profile, lacking a process at the lacrimal suture. There are considerable differences among the dwarf varanids of the *Odatria* group. *Varanus eremius* has a maxillary morphology very similar to *acanthurus*, and *storri* intermediate between the two (mL), while the condition in *gilleni* is most similar to west African *griseus*. The remaining *Odatria* species are divided among morphology types mL and mP.

Species-rich New Guinea and associated islands (e.g., Halmahera, Aru, Solomons) contain types mA and mC, though the latter are, at least as juveniles, difficult to distinguish from the mA species. The large *salvadorii* is similar to *komodoensis* in shape, but instead of having straight dorsal and ventral edges, the entire maxilla is arched such that the lower (tooth-bearing) surface is distinctly concave, the dorsal less so. Also like *komodoensis*, the posterior edge is nearly vertical, but the short maxillary-frontal process is shorter (versus subequal) than the posterior orbital process. Most members of the *indicus* group have maxillae very similar to *giganteus*, while *prasinus* is similar to *tristis*, having the mP morphology plus a more elongate and curved posterior orbital process. The similarity is extremely pronounced in *prasinus* and some of the western *indicus* taxa to the point of being nearly identical.

Based on an analysis of the distribution patterns of varanid maxillary morphology, seven concepts can be inferred (Map 5.2 and Table 5.2):

1. The morphotypes are not finely demarcated, but form a near continuum. Therefore, the use of any descriptive terms (e.g., platyprosopic, mL) is merely a convenient way of referring to a part of the spectrum of types.
2. All the largest species (TL > 200 mm) of *Varanus* share similar type mC maxillae, and include *albigularis*, *giganteus*, *komodoensis*, some *indicus*, *salvadorii*, *salvator*, and *varius*. Though all mC species are large (total length > 122 cm), not all large species are mC (e.g., *albigularis, bengalensis, niloticus, rudicollis*).
3. The mA is the only universally distributed morphotype, found in all distribution provinces, and prevalent in xeric, often sandy, habitats.
4. Type mP maxillae are found among all varanids that are largely durophagous, at least later in life. This may be related to the need for broader muscle attachments that allow crushing of very hard prey. Included are *albigularis*, *bengalensis*, *niloticus*, and *olivaceus*.
5. Pattern shows convergence of morpho-ecotypes, and not necessarily relationships, vide Losos (2009, 2017). Specifically, northern taxa tend to have some variant of the mL maxillary morphology, southern taxa the mB variant, and going west – east from Sri Lanka to the Solomons in a nearly straight line are the mC varieties.
6. The pattern seen in (5) here allows recognition of four north-to-south biological zones of closely repeating morphological patterns, e.g., (1) African-Middle Eastern; (2) southern central Asia; (3) largely forested southeastern Asia, and (4) Australian-Oceanian, from Wallace's line east to the Solomon Islands.
7. Some species experience considerable ontogenetic change in maxillary morphology (Andrews et al. 2017), sometimes drastic enough to develop from one morphotype to another. Furthermore, the process tends to elongate the snout forward, away from the orbit, in agreement with the suggestion that "the long rostrum of the varanids is a secondary elongation superimposed upon a blunted muzzle, rather than a heritage from ancestral anguinids" (McDowell and Bogert 1954:52).

In each zone, the northernmost taxon possesses the mL-form maxilla and, fitting this character in with phylogenetic analyses of the genus corroborates *V. griseus* as basal to living *Varanus*, and supports the hypothesis that the earliest *Varanus* was very similar in morphology to *griseus* (Openshaw et al. 2016, Sprackland 1991).

5.7 THE VERTEBRAE

My review of the varanid spine is brief, because this subject was only a minor focus of my morphological work. However, the axial skeletons of varanids have been studied by several anatomists (Cieri 2018, Conrad 2015, Garberoglio et al. 2019, Hoffstetter and Gasc 1969, Roberts 2022), particularly for comparison with putative prehistoric monitors (Fejérváry 1918, 1935, Gilmore 1928, Hooijer 1972). It is composed of numerous strong, constricted, procoelous vertebrae, each centrum ending in a large ventral condyle that articulates with a corresponding anterior condylar fossa (Fig. 5.32). Characteristic of varanoid vertebrae is the distinct pericondylar flange, anterior

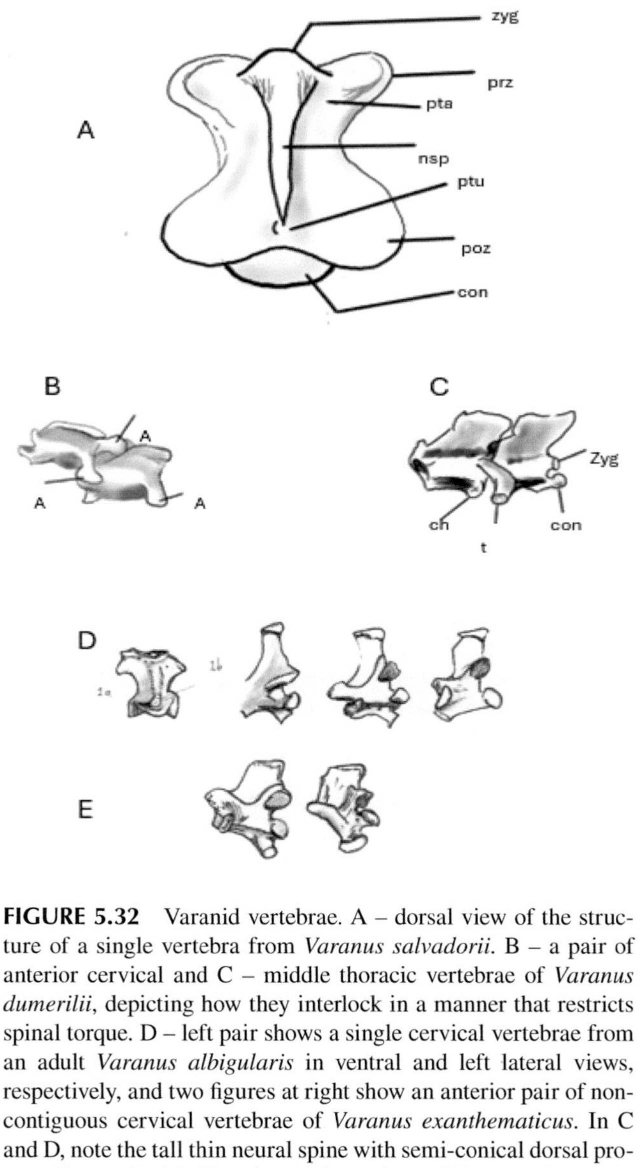

FIGURE 5.32 Varanid vertebrae. A – dorsal view of the structure of a single vertebra from *Varanus salvadorii*. B – a pair of anterior cervical and C – middle thoracic vertebrae of *Varanus dumerilii*, depicting how they interlock in a manner that restricts spinal torque. D – left pair shows a single cervical vertebrae from an adult *Varanus albigularis* in ventral and left lateral views, respectively, and two figures at right show an anterior pair of non-contiguous cervical vertebrae of *Varanus exanthematicus*. In C and D, note the tall thin neural spine with semi-conical dorsal process. E – cervical (left) and caudal vertebrae of *Varanus dumerilii*, with less prominent apical structures. In D, the neural processes are much thinner than in the corresponding processes in *Varanus dumerilii*. B through E roughly to scale. Abbreviations: A – articulation process; ch – hypophysis for caudal chevron; con – condyle; nsp – neural spine; poz – postzygapophysis; prz – prezygapophysis; pta – pars tectiformes arcus vertebra; ptu – posterior tubercle; t – transverse process of centrum; zyg – zygosphene.

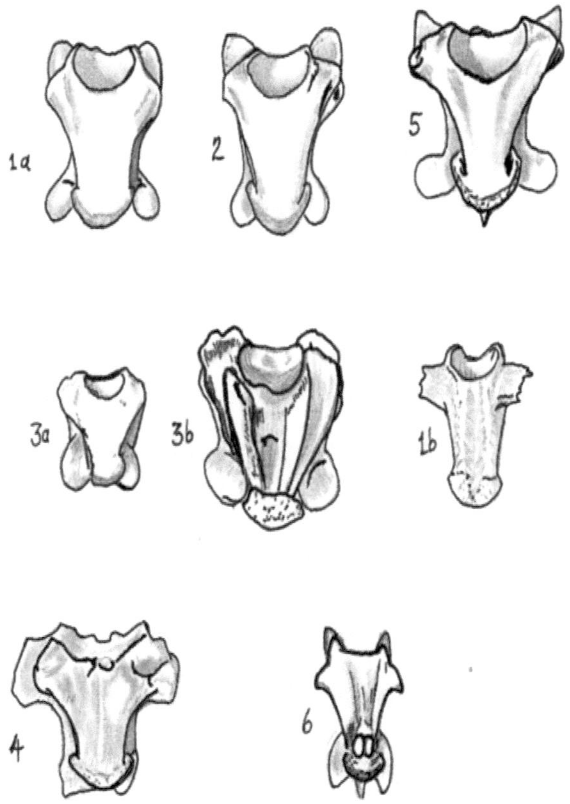

FIGURE 5.33 Comparison of ventral morphology of vertebrae from fossil and extant varanids, showing the shared enlarged posterior condyles and considerably constricted bases. 1 – †*Saniwa ensidens* (USNM 2185, holotype) from a) thoracic and b) anterior caudal regions. 2 – †*Saniwa crassa*. 3 – †*Palaeosaniwa canadensis* (USNM 10864, holotype) showing a) thoracic and 2) lumbar examples. 4 – †*Varanus sivalensis*, with broadly expanded anterolateral processes similar to 5 – extant *Varanus niloticus*. 6 – As with many other features of its morphology, the thoracic vertebrae of *Varanus salvadorii* differ in the shape of the centra from other varanids. Note that the anterior articulating fossa are large, accommodating the enlarged posterior condyles. From on-site sketches by author.

to which the condyle is narrower in diameter. These features are synapomorphic to Platynota. Also typical of the varanoid post axis centrum is a distinct ventral Y configuration (Fig. 5.33), the lower portion comprised of the convex body and condyle, the upper by the diverging anterior ventral processes. Within the fork of the Y is the partially downward-facing condylar fossa; this structure is always visible in ventral view. The relative depth of indentation of the upper part of the Y shape is useful in distinguishing among genera. In †*Saniwa ensidens* and †*S. crassa*, the Y has a short main body, the V section is shallow, and lateral constriction of the centrum is minor (Fig. 5.33 1a-b, 2). *Lanthanotus borneensis* has centra that are extremely similar to those in *Saniwa* save the postzygapophyses in *Lanthanotus* are considerably larger, and the condylar fossa is not as exposed ventrally. In †*Palaeosaniwa canadensis*, the entire ventral surface more closely resembles a V than a Y (Fig. 5.33 3a-b); the longitudinal body is short, the condylar fossa is shallow, and the only appreciable constriction lies immediately anterior to the condyle. I concur with Balsai (2001) that *Palaeosaniwa*, while not varanid, is varanoid. †*Varanus sivalensis* has a shape that is more slender (Fig. 5.33 4), with a longer distinct central body and more laterally facing anterior processes than *Saniwa*, a morphology very close to contemporary *V. bengalensis, V. nebulosus*, and *V. niloticus*. The highly modified *V. salvadorii* has centra in which the Y is T-shaped (Fig. 5.33 6), having a long straight central body, and anterior processes that jut laterally at 45-degree angles from the midline. In this species, too, the condylar fossa, though visible from beneath, is less so than in any other varanid.

The articular contacts are broad and give considerable strength to keep the spine intact when the lizards, for example, engage in "feeding rolls," an action also seen in crocodylians, where prey is seized and shaken to death as the reptile rapidly spins longitudinally. In several of the largest monitors, there is a shallow dorsal fossa of the prezygapophyis and the larger articulating postzygapophysis that, combined, provide protection from axial torque, making the joint both strong and considerably flexible.

6 Monitor Anatomy III
Internal – the Soft Anatomy

6.1 GENERAL FEATURES OF THE SOFT ANATOMY

The soft internal organs, or viscera, of monitors are enclosed within a single-cell-thick continuous layer of transparent tissue, mesothelium, itself an extension of the body wall lining of similarly thin tissue, the peritoneum. Internal organs are suspended from the outer layer by sheets of mesentery tissue that also provide cushioning for the organs. Though there is no direct equivalent to the mammalian diaphragm, the varanid lungs are enveloped within a pulmonary sac of post-pulmonary mesentery. A similar arrangement is seen around the liver. The organs are attached to each other by sheets of clear peritoneal tissue. This subdivision into peritoneal envelopes is known only from varanids and tegus, taxa that are very distantly related. Also, the two groups contain the largest of the carnivorous lizards and have numerous life history convergences that include habitats, thermal tolerances, and ecologies. Investigation of links between the anatomy and the size of these animals would be useful in better understanding convergence.

Obese monitors will have masses of reticulate matter that partially or wholly surround the ventral and lateral coelom, covering organs, a condition seen more often in captive than field specimens. Viscera have well-defined shapes, and none is so large as to be confused with any but the smallest fat bodies.

6.2 CRANIAL ANATOMY

6.2.1 Soft Palate, Tongue, Buccal Cavity, and Pharynx

The soft palate, tongue, buccal cavity, and pharyngeal region constitute the interior of the mouth (Figs. 6.1 and 6.2). The bony palate and dentaries are covered in Chapter 5. The soft palate has a corrugated exterior, its form closely following the underlying bony palate. The center of the palate varies in size by species and often by age. The most anterior of the openings is the vomeronasal fenestrae or septomaxillary fossa, the opening from the buccal cavity to the cavity ventral to the septomaxilla where the vomeronasal/Jacobson's organ is located (Fig. 6.3). The fenestrae only receive air samples, and there is no other channel to the exterior. Air samples collected by the tines of the tongue are analyzed at the molecular level. Just posterior to the much larger nostril openings are the choanae (see below).

6.2.2 The Tongue

The tongue is a muscular structure that serves as a primary environmental sensor for monitors of all sizes. Unlike many lizards (but not snakes) the varanid's tongue is smooth dorsally. The long, deeply bifid and protrusible varanid tongue is characteristic of the genus *Varanus*, and possibly its fossil family representatives, but differs in length and breadth from the other platynotans, and from the fork-tongued tegus (*Salvator, Aspidoscelis*) in being thinner. In general morphology, varanid tongues consist of three distinct sections (Fig. 6.2). The anterior portion of the tongue is both thin and deeply bifid, the individual tines capable of diverging when extended up to 180°. The central section, the main body, is highly elastic and the primary section that engages in tongue elongation during the flicking of the tongue outside the mouth. The most posterior is the thicker base, an anchor region that attaches the tongue to the hyoid apparatus that makes tongue extrusion possible. Helodermatids have protrusible tongues, but these are much broader than in *Varanus*,

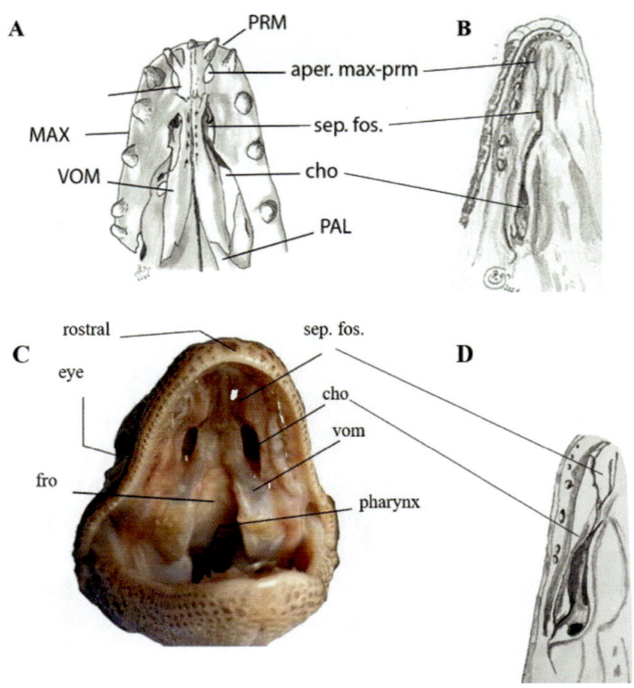

FIGURE 6.1 Palates in *Varanus*. (A) Bony palate of *Varanus indicus* for comparison with soft palate topography. (B) Soft palate of *Varanus nuchalis* (USNM 305982), with narrow fossae, showing correspondence with the bony palate. (C) *Varanus orientalis* showing soft palate with broad fossae, in anterior perspective. (D) Same, diagrammatic for comparison with B.

DOI: 10.1201/9781003463528-8

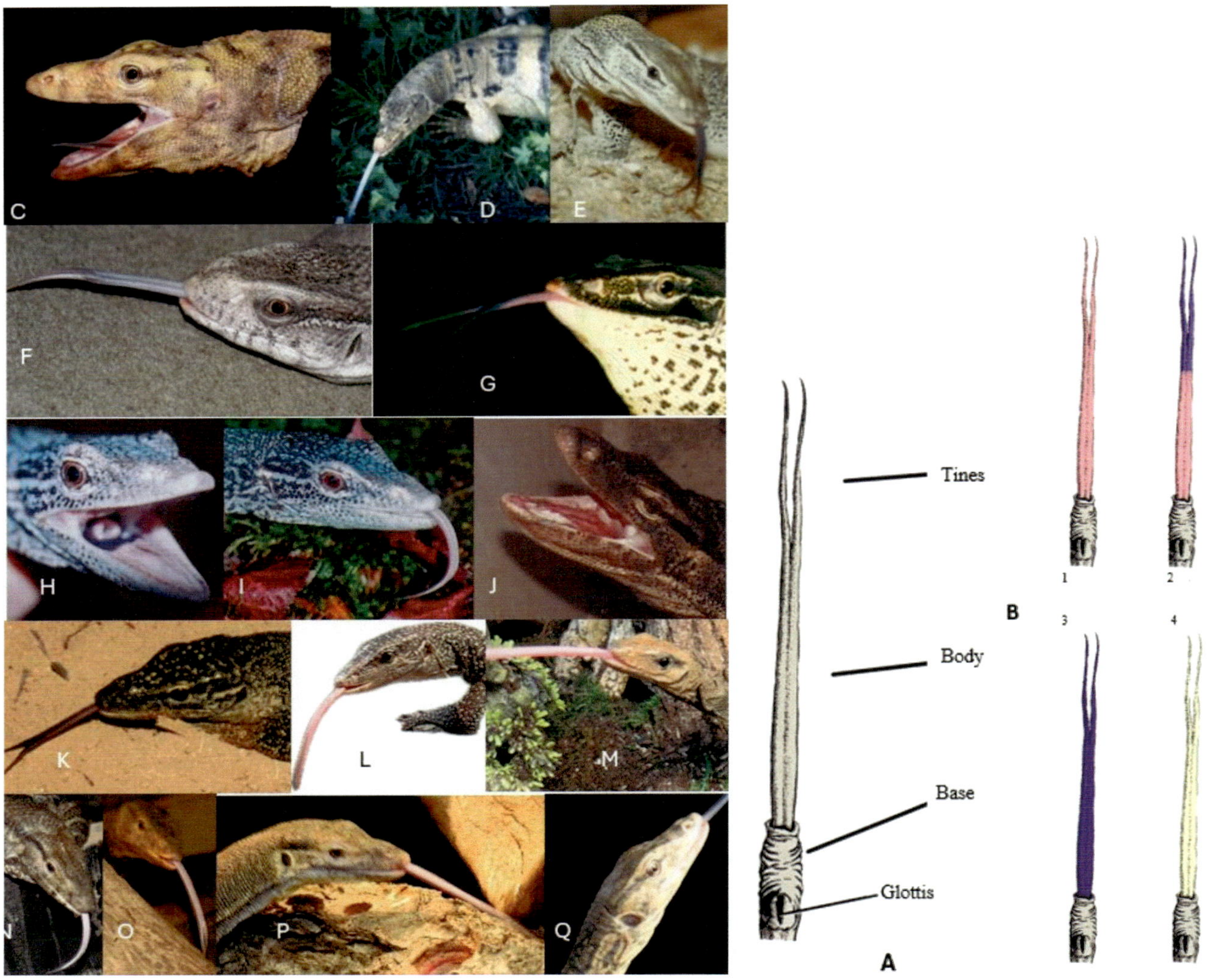

FIGURE 6.2 Tongues of *Varanus* are morphologically similar but vary in color. (A) General morphology of the protrusible tongue. (B) Four of the color morphs were displayed among different monitors. Pink (1) is the most common color, and bicolored (2) is widely distributed. Purple and blue (3) are less common and rarely occur as part of a bicolored tongue, and yellow (4) is found in only a few species. Tongue variety displayed in monitors: (C and D) *Varanus bivittatus*, "sulfur" color morph; (E) *Varanus flavirufus*; (F) *Varanus exanthematicus*; (G) *Varanus panoptes*; (H) and (I) *Varanus macraei*, with the fully opened glottis visible in H; (J)*Varanus macromaculatus*; (K) *Varanus indicus* from western Indonesia; (L) *V. indicus*; (M) *Varanus spinulosus*; (N) *Varanus dumerilii*; (O) *Varanus olivaceus*; (P) *Varanus mertensi*; (Q) *Varanus salvator*. Several photos show the extreme length of fully extended tongues. In G the tongue is bicolored, with pink on both the ventral and posterior (not visible) surfaces, while in K, the entire dorsal surface is dark, and the ventral surface is pink. A completely blue tongue (D, F, Q) is relatively uncommon, while the yellow tongue is restricted to a few species from Indo-Australia (e.g., *Varanus komodoensis*, *Varanus doreanus*, some *Varanus varius*).

and the tongue of the former is unlikely to be confused with that of the latter. In *Lanthanotus* the tongue may extend about the length of the skull and is intermediate in width between *Varanus* and *Heloderma*. In *Varanus* the tongue fits into a narrow epiglottal sheath (Fig. 6.2), the anchor, such that when the tongue is withdrawn, all but the long tines may be encased therein. Unlike the situation in snakes, varanids do not extend the glottis out of the mouth while feeding on large objects, but instead may engage in deep exhalations that dilate the glottis and, sometimes, may assist in moving food back into the throat. Both *Lanthanotus* and *Varanus* that forage underwater are fully able to use their tongue to detect and track prey such as fish, shrimp, and worms.

No taste buds have been found in the tongues of *Varanus indicus* (Schwenk 1984, 1985), *Varanus bengalensis* (Pathak 2015), *Varanus niloticus* (Al-Zahaby et al. 2018), and 12 other species (Young 1997). The vomeronasal organ seems to combine both scent and taste functions. Monitors will avoid certain potential foods after a simple tongue-flick investigation, but they may also take an item in the mouth and reject it after a few seconds. I have repeatedly and successfully fed bits of banana to specimens of *Varanus prasinus* and *Varanus beccarii*, but the fruit was taken and quickly spit out by *Varanus spinulosus*. It seems that varanids have lost the ability to taste with the tongue and replaced it with greater sensitivity and use of the vomeronasal organ, which

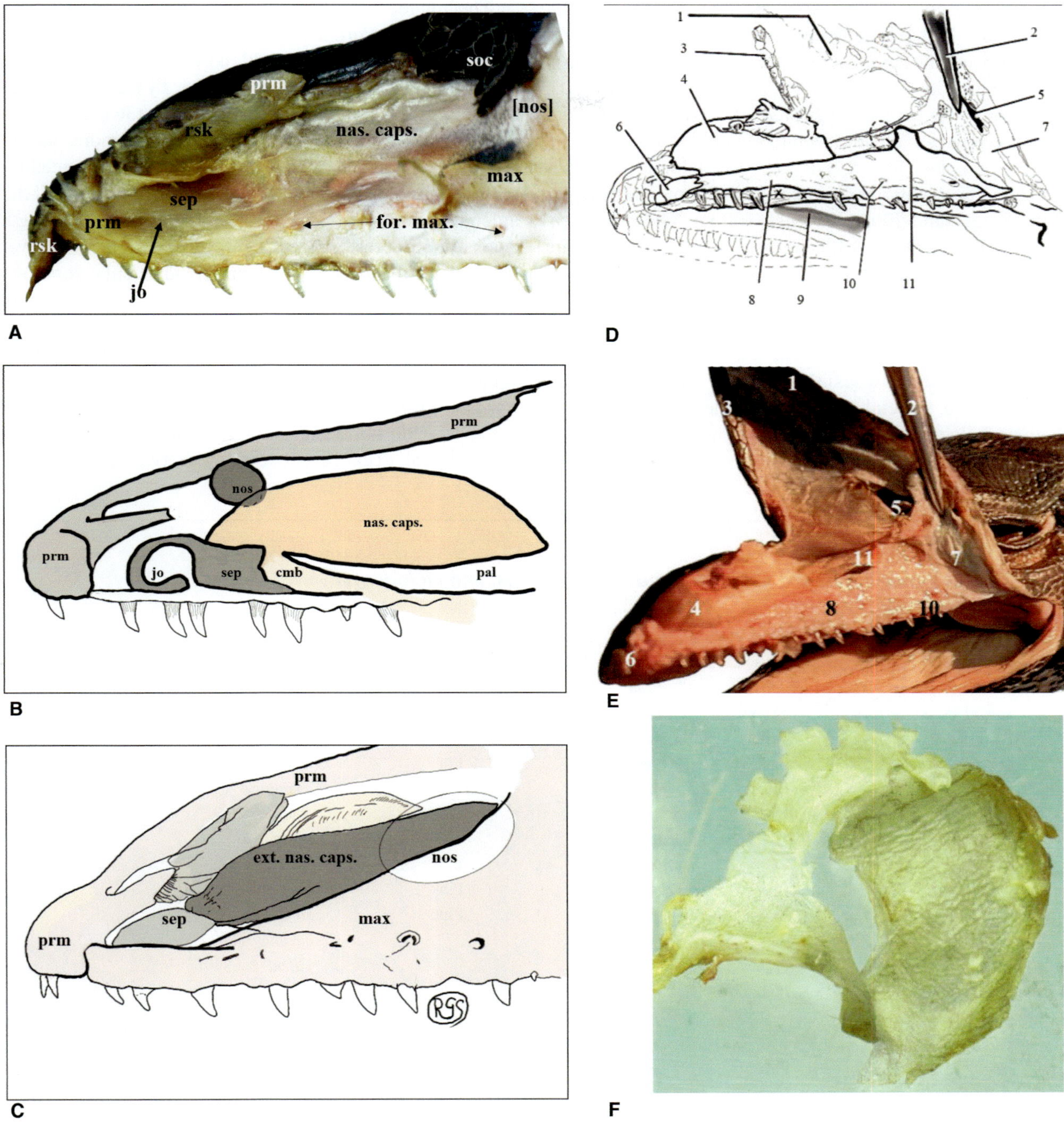

FIGURE 6.3 Internal arrangement of the snout showing location of the nasal capsule relative to the septomaxilla and the underlying Jacobson's organ. (A and C, D and E) *Varanus rudicollis*, adult, from the same specimen, with lateral skin pulled dorsally (rsk) to show underlying features. (B) *Varanus prasinus*, adult, diagrammatic representation of medial section with exterior structures (bone and skin) removed for clarity. Abbreviations: cmb – nasal choana; ext. nas. caps. – dorsal membrane covering nasal sac; for. max. – maxillary foramina; nas. caps. – nasal chamber with membrane cover removed; nos, [nos] – nostril position; pal – palatine bone; prm – premaxilla; rsk – reflected skin; sep – septomaxilla; soc – supraoccipital scales. The thick arrow in a shows the position under which lies Jacobson's organ. (D and E) Nasal chamber and associated structures in *V. rudicollis*, left ventrolateral view. (1) Medial surface of turned-back maxillary skin. (2) Probe. (3) Lateral surface of maxillary scales. (4) Enveloping nasal capsule of the nasal chamber. (5) Medial view of nostril; Note the length, which is comparatively longer than in any other lizard. (6) Lateral face and maxillary process of the premaxilla. (7) Sheet of fascia that bonds skin to maxilla. (8) Anterior of maxilla; The x indicates missing tooth. This species has both the smallest teeth and the greatest number of teeth in *Varanus*. (9) Dorsal surface of tongue (retracted). (10) Maxillary nutrient foramina. (11) Location of nostril. (F) Medial view of left nasal chamber (4) after removal, anterior at left, and the large dorsal dome at right. The tissue is reticulated with small circulatory vessels and nerve fibers.

occupies considerable space in the maxilla-septomaxilla area of the skull. Three cranial nerves have motor control of the tongue: the sensorimotor glossopharyngeal (CN IX), motor vagus (CN X), and motor glossopharyngeal (CN XII). Extrusion of the tongue is achieved by flexing its hyolingual muscles (Herrel et al. 2001, Smith 1986).

Tongue color (Fig. 6.2) is a key taxonomic character among some cryptic varanid species (Böhme and Ziegler 1993, Sprackland 1995a, b, 1999, 2000a). The diagnostic value of this character will be most appreciated by workers with fresh or living material. The species-specific nature of tongue color will thus be of great benefit to workers trying to establish breeding groups, study the ecology of sympatric cryptic species, or make positive identifications for legal reasons. The tongue loses its diagnostic value when dealing with dried, skeletal, or fossil specimens. In these cases, the morphology of the septomaxilla may be more useful in making an identification.

The tongue in platynotans (certainly in the polytypic *Heloderma* and *Varanus*; *Lanthanotus* is monotypic) varies in color among species (Sprackland 1995, 1999b, c; Fig. 6.2; Table 6.1). The range of colors includes, by frequency for 56 species in Table 6.1: pink (52 percent), blue (36 percent), yellow (9 percent), and purple (4 percent). Furthermore, hue and intensity within each color vary, e.g., purple can be medium to nearly black in hue. Sprackland (1999b, c) suggested that among sibling species of the *V. indicus* complex, tongue color might serve as a species recognition signal. However, the considerable overlap of seemingly identical colors (possibly differentiated in the ultraviolet wavelengths) among many sympatric monitors, notably in Australia, negates the universality of tongue color as a species identifier. In some cases, between very similar-looking purple-tongued *Varanus primordius* and the pink-tongued *Varanus storri*, the species do not occur naturally and sympatrically. The geographical distance between those two species is not

TABLE 6.1
Tongue Color across the Genus *Varanus, Lanthanotus*, and *Heloderma*

acanthurus		*insulanicus*		*brevicauda*	
giganteus	Pink	*gouldii*	Blue	*flavirufus*	Blue
rosenbergi		*panoptes*	Blue	*rubidus*	
hammersleyensis		*bushi*		*sparnus*	
komodoensis	Yellow	*mertensi*	Blue/purple	*spenceri*	
varius a	Yellow	*bengalensis*		*nebulosus*	
varius b	Pink	*dumerilii*	Pink	*rudicollis*	Pink
varius c	Blue	*flavescens*		*spinulosus*	Pink
indicus	Pink	*jobiensis*	Pink	*caerulivirens*	
cerambonensis		*doreanus*	Yellow	*juxtindicus*	
finschi		*melinus*	Pink	*yuwonoi*	Blue
prasinus	Pink	*beccarii*	Pink	*bogerti*	
macraei	Pink	*keithhornei*	Pink	*kordensis*	
telenesetes	Pink	*baritji*		*citrinus*	
primordius	Purple	*caudolineatus*		*gilleni*	
storri	Pink	*ochreatus*		*glauerti*	
glebopalma		*kingorum*		*mitchelli*	Pink
pilbarensis		*semiremex*	Pink; blue tip	*timorensis*	Pink
tristis	Pink	*orientalis*	Pink	*scalaris*	Pink
auffenberg	Pink	*similis*	Pink	*pengilleyi*	
mabitang		*olivaceus*	Pink	*salvadorii*	Yellow
exanthematicus	Pink	*albigularis*	Blue	*angolensis*	
microstictus		*yemenensis*		*nesterovi*	Pink
griseus	Pink	*caspius*		*koniecznyi*	
niloticus	Blue	*ocellatus*	Blue	*ornatus*	
salvator	Blue/pink	*andamanensis*		*bivittatus*	Blue
cumingi	Blue	*komaini*	Blue	*marmoratus*	Blue
nuchalis	Blue	*togianus*	Blue	*bangonorum*	Blue
bennetti		*bitatawa*	Blue	*boehmei*	
chlorostigma	Pink	*dalubhasa*		*douarrha*	
eremius		*louisiadensis*		*obor*	
palawanensis	Blue	*rainerguentheri*	Pink	*rasmusseni*	
reisingeri	Pink	*samarensis*		*semotus*	
tanimbar		*tsukamotoi*		*zugorum*	
macromaculatus	Blue				
Lanthanotus borneensis	Pink	*Heloderma horridum*	Pink	*Heloderma suspectum*	Purple

considerable, so it is possible that the tongue colors did serve as differentiators at an earlier time in their history. An interesting case of tongue color variation occurs in the eastern Australian species, or species group, of the lace monitors, in which purple, yellow, and blue tongue colors are represented in distinct geographical areas, which may indicate yet to be described taxa. So, too, are there different tongue colors in different populations of *Varanus salvadorii* and among members of the large water monitor complex.

Monitor lizards very rarely use the tongue to help manipulate food in the mouth, but on occasion, they may use a strong retraction of the tongue to help press food into the gullet. This behavior is more likely to take place when the lizard has engulfed a prey item at or near the limit of their capacity to swallow (Montuelle et al. 2011, 2012, Reippel 1978, Schaerlaeken et al. 2012, Smith 2011), and may represent a passive mechanical occurrence.

6.2.3 Jacobson's Organ and Nostrils

The anterior third of a varanid snout (Fig. 6.3) contains the structures that (1) allow tongue protrusion, (2) analyze molecules obtained by the tongue, (3) permit inspiration and expiration of air to the lungs, (4) seize food, and (5) initiate the swallowing process. It is (2) and (3) that take a proportionately larger role in monitors than in most other squamates excepting snakes. Though the tongue collects aerial molecules, it is the vomeronasal (Jacobson's) organ that does the actual analysis to determine, for example, if an object is food, conspecific, or potential predator (Cooper 1989). There has been suspicion that this process is incomplete, based on a study of snakes and using isotopes to track particles from tongue to mouth, in which no marked molecules were found in the vomeronasal organ (Young 1990), but the long-standing model is still accepted by most zoologists.

Jacobson's organ is a partially coiled structure that lies in a cavity beneath the external nares and connects with the anterior palate via paired septomaxillary fossae (Figs. 6.1 and 6.3). The fossae are separated and thus provide discrete apertures and organs for each tine of the tongue. The intensity of sensory input by the tines can help the monitor in determining the direction of an odor source. Cooper (1989) performed the classic experiment demonstrating odor-based food selection in *Varanus exanthematicus* and *Heloderma suspectum*, and Sprackland (1989d) demonstrated that *Varanus jobiensis* and *V. exanthematicus* discriminate potential prey items by taste.

The size, shape, and position of nostrils in monitors are three of the most variable and noted characteristics of the genus (Fig. 6.4). In hatchlings of most, if not all, species, the nostril is round, small, and positioned roughly equidistant between the snout tip and anterior edge of the orbit (Fig. 6.4).

FIGURE 6.4 Representative variation in external nasal openings among *Varanus*. The photos at the top of young (left) and adult *Varanus rudicollis*, showing the difference in nostril shape and length during growth. The upper row of illustrations shows lizards with nostrils near the orbits. (A) *griseus* (Morocco). (B) *rudicollis*. (C) *albigularis*. The middle row shows lizards with centrally located nostrils. (D) *dumerilii*. (E) *exanthematicus*, young. (F) Same, Adult. The lower row shows lizards with anterior or anterior-facing (G) nostrils. (G) *glauerti*. (H) *timorensis*. (I) *macromaculatus* (Java).

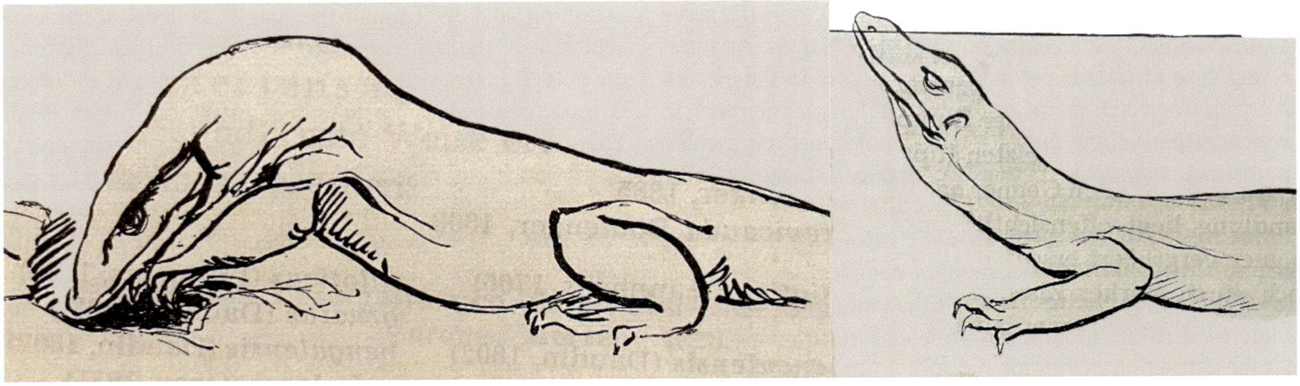

FIGURE 6.5 Rotter (1963) suggested that the nostril positions found in *Varanus* are directly correlated with special foraging habits, such as excavation by *Varanus bengalensis* (left) and snorkeling in *Varanus salvator*, while this explanation may have some validity, it does not hold as a rule across the Varanidae, there being many exceptions (see text). (Illustrations from Rotter [1963].)

As the lizard grows, so does the size of the nasal aperture and, in about half the species, its position on the snout. Rotter (1963) linked nostril shape and position to ecology, such that a posterior nostril is found in burrowing foragers and anterior positions occur in aquatic species (Fig. 6.5). Species that spend considerable time in the water tend to have the most anterior and dorsally situated nostrils (e.g., *cumingi, komodoensis, mertensi, macromaculatus, mitchelli*, and *salvator*), though *Varanus dumerilii* is a notable exception with its centrally situated lateral nostrils. In each of those taxa, the nostril is round or slightly oval. The primarily arboreal *V. prasinus* group possesses round nostrils placed only slightly closer to the rostral than to the eye. The large *Odatria* group also tends to have round, centrally situated nostrils, combined with having the proportionately smallest nasal openings in the genus. At the extreme of nostril morphology are two species with greatly different habits, the dry forest and grassland *Varanus albigularis*, and the wet forest dweller *Varanus rudicollis*, which tends to be the outlier for many varanid traits. In each of these, the nostril is a long slit, much more elongate in adults than in juveniles. In *albigularis*, it lies almost in contact with the orbit (Fig. 6.4A), while in *rudicollis,* it forms a slit nearly half the distance between rostral and eye (Fig. 6.4B and upper), though only the posterior half actually contains the opening to the nasal chamber. In all other features, the two species differ considerably, the cube-headed, stocky, comparatively short-tailed *albigularis* versus the flattened arrow-headed, slender, long-tailed *rudicollis*. With regard to nostril, shape, size, and position, *albigularis* is very similar to *griseus* and *bengalensis*.

Obviously, dorsal nostrils correlate well with aquatic species, but do the other forms also link with aspects of ecology? I believe that for most monitors, the answer is yes. *Odatria* species are predominantly residents of extremely dry and sandy habitats where water is almost completely metabolized from prey and is thus particularly difficult to conserve. Small nostrils reduce the flow of exhaled air and thus help retain body moisture (Bourke and Witmer 2023) while also barring all but the smallest soil particles from entering. The similar configurations of *V. albigularis* and the desert-dwelling *Varanus caspius, Varanus griseus*, and *Varanus nesterovi* are similar in two important features, e.g., the oval nostrils practically about the orbit, and the interior nasal chamber is long and folded in a way that doubles back on itself, again helping retain water content. All of these species excavate their own burrows, often in hard-packed soil, and I suggest that the nostril position keeps it away from the tip of the snout, maintaining uninhibited airflow, and the posterior orientation of the opening helps keep soil from lodging in the nostril. In the ecologically similar *V. bengalensis* and *Varanus nebulosus,* the oval nostril lies distinctly forward of the orbit, but the internal nares are nearly identical in form (Bellairs 1949; pers. obs.).

Nasal flaps occur among some monitors that spend considerable time foraging in the water (Sprackland 1972, 1976, 1989a, 1992, 2009, McDowell, pers. comm.), but not all. In *V. dumerilii* there is an extension of the most posterior nasal scale that folds inwards to a fully medial position while the lizard is not in water and mechanically shuts like a door while submerged (Sprackland 1976, 1992), and so far as I have observed, it is unique. In *Varanus mertensi* there is a thin sheet of skin located on the floor of the internal naris, above the septomaxilla that is passively raised by pharyngeal pressure when submerged, thus closing the nostril, at least in young specimens; the feature seems variable in adults (pers. obs.) which may be evidence of there being two different taxa involved (see Chapter 11). Though *Lanthanotus* has a small cutaneous tissue arrangement like that of *V. mertensi*, it does not seem to be functional for closing the nostril. Observation of living specimens shows, instead, the lizards seem to have only air pressure forcing a small bubble from the nasal chamber forward to perform a sealing action (pers. obs.).

6.2.4 Eyes

Lizards, including varanids, possess two distinctly different types of optic organs. Nearly all vertebrates have family-level taxa that possess cranial organs involved in detecting a range of radiation wavelengths that are classified as light, including several invisible to humans. Paired eyes are ubiquitous among

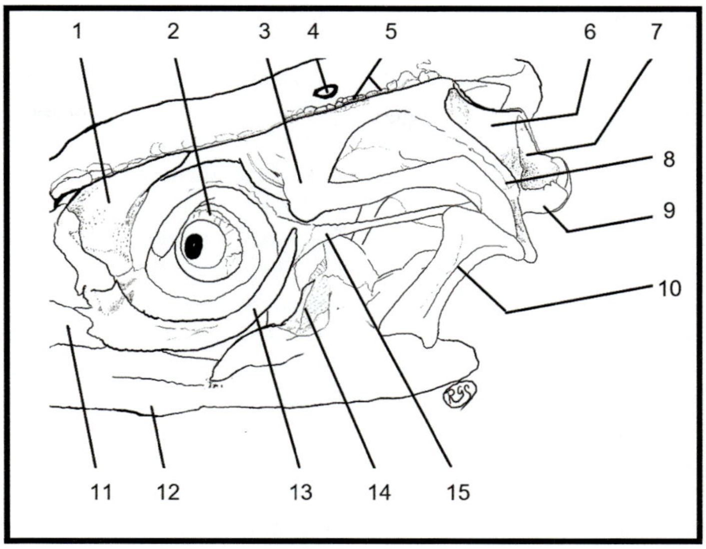

FIGURE 6.6 The orbit and periorbital anatomy of *Varanus rudicollis*. key: (1) Frontal. (2) Scleral ring. (3) Postorbital–postfrontal. (4) Parietal foramen. (5) Cranial scales. (6) Dorsal portion of supraoccipital. (7) Foramen magnum. (8) Parietal process. (9) Left basi-occipital. (10) Quadrate. (11) Posterior edge of maxilla. (12) Dentary. (13) Jugal. (14) Anterior of coronoid process. (15) Optic nerve.

reptiles, though many burrowing genera may have the organs greatly reduced and covered by dermis, e.g., the Dibamidae, many Scincidae, and Typhlopidae. The paired eyes are used to collect light and transfer the stimuli into afferent neuronal messages to be interpreted by the brain (q.v.), where the largest section is typically the optic tectum.

Additionally, there is a much smaller organ situated mid-parietal that contains several features of the "seeing" eyes, but is incapable of transmitting images to the brain (Figs. 6.6 and 6.7). This latter has gone by the names pineal (Foá 1991, Ralph et al. 1979a, b) and parietal (Spencer 1885, Tracholt 1997) organ/eye. Anatomical and physiological

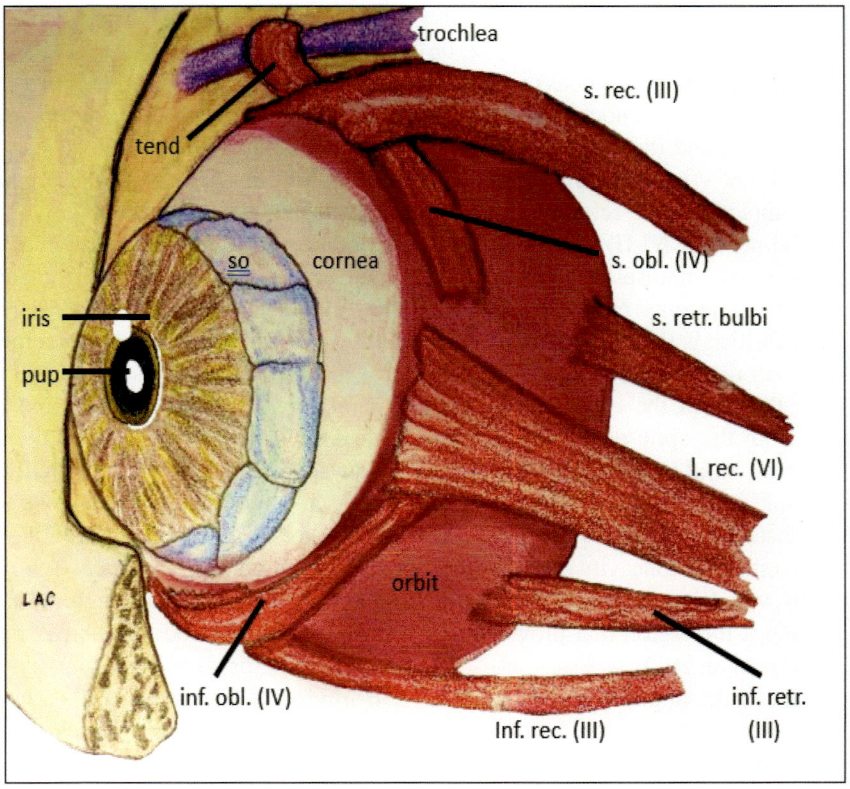

FIGURE 6.7 The left eye of *Varanus prasinus* showing structures and associated muscles and their cranial nerve innervations. Abbreviations: inf. obl. – inferior oblique, inf. rec. – inferior rectus, lac -lacrimal bone, l. rec. -lateral rectus, pup -pupil, sso -scleral ossicle, s. obl. -superior oblique, s. rec. – superior rectus, s. retr. Bulbi – superior retractor bubi, tend tendon of trochlea.

research has shown that the pineal gland is a portion of the brain that produces and regulates several essential hormones (Lillywhite 2008). The parietal eye has embryonic origins that largely parallel the larger eyes and develops a cornea, lens, rudimentary retina, and a thin optic nerve type extension that extends to the optic tectum. The structures are simple and not functional for sight. Because of these differences, the preferred term for this organ is parietal eye and is used throughout this monograph as such. Further details are given below.

6.2.4.1 Orbits

Monitor eyes (Figs. 6.6 and 6.7) are spherical and protected by an external scleral ring composed of 15 thin, slightly overlapping cartilaginous plates that closely form the central (anterior) perimeter of the retina (Underwood 1970). Like other squamates, monitors have a single layer of retinal cells. The rear of the eyeball has a prominent conus that originates proximal to the fovea and terminates near the innermost part of the lens. Like all other examined diurnal lizard species, the retinas of varanids contain only color-detecting cones, which would seem to make the lizards acutely night blind. Most diurnal lizards have color vision enhanced by the presence of four opsin types in their cones (Romero and de Souza 2025): long-wavelength-sensitive (LWS), short-wavelength-sensitive type 1 (SWS1), SWS type 2 (SWS2), and rhodopsin-like type 2 (RH2), sensitive to red, ultraviolet, blue, and green light wavelengths, respectively. Varanids have only three types (Romero and de Souza 2025). The absence of light intensity detecting rods seems to have no detrimental effect on following motion or hunting in the low light conditions of twilight, as evidenced by monitors that have been observed in nocturnal activity. Their visual acuity is excellent, and the eyes are probably as important as olfaction in detecting prey, predators, and mates; they are certainly employed to detect motion.

The optic nerve (cranial nerve [CN] II) transmits light-activated neuronal receptor cell impulses (the cones) to a cross-over structure, the optic miasma, ventral to the brain, thence to the optic tectum, which is the largest structure of their brains. The anterior terminus is the optic disc, a tight cluster of ganglion cells that form the central collection point for retinal signals from the optic nerve. The region lacks cones and forms a small blind spot. It is also the entry and exit point for arteries and veins that serve the interior of the orbit. Near the optic disc, and only a few degrees off a direct line to the lens, lies a shallow concavity on the retinal surface, the fovea centralis, which has the highest density of cones in the retina, making it the region that provides the greatest visual acuity (Rasys et al. 2021).

In Chapter 5, I introduced the angle of the snout as a diagnostic character, that is, the angle formed by two lines running from the orbit to the snout tip. The resulting angle is, to some extent, diagnostic for differentiating species. The resultant angle in degrees runs from 26° to 40°, in which 26° has a narrow snout and 40° has a broad box-like snout. As is so often the case in *Varanus, V. rudicollis* is the outlier with a narrow 26°. This number has significance because the narrower the snout, the more the eyes get a broad overlapping field of view, and that gives *rudicollis* the best depth of field of the monitors. At the other end of the scale with a 40° angle is *exanthematicus*. The cube-like head shape results in *exanthematicus* having limited forward vision (unless it turns its head in bird fashion) and little depth of field, but has excellent lateral views.

Eye movement is achieved by several thin muscles innervated by three sets of cranial nerves. The oculomotor nerve (CN III) moves the orbit laterally via the inferior rectus, inferior retractor, superior retractor, and superior rectus muscles (Fig. 6.6), and also controls visual tracking by the eye, focusing the lens, and responding to light intensity by pupil dilation and contraction. The trochlear nerve (CN IV) controls the superior and inferior oblique muscles that allow the orbit to rotate and look downwards. The abducens nerve (CN VI) controls the lateral rectus muscles, which move the orbit posterolaterally.

Varanid eyes, like their tongues, come in different colors, from copper to black. There is no single published treatise about varanid eyes, but useful information is found in Hoffmann (1890), Underwood (1970), and more general accounts in Lawler (2019) and Montiani-Ferreira et al. (2022).

6.2.4.2 Parietal Eye

There is some confusion in the literature as regard the terms “parietal eye” and “pineal eye,” in many cases, the word “organ” replaces “eye.” The structure was well described and illustrated by Spencer (1885, see also Foá 1991, Ralph et al. 1979a, b), who used the term “pineal,” while Eakin (1973) used the term “third eye,” as do popular and non-scientific publications.

The light-sensitive, eye-like organ located in the upper skull of many vertebrates (Fig. 6.6) is the parietal eye or parietal organ (Eakin 1973, Irwin et al. 1994, Leydig 1890, Spencer 1887). Like the eyes, the parietal eye is an extension of the forebrain that rests dorsally, posterior to the paired eyes (Fig. 6.8) and lies in a foramen within the parietal, differing from most non-anguimorphs that have the foramen between the frontal(s) and parietal bones. A thin pineal nerve runs from the organ thence posteriorly past the epiphysis (with which it has a small neuronal contact), and finally into the diencephalon close to the junction for the optic nerves. Similar organs are found in sharks, some teleost fishes, amphibians, *Sphenodon*, and most lizards (e.g., Iguania, Scincomorpha, Anguimorpha), but are absent in snakes, turtles, crocodilians, birds, mammals (Eakin 1973), and most nocturnal or burrowing lizards (Teiidae, Gekkota, Dibamidae, Chamaeleonidae). Though *Varanus* retains an ancestral parietal eye, its closest relatives, *Heloderma* and *Lanthanotus*, lack this organ and its associated foramen in the parietal bone. Gundy and Wurst (1976) found seven distinct morphological types of pineal complex in the examination of 85 lizard species, varying among arboreal, terrestrial, and fossorial species.

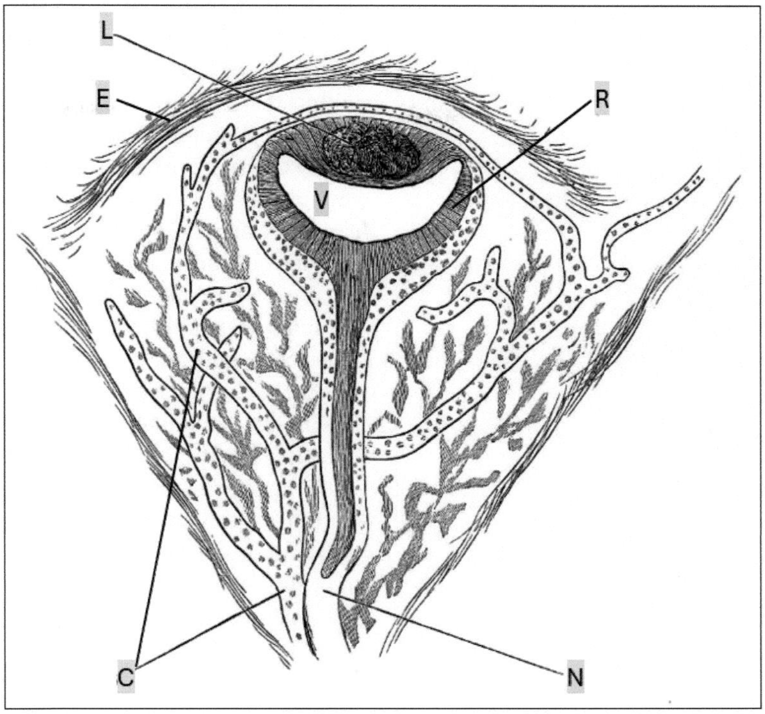

FIGURE 6.8 The parietal eye and its associated structures in *Varanus giganteus*. C – capillaries, E – translucent epidermal layer, L – lens (occluded), N – parapineal nerve, R – retina, V – vitreous. After Spencer (1887).

The organ is sometimes considered a degenerate eye, containing structures common to, but not as complex as, the typical eye. It contains a foveal cavity, a distinct occluded lens, and a direct parietal nerve that is far smaller in diameter than the optic nerves and links to the brain (Eakin 1973, Hoffmann 1890, Leydig 1890, Locy 1894, Solessio and Engbretson 1999, Spencer 1887). Unusual among squamates, the parietal nerve stalk in *Varanus giganteus* is composed of three strands (Spencer 1885) instead of one that meets the lateral and ventral surfaces of the organ to form a single nerve stalk between the epiphysis and pineal gland. The single-layer retina contains cones, particularly rich in green- and blue-sensitive cells. A dorsal corneal layer lies proximal to the epidermal covering scale, and a fluid-filled conjunctival cavity separates the cornea from the lens, the latter spherical and pale white. There is no scleral ring. The parietal eye is an extension of the epithalamus that, in varanids, originates on the anterior surface of the pineal gland and extends dorsally via the thin pineal nerve, analogous to the optic nerves. It then terminates as an eye-like organ that lies in a medial foramen of the parietal bone and is overlain by a usually translucent single scale composed of a thin keratinous epidermal layer. Some structures homologous to those of the orbits are present but modified. The lens is immobile, opaque to varying degrees, and is nearly transparent in neonate monitors but becomes less so with age. Being both opaque and immobile, the lens cannot focus light, but short-wavelength light in the blue to ultraviolet range is transmitted to cones that are sensitive to even small doses of light at those wavelengths (Romero and de Souza 2025). Engbretson and Lent (1976) confirmed the link between the parietal eye and the pineal gland. When light hits the parietal eye, afferent impulses release the neurotransmitter norepinephrine to the pineal gland, enhancing the eye's sensitivity to light. In the absence of light, the parietal stops transmitting norepinephrine and instead receives the neurotransmitter, stopping the eye's secretion of norepinephrine (Arushanian and Ovanosov 1995). The epiphysis is a structure linked to the parietal eye via the thin parietal nerve, and the nerve terminates in the midbrain, including the pineal gland. There is no other physically close association between the parietal eye and the pineal gland.

Several functions have been linked to the parietal eye, including influencing or regulating behavior (Tosini 1997, Tracholt 1997), including aggressive behavior (Phillips and Howes 1988) and circadian rhythms (Berry 2012, Foá et al. 2018, Janik and Menaker 1990, Tracholt 1997), orientation (Beltrami et al. 2010, Foá et al. 2018), ultraviolet radiation dosimeter that regulates hormonal release and inhibition of neurotransmitters, particularly of the diencephalon (Jenison and Nolte 1980, Nishimura 2020), melatonin catabolism (Grace and Beshare 1994), and thermoregulation (Avery 1979, Firth and Turner 1982, Hutchinson 1974, Phillips and Howes 1988, Ralph et al. 1979a, b). Though it likely serves more than one function, the parietal–pineal complex in varanids has been experimentally demonstrated to be an important ultraviolet radiation dosimeter. Physiological functions of the parietal–pineal complex are covered in Chapters 7 and 8. The reported fourth eye in †*Saniwa* is covered in Chapter 11.

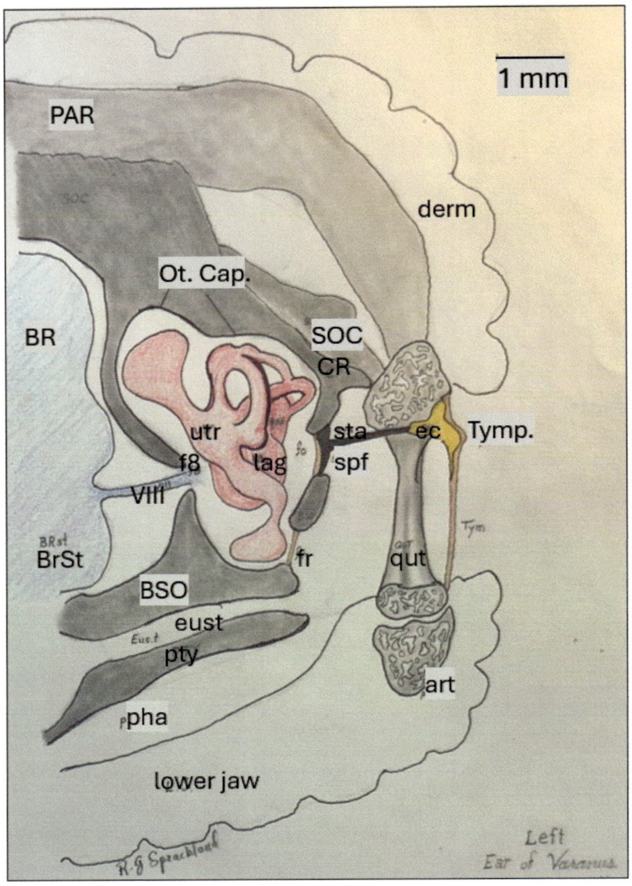

FIGURE 6.9 Configuration of the right ear in *Varanus bivitattus* in posterior view Abbreviations: art – articular bone, BR – brain, BrSt – brainstem, BSO – basioccipital, CR – crista prootica, derm – dermis/epidermis, ec – extracolumella, eust. – eustachian tube, f8 – foramin for CN VIII, fo – foramin ovalis, fr – foramin of recessus scala tympani, lag – laguna of inner ear, Ot. Cap. – otic capsule, PAR – parietal, pha – pharynx, pty – pterygoid, qut – quadrate, SOC – supraoccipital, spf – stapedal footplate, sta – stapes, Tymp. – tympanic membrane, utr – utriculus of inner ear, VIII – auditory nerve CN VIII.

6.2.4.3 Ears

The skeletal properties and internal structure of varanid ears are covered in Chapter 5. Externally, the visible portions are the tympanum (eardrum), extracolumella, and often the posterior edge of the quadrate bone (Figs. 5.24 and 6.9). The tympanum is translucent and functionally comprises two sections. The larger upper portion is flexible, the lower less so (Han and Young 2018). Very small tympanus muscles attach to the edges of the tympanum and the depressor mandibuli and can be flexed, thus causing a tightening, less pliable eardrum, or be at rest, allowing the greatest mobility of the tissue. This flexibility allows monitors to fine-tune the amplitudes of external sounds, a feature that is useful terrestrially and in water (Han and Young 2018, Vedurmundi et al. 2020). Han and Young (2018) suggest that in allowing the tympanum to respond to different frequencies and amplitudes, varanids obtain a way to detect the direction of sound sources. Based on the few experiments looking at varanid hearing, monitor lizards have a hearing range of 40 Hz to 5000 Hz and of –10 dB to 40 dB sound pressure levels (EOL 2025), well within the limits of human hearing and enough to hear a potential predator or prey at least several meters away. The acuity of that hearing is still largely unknown. For example, several people who work with varanids claim that the lizards can recognize their names and come when called. Videos abound on YouTube and other Internet sites, and some of those more reliable sources have shown videos of their varanids responding (e.g., Tom Crutchfield). Experiments must be conducted to see if the lizards are performing a variant of Clever Hans, the horse (Samhita and Gross 2013), or Chaser, the border collie (Pilley and Hinzmann 2014). (These animals were publicized to be highly intelligent, with Hans able to do simple math addition and subtraction. Careful investigation discovered his owner giving a barely perceptible movement when Hans was to stop tapping his hoof to give a correct answer. This alone argues for some intelligence, if not what was claimed, in the horse learning a subtle cue that seemingly he was not specifically trained to understand. Chaser, however, repeatedly showed his excellent memory and ability to recognize and retrieve one of over 100 toys by name.)

6.2.5 The Brain

Lizard brains are not particularly well-studied organs, though the brains of some of the larger varanids are better known than the smaller ones. Varanids, teiids, and iguanians include the largest living lizards (but see Stroud 2019), have the largest brain to body mass ratios among living reptiles (Ebbesson and Northcutt 1976, Platel 1979), and probably the most complex brains of living lizards (ten Donkelaar et al. 1987). Because of the downward lateral processes of the frontal and parietal bones, the varanid brain has more osseous protection than any other lizards (possibly excepting dibamids). In general configuration, there are five distinct regions or segments: anterior olfactory stem, telencephalon, tectum (or optic tectum), cerebellum, and brain stem (Figs. 6.10 and 6.12). In this regard, the brain fits a generalized reptilian form, though it differs in fine morphology (Ebbesson and Northcutt 1976, Platel 1979, ten Donkelaar 1987, ten Donkelaar and Bangma 1987, ten Donkelaar and Huizen 1984, ten Donkelaar et al. 1987; see below). The telencephalon of varanids is comparatively as large as that of basal birds and mammals (Northcutt 2013) but not the largest complete brain size among squamates (De Meester et al. 2019). That particular distinction belongs to a small iguanian, *Anolis stratulus* (Stroud 2019).

The olfactory stem contains a pair of greatly enlarged olfactory nerve bundles (CN I) that innervate the interior of the nasal capsule and vomeronasal organ. Posteriorly, they meet the anterior telencephalon as narrow cylinders (Figs. 6.10–6.12). Anterior to the sub-frontal conduit, they become laterally compressed, taller, and then form small branches within the nasal chamber. The telencephalon, like

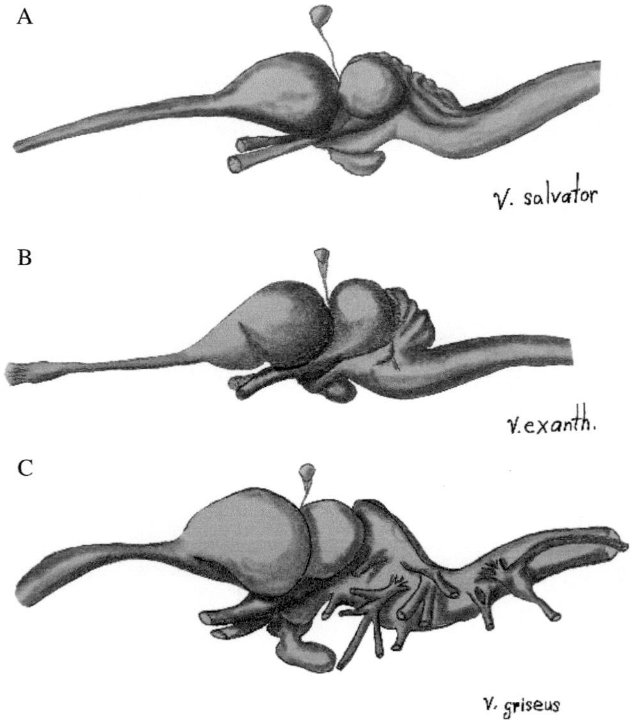

FIGURE 6.10 Brains of different *Varanus*. Left lateral views of (A) *salvator* and (B) *exanthematicus, with cranial nerves removed, and* (C) *griseus*. Though the species are morphologically distinct, their brains are very similar in appearance. See also Fig. 6.11.

the olfactory stem and tectum, is divided into left and right halves. Inferior to the telencephalon is the broad optic chiasma and roots for the optic nerves (CN II). Beneath the chiasma and partially underlying the tectum are the hypothalamus and pituitary structure. The telencephalon partly overlaps the anterior tectum, but the tectum and cerebellum are only in minor contact. Inferior to the mid-tectum and extending caudad beneath the cerebellum is the brain stem, from which all cranial nerves except I and II emerge (Fig. 6.10).

The brain stem is a barrel-shaped nerve bundle subdivided into a forebrain (diencephalon), midbrain (mesencephalon), and hindbrain (rhombencephalon). As in mammals, the diencephalon contains the medial thalamus and hypothalamus. Beneath the hypothalamus is the nerve mass that is the optic chiasma; the optic nerves (CN II) enter anteriorly. A narrow infundibulum links the hypothalamus to the adenohypophysis and neurohypophysis, which together form the pituitary gland. The midbrain lies both medial and inferior to the tectum and contains neuronal tracts that coordinate impulses of the oculomotor (CN III) and abducens (CN VI) nerves (Fig. 6.11). According to ten Donkelaar et al. (1985), neurons of these pathways control the precision involved in coordinated eye-head and paired eye movements.

6.2.5.1 The Pituitary

The pituitary gland is the encompassing term for a small oval area beneath the tectum that in humans is about the size of a pea and weighs about one gram. In varanids it is comparatively about as small, making it difficult to see without microscopy. Nevertheless, varanid pituitaries are sufficiently large to fit into the well-developed bony sella turcica. The organ includes two functionally distinct sections. The neurohypophysis lies dorsal to the adenohypophysis. There are no data for specific hormones of the varanid pituitary, but a study across three snake families was consistent in finding Ile_s-oxytocin and Arg_s-oxytocin (Acher et al. 1969). Furthermore, the oxytocins "showed no variation" between the snakes and amphibians (Acher et al. 1969), making it highly likely that *Varanus* produces the same chemicals. Among the hormones produced or regulated by the pituitary are adrenocorticotropic, growth, luteinizing, prolactin, and thyroid-stimulating hormones, which control the development and condition of the ovaries and testes. The pituitary complex connects dorsally to the hypothalamus via the short medial stalk, the infundibulum. Extensive reviews of reptilian pituitaries can be found in Pearson (1985) and Saint Girons (1970).

6.2.5.2 Cranial Nerves

The majority of the cranial nerves originate in the rhombencephalon. Varanids have 11 pairs of cranial nerves (CN), lacking CN XI (accessory) found in many other amniotes (ten Donkelaar et al. 1984:10) (Figs. 5.15 and 6.11). CN I originates as a thick extension of the anterior telencephalon. CN II emerges from the anteroventral portion of the optic tectum and is also a thick nerve. The nerves converge posterior to the orbits, their tracts bringing afferent signals primarily to the contralateral side of the optic tectum. In mammals, each nerve divides equally, sending signals from each eye bilaterally to each hemisphere of the cerebrum, but in varanids, a smaller branch of the optic nerve routes to the opposite hemisphere of the tectum. These two nerves are unique in not exiting the braincase via bony foramina and not originating from the mesencephalon or rhombencephalon (brainstem). The remaining nine cranial nerves emerge laterally from the proötic, those most posterior (CN X and CN XII) from the exoccipital, and the most anterior CN III (oculomotor) (Fig. 6.11). CN III innervates the eye's medial rectus muscles that control movement of the eye rostrally (looking inwards) and, when so pointed, allows the eyes to look upward. Of extreme importance for predators of swift prey, CN III operates the visual convergence function that allows the eyes for depth perception. CN III also controls the aperture width of the pupil and opening of the eyelid. Additional control of the eye includes CN IV (trochlear), which helps the eye look down and rostrally, and CN VI (abducens), which innervates the lateral rectus muscle and is responsible for coordinating eye movements toward the outside, e.g., anti-rostrally.

The trigeminal (CN V) begins as an enlarged, the largest in some species, proximal root that branches into four much thinner nerves, from anterior to posterior: ophthalmic, maxillary, mandibular, and ramus muscularis. It is a sensory nerve that receives touch stimuli of the face and

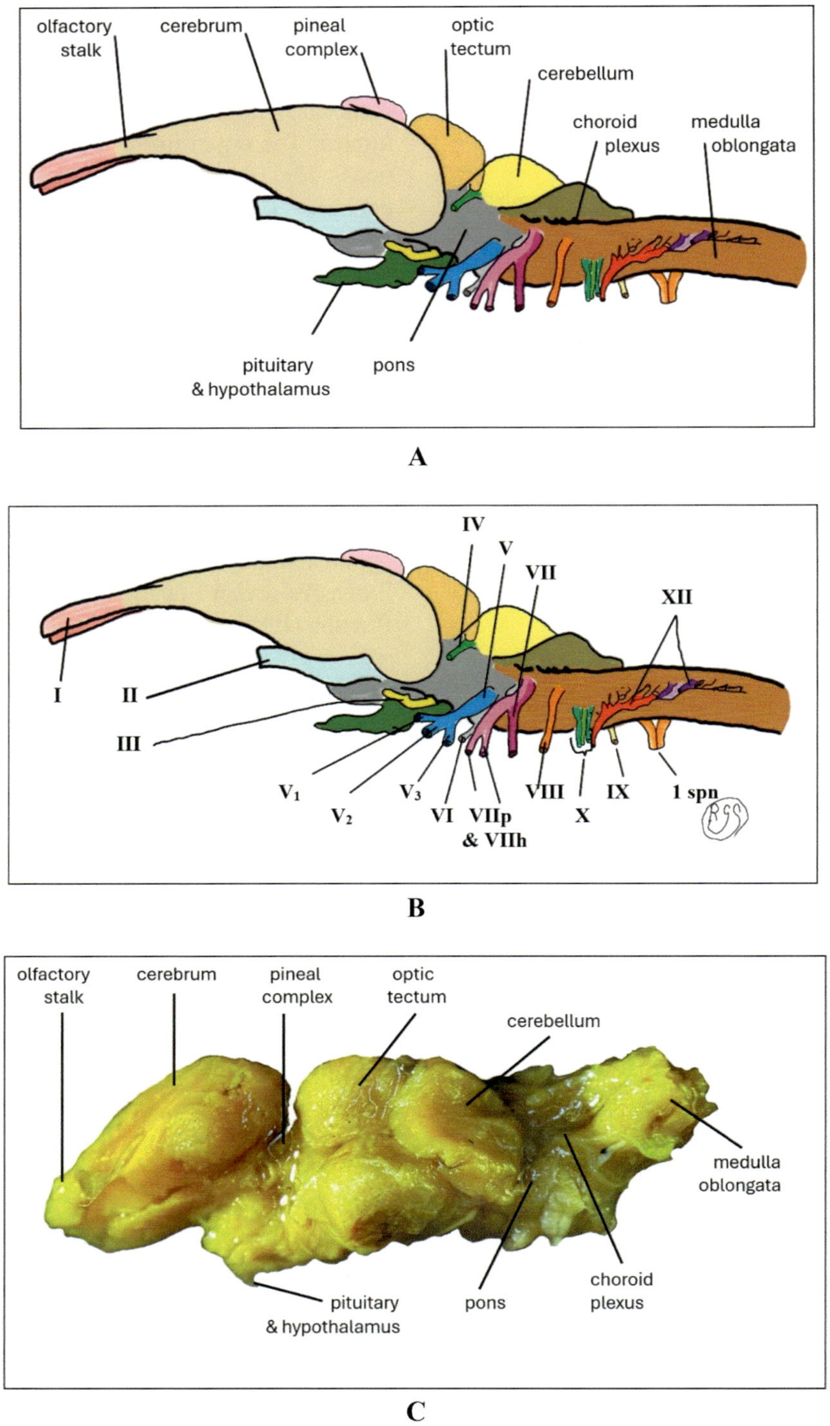

FIGURE 6.11 Left lateral view of brain of (A and B) *Varanus bivitattus* showing (A) major structures, and (B) the origins of the cranial nerves. Note that *Varanus* lacks CN XI. CN VIII does not exit the skull but enters the inner ear. (C) Same view of brain of *V. prasinus*, with left cerebrum excised to show the position of pineal complex.

mouth that include heat, cold, and pain, and a motor function that controls dentary movement. The facial (CN VII) emerges from the brain stem inferiorly and adjacent to the forked vestibulocochlear, and its branches extend to the supraorbital, facial, and mandibular regions. This nerve also has sensory and motor functions. Taste is a significant stimulus to varanids, and the nerve controls the release of saliva. Tear production and flow are also controlled by this nerve. There is also a role for CN VII in controlling fine muscles that attend the tympanum and thereby influence sound perception (see Ear above). A central extension of this ubiquitous nerve controls closing the eyelid. Sensory neurons form a solitary tract nucleus that enters the pons via a parabrachial nucleus. From the pons, signals are

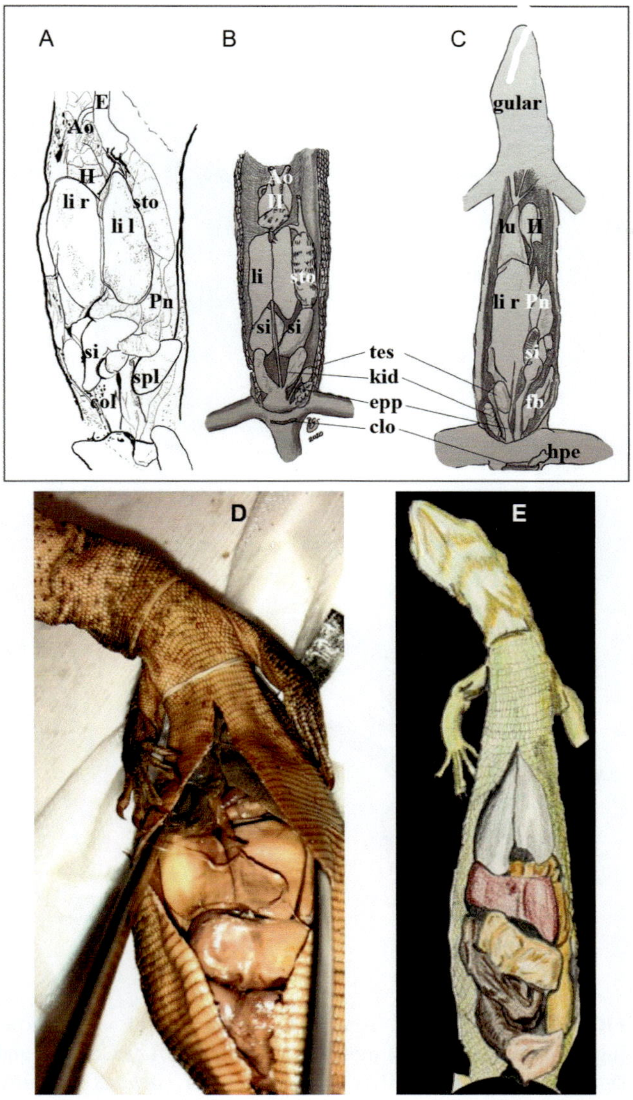

FIGURE 6.12 Variation in the arrangement of the visceral anatomy of *Varanus*. (A) Dissection of *Varanus macromaculatus* male. Modified from Srichairat et al. (2018). Diagrammatic representations of (B) *Varanus prasinus* male, (RGS 42) lungs removed. (C) *Varanus indicus* male, (CAS, uncat.) left lung, left testis and kidney, and stomach removed. (D) Dissection of *timorensis* (USNM 128574, female). (E) Diagram of the same. Abbreviations: Ao – aorta; clo – cloaca; col – large intestine; E – esophagus; epp – epididymis; fb – fat bodies; gular – gular/throat region; H – heart; hpe – hemipenis; kid – kidney; li – liver; li l – liver, left lobe; li r – liver, right lobe; lu – lung (right); Pn – pancreas; si – small intestine; spl – spleen; sto – stomach; tes – testis. In both B and C, a large thin mass of abdominal fat bodies was removed to show the organs.

relayed to taste-identification regions of the forebrain along a medial parabrachial tract.

The vestibulocochlear (formerly acoustic and auditory) CN VIII nerve has an anterior vestibular branch that connects to the ear's semicircular canal system (which recognizes X, Y, and Z coordinates) and relays information to the balance processing portions of the brain. Part of the nerve that connects to the inner ear is unique in being wholly confined within the bony braincase. The posterior cochlear branch is the acoustic nerve that conducts afferent neuronal signals to the hearing center of the brain. Like the oculomotor, the abducens, and vagus nerves come from the ventral part of the brain stem. The glossopharyngeal (CN IX), like the facial, is involved in taste, and its motor function includes tongue movement and pharyngeal muscle actions that include swallowing.

The vagus nerve (CN X) is the only cranial nerve to regulate organs beyond the skull, specifically parasympathetic actions of the heart and digestive systems. It complements the glossopharyngeal for swallowing and additionally regulates breathing and heart rate. The longest branch of the vagus nerve connects with the stomach and regulates part of the digestion (and vomiting) processes. Varanids lack cranial nerve XI, the accessory; at least it is absent in *V. exanthematicus*, *V. salvadorii*, and *Varanus glauerti* and *V. prasinus* (pers. obs.), and in studies that have listed said nerve, it may have been a misidentification of the most posterior root of the vagus (CN X) nerve (ten Donkelaar et al. 1987). In studying the rubrospinal tract among several reptile taxa, ten Donkelaar (1987) found that while lizards, including *Varanus*, have a well-developed tract and associated red nuclei, the structure is absent in pythons, leading them to suppose that all limbless reptiles lack that tract. ten Donkelaar hypothesized that the nerve tract was important in neuronal control of limbs, hence its presence in quadrupedal but not limbless reptiles. This idea is corroborated in part by a study showing that lizards with limbs have larger brains relative to body mass than limbless species, as do non-social species (De Meester et al. 2019, Platel 1979). The final cranial nerve, and one with a thick proximal root, is CN XII (hypoglossal), the efferent nerves controlling tongue movement via manipulation of the hyoid apparatus.

6.3 POST CRANIAL SOFT ANATOMY

Varanid viscera do not differ appreciably in content from those of other tetrapods, having paired lungs, kidneys, and gonads (Figs. 6.12 and 6.13). Organs may be paired and asymmetrical, or asymmetrical, and morphologically variable among taxa. The gastrointestinal tract, too, is asymmetrical and shows considerable diversity among species. The compact and roughly spherical heart is just slightly left of the midsagittal plane and closely surrounded by the lungs and liver. In gravid females, the oviducts and their eggs are undisputedly the largest of internal organs.

6.3.1 The Heart

The structure of squamate hearts has been well reviewed for many taxa, notably *Sphenodon* and several lizard species (Mathur 1944, Piskovská 2019, Rajani et al. 2021), though there are few genus or family-wide reviews (Figs. 6.14–6.16). For most of the history of modern herpetology, it has been known that crocodilians possess a ventricle with a

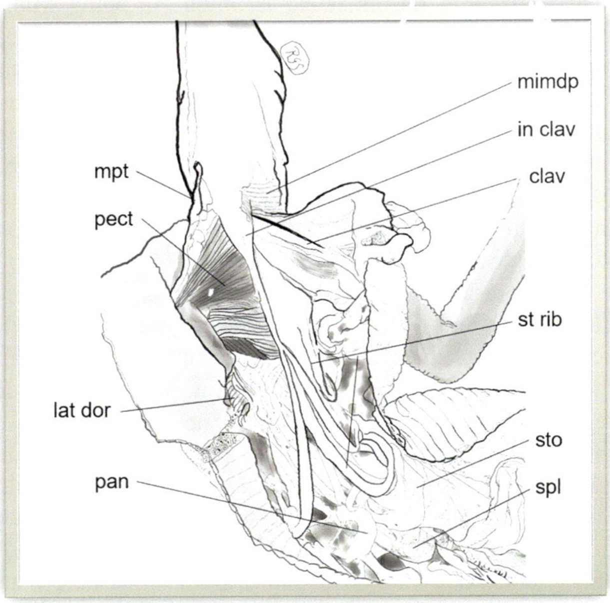

FIGURE 6.13 Viscera of *Varanus prasinus* (RGS 158). Abbreviations: clav – clavicle; in clav – Interclavicle; lat dor – M. latissimus dorsi; mimdp – M. intermandibularis posterior; mpt – M. cervicomandibularis (part); pan – pancreas; pect – M. pectoralis; spl – spleen; sto – stomach; st rib – sternal rib. The pancreas has been moved to the left to display the stomach.

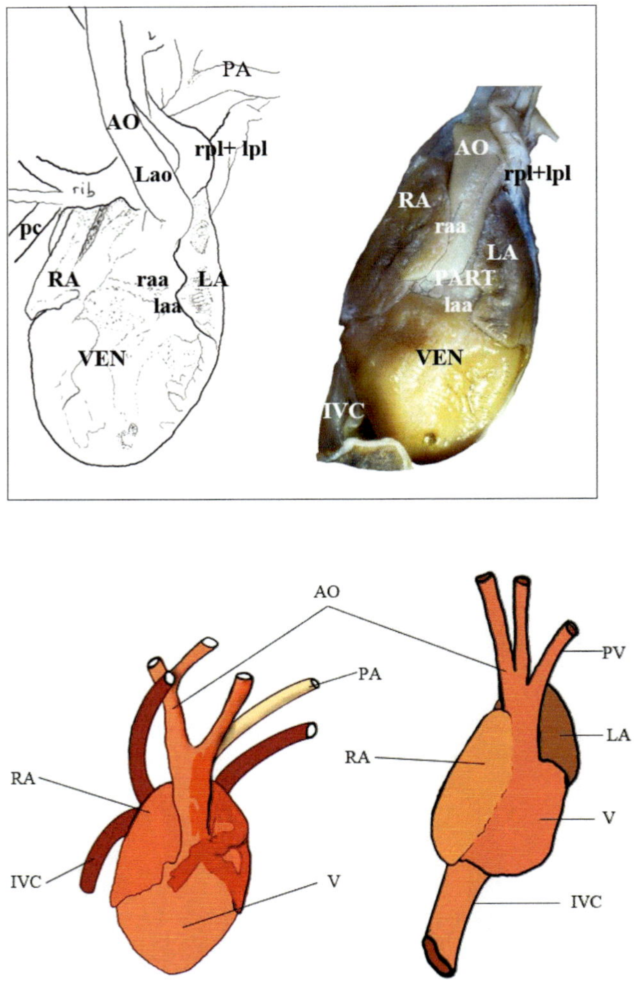

FIGURE 6.14 Hearts from *Varanus*, in ventral view, showing variation in external morphology. Upper left (RGS 158) and right (RGS 42), from different specimens of *Varanus prasinus*. Lower left, *V. prasinus* (CAS uncat); Right, *Varanus indicus* (CAS uncat.). Abbreviations: Bca, brachycephalic artery; Cpl, common pulmonary artery; Laa, left aortic arch; LA, left atrium; Lao, left aorta; Lpl, left pulmonary arch; Lsa, left systemic arch; Lsc, left subclavian artery; PART, base of pulmonary artery; Pc, postcaval vein; RA, right atrium; Raa, right aortic arch; Rao, right aorta; Rpl, right pulmonary arch; Rsa, right systemic arch; VEN, ventricle.

considerable longitudinal septum that effectively turns that chamber into the functional equivalent of the two chambers seen in the hearts of mammals and birds. So important is this septum that it was long used as an important distinguishing character for diagnosing crocodilians from other reptiles. In actuality, the crocodilian ventricle not only has a complete intraventricular septum derived from the longitudinal muscular ridge, but each half also contains several incompletely separated smaller sacs. The entirety of lizards and snakes were considered to have simple three-chambered hearts with at best an insignificant partial ventricular septum. It has subsequently been learned that some lizards, e.g., the teiid genus *salvator* (Gregorovicova et al. 2022) and *Varanus*, possess a long ventricular septum nearly the equivalent of that seen in crocodiles, along with two other intraventricular partial septa (Hanemaaijer et al. 2019a, b), but only the Varanidae have a functionally divided ventricular septum (Hanemaaijer et al. 2019a). Although this is still an incomplete septum, it very effectively functions to divide ventricular blood flow with great efficiency. In *Varanus* the ventricle encloses three small and partially separated chambers (Fig. 6.16). The largest of these is the cavum arteriosum, which occupies nearly the entire left side of the ventricle. Oxygenated blood from the pulmonary veins enters the left atrium and enters the cavum arteriosum from which it is pumped into the small lateral cavum venosum. In addition to oxygenated blood from the cavum arteriosum, the cavum venosum also receives deoxygenated blood from the right atrium. Mixed blood from the right atrium and cavum venosum passes into the cavum pulmonale. That is, blood flow dynamics, guided in large part by the septum, give the circulatory system an advantage in keeping oxygenated and deoxygenated blood streams separate (Bickler and Anderson 1976, Burggren and Johansen 1982, Gregorovicova et al. 2022, Hanemaaijer et al. 2019, Heisler et al. 1983, Ishimatsu et al. 1988, Mathur 1944, Piskovská 2019, Webb et al. 1971, Wood et al. 1978). The efficiency of blood separation is enhanced in part by the cavum venosum, which is quite reduced in varanids. Intraventricular separation seems to be a characteristic of lizards that have highly active, energetic lives.

In external view the ventricle is considerably larger than the combined atria, and the ventricular wall is several times thicker than that of the atria (Figs. 6.14 and 6.16). In addition to the nearly complete ventricular septum, the lumen of the ventricle is lined with numerous thin trabeculae,

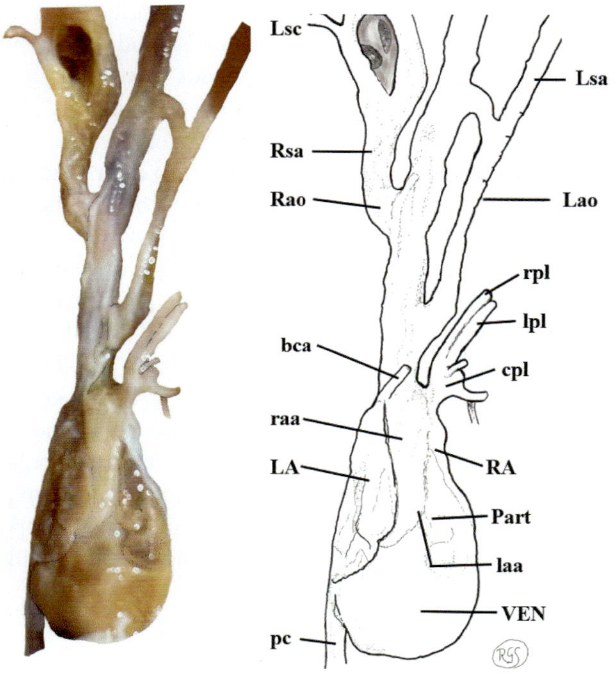

FIGURE 6.15 Hearts from *Varanus*, in ventral view, showing variation in external morphology. Upper left (RGS 158) and right (RGS 42), from different specimens of *Varanus prasinus*. Abbreviations: bca, brachycephalic artery; cpl, common pulmonary artery; laa, left aortic arch; LA, left atrium; Lao, left aorta; lpl, left pulmonary arch; Lsa, left systemic arch; Lsc, left subclavian artery; PART, base of pulmonary artery; pc, postcaval vein; RA, right atrium; raa, right aortic arch; Rao, right aorta; rpl, right pulmonary arch; Rsa, right systemic arch; VEN, ventricle.

finger-shaped structures that subdivide the lumen into many small chambers, increasing the internal surface area. There is a sac called the sinus venosus attached to the dorsal surface of the ventricle, and this contains the pacemaker. Deoxygenated blood enters the sinus via a pair of cranial vena cava, caudal vena cava, and hepatic vein, and drains through the sinoatrial foramen into the right atrium (Ishimatsu et al. 1988). The atria lie on the upper (anterior) portion of the ventricle, where they are flap-like in structure and can be separated from the rest of the heart except around the aorta and pulmonary artery. The capacity of the right atrium is greater than that of the left, which is the receptacle for oxygenated blood from the lungs.

In most lizards, the heart is enveloped by the sternum and ribs, but in varanids, it lies more posterior, at the anterior end of the coelomic cavity (Fig. 6.12A–C). The heart is surrounded by a thin, transparent, and strong pericardium. During dissection, the lower portion of the ventricle is clearly visible. External to the pericardium, the heart is bracketed laterally by the lungs and the large right lobe of the liver (Fig. 6.13). The posterior location may facilitate necessary flexibility of the heart, which, along with efficient intraventricular hemodynamics, allows for the exceptional respiratory physiology of such active lizards (Owerkowitz et al. 1999, Wood et al. 1978).

In summary, the hearts of *Varanus* species differ from almost all other lizards (non-snake squamates) in possession of the following suite of characters: (1) a longitudinal muscle-derived ventricular septum that physically divides the left and right ventricular regions with (2) a transverse

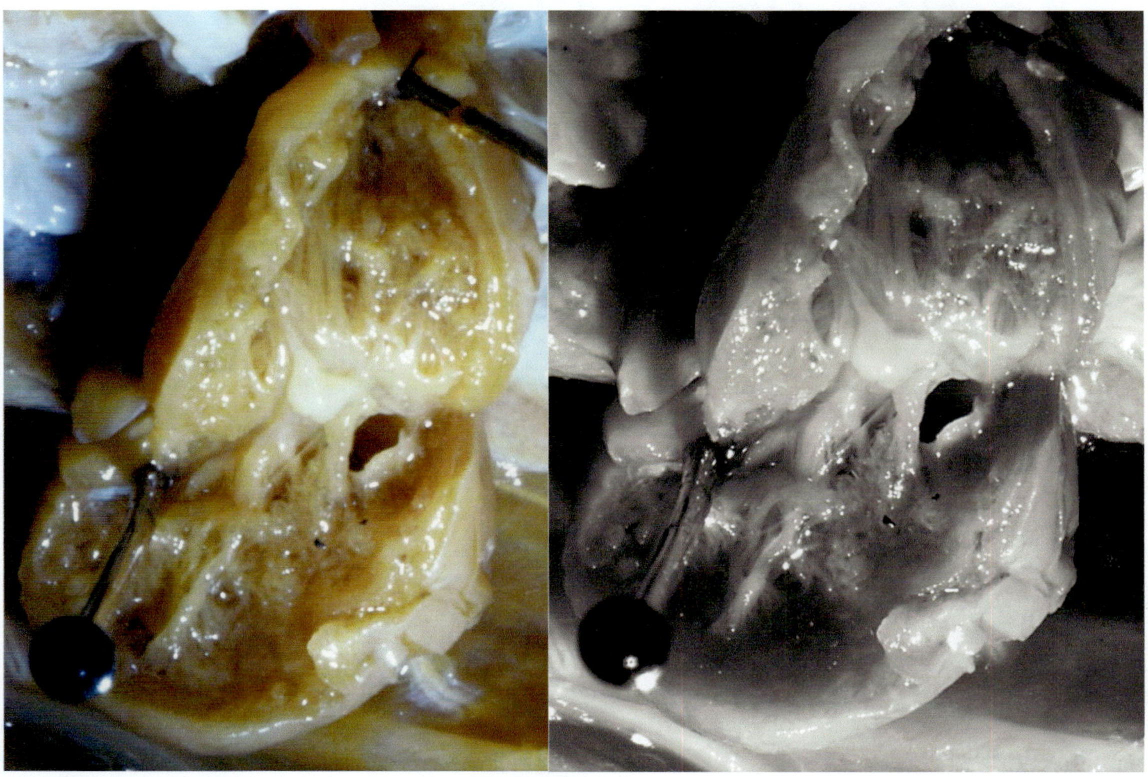

FIGURE 6.16 Detail ventral view of heart of *Varanus prasinus* (RGS 42), internal view of ventricle, dorsal section reflected at top, sagittal dissection. The thick intraventricular septum is seen starting just right of the pin head. Abbreviations as in Fig. 6.16.

aperture, the cavum venosum, that perforates the septum and allows left and right atrial blood to mix, but (3) effectively and demonstrably separates systemic and pulmonary circulation; (4) the small cavum venosum induces higher blood pressure than that in the cavum arteriosum; (5) the heart lies posterior to the sternum where (6) it is more affected by peripheral muscular movements during locomotor activities, providing enhanced endurance; (7) has a thin pericardium to which the ventricle does not attach; (8) possesses a large dorsal sinus venosus that receives deoxygenated blood and contains the pacemaker; and (9) is surrounded anterolaterally by the lungs and posterolaterally by the liver.

6.3.2 Lungs

Vertebrate organ systems work unidirectionally, each component dealing with one-way traffic (Fig. 6.17). Neurons are afferent or efferent, not both; blood vessels transport blood in one direction, either toward or away from the heart, not both; and the gastrointestinal tract moves food from mouth to cloaca (excepting unfortunate periods of vomiting), not the other way. The exception – for biology is nothing if not the science of exceptions – is the respiratory system. Here is the single case where the component (air) enters and exits through the same conduits of trachea, bronchi, and the lung itself. Of these organ structures, it is the lungs that show the most variability, being asymmetrical, differing in size and shape, and having lumena of single, di-, or multichambers.

Pulmonary anatomy within varanoids is variable at the species level and of phylogenetic significance (Becker et al. 1989, Becker 1991). Of the several types of squamate lung morphologies (Perry 1998), only two families of platynotans, Helodermatidae and Varanidae, have multichambered organs (Becker et al. 1989). Varanids and *Lanthanotus* are unique in having secondary bronchi with cartilaginous support (Kirschfeld 1970, Perry 1998). Their lungs are thin-walled

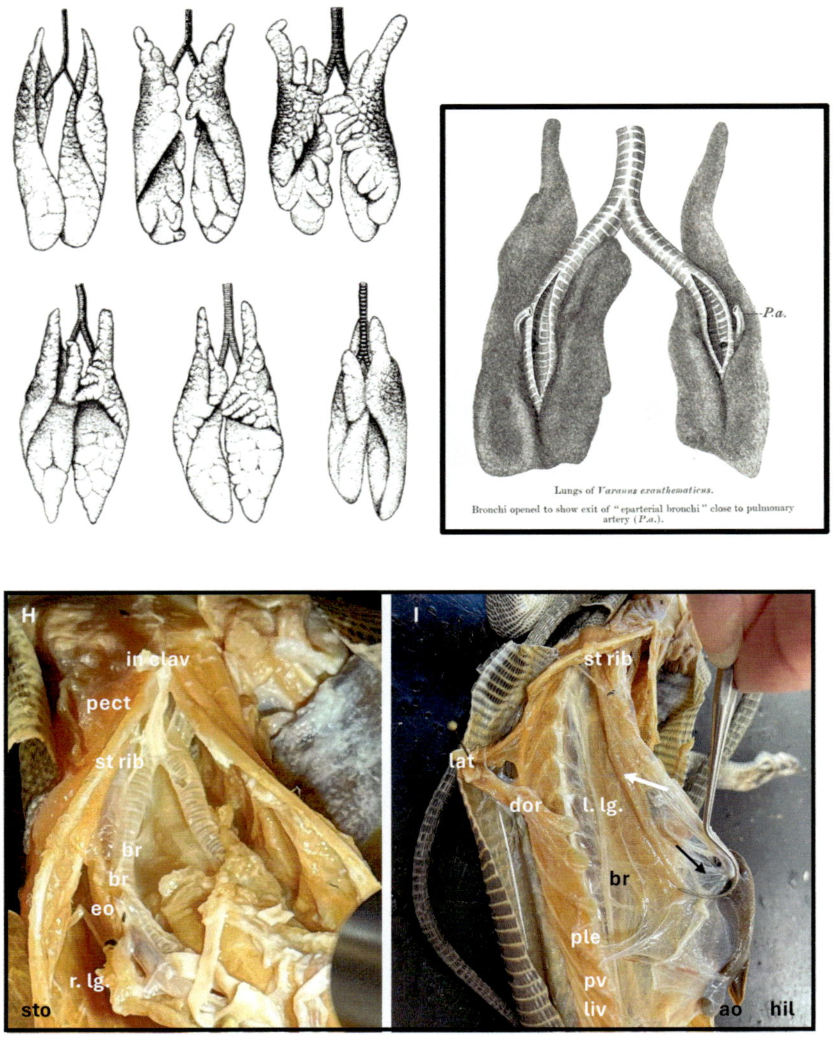

FIGURE 6.17 The lungs in *Varanus*. *Varanus prasinus* (RGS 42). A clearly shows the bronchi that separate anterior to the opening between the sternal ribs, and extend nearly halfway down the length of the abdomen. B shows the left lung, which is long, laterally compressed, and flaccid. The bronchial hilus (ple) is at the tip of the probe. Abbreviations: ao – aorta; br – bronchus; hil – bronchial hilus; eo – esophagus; in clav – interclavicle; lat dor – m. latissimus dorsi; liv – liver (reflected to left); l. lg. – left lung; P.a. – pulmonary artery; pect – m. pectoralis; ple – peripleural fascia; pv – dorsal portal vein; r. lg. – right lung; si – small intestine; sto – stomach; st rib – sternal rib.

sacs restricted to the anterior half of the body cavity, anterior to the liver and adjacent to the heart, the left lung slightly longer than the right. They are long and arrow-shaped, with most of each organ protected by the ribs. The interior (alveolar) sacs are comparatively larger than the spongier form seen in mammals and give comparatively less surface area for respiratory gas exchange. Lung shape parallels body form, where long-bodied species, such as *V. prasinus*, have long, oblong, thin lungs, and stout-bodied species, such as *V. exanthematicus*, have shorter, more oval organs.

Varanids possess a typical vertebrate trachea, a slightly flexible tube originating in the throat as the larynx, the glottis being the anterior (buccal) opening. Tracheal support and protection come from thin, evenly spaced, hyaline cartilaginous rings, connective tissue, and smooth muscle. Perry (1998) states that the rings may be "of great taxonomic significance, as this character is independent of the size of the animal" (Perry 1998:32), probably at taxonomic levels of family or higher. The cartilaginous rings do not meet dorsally. The tube is thus firm yet flexible. The caudal portion of the trachea bifurcates into bronchi just anterior and slightly dorsal to the lungs, the point of connection (hilus) to the lungs varying in position (see below). There are no bronchioles.

Pulmonary circulation has been reviewed by Burggren and Johansen (1982) and Owerkowitz et al. (1999), the latter describing the possibly unique mechanics of gular pumping and lung ventilation in *V. exanthematicus*. Though most lizards engage costal muscles to operate the lungs, monitors employ both body motion and gular pumping to attain movement of air. The pump, inducing use of alternate lungs, allows increased periods of activity by providing a near-continuous flow of air across pulmonary gas exchange surfaces, not unlike the flow system of birds. The mechanism is complementary to the models of Bartholomew and Tucker (1964) and Ishimatsu et al. (1988) for explaining the often described "mammal-like physiology" of varanids, which are able to stay more active and for longer periods than other lizard species of similar size. The ability to maintain a high level of aerobic metabolism is an important factor in allowing monitors to stay active enough to be the active high-level predators that they are. The physiology of gas exchange, oxygen consumption, oxygen debt, and activity in *Varanus* have long been areas of intense study (Bartholomew and Tucker 1964, Bennett 1972, Bickler and Anderson 1982, Earll 1982, Christian and Conly 1984, Christian and Weavers 1994, Gleeson 1981, Wood et al. 1978).

The lungs vary considerably among taxa (Becker et al. 1989), with intrageneric variation in where the bronchi enter the lungs (Becker et al. 1989). In the African species examined (*V. exanthematicus, V. niloticus, Varanus yemenensis*), plus Australian *V. giganteus* and *Varanus mitchelli*, the bronchial hilus enters the lungs anterior to the first ribs, while the Australian and New Guinean taxa *V. prasinus, Varanus salvator, Varanus glebopalma, V. indicus, V. salvadorii*, and *Varanus varius*, plus *V. bengalensis*, the entry point is distinctly posterior to the ribs. Subsequent to the study of Becker et al. (1989) the *V. indicus* group was taxonomically divided into several taxa (Böhme et al. 1994, Harvey and Barker 1989, Weijola and Sweet 2010, Weijola et al. 2016), so it is possible there is variation in hilus location in the different species of that widespread complex.

6.3.3 Liver, Gall Bladder, Spleen, and Pancreas

Certain aspects of reptilian anatomy generally, and of squamates in particular, remain very incompletely known despite more than two centuries of scientific interest (Fig. 6.18). These include the pancreas (see below) and

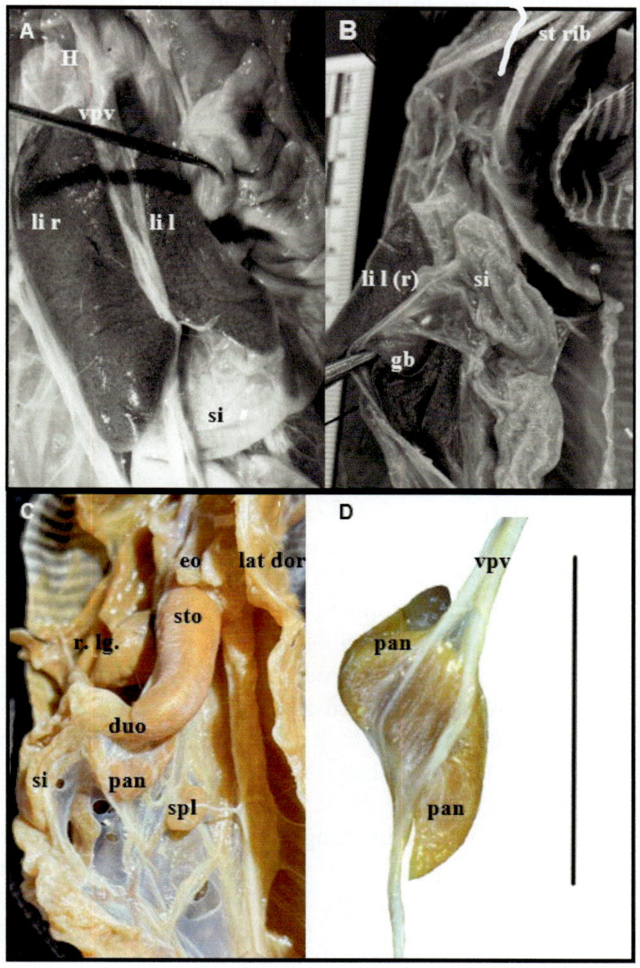

FIGURE 6.18 The liver and gall bladder in *Varanus prasinus*. (A) shows paired lobes of liver in situ in ventral view, firmly positioned between the anterior heart and the lateral and posterior stomach and small intestine. (B) shows the reflected right lung with an embedded gall bladder next to the probe. The clear fascia normally surrounds the entire organ, but has been partially removed for clarity. (C) The main components of the gastrointestinal tract are reflected to the right to expose underlying organs, e.g., pancreas and spleen. (D) Enlargement of the pancreas in the ventral view. Abbreviations: duo – duodenum; eo – esophagus; gb – gall bladder; H – heart; lat dor – m. latissimus dorsi; li l – liver lobe, left; li r – liver lobe, right; li l(r) – left liver lobe, reflected to show dorsal aspect; lg. – right lung; pan – pancreas; si – small intestine; spl – spleen; sto – stomach; st rib – sternal rib; vpv – ventral portal vein.

the largest visceral organ, the liver, about which Shaffner (1998:486) wrote

> The data available on the biology of the reptilian liver are truly inadequate.
> Consequently, this chapter relies heavily on observations in mammals.

An unusual feature of vertebrate livers is their lack of symmetry, the left side generally thin with a posteriorly pointed apex, and the right being broad and largely transverse in the anterior thorax. The liver in varanids is located more posteriorly than in other lizards, lying caudad to the heart and medial to the lungs. The lobes form a distinct border separating the heart and lungs from the gastrointestinal tract and are considerably larger and denser than the lungs. A clear, thin diaphragm-like membrane separates the liver from the heart and lungs. Within a fold of the right posterior lobe of the liver lies the gall bladder, which may lie completely enfolded within the liver or be largely exposed; the latter condition is more likely in juveniles. The pea-shaped gall bladder has a distinct and relatively wide bile duct leading to the duodenum. Analysis by thin-layer chromatography showed the bile acids in *Varanus "monitor"* (likely *bengalensis*; see Sprackland 1982a) are steroid derivatives of the amino acid taurine (Ali et al. 1982, Monte et al. 2009). The primary constituents are bile salts, which have both hydrophobic (water-repelling) and hydrophilic (water-attracting) properties, cholesterol, phospholipids, and biliverdin. Mammals secrete the yellowish bilirubin, but varanids instead produce biliverdin, which, as the name suggests, is green in color. The characteristic greenish postmortem ventral discoloration seen in reptiles is caused by the denaturation of hemoglobin into the bile salt biliverdin when it reacts with a by-product of decay, hydrogen cyanide. Biliverdin is produced in the liver from dead erythrocytes and functions as both an anti-inflammatory and an antioxidant. Bile is essential for the pre-digestive catabolism of dietary lipids in all vertebrates.

Livers are unique in being the organs that receive the entirety of intestinal absorption, and thereby the first stop in processing nutrients. The liver synthesizes blood-clotting factor, metabolizes fats, produces uric acid and bile, detoxifies the blood, converts precursor vitamin D into activated D_3, and stores lipids, among other things. The interior is a spongy mass of blood and lymphatic vessels, with blood arriving via several arteries. Both the hepatic artery and vein are among the broadest vessels in the body, the vein transporting blood directly from the liver to the heart. The hepatic vascular system has been described by Yeager (1973).

The spleen has more of an immunological role in reptiles than in mammals, acting in lieu of lymph nodes. As in other amniotes, the reptilian spleen is composed of red and white pulp cells and is a production and storage site for leucocytes and erythrocytes. In this regard, the reptilian spleen is not an "optional" organ, as it is in mammals that can survive splenectomy. In some squamates, notably in snakes, the spleen and pancreas are joined into a single composite organ; in varanids, the organs remain distinct, both loosely attached to the gastrointestinal system near the union of stomach, intestine, and liver. Varanid spleens are oval or bean-shaped, often with a flat side opposite a convex side, and attach via connective tissues to the stomach, left liver lobe, and sometimes the costal muscles.

Certain organs of reptiles remain poorly studied even after many others have attracted great interest and innumerable publications. Of the pancreas, Miller and Lagios (1970: 319) wrote:

> Very little has been published specifically pertinent to the gross anatomy or detailed features of the reptilian pancreas.

The pancreas of *Varanus* resembles that of other lizards in being trilobed and is attached to the spleen, gall bladder, and small intestine (Figs. 6.12A, C and 6.18) and has both exocrine and endocrine functions. Pancreatic endocrine cells produce enzymes that regulate blood glucose levels. Exocrine products include proteolytic, amylolytic, lipolytic, and chitinolytic digestive enzymes. The pancreas also produces an alkaline "antacid" that neutralizes stomach acids as they exit the pyloric valve of the stomach. Unlike other lizards, the greatest concentration of islet (isle of Langerhans) tissue in varanids is in a distinct elongated islet structure attached to the dorsal pancreatic lobe by a thin conduit linking the pancreas to the spleen. This juxtasplenic body is not found in other lizards, but somewhat resembles the pancreas of "primitive snakes" (Stahl 2003). The trilobed pancreas has been considered evidence of a close phylogenetic relationship between monitors and snakes, but numerous subsequent studies of morphology and molecules refute that hypothesis.

6.3.4 Digestive System

The gastrointestinal (GI) tract in most varanids consists of a long muscular esophagus, a large thick-walled stomach, and a highly coiled intestine that terminates at the cardiac sphincter, which separates esophagus from stomach (Figs. 6.19–6.21). The stomach has a nearly parallel orientation to the vertebral column, is situated to the left of the midline, and is some 0.2 SVL. The anterior cardiac section (fundus) is narrower and thinner than the broad posterior (pyloric) portion. The fundus primarily releases digestive enzymes that soften food, and the thicker-walled pyloric region does the majority of pre-intestinal digestion. Internal folding is distinct, in parallel rings of thick ridges of rugae that pulverize food, and large regions of soft, textured tissue (Figs. 6.21) that provide a large surface area to bathe food in digestive juices, largely hydrochloric acid, and absorb some of the products of digestion. The pyloric sphincter limits the flow of partly digested food from entering the small intestine, allowing controlled emptying into the duodenum. It is here that bile salts and pancreatic enzymes act, in part,

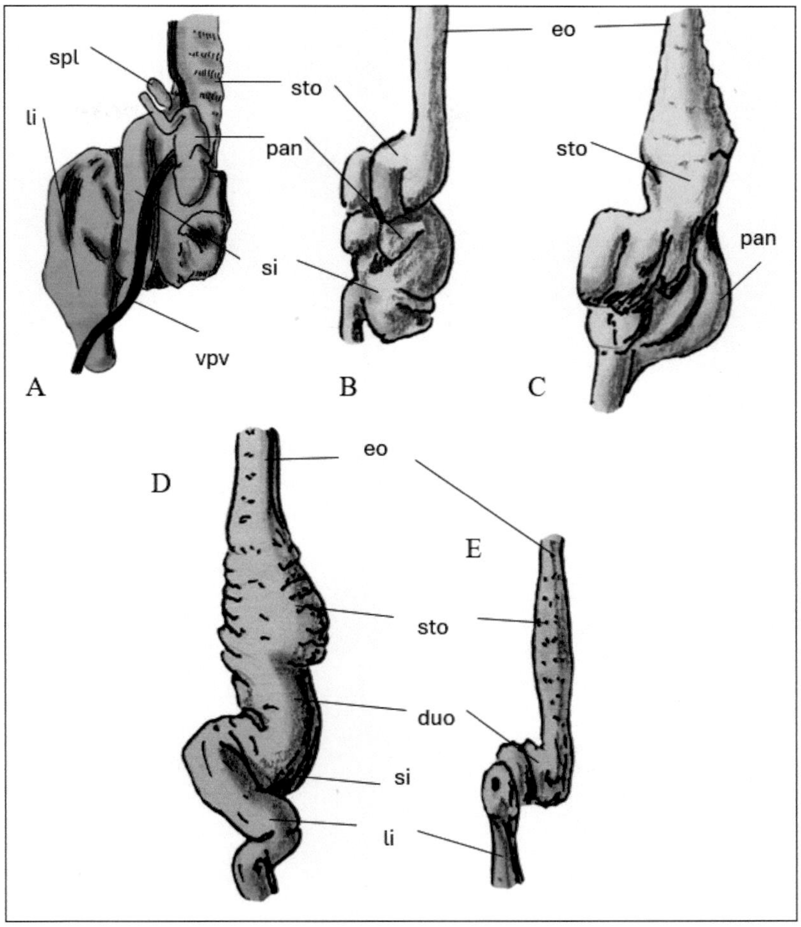

FIGURE 6.19 Comparison of the digestive tracts of five platynotans showing the morphological variation within the group. The liver and peripheral organs are omitted. (A) *Varanus exanthematicus* (uncat.). (B) *Varanus bengalensis* (uncat.). (C) *Varanus indicus* (RGS 38). (D) *Varanus prasinus* (RGS 37). (E) *Lanthanotus borneensis* (AMNH 87375). Symbols: dor – dorsal portion of small intestine; duo – duodenum; eo – esophagus; li – large intestine; pan – pancreas; si – small intestine; spl – spleen; sto – stomach; vpv – ventral portal vein. From Sprackland (1982). Not to scale.

to increase the pH of stomach contents, an essential step in preventing self-digestion of the intestinal tract (Monte et al. 2009). The small intestine is the longest segment of the GI tract, generally 1.5–2.8 times SVL, and occupies most of the posterior half of the post-liver coelom. The lumen is smooth, or with a few spaced patches of small irregularities, but is in no way rugose (Fig. 6.21B). The intestinal tissue is highly pliable and capable of distension several times its diameter, a trait essential in species that consume large food items. The large intestine is wider in diameter and thicker in structure than the small intestine. The majority of the organ runs dorsal to the small intestine, excepting the posterior descending process that empties into the coprodeum.

There is significant variation among varanid species in digestive tract morphology (Fig. 6.19, Sprackland 1982b, Srichairat et al. 2018, Surahya 1989). The well-studied Bengal Monitor (*V. bengalensis*) has a "typical" carnivorous lacertilian gut, including a narrow small intestine convoluted into four subequal chambers (Fig. 6.19B). The organ may distend by 300 percent when processing food. The large intestine occupies more space and is 50 percent to 110 percent wider than the small intestine. Prey of the Bengal Monitor is highly varied, including insects, worms, eggs, small mammals and birds, lizards, snakes, and clams, so its gut morphology is appropriately unspecialized. The GI system of *V. salvator*, another large opportunistic feeder, is also unspecialized (Srichairat et al. 2018), though the liver is somewhat larger than in other monitors of similar size. The more aquatic Mangrove Monitors (*V. indicus* group, Fig. 6.19C), which also take a wide variety of foods, have similar gut morphology, though proportionately larger organs, while *V. dumerilii*, which includes crabs as a major dietary component, has even larger organs. In tree monitors (*V. prasinus* complex, Fig. 6.19D) and Earless Monitors (*Lanthanotus borneensis*, Fig. 6.19E), species that largely consume small, soft-bodied prey, the gut is abbreviated in comparison to other varanids; in the latter, the gut shows less distinction between constituent organs. In the tree monitors, however, there are notable modifications from the "typical" varanid gut morphology. The stomach is slightly conical, the narrow end at the esophagus, and consists of five to eight narrow muscular bands. The duodenum and

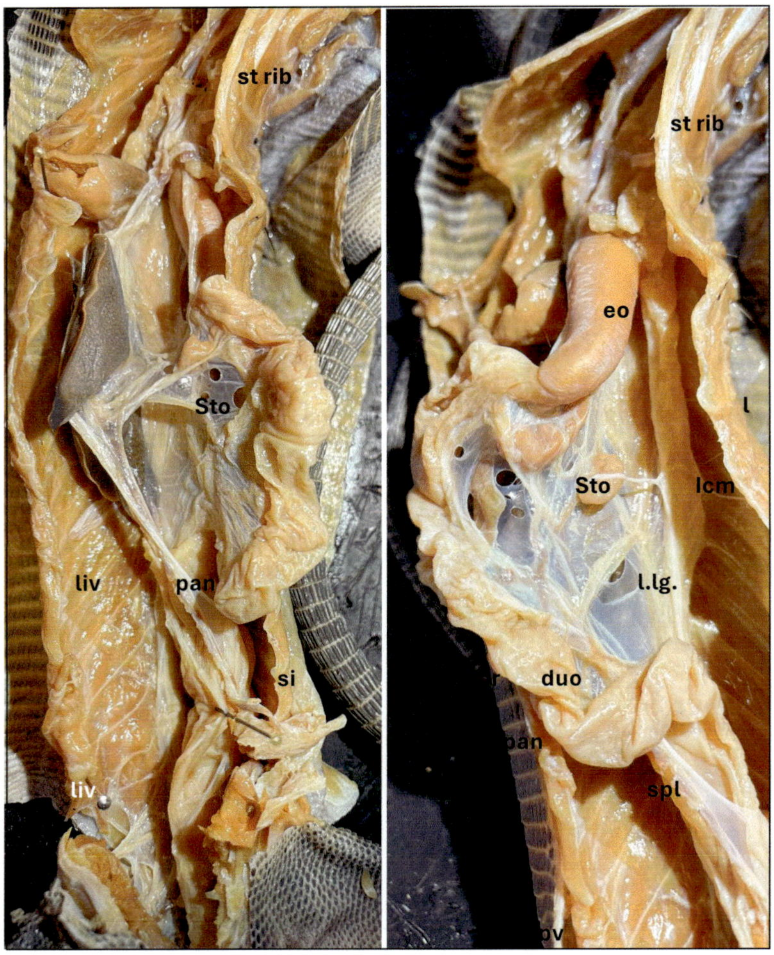

A B

FIGURE 6.20 Gastrointestinal tract of *Varanus prasinus* showing (A) intestine pulled to show underlying organs and (B) intestine reflected to the right to expose major vessels, spleen, and pancreas. Symbols: dor – dorsal portion of small intestine; duo – duodenum; eo – esophagus; fb – fat body; Icm – intercostal muscles; li – large intestine; liv – liver; l. lg. – left lung; pan – pancreas; si – small intestine; spl – spleen; sto – stomach; st rib – sternal rib; vpv – ventral portal vein.

small intestine are wide, 2.2–3.5 times wider than the anterior colon, and of subequal length.

6.3.5 Kidneys and Urinary System

Kidneys are essential organs that filter waste products and ions from the bloodstream, regulate blood pressure, and reduce water loss, and, though simple in appearance, are complex in structure (Fig. 6.22). They work in concert with the large intestine, cloaca, and salt-excreting glands in the reabsorption of body water and countering dangerous levels of dehydration. Urinary bladders, which also act to reabsorb water, are absent in varanids, but their particular physiological process of resorption is partially performed by the cloaca and lower large intestine. (There have been conflicting published comments on whether varanids possess urinary bladders. Most reports deny their presence, and I confirm this based on my dissections of numerous individuals of many taxa. I suggest that some researchers mistook rounded precloacal fat bodies for bladders.) Varanid kidneys are paired organs that lie dorsal to the GI tract, in a dorsolateral position just dorsal to the posterior tips of the liver. The highly corrugated surface makes recognition easy, as does the accompanying long, thick ureter. Each kidney is bean-shaped and depressed, divided into several lobes (Fig. 6.22) and covered by a smooth transparent capsule. In males, the distinct, round, yellowish testes and epididymis are closely associated with the kidneys, the latter attached to the ureter (Fig. 6.22). In females, the ovaries form a greyish sac that originates well anterior to and extends posterior to each kidney. Ova are seen as grey-white structures of varying sizes in the oviducts.

Reptiles lack the loop of Henle tubule of mammals and are thus unable to produce urine more concentrated than blood. Without this process, dehydration increases with potentially fatal results. Instead, squamates are uricotelic, mixing nitrogenous wastes and ions, mainly salt (NaCl), into

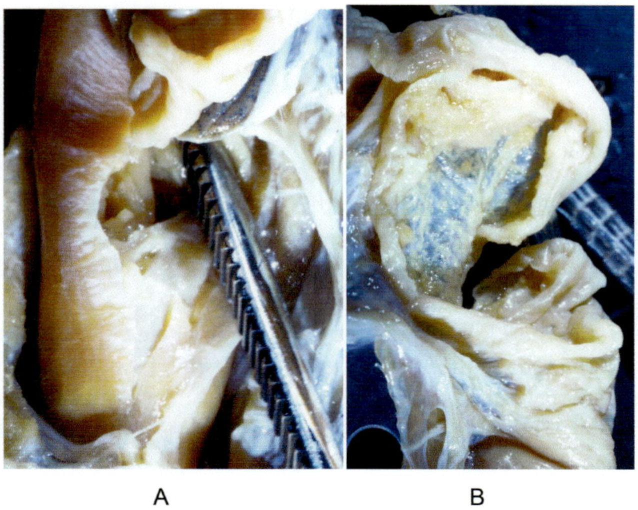

FIGURE 6.21 Interior surfaces of (A) stomach and (B) small intestine of *Varanus prasinus*. The stomach has a common, highly infolded lumen, with tough, fibrous regions that help grind food, and a greatly enhanced surface area that provides for highly efficient nutrient absorption. The intestine, however, is considerably smoother and thinner than in most other lizards, which suggests a diet high in small and/or soft-bodied prey.

uric acid, which is insoluble in water, partially performing the loop of Henle's function. Well-hydrated monitors excrete more uric acid than dehydrated lizards, but additional methods of water retention are necessary. In reptiles with urinary bladders, that organ performs a water reabsorption function, but because monitor lizards lack urinary bladders, alternative mechanisms have evolved that compensate. The kidney tubule products empty into the coprodaeum, the anterior chamber of the cloaca, and are stored with the end products of the intestinal tract. This means that lizards do not have separate urinary and fecal elimination functions, but they occur simultaneously. The matter so excreted, liquid at first but quickly drying into a powdery white substance, is uric acid. Nevertheless, the loss of ions is disadvantageous to the lizards, both in terms of the energy used in creating uric acid and losing body water to the external environment. Resolving both issues are two extrarenal organs, the cloaca and salt glands (see below).

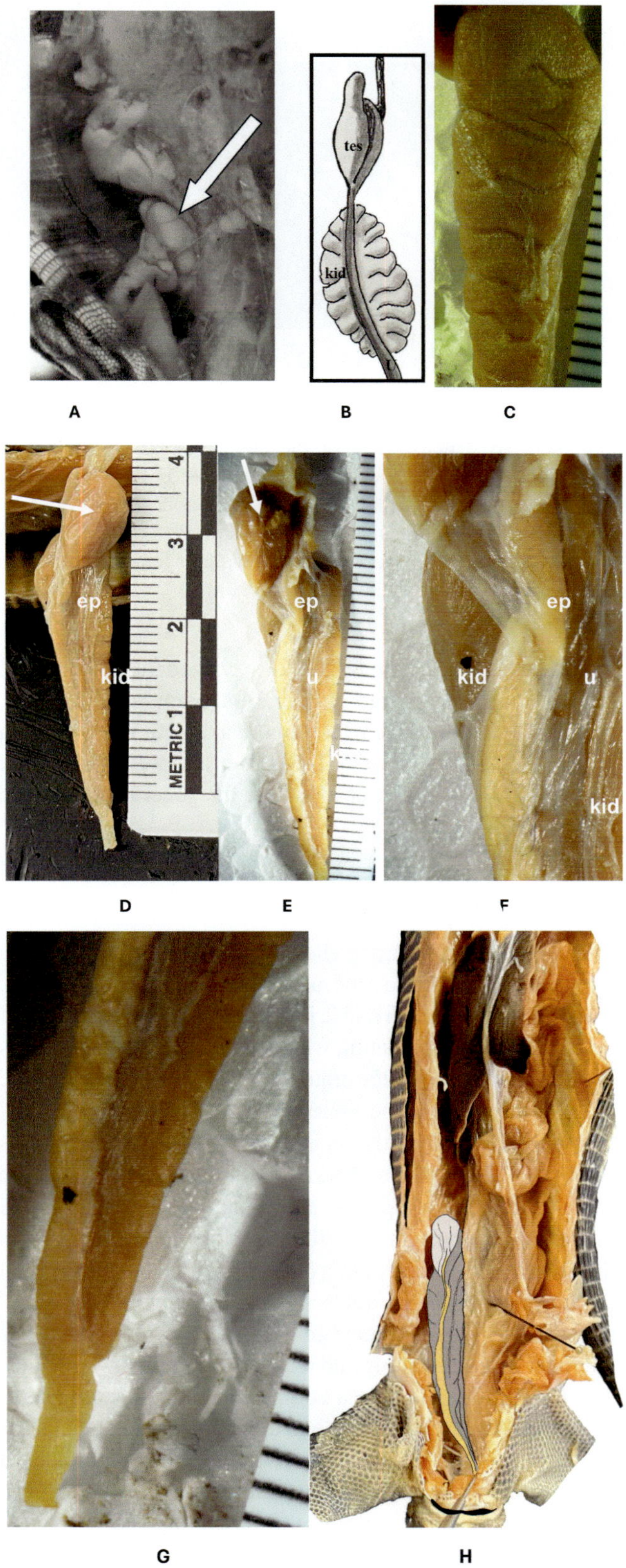

FIGURE 6.22 Kidneys and the urinary system of *Varanus*. (A) Five lobes of the right kidney (arrow) are clearly visible in a young *Varanus glauerti*, and its descending ureter to the left. Above the kidney is a group of fat bodies, and enveloping connective tissue is clearly seen throughout. (B) A diagrammatic illustration of a generalized varanid right kidney and testis based on an oft-published illustration (e.g., in Porter [1972] and Romer [1963]). (C)–(H) The right kidney of a *Varanus prasinus* male. (C) Dorsal surface showing clear demarcation of lobe sections. (D) Same kidney in ventral view (with testis [arrow] moved to right) showing anterior section. (E) Same, with testis in normal position relative to the kidney, and showing parallel arrangement of the epididymis and the ureter. (F) Close-up of E to show the complex folding of the epididymis and the clear fascia that encloses the testes and the kidney as one. (G) Close-up of the terminus of the epididymis (left) and the ureter (at right, under the fascia). (H) Positions of the right testis (light gray), epididymis (yellow), and kidney (dark gray) in situ in the abdomen. The epididymis and the ureter both empty into a common cloacal chamber, the urodaeum. Note the absence of bladder. Abbreviations: ep – epididymis; kid – kidney; u – ureter; tes – testis.

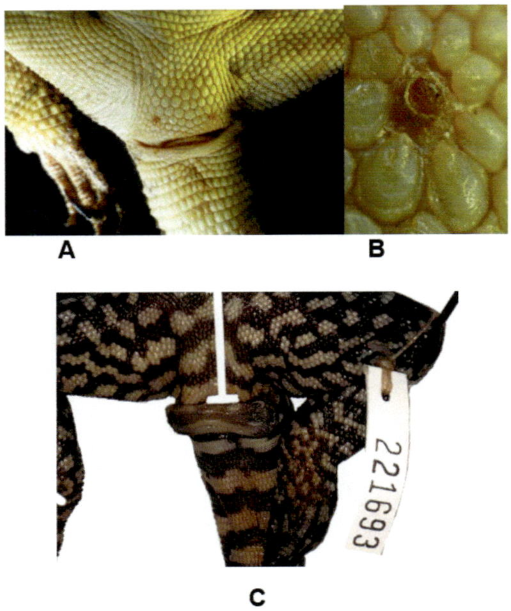

FIGURE 6.23 (A) Cloaca and preanal pores of an adult *Varanus dumerilii.* (B) Enlargement of the right preanal pore. Though common in many lizards, notably gekkonids and iguanines, few varanids possess the pores or their subcutaneous glands. (C) Cloacal region of *Varanus rainerguentheri* (USNM 221693) showing complete absence of either preanal or femoral pores, the condition most common among *Varanus.*

The rate of filtration by the kidneys is driven by several factors, including degree of hydration, contents of the diet, temperature, reproductive condition, and activity (Christian et al. 1996, Dantzler 1976, Dryden et al. 1992, Green and Christian 2007, Green et al. 1991; for general review see King and Green 1999).

6.3.6 Cloaca

The cloaca is the common chamber at the termini where the digestive, reproductive, and urinary systems converge, and, like Caesar's Gaul, is divisible into three parts (Fig. 6.23). Most anterior is the coprodaeum, which is bracketed by sphincters separating it from the terminus of the large intestine and the anterior urodaeum. The lumen of the coprodaeum contains many ridges of folded tissue, increasing its surface area. This is where water is reabsorbed from the feces. The urodaeum is the receptacle for digestive and urinary products, e.g., fecal matter passed through from the coprodaeum and urine from the ureters. The lumen of this region is smooth and is the final reabsorption point for water and high levels of Na^+ and K^+ ions that were not extracted in the nephrons. The cloaca has the function of extracting and reabsorbing them into the body. The terminus of the cloaca is the short proctodaeum chamber, from which fecal matter and uric acid are excreted. It has a sphincter separating it from the urodaeum, and another at the opening of the vent (cloacal opening).

In addition to its role as the receiving and final processing area for urinary and digestive tract products, the urodaeum and proctodaeum in females also serve as the entry for sperm and the exit for eggs.

6.3.7 Glands

Varanids have numerous glands, organs that produce and regulate hormones, and other proteinaceous molecules. The pituitary has been discussed under the section on the brain, above. Of particular note in varanids are the salt and venom glands.

Reptile kidneys lack the evolutionary refinements seen in mammals that reduce salts from the blood through a unique process that uses more water than a reptile can spare. Therefore, their urine cannot be more concentrated than the blood (i.e., high in filtered molecules such as salts and amines). Converting urine into uric acid in the kidney is a partial solution, and additional salt excretion elsewhere is another. Many lizard species possess nasal glands, including non-varanids (e.g., Braysher 1971, 1986, Hillman and Pough 1976), that complement the work of the kidneys by excreting excess salt from the blood, then releasing it with little water from around the eyes or nostrils. The glands are highly effective and are seen in many desert and marine/brackish water residents. They are widely present in varanids, possibly the majority (Table 6.2). The partially aquatic *Varanus semiremex* lives in a salt-rich environment (mangroves and mud flats) and consumes prey from fresh and brackish water, increasing its NaCl intake over other varanids. It is the second fastest lizard in terms of salt excretion rate (Dunson 1974, 1976). The larger Mertens Monitor, *V. mertensi*, has similar habits and diet, but entirely in freshwater locations, and lacks the glands (Christian et al. 1996). Most lizards excrete more potassium than salt, but *Varanus gouldii* is an exception (Dunson 1976). The excreted fluid, like uric acid, emerges moist but quickly dries into small white crystals. In many desert lizards (e.g., *Crotaphytus collaris, Dipsosaurus dorsalis, Sceloporus occidentalis*),

TABLE 6.2
Species of *Varanus* Confirmed to Possess Nasal Salt-Excreting Glands

Species	Environment	Reference
Varanus acanthurus	Desert	Dryden et al. (1990)
Varanus bennetti	Mangroves	Pers. Obs.
Varanus flavescens	Marshes	
Varanus giganteus	Desert	Green et al. (1986)
Varanus gouldii	Desert	Green (1972)
Varanus griseus	Desert	Vernet et al. (1988)
Varanus indicus	Mangrove	Pers. Obs.
Varanus komodoensis	Arid, marine	Green et al. (1991)
Varanus rosenbergi	Arid	Green and King (1993)
Varanus salvator	Riparian, mangrove	
Varanus semiremex	Riparian, mangrove	Dunson (1974)
Varanus tristis	Desert	
Varanus tsukamotoi	Island	

additional salt excretion emerges from the nostrils; I am unaware of this process in varanids.

A point of clarification regarding the salt glands in *V. indicus*. Dunson (1974) referred to the eastern Australian *V. semiremex* as "a mangrove monitor," a vernacular name almost entirely restricted to *V. indicus* and its close allies. This has led to some accounts referencing Dunson, in error, as the source of claiming the glands in *V. indicus*. Between 1990 and 2023, "*V. indicus*" underwent several taxonomic revisions, establishing numerous species. Not all have been examined to determine the presence of salt glands; however, specimens from Palau (now *Varanus bennetti*), western New Guinea (*V. indicus*), and northeastern Australia (sp. undet.) do have such glands (per. obs.). Each of these taxa, like *V. semiremex*, inhabits environments with brackish water.

Salt glands lie within the large nasal cavity, lateral and external to the nasal chamber (Fig. 6.24), and extend between the orbit and septomaxilla, the precise extent depending on species and age.

The discovery of mandibular venom-secreting glands in *Varanus* and *Lanthanotus* has been met with both controversy and misunderstanding (see Venom_Doc online; Sweet 2016; numerous papers by Fry et al. 2005, 2006, 2009), but the reports of envenomation by monitors date back many years (Sopiev et al. 1987). Because the glands are long and very thin, they have likely been misinterpreted as part of the dentary musculature (Fig. 6.24). However, subsequent research has confirmed such rudimentary venom glands throughout the Anguimorpha and other lizard groups as well (Fry and Scheib 2007, Fry et al. 2006, 2009, 2010, Kaludarov et al. 2017). The crux of the issue is the use of the term "venom system," used in the first description (Fry et al. 2006) instead of "venom," which the authors stress has distinct meanings. Specifically, a venomous animal can actively inject toxin into another creature; having a venom system does not necessarily imply the ability to deliver toxin. In the case of varanids, both terms have some relevance. It is noteworthy that the long-stated opinion that infection-causing bites from Komodo Dragons, and presumably other varanids, were believed to be the result of toxic bacteria from the lizards' mouths entering the wound, has been discredited by Dobson et al. (2019) and Fry et al. (2009, 2019).

Though helodermatid venom is a conservative set of proteins within the genus, varanid venoms show considerable diversity (Arbuckle 2009, Chandrasekara et al. 2024, Fry et al. 2007, 2009, 2010). Several records of envenomation leading to toxic effects – paralysis in birds and rodents, dizziness, tightness of the chest, muscle soreness in humans – are known (mostly anecdotally) from *V. griseus* (Fry Venom_Doc), and one fatality has been published on a human death following a bite by *V. bengalensis* (Vikrant and Verma 2014). Reports of non-lethal envenomation in humans are also reported for *V. albigularis* (three cases), *Varanus komodoensis, Varanus kordensis* (four cases), *V. salvadorii* (one case), *Varanus scalaris* (one case), and *V. varius* (five cases). In *V. komodoensis, V. salvadorii*, and *V. varius*, symptoms included pronounced bleeding, suggesting the venom has anti-coagulant properties, but as these represent three of the largest species of *Varanus*, the sheer size of the lizard and consequent power and depth of a bite may have significant effects as well. Fry et al. (2009) demonstrated that the bite of *V. komodoensis* produces a deep wound capable of delivering a lethal dose of venom, and argue that venom, and not oral microbes, cause the toxic effect of the lizard's bite. In the half-century that I have studied varanids, I have heard of no bites with notable effects that seem venom-induced. *Lanthanotus* has a venom delivery system evolutionarily close to that in *Varanus*, along with a history of being suspected of having a venomous bite. This hypothesis was based on the supposed physical similarity in

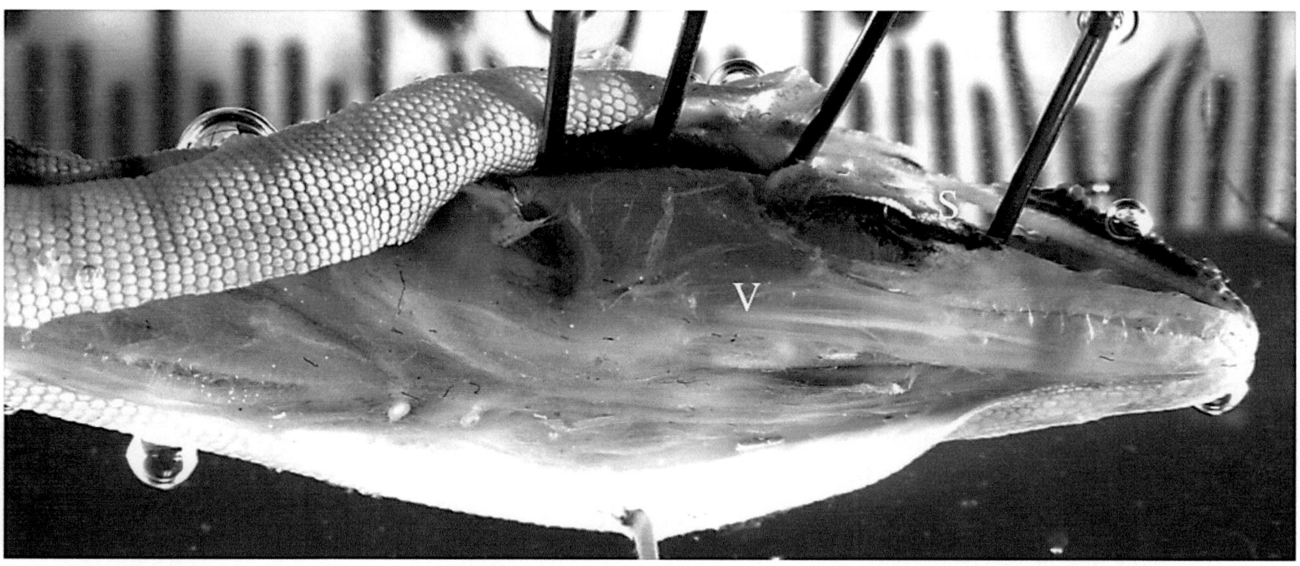

FIGURE 6.24 Location of salt (s) and venom (v) glands of *Varanus* and *Varanus glauerti* juvenile. The venom gland is elongated and quite thin. Varanid venom has been implicated in rare human deaths, but the genus is very rarely capable of harming humans.

appearance of *Lanthanotus* with *Heloderma*, and was long considered a helodermatid (Steindachner 1977, Boulenger 1885). *Lanthanotus* was firmly established as being only distantly related to *Heloderma* (McDowell and Bogert 1954), and no known bites have included any symptoms of note (pers. obs.). Nevertheless, the species does possess a mandibular venom system.

Other studies compared the effects of the venom of different species of *Varanus* on birds and mammals (Koludarov et al. 2017, Dobson et al. 2019, 2021). Does the venom serve defensive, offensive, or prey-subduing roles (Arbuckle 2009)? So far the evidence supports the latter suggestion. Dobson et al. (2021) found that venom from *V. salvadorii* and *V. varius* had no effect on chickens (*Gallus*), but did bind to ion channels in mammals, causing anticoagulation. I find this interesting because both species are known predators of eggs and, much less commonly, birds. However, susceptibility to venoms varies greatly among and even within species, so the immunity displayed by chickens may have nothing to do with susceptibility to varanid venoms in other birds. In mammals, Dobson et al. (2021) noted that the tested varanid venoms may have been evolutionarily selected for based on the prey of varanids (and the allied *Heloderma*), which favors mammals over birds. Rodents experimentally injected with venom from *V. varius* caused extreme smooth muscle contraction, a tetanic response associated with neurotoxins. In humans, the bites have caused lethargy and paralysis (Dobson et al. 2021). Injection with venom from *V. griseus* caused immediate paralysis in rodents (Zima 2019), which is expected when a predator must quickly subdue prey in a prey-poor environment and where the prey is capable of swift escape. Anecdotal reports of *Varanus* bites almost universally tell of short-term anticoagulatory symptoms, bleeding taking "much longer" to stop than from similar or worse bloodletting injuries.

Anguimorph venom glands are all mandibular (Fry et al. 2009, 2010), extending laterally along the dentary as far posterior as the intramandibular joint, and are parallel to the lingual mucous glands. The venom glands have numerous ducts that open between dentary teeth, providing an efficient venom delivery system (Fry et al. 2009). The gland is laterally compressed and covered by the dentary portion of the m. pterygoideus. The mode of excretion of venom is still unclear, but it is probably linked to muscular contractions of the surrounding muscle under the control of the mandibular branch of CN V.

6.3.8 Reproductive Systems

Varanid reproductive systems follow a common amniote Bauplan, the two sexes being morphologically distinct (Figs. 6.12, 6.23, and 6.26). Sex is genetically determined, with females being heterogametic ZW and males homogametic ZZ (Fig. 6.25). Sexual dimorphism ranges from subtle to conspicuous and is covered in Chapter 4. Both sexes have paired gonads that lie lateral to the large intestine

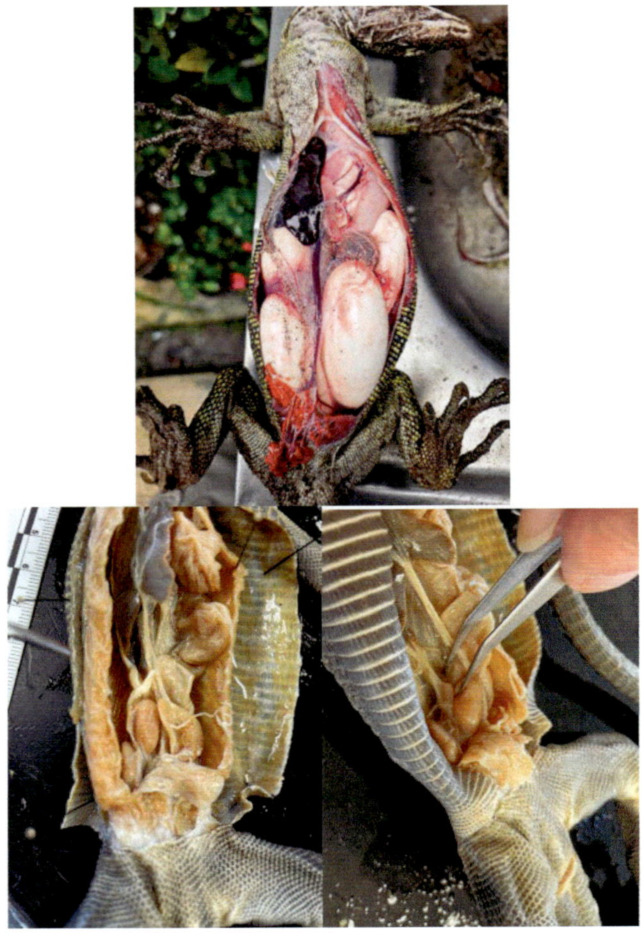

FIGURE 6.25 Reproductive systems of (Upper) Female *Varanus colei* with oviducts full of eggs (courtesy Chris Aplin) and (Lower) A male *Varanus prasinus* showing positions of reproductive organs. Int – intestine, Kid – kidney, Liv – liver, Sto – stomach, tes – testis.

and in close association with the kidneys (Figs. 6.12, 6.23, and 6.26).

6.3.8.1 Female

The ovaries and oviducts vary so much through different stages of ovulation that a single description is not possible. The ovaries comprise tightly compressed microtubules that, in the pre-ovulation state, may resemble testes but lack the clearly defined smooth cortex of the latter. As ova develop, the ovaries enlarge, and with egg growth and movement caudad into the oviducts, tend to lose clear definition. The oviduct is a slightly broader structure than the male's epididymis throughout the year and lacks the latter's complex coiled morphology. The width of the oviduct increases with the growth of the eggs, and to such a degree that it is nearly invisible (Fig. 6.26A). Egg size becomes so great as to obscure more dorsal organs and systems, and to a degree that they are quite visible through the skin when viewed ventrally. Mature eggs remain in the oviducts until time for parturition, at which time they are moved caudad into the urodaeum and expelled from the proctodaeum by

A

B

FIGURE 6.26 (A) Extreme 85° gape of *Varanus salvator* (= *komaini*), possible in part because of the ability of the hyoid structure to both drop and maintain a well-supported pharyngeal–esophageal opening. The major muscle involved in opening the mouth is the m. pterygoideus. Photo courtesy Katie Bruce, the legend lizards. (B) A captive *Varanus tristis* swallowing an adult *Sceloporus undulatus* entire. The major muscles closing the jaws and maintaining the grip on prey are the m. superficialis adductor mandibulae externus and m. levator anguli oris. The degree of gape seen in A was not necessary, but note the vertical and horizontal extension of the throat that allows such a large item to be consumed. During prey manipulation, the gape did not exceed 20°. Note that the conspicuous lateral post-aural and anterior two of the three gular folds in A, which allow the extension during swallowing, are seen in B, where only one vertical fold remains conspicuous. The trans-sternal gular fold is universal in *Varanus*, where it forms a straight line that is almost entirely ventral in position. Unlike the anterior folds, the trans-sternal fold is not an accessory to gular enlargement or expansion. Instead, it acts as a reinforced hinge that provides muscular support to the throat when large food items pass through. It moves passively in the vertical plane, the degree of movement determined by anterior glenoid enlargement. As such it also acts as a regulator; if food cannot pass the trans-sternal gular region, it is disgorged, at which time the monitor will (1) make a second swallowing attempt, (2) claw the object into smaller pieces, or (3) abandon the food.

muscle contractions. Because each ovary is fertilized independently of the other, it is possible that only one ovary will develop viable eggs; total egg number for a given clutch depends on whether one or both ovaries have received viable sperm.

6.3.8.2 Male

Male varanids have round or oval testes that lie along the anterior portion of each kidney (Figs. 6.23 and 6.26), and unlike ovaries retain their shape throughout the reproductive cycle. A thin fascia covers both the testes and the kidney, keeping the organs in constant close proximity.

An interesting feature of several lepidosaurian males is the presence of paired hemibacula, thin fibrous or cartilaginous rods that help support the everted hemipenis. The structures are not universal in *Varanus*, but when present, they show up as faint lateral post-cloacal images on X-rays.

6.3.9 Cranial and Throat Muscles

Varanid musculature has been studied in considerable detail and, probably because of their size, has been used for standard anatomical reference figures in many publications (Cieri 2008, Porter 1972, Ritter 1995, Romer 1963). Important studies on lizard musculature include Abel and Werneburg (2021), Camp (1923), Daza et al. (2011), Dick et al. (2016), Hoffmann (1890), Kępa et al. (2023), Muzammil (1949), Oelrich (1956), Reippel (1978, 1980, 1981), Ritter (1995), Romer (1956), Surahya (1989), ten Donkelaar and Huizen (1984), and Wilken et al. (2019). Those focused particularly on *Varanus* are Cieri (axial 2018), Reippel (cranial 1980), and Watkinson (cranial 1906). The terminology of reptile muscles varies in the literature. Herein, I largely follow the terminology of Johnson (2014) and Reippel (1980).

The bite force of monitors has been documented and demonstrated to vary significantly among species (Herell et al. 2001, Smith and Hylander 1985), and in large varanids (>1.25 m), the jaw power and mechanics are similar to those of *Tiliqua rugosa* (Gans et al. 1985). Force not related to the absolute size of jaws depends in large part on the hyoid structures of the species (Daza et al. 2011, Herell et al. 2001, Montuelle et al. 2011, 2012, Rieppel 1978, 1981, Schaerlaeken et al. 2012, Sinclair 1983, Wilken et al. 2019), which delimit the maximum extent of vertical gape capable of admitting prey. Gape is enhanced by the lateral and vertical integumentary folds (Fig. 6.25) that allow the great increase in gular capacity to safely contain prey. Maximum prey size does have a finite limit, restricted by maximum gape size and the extent of trans-sternal gular fold ability to accommodate the passage of food. In large species (e.g., *V. komodoensis, Varanus macromaculatus, V. salvator, V. albigularis*, and presumably *V. salvadorii*), prey is often torn into manageable pieces, made possible by specialized teeth, a strong and flexible hyoid apparatus (Chapter 5; Rieppel 1981, Schaerlaeken et al. 2012, Smith 1986, 2011), and strong muscles. This behavior is also common among the large arboreal specialists such as *V. beccarii, V. prasinus*, and *V. rudicollis*.

Major cranial muscles are concentrated between the jugal and occiput in three distinct layers (Figs. 6.28 and 6.29). The m. depressor mandibularis is visible at all three layers and is nearly cylindrical. It lies posterior to the ear, its anterior edge forming the posterior border of the tympanic membrane. The origin lies on the posterior surface of the parietal, squamosal, and supraoccipital, and its insertion is on the articular and posterior of the angular of the

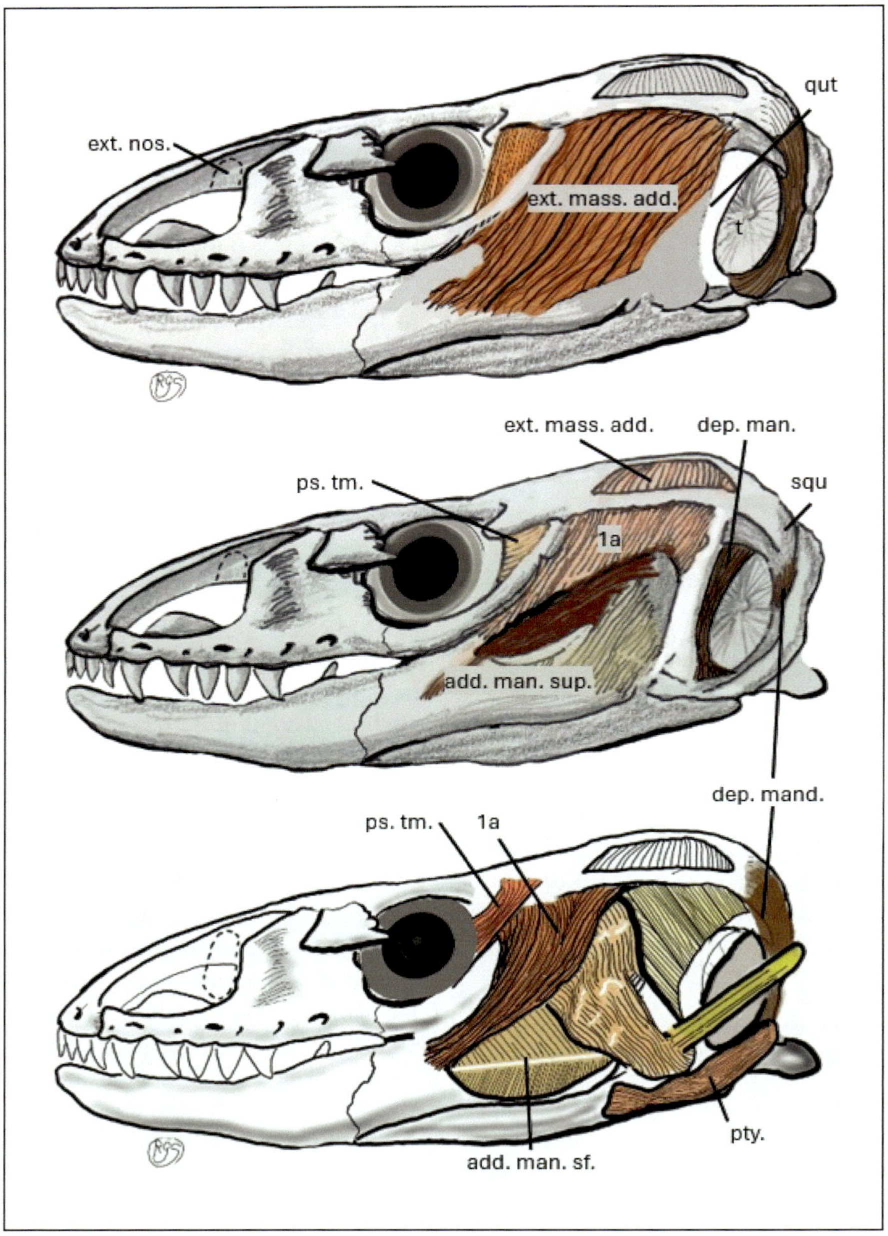

FIGURE 6.27 Successive layers of varanid temporal and jaw musculature based on half-grown *Varanus exanthematicus* (RGS 205). Abbreviations: 1a – 1a levator anguli oris; 1b – ; add. man. sup. – adductor mandibularis superior; add. man; den – dentary; cor – coronoid bone; dep. mand. – depressor mandibularis; exc – extracolumella of stapes; ext. mass. add. – exterior masseter adductor; ext. nos. – position of exterior nostril; jug – jugal; ps. tm. – pseudotemporalis; pty – pterygoideus, qut – quadrate, squ – squamate; sf. – adductor mandibularis externus superficialis; ten – tendon.

mandible. The most exterior layer contains primary biting/grasping muscles and is oriented vertically. The m. superficialis adductor mandibulae externus (MAME) fills most of the lateral region between the jugal and quadrate, forming a large muscle that broadly originates on the descending lateral processes of the parietal, a surface area that gives this muscle considerable strength, both in contraction and biting force. Its insertion is lateral on the mandible, including most of the lateral and medial coronoid, lateral surangular, and may cross the intramandibular joint to the extreme posterior dentary. In some species (e.g., *V. albigularis*, *V. dumerilii*), the posterior edge of the MAME is thicker and extends slightly caudad to cover the articular and part of the upper angular, presumably increasing bite force.

Among *Varanus timorensis* and the *Odatria* group, it appears that the ability to retain a grip on prey favors them having well-developed MAMEs with comparatively smaller and thinner MPs. This is an area worth further study from a functional morphology perspective (Gray et al. 2019, McCurry et al. 2015, Schaerlaeken et al. 2012, Wilken et al. 2019). Aside from their kinetic functions, the temporal muscles, together with the proötic, basisphenoid, and

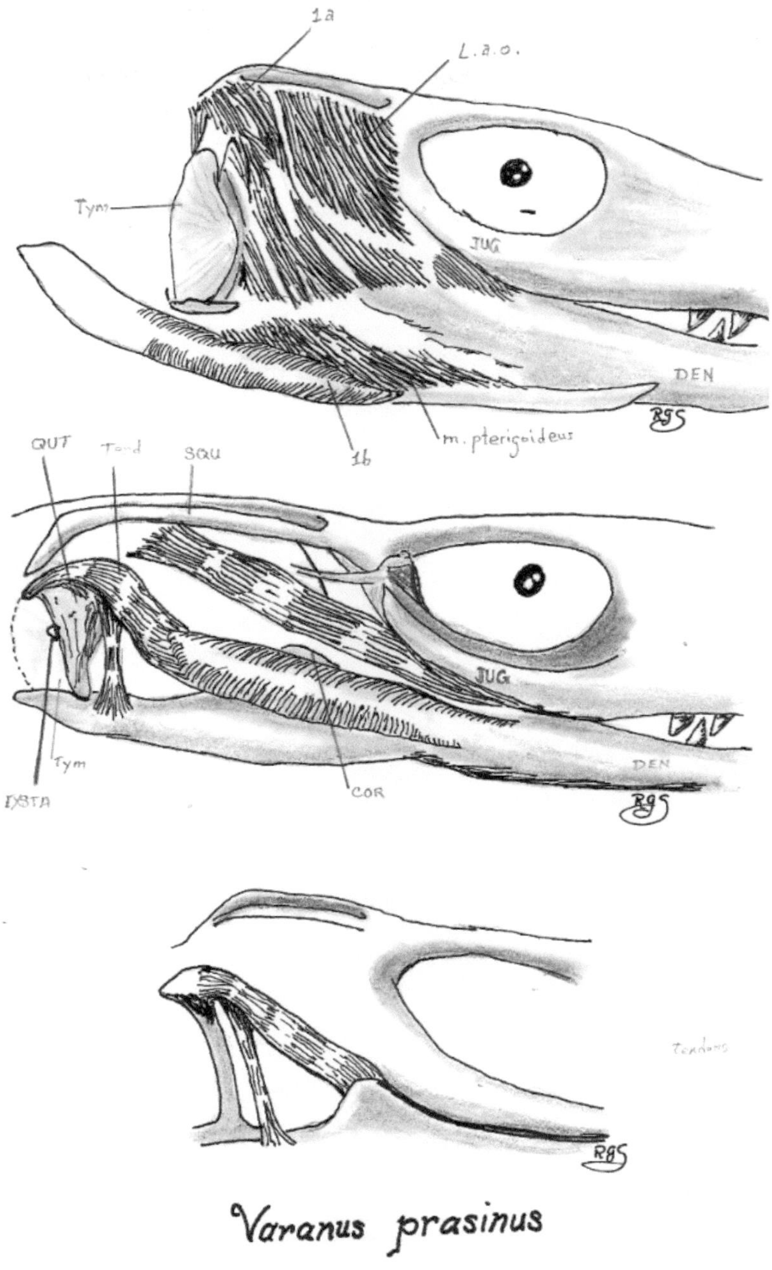

FIGURE 6.28 Successive layers of varanid temporal jaw musculature based on adult *Varanus prasinus* (CAS uncat.). Abbreviations as in Fig. 6.27.

supraoccipital, form a thick and strong enveloping cushion that protects the brain from pressures resulting from swallowing large or hard prey.

The neck in varanids is relatively and often absolutely more elongated, and often the prey to gape ratio is greater than in other limbed lizards. This generally requires a swallowing behavior that is replicated in limbless lizards, including snakes. Two actions particularly warrant attention: lunging and sinuous movement. Lunging occurs when the lizard has prey in its mouth and, to facilitate moving it caudad, loosens the bite grip, lunges the head forward rapidly such that the prey barely moves while "in mid-air," and then clamps the jaws at a point that had been more distal before the lunge. It is a swallow-facilitating action seen in many long-necked predators, including some turtles and birds. Sinuous, or sinusoidal, movement occurs as the neck is moving in strongly curving motions, most especially on the lateral (relative to the spine) plane. The motion is important in moving large food items past the pharynx to a point where esophageal peristalsis can continue the swallowing action. If the food is large enough, the sinuous motion uses neck muscles to forcefully push food toward the stomach. Accordingly, varanids have evolved strong and elongated neck and throat muscles (Fig. 6.29) that perform this task, predominantly the long, broad constrictor colli, an exterior muscular tube with transversely oriented fibers. The muscle

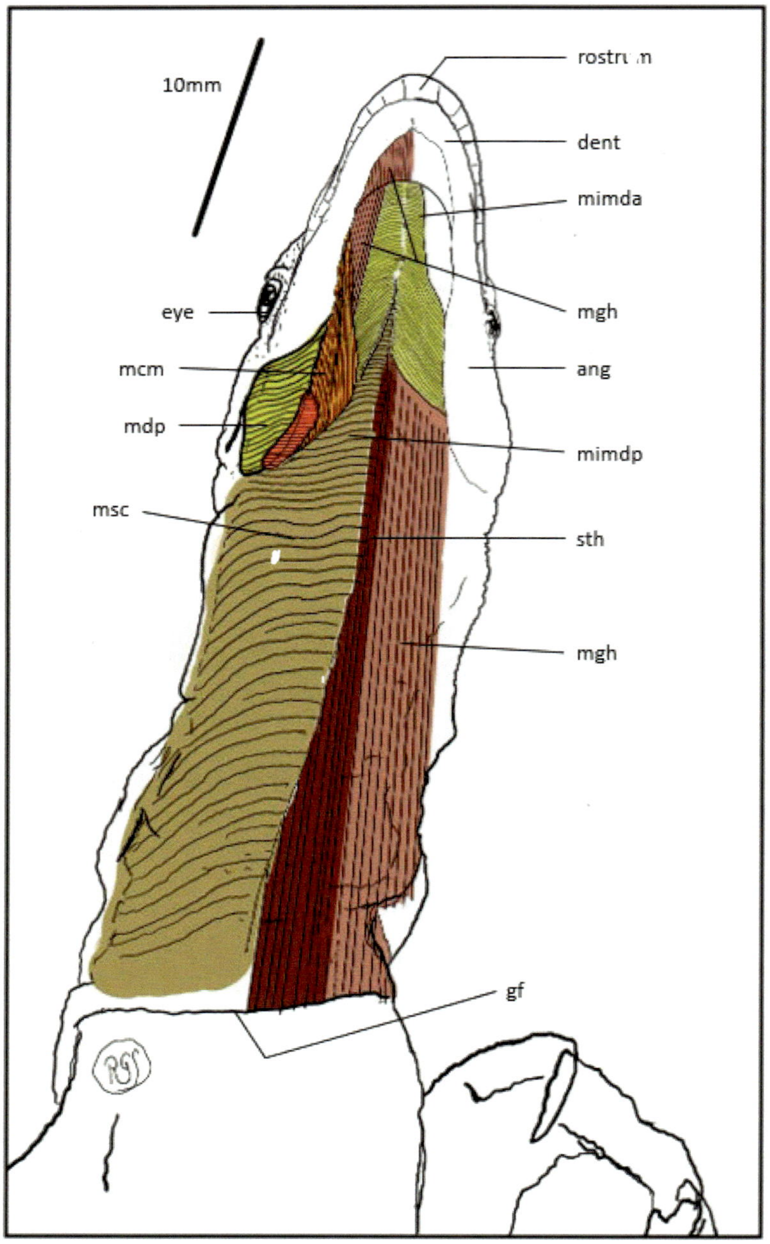

FIGURE 6.29 Throat musculature. Abbreviations: ang – angular; den – dentary; gf – gular fold; mcm – cervicomandibularis; mdp – depressor mandibularis; mgh – geniohyoideus; mimda – anterior mylohyoideus; mimdp – posterior mylohyoideus; sth – sternohyoideus.

originates ventral to the glottal opening and extends ventrolaterally for the length of the neck to the clavicle. The constrictor colli in *Varanus* is much longer than the homolog in most other lizards, including many limbless species. The genioglossus muscles (CN IX), which govern tongue movement, are deep to and segments nearly as long as the constrictor colli but are divided into two pairs. Dorsal to them lie the tongue-retracting hyoglossus (CN XII). The genioglossus, an anterior muscle linking the tongue and medial dentary, is used in the rotation of the tongue along a longitudinal axis and is involved with tongue protrusion, and in varanids, it is quite small. The hyoglossus also controls tongue flicking, and muscles responsible for glottic pumping.

6.4 VARIATION AND ECOLOGICAL CONSIDERATIONS

Varanids possess several synapomorphies related to soft anatomy:

1. The palate contains distinct, unconnected apertures for the nasal channel (choanae) and an opening for the vomeronasal organ.
2. A distinct tongue that is slender, bifid, highly protrusible to a length equal to or greater than the length of the skull, that when retracted rests in a medial groove of soft sublingual tissue. Tongue and associated structures lack taste buds.

3. A parietal eye that sits in a foramen entirely within the parietal bone.
4. Excepting absolute size, all varanid hearts have a nearly complete vertical intraventricular septum that causes enhanced cardiac flow and separation of oxygenated from deoxygenated blood in a way closer to the crocodilian and mammalian hearts than in any other lepidosaurs. This gives varanids greater endurance than reptiles lacking the complete septum.
5. Ventricles not attached to pericardium or nearby organs; the auricles are so attached.
6. Multicameral lungs that contain a single large lumen and numerous smaller diverticulae. These do not approach the respiratory surface area of bird and mammal hearts but exceed that of other reptiles.
7. With *Lanthanotus*, the only squamates with secondary bronchi supported by cartilaginous rings.
8. The liver is situated posterior to the heart, except for a very short pair of anterior extensions of its lobes, upon which the dorsal ventricular surface rests. The gall bladder is contained within the left lobe of the liver, exposed only by a small opening.
9. The gastrointestinal tract is distensible to a degree not seen in other non-snake lizards. The stomach contains numerous tough, fibrous longitudinal rugae, and the intestines have a predominantly smooth lumen.
10. The pancreas has a trilobed shape, with the individual points in contact via fascia to the pyloric region of the stomach, the duodenum, and part of the small intestine.
11. An elongated process of the spleen is semi-independent and attaches the spleen to the pancreas via a unique narrow infundibulum.
12. There is no urinary bladder.
13. In addition to standard vertebrate endocrine and exocrine glands (e.g., thyroid, parathyroid, pineal, lacrimal, adrenal, and salivary), varanids tend to possess antorbital nasal salt-excreting glands (which may not be universal) and mandibular venom glands (which are).

However, the variation of the soft anatomy within the Varanidae can be considerable:

1. The width of the post-vomer palatal region may be narrow, the two sides nearly touching, or widely separated into a V shape. Species with the broader palatal space and have the greatest possible gape are the species that consume large or hard-bodied prey. In contrast, the small primarily insectivorous species (*Odatria* group) have the smallest gape and narrowest gap between palatal bones.
2. The angle formed by the maxillary bones determines acuity of the snout and is closely linked to preferred prey types, e.g., broad-snouted monitors can consume larger prey than narrow-snouted species, and where they are consuming an item of equal size, the broad-snouted lizards do it with noticeably lesser difficulty.
3. Tongue color varies among species and consists of pale yellow, pink, purple, or blue shades, either uni- or bicolored. In many cases, sympatric species of a closely related species group (e.g., the *V. indicus*, *V. niloticus*, and *V. albigularis-exanthematicus* groups) have tongues of different colors that may serve as intraspecific recognition flashes, where a single species sports different tongue colors (*V. varius, V. salvadorii*), cryptic species may await recognition.
4. Nostrils vary widely in shape, size, and position on the snout. Additionally, these nostril features may differ based on the ontological stage of the monitor. The ancestral nostril position is probably midway between the rostral and orbit. Truly anterior nostrils are features of partially aquatic species, as is the more dorsal orientation of the openings (e.g., in *Varanus bivittatus, V. komodoensis, V. mertensi, and V. macromaculatus*).
5. Some of the more highly aquatic species possess an interior nasal flap (*V. dumerilii, V. mertensi*), capable of closing the nostril when submerged. The flaps function passively as buccal air pressure increases.
6. The degree of forward vision, and thus depth perception, varies from minor to considerable, often correlated with age. Acutely pointed snouts correlate with species that eat small, highly rapid-moving prey, the capture of which benefits from increased visual depth of field that is enhanced by the optical overlap afforded by an acute snout.
7. Relative size and symmetry of the lungs and bronchial hilus location vary considerably and may be species or species group diagnostic.
8. The shape, rugosity, and thickness of the stomach are variable, as are the morphologies of pyloric and cardiac sections. Visceral structure coordinates closely with maximum gape and size and hardness of food, likely linked to palatal morphology.
9. Kidneys vary in shape from somewhat ovoid to triangular and in size from slightly larger to much larger than the testes. Further morphological and physiological studies are needed.
10. Preanal pores are rare in varanids and come as a single pair when present. They are less conspicuous or absent in females.

The preponderance of the biological links of these characters are broken down as follows: respiration (nos. 1, 4–8, 13, 17, 20), feeding and digestion (2, 11, 9–11, 13, 14, 18), hormones and behavior (3, 13), and urinary (12), while some (12–16) are involved in more than one significant role. The morphologies of the respiratory system enhance the stamina, and thus ability to pursue and overcome prey, and thus it is not surprising that 16 of the 20 characters noted above are enhancements related to prey capture.

Part III

Ecology and Behavior

The business of being a living monitor lizard is this section's subject. How do they regulate physiology, adapt to a habitat, find food and mates, and reproduce? How do they adapt to the environment and respond to other animals? The subject of physiology is examined directly in ecological contexts. Biogeographical patterns and causes are discussed, including conditions that affect small/island populations. Links between morphology and ecology are presented.

DOI: 10.1201/9781003463528-9

7 Behavior

7.1 INTRODUCTION

Behavior, intelligence, and learning are grouped together because they are all individual versus population-level actions of individuals rather than populations and thus are fully biology-based. Behavior represents the entirety of actions and inactions that are taken by organisms and includes conscious and unconscious mental responses to their environments. Investigations into varanid intelligence are a recent development (Getsy et al. 2022, Howard and Freeman 2022, Huo et al. 2024, Kane et al. 2019, Mendyk and Augustine 2023, Waterman et al. 2021). This chapter looks at the individual lizard, while Chapter 8 continues to review the more complex relationships among numerous individuals and the net interactions between the monitors and their environment. Gans and Tinkle (1977:vii) wrote that the study of reptile behavior has generated an "... enormous mass of random observations..." that requires critique and organization. Baum (2013) proposed that for something to be considered a behavior, it must include four elements. First, only the organism as a whole can display a behavior. Things like breathing, heartbeat, and thirst are not behaviors. Second, a behavior must have a purpose, whether it is essential for survival or provides entertainment; it is goal-oriented. Using the examples from above, holding breath and drinking qualify as behaviors, as does freezing in place. Third, a behavior takes time to become ingrained and repeated because the interplay is between organism and environment, including other organisms, and initially might not be what it appears to an observer. This partially explains why no two organisms in identical situations can ever be seen to respond in *exactly* the same way. Fourth, behavior is based on a choice: the organism decides from two or more possible reactions to its environment. The four elements of behavior must work simultaneously. The commonality that links Baum's (2013) hypothesis is that a behavior is a response of the voluntary nervous system (CNS). The actions of the heart and skin are directed by the CNS and cannot function for long without such guidance but can perform their functions independent of most (or, rarely, all) of the rest of the organism. In comparison, chasing after a potential mate does require involvement of the entire organism. It also is goal-oriented, takes time, and is the result of a choice. The topic of behavior and intelligence has long been controversial and remains so now. Delving deeper into these subjects is not part of the goal of this book; there are thousands of resources about them elsewhere. In that mass of information, only a tiny fraction focuses on reptiles and specifically monitors. It is support from those sources that forms the core of this chapter.

Lizards have become standard models for a variety of behavioral studies and are well represented in the literature (see Bels and Russell 2019, Gans and Dawson 1976, Gans and Tinkle 1977, Reilly et al. 2007, Vitt and Pianka 2014, Whiting et al. 2006, and references therein). Unfortunately, few of these studies focus on varanids, and of those many are based on observations of fewer than ten specimens (Attum et al. 2000, Cooper et al. 2020, Davis et al. 1986, Irwin 1994, Manrod et al. 2008, Thompson et al. 1995). Understandably, most studies have used smaller lizards because of availability, the number of specimens that can be observed, size and husbandry considerations for laboratory work, cost of care and feeding, permitting bureaucracy, ease of handling, and the costs associated with all of these. Nevertheless, studies that employed varanids have added considerably to our understanding of their behavior, and studies that involved other lizard taxa provide a valuable set of comparative data.

In concert with the concepts of Baum (2013), I have included specific activities of the monitors to be grouped under the umbrella term "behavior." These activities are (1) displays of intelligence, (2) choice of locomotion, (3) bipedalism, (4) modes of foraging and feeding, (5) thermoregulation, and (6) agonistics.

7.2 INTELLIGENCE

Ask anyone who has spent time observing *Varanus* if they consider the lizards intelligent, and the answer is almost consistently "yes." In my early days in herpetology, I asked two different and highly regarded mentors with considerable knowledge of monitor lizards if, in their opinion, varanids are intelligent, well beyond other lizards. The first, a morphologist, said yes, that live monitors he had seen seemed, to him, more inquisitive than other lizards, but his impression, he added, was mostly a feeling. Not being his area of experience, he said that he did not know how to go about testing the hypothesis, but someday someone probably would follow up. The second, a field ecologist with extensive experience working with live monitors, also answered "yes." In fact, he followed that up by saying "of course." But, he warned, we had no way yet to define lizard "smarts," and he felt that anyone trying to research the question might be making what Silicon Valley folk called "a career-limiting move," which until recently was certainly the case for dogs (Horowitz 2009). Back then, reptiles were rarely considered to have "advanced" mental abilities, and what few studies that were published focused on turtles (Burghardt 1978, Szabo et al. 2021); "back then" was also 50 years ago, and a lot of new avenues of research (an acceptance of studying a few "out there" questions)

DOI: 10.1201/9781003463528-10

have become more widely accepted. Even so, a few major limitations to such studies are indeed formidable, not the least of which is, how should we define intelligence? Over 100 years of psychological research and debate have failed to accept a definition of intelligence for that most well-studied vertebrate, humans, so it is not unexpected that we lack a useful way to apply the term to other species. There is also the problem that the terms “sentience” and “intelligence” are often used interchangeably, though the former is more commonly associated with “feelings” such as anxiety, fear, stress, excitement, and pain (Lambert et al. 2019).

However daunting the task of defining intelligence, scientists have accepted certain markers that they feel must be part of the larger definition. For example, teaching an ape to talk via sign language assumes some intelligence; for the ape to learn sign language argues for considerable intelligence. A dog that knows all of its toys by name, sometimes 100 or more, or can monitor a flock of sheep without human direction, is considered to have intelligence (see Horowitz 2009, Pilley 2013, Sprackland 2015 for reviews). The ability of an octopus to recognize humans, in diving gear or not, seems to imply considerable intelligence, as does their ability to solve puzzles and learn to unscrew a jar by observing another octopus doing so (Cousteau 1972, Foster 2020, Sprackland 2015). These behaviors are not the activities that equate with one learning not to touch a hot stove by having been burnt once. These activities require memorization and choice that involve accessing a large store of mental recollections. A question being studied now is about kinds of intelligence, including categories of social intelligence used in human psychology (e.g., spatial, numerical, sociability, auditory, and tactile) and whatever “native” intelligence that different taxa possess. Octopuses, dogs, and lizards see their world through eyes and mental processing different from humans, and the biggest difficulty is trying to decipher the methods needed for researchers to understand how to see through non-human eyes. In this quandary I am reminded of the two Chinese philosophers discussing the dreams of one of them. “I often dream that I am one of our goldfish,” says the one. “Ah, are you sure you are a human dreaming that you are a fish? Might you not be a fish now dreaming it is a human?” The difficulty of understanding perspective, whether of fish, dog, friend, or the opposite sex, has been with us for millennia. Quantifying remains elusive.

Do reptiles learn in ways acknowledged by human observers? Published evidence confirms that lizards do learn to act and behave in ways outside basic instincts (Howard and Freeman 2022, Huo et al. 2024, Kirkish et al. 1979, Kis et al. 2015, Krekorian et al. 1968, LaDage et al. 2017, Mendyk and Augustine 2023, Noble et al. 2014, Szabo et al. 2019, 2021; see Burghardt 1978 and Szabo et al. 2021 for review). In some instances, lizards may be strongly influenced by abiotic as well as biotic factors (Batabval and Thakar 2019, Clark et al. 2014, Riley et al. 2017). As for *Varanus*, there is now considerable consensus that they are mentally flexible enough to learn more than just basic behaviors (Cooper et al. 2019, Howard and Freeman 2022, Huo et al. 2024, Kirkish et al. 1979, Loop 1976, Manrod et al. 2008). This view is in part due to tasks performed and then observed by researchers. Varanid brains are comparatively slightly larger, but morphologically quite similar to brains of other lizards. It has been proposed that the lengthy incubation time required for varanid eggs is due to a late period for developing a large brain and braincase (for development account see Chapter 9; for braincase and brain see Chapters 5 and 6, respectively). Though lizards do exhibit several instances of intelligence, their abilities may be limited by the number of neurons in the brain, about 20 times fewer than in mammals or birds of similar size (Kverková et al. 2022). The squamate with the largest number of brain neurons in the study by Kverková et al. (2023) was a 3883 g Asian water monitor (Table 7.2), the brain of which held 78 million neurons, far fewer than the 84 million in a 168 g mammal or 80 million in a 44 g bird (Table 7.1). Only four species of *Varanus* were assessed by Kverková et al. (2022). An interesting finding is that the *Varanus gouldii* (6605.30 g) of twice the mass of the *Varanus salvator* (3883.00 g) had nearly 28 million fewer neurons (Table 7.1). There are no published data proposing that *V. salvator* is “smarter” than *V. gouldii* nor the 400 g *Varanus macraei*. In fact, *V. macraei* belongs to the *prasinus*-group, judged to be among the most intelligent of varanids (Mendyk and Augustine 2023, Mendyk and Horn 2011). The mass of a monitor and number of brain neurons do not necessarily show positive correlation (Table 7.1 graph). When calculating mass to number of neurons, the range for the four species studied by Kverková et al. (2022) goes from 20,147 neurons per gram (*salvator*) to 82,727.50 neurons per gram (*macraei*). While these numbers are only a small fraction of the equivalent measures for birds and mammals, they are ambiguous in relation to determining intelligence based on neuron capacity of a varanid's brain. The high number given for the arboreal *V. macraei* is expected; the number given for *V. salvator*, which has been observed doing similar tasks as *V. macraei*, is not.

Does the number of neurons to body mass data (Table 7.2) tell us anything about intelligence in varanids? It could be argued that arboreal species may need better visual acuity and recognition patterns and therefore larger brains to handle the information processing. Does the difference between *V. salvator* and *V. gouldii* reflect on their diets or habitats? Or is the difference aligned with being a generalist or a specialist? Such questions will require specific knowledge of the behavior of a species and examination of that species' brain, and not one from a different species.

Why should we study lizard behavior for indications of intelligence? The question has nothing to do with the discredited concept of “the lizard brain” or “the reptile brain” of popular psychology, but with the more fundamental questions about brain evolution and the foundations of cognition and memory. Researching the development of intelligence among organisms allows us to discover and test links among sociality, environment, development, and memory that could have important implications for both

TABLE 7.1
Comparison of the Number of Neurons in Four Species of *Varanus* Studied by Kverková et al. (2022)

Species	Grams	Neurons (millions)	Habitat
acanthurus	308.7	40,201,000	Saxicolous
macraei	400.0	70,709,000	Arboreal
salvator	3883.0	188,500,000	Generalist
gouldii	6605.3	167,448,000	Open scrub, forest edge

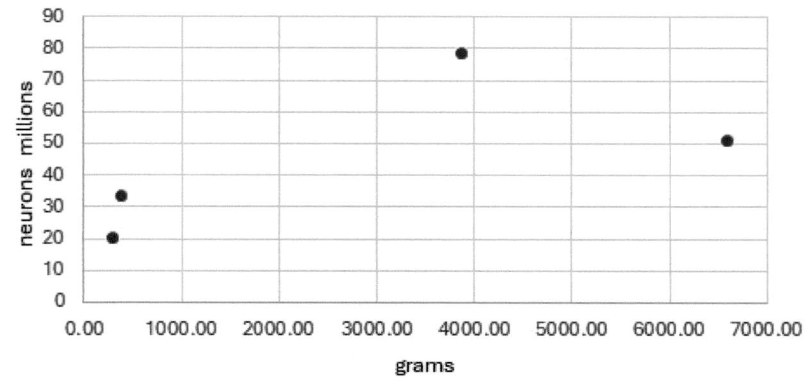

Note: Habitat preferences are saxicolous (*acanthurus*), open scrub and forest edge (*gouldii*), specialized arboreal (*macreai*), and generalist (*salvator*).

zoology and mental health. Lest we dismiss activities such as behavioral thermoregulation as an especially reptilian phenomenon, consider the very common responses of, for example, beach-going humans on a hot day, shuttling from "getting their rays" to seeking shade or the cooling effects of entering the water. These responses have definite affinities with brain function and at least basal links to intelligence. Many "higher level" behaviors might be overlooked without our having access to more ancestral beings and their obvious responses to their environments. Presently, our most challenging task in this regard remains the defining of "intelligence" (see below).

TABLE 7.2
Terms for Specific Actions Taken by Lizards in Physically Performing Thermoregulation

Thermoprotective Postures	Thermoregulatory Postures
All feet on substrate	Torso on substrate; head and neck may be raised; legs outstretched
Digits of 1+ feet lifted off substrate	Head and anterior body lifted; posterior torso, hips, and tail on substrate
1+ feet off substrate	Body raised above substrate by all four limbs; tail raised or on substrate
1 leg off substrate	
"2" + 1 additional feet off substrate	
2 legs off substrate	
"3" + 1 feet off substrate	
Abdomen on substrate, all legs lifted	

Source: From Wilén (2011).
Note: Not the distinction between postures that are to protect the from heat versus those that are for acting thermostatically to control body temperature.

It is difficult, if not impossible, to measure a quantity without having an accepted definition for what is to be measured, in this case, intelligence (Cooper et al. 2019, Whiting and Noble 2018). There has been evidence that viviparity may convey an intellectual advantage (Halliwell et al. 2017, Szabo et al. 2021, Whiting and Noble 2018), presumably because the female must care for her developing young for a longer, and thus more vulnerable time, than an egg-layer. The 21st century has seen varanid intelligence become an important field of study (Burghardt 2013, Cooper et al. 2019, 2020, Cross et al. 2020, Doody et al. 2013, Manrod et al. 2008, Mendyk and Horn 2011, Roth et al. 2019) that has included such aspects as play (Eagan 2019, Gaalema 2011, Howard and Freeman 2021, Huo et al. 2024, Kane et al. 2019, Khandahar et al. 2020), an activity almost universally agreed to imply a level of intelligence.

Based on these studies, and my own observations (Sprackland 2015), I understand intelligence, as mental activity having two aspects. First, there is instinctive behavior, in which an animal engages in an activity automatically, without thought (or its equivalent). The mental programming needed for a neonate to escape an egg, the fixation and chasing of certain moving objects, and the inclination to run from large objects moving in their direction fall into this category. So, too, do mating ritual, copulation, and egg-laying. These are instinctual in that a lizard raised in the isolation of captivity still has the ability to perform these activities without any learning from or observation of other members of its species. Some of these actions are rather complex and may be varanid or even species-specific; consider the tail lashing and lateral positioning of a defensive monitor, or the spectacular three-meter spiral nests and communal egg-laying of at least one Australian species (see Chapter 9). Instinctual knowledge is not necessarily simple. And all of these activities involve the entire animal.

The second broad category of intelligence is learned behavior. Experience teaches varanids what potential prey may fight back, that a branch swaying in a breeze is not a predator, and where to go to find shelter from the sun. While many monitors will flee from a person while still separated by a score of meters, others will quite deliberately walk up to a picnic table and hunt or "beg" for handouts of human foods. Both behaviors have been seen in the same species. Mendyk and Horn (2011) reported on an interesting method used by captive *Varanus beccarii*, a highly arboreal species, to acquire hidden insects. It would extend its claw into a small hole in a tree to pull out an insect that was hiding there. The procedure was repeated on later occasions, which suggests it had a learned and remembered experience. I was later able to observe the same species in a different facility perform the same action, as did *Varanus prasinus* in a nearby enclosure. The lizards could not see each other, so the one monitor did not learn the probe-and-catch technique by watching the other (as I had observed among Pacific octopuses, Sprackland 1995). A question arises: does play evolve from foraging instinct, or did they co-evolve? Both species have been reported to play in captivity, as has the related *V. macraei* (Kane et al. 2019). The same test was given to *Varanus mertensi*, a much larger and non-arboreal species, that did nothing to attempt extraction of a preferred prey from a hole in a log (pers. obs.). *Varanus indicus* is a terrestrial, somewhat arboreal, and largely aquatic monitor that seems to have intelligence in line with the two tree monitor species (pers. obs.). Hediger (1934) reported that while he was conducting field studies on the New Guinean island of New Britain, he had daily contact with a *V. indicus* (probably = *V. finschi*) when he went to the beach for his daily swim. Though the lizard did not closely approach him, it did, he wrote, seem to wait for him each day, entering and leaving the surf when he did. I have kept a *V. indicus* that was an expert at escaping its terrarium. On one occasion it found a place where the ventilation screen was slightly less taut, and it would suspend itself from the screen and use its weight and periodic clawing and swinging motions until the edge of the screen started to rip from its frame, thus providing an escape route. That same lizard used a more ingenious method about a year later; it had figured out how the lock-latch on the enclosure's front door worked, and, like the tree monitors reaching for food in a log, used its claws to reach into the interior opening of the lock and managed to open the door and get out. The lizard had done that when I wasn't there to observe. On the last occasion, though, it immediately began to pick the lock when I placed it back in its terrarium. I was able to videotape this second attempt, then found a different way of securing the lock (which he has yet to figure out).

An important question regarding perceived learning is how to differentiate between habituation and something more random. Koi and guppies will learn to congregate near the surface of the water when someone approaches to feed them (Brown et al. 2011), as will most reptiles that I have worked with, from *Terrapene* to *Eunectes* to *Crotaphytus* to *Varanus*. True, some species and individuals are less susceptible to conditioning, and as they include no varanids, they remain beyond the scope of this chapter. In each of these cases, the learned behaviors have a positive outcome regarding survival value tasks. Rarely addressed is the question of how much does human observation affect the behaviors of the objects of study (Langkilde and Shine 2006)? Aside from the considerable problem of objectively defining intelligence, there are problems relating to factors that may dramatically alter that quality that are independent of learning through experience, among which are the effects of egg incubation temperature (Amiel et al. 2014, Amiel et al. 2017, Amiel and Shine 2012, Clark et al. 2014), interspecific differences in habitat (Batabyal and Thaker 2019, Bezzina et al. 2014), and age (Noble et al. 2014). As much as varanids are lauded for their alleged intelligence, they have been the subjects of rather few experiments to document that behavior. Skinks, particularly the Australian genera *Bassiana* (Amiel et al. 2014, 2017, Amiel and Shine 2012, Clark et al. 2014), *Egernia* (Riley et al. 2017, Szabo et al. 2020), *Eulamprus* (Langkilde and Shine 2006, Noble et al. 2014, Szabo et al. 2020), *Lampropholis* (Bezzina et al. 2014), and *Tiliqua* (Szabo et al. 2019, 2021), have been the primary lacertilian recipients of researchers' attentions, with other taxa less well represented (see Burghardt 1978, Cooper et al. 2019, Kirkish et al. 1979, Kis et al. 2015, Krekorian et al. 1968, LaDage et al. 2017, Wilkinson and Huber 2012, Whiting and Noble 2018 for comprehensive reviews). Nevertheless, what has been observed in *Varanus* has been informative and, in most cases, worthy of continued investigation (Burden 1927c, Cooper et al. 2019, De Meester et al. 2019, Gallema 2011, Manrod et al. 2008, Tsellarius and Men'shikov 1994, Tsellarius and Tsellarius 1997). There is a good deal of information from which to build experiments that can help quantify varanid behavior and intelligence with many other squamates. Also, because monitor lizards engage in many behaviors that are not shared by other lizards, the behaviors should be considered phylogenetically important.

If we consider how often monitors are credited as intelligent in the literature, analysis shows that most of the quotes refer back to very few published data. It is the fact that there have been so few experiments conducted with these "very intelligent" lizards that it is curious, for although size alone presents difficulties for lab work with large lizards, at least 30 percent of known *Varanus* species are close to the sizes of many lizards that have been tested. Experiments in which three varanid species, *V. beccarii, V. mertensi,* and *V. prasinus,* were given food puzzles provided evidence that the lizards could learn to obtain food repeatedly (Cooper et al. 2019, Mendyk and Horn 2011). When retested a year later, the first two species figured out the puzzle faster than they had when previously exposed (Cooper et al. 2020). In specimens of *prasinus* and *mertensi* observed at the National Zoological Park, the former was quickly able to find and extract food placed inside a narrow hole in a tree, while the latter showed no interest in the food (pers. obs.). Two further species of monitor that I tested (*V. indicus* and *Varanus spinulosus*) were able to solve a similar puzzle feeder (Sunglow Claw Puzzle, Fig. 7.1), but with differing results. On first trial the lock-picking *indicus* mentioned above was able to solve the puzzle and take all the shrimp pieces in it within a half-hour. The first discovery of food was accidental, when the lizard bumped the puzzle and one door opened slightly. After eating that shrimp, the lizard explored the rest of the puzzle for several minutes before intentionally opening a door by placing a forelimb on it and pulling back. On repeated occasions, the same lizard would either solve the puzzle within about 20–30 minutes or would ignore the puzzle completely. The other species, *spinulosus*, though phylogenetically close to the *indicus* group, was much slower in solving the puzzle and seemed to have to relearn the process each time the test was administered. Admittedly, these unpublished observations were limited in time and repetitions and are consequently insufficient to form a decent hypothesis or paper.

At about the same time as the Mendyk and Horn (2011) report, Gaalema (2011) published her results regarding visual discrimination and reversal learning based on two specimens of *Varanus rudicollis*. Reversal learning is defined as the ability of an organism to change its reaction to a reward-based choice (e.g., food on white but not black dish) to the opposite (e.g., food on black but not white dish). In this case, the reaction is to the color of the dish, the food possibly hidden beneath it and out of sight. Gaalema correctly noted that the study was limited in number of subjects (one) and trials, and that "given the small sample the… discussion is purely speculative," but that at least one *rudicollis* has the ability for discrimination and to learn reversals.

Problem-solving was the trait sought in the large African species *Varanus albigularis* (Kaufman et al. 1996, Manrod et al. 2008). Juvenile *albigularis* were offered food in a clear tube; the lizard had to figure out how to get to the food (Manrod et al. 2008). The eight lizards managed to access the food within 10 minutes on the first trial, and more quickly in subsequent tests. Additionally, the lizards quickly eschewed failed methods more often in later trials. Specimens at the San Diego Zoo were given snails as prey in tests to see if the lizards could keep track of the number offered. In unpublished studies by John Phillips, it was shown that the lizards had some ability to keep track of the number of prey items. Given just four snails in an enclosed feeding compartment, the monitors learned to eat the four and then enter the next compartment for another four snails. If a snail was removed from a compartment, the lizard would eat the three snails offered and then wait for the expected fourth snail (Murphy et al. 2019, Pianka and Vitt 2003). Additional testing showed that the lizards could keep track of six snails and then move to the next compartment for six more, but above that number they fed indiscriminately on whatever number above six was offered.

Among the large community of herpetoculturists (private reptile keepers), there have been several online discussions and videos purporting to demonstrate intelligence in *Varanus*, but such documentation varies from incomplete (particularly in lacking controls or replicable results) to questionable. Gaalema (2011) comments on the widely disseminated view that varanids are mammal-like (King and Green 1999) based on field and captive observations that

FIGURE 7.1 *Varanus indicus* figuring out how to extract food (shrimp pieces) from a Sunglow Claw Puzzle set up by the author. (A) Approaching the one slightly open compartment used to start the monitor probing. (B) Probing an adjacent sliding door. (C) Opening the door with a combination of pushing with its snout and clawing it to the right. (D) Monitor consuming shrimp from second compartment. Testing using this or a similar device is frequently videotaped and uploaded to Internet sites, notably YouTube.

"have yet to be examined in the peer-reviewed literature." Of the several video presentations I watched, most fall into one of two categories. First are those that show little, if any, activity by the monitors to confirm what the on-screen presenter claims. Viewers are given lectures that are objectively unsubstantiated by presenters with a poor knowledge of conducting an experiment, or how to use video to convey meaningful content. From this group it remains to be seen how many sources actually have something objective to offer investigators. The second group makes an effort to explain their intent, show their experimental setup, and have the testing conducted on camera. There are faults here, as well, the most frequent being the presenter "cheating," showing the lizard where food is placed (though not how mechanisms, such as the Sunglow Claw Puzzle, work), or partially opening closed slots and thus providing additional cuing. While it is possible that the lizards can perform the task, the video material does not provide sufficient evidence to prove it. Alternately, presentations are marred by poor editing. An example is when a group of excited testers provide the lizards with the experimental materials and, just as the monitor appears to be performing the task, the edit cuts to a view of the excited boisterous observers. Such videos seem as focused on aggrandizing the people as (or more than) observing demonstration of varanid intellect. In spite of these objections, many other videos that are not presenting material to show intelligence sometimes do have monitors behaving in ways, and repeatedly, that suggest intelligence and seem to warrant further, more formal, study. I request academic colleagues to stay focused on the activities of the monitors when watching these programs; indeed, I now watch them with the volume off. Other ambiguous presentations attempt to show varanids come when called by name or respond to a clicker, such as that used in training dogs. Having witnessed the former at several facilities I believe that more rigorous and controlled testing is warranted, but at present the offered evidence is overwhelmingly anecdotal. Anecdotal does not mean worthless. Several experienced varanid keepers have shared considerable video evidence that shows species such as *Varanus macromaculatus*, *Varanus bivittatus*, *Varanus niloticus*, and *Varanus salvadorii* responding to being verbally summoned in the absence of food offerings or physical signaling by the keeper (see Fig. 7.2). Tom Crutchfield has been working with monitors for over three decades and claims he has been able to "tame" newly imported *V. salvadorii* and get them to come to him

FIGURE 7.2 Examples of large adult monitors that have been acclimatized to being handled. (A and B) *Varanus salvadorii*, widely considered by herpetologists and keepers as the most aggressive of varanids, including the Komodo Dragon. They are agile, alert, top-level predators in their native New Guinea. In (B), keeper Jeremy Westerman compares the length of his specimen to his own height. (A) Courtesy Nick Quinones; (B) courtesy Katie Bruce; (C) source unknown; (D) courtesy Urrutia Mario and Judy Hoffman.

when called by name (documented on videotape). This species is particularly known for aggressive behavior, and both its large adult size (total length 2+ m) and unique dentition (see Chapter 5) make casual handling of specimens potentially foolish and extremely dangerous. Nevertheless, he and other patient keepers have been able to help the more aggressive varanids adjust to captivity and become tractable. Other examples of large varanids that allegedly respond to being called include *V. bivittatus* (pers. comm. Mark Bayless, Gin Black), *Varanus komaini,* and *V. salvadorii* (Fig. 7.2). Because responding to one's name requires some level of intelligence, and because the monitors must become familiar and comfortable (e.g., specifically recognize) with keeper and signal, a certain level of intelligence beyond instinct seems required. These examples, however, will require much more rigorous testing, isolating the lizard away from background noises, the keeper refraining from making any gestures, and, ideally, not being in sight of the lizard. From the examples that have been publicized, it seems that the subject is worthy of experimentation.

For any evaluation of varanid intelligence, an important criterion is to have the required action repeatable with less time between subsequent attempts, coupled with some ability to retain the learned pattern at a later date. We must also accept that any testing, especially under captive conditions, is likely to give questionable results. An example covered in Chapter 9 shows at least two large Australian varanid species excavate incredibly complex, deep nesting tunnels. It is not uncommon to find lizards that share the same nesting place, notably among geckos, but the complexity of the varanid subterranean lair at least hints that cooperation could be involved in its excavation. Evaluating their reproductive behavior in captivity will not let us discover such a potentially unpredictable behavior. Even the best zoo or well-funded private keeper would not have constructed an enclosure with 3m of soil for what was then considered an extraordinary and unpredictable "new" behavior of digging a deep, spiral nest. Some lizards may thrive under captive conditions but not in seemingly identical conditions in the field: given the same temperature, ultraviolet exposure, and humidity, a captive lizard should bask and feed while in the field the addition of a minor breeze may keep the same lizard sequestered in its den. Testing is affected by human-linked biases that may seem insignificant but might not apply at all in the field.

If there exists cooperative behavior, where, then, does play fit in the intelligence taxonomy? A definition for play that I find useful is: "Play is defined as an activity in which an animal engages for pleasure or enjoyment rather than for a serious or practical purpose" (Khandakar et al. 2020: 257). Play behavior has been documented for various reptile groups, including turtles (Burghardt et al. 1996), crocodilians (Dinets 2015), geckos (Barabanov et al. 2015), skinks (Sprackland 1972b), and varanids (Burghardt et al. 2002, Gaalema 2011, Howard and Freeman 2022, Kane et al. 2019, Khandakar et al. 2020, Manrod et al. 2008), though there are still very few detailed publications on the subject for varanids. Those who research reptilian play behavior frequently consider it a function of intelligence beyond a simple survival necessity. A relatively new aspect of captive animal management is enrichment, whereby the animal's living area is provided with objects that are supposed to provide a way to perform "normal" activities or play. The ability to engage in such activities has been widely accepted as important for captive animal welfare, including reptiles (Burghardt 2013, Burghardt et al. 1996, Eagan 2019, Lambert et al. 2019, Meester and Baeckens 2021), but again, much of what we know is based on anecdotal or minimal objective data.

7.3 LOCOMOTION AND POSTURE

Prior to the late 20th century, most studies on varanids were based on anatomical work, coupled with scant field data from expeditions. When Douglas Burden undertook his expedition for the American Museum of Natural History to collect Komodo Dragons, his comments reveal his lack of knowledge of monitors generally. Of their locomotion Burden (1927) wrote,

> ...is known for its moving about in a way quite different to that observable in other Varanians: for it does not creep, but keeps up high on its extremities, just like most of our Mammalians do, or like the orthopodus Dinosaurs did, the belly of the moving animal never touching the floor.

In the few years after the discovery of Komodo Dragons, information about them was extremely limited, sometimes based on anecdotes, and misleading. De Rooij (1915: 151) stated that "total length reaches 7 M," about 4m too many. Burden had gone to Komodo 15 years after discovery of the giant lizards – safely after the dust from the Great War had settled, and a time when he had no idea how big the lizards might actually be. He brought specimens back, some going to zoos. At the time, herpetoculture was in its infancy, when even the best zoos often provided temperatures to their charges that were suboptimum, sometimes even fatal. The first western zoos kept Komodo Dragons on concrete floors, heat coming largely from hot water pipes under the concrete. Thus, seeing a captive varanid, or almost any other lizard, in conditions suitable for it to walk or run as it would in nature was extraordinarily rare. As we know now, monitors and a great many other lizards, when healthy and properly heated, typically "keeps up high on its extremities."

7.3.1 Bipedality

The ability to stand or run upright on just the rear legs, bipedality, occurs among lizards as either a dynamic (running) or a static (tripoding) behavior (Fig. 7.3). There has been considerable interest in the bipedal behaviors of lizards (Aerts et al. 2003, Ballinger et al. 1979, Clemente 2014, Clemente et al. 2009, Fedak and Seeherman 1979, Hertz et al. 1988, Hsieh 2003, Lee et al. 2018, Olberding et al. 2012, Schuett et al. 2009, Suzuki et al. 1991), though considerably less on varanid bipedality (Clemente et al. 2009, Neugebauer

FIGURE 7.3 Bipedality n varanids is not often associated with locomotion as it is for other lizards. (Left) *Varanus indicus* in upright posture assumed as part of a threat display, accompanied by long hissing and occasional tail lashing. (Right) *Varanus panoptes* in Papua New Guinea, taking a tripod stance in reaction to the photographer's proximity.

1976, Schuett et al. 2009, Suzuki et al. 1991). Dynamic bipedality was first studied in basilisks (Snyder 1949) by analysis of high-speed movie images. Dynamic bipedality is found among the Iguania (*Basiliscus, Callisaurus, Crotaphytus, Dipsosaurus, Holbrookia, Iguana, Liolaemus, Sceloporus, Tropidurus, Uma*), Agamidae (*Agasthyagama, Chlamydosaurus, Ctenophore's, Hydrosaurus, Intellagama, Lophognathus, Otocryptis, Pogona, Sitana, Tympanocryptis*), Lacertidae (*Acanthodactylus, Takydromus*), Teiidae (*Aspidoscelis, Salvator, Tupinambis*), Anguidae (*Gerrhonotus*), Lanthanotidae (*Lanthanotus*), and Varanidae. However, for most of those taxa, bipedality must be the result of running swiftly at maximum or near maximum speed. Basilisks do not walk on water, they run on it, or they sink. In contrast, static bipedality (Fig. 7.3), the ability to tripod (or periscope) – that is, to raise the body at a near right angle to the ground, raising the head high enough to see over objects – differs from dynamic bipedality in that the lizard uses, in addition to its hind limbs, its tail to form a well-balanced tripod (Schuett et al. 2009, Suzuki et al. 1991) and remains in one spot. Tripoding ability is less common than bipedal locomotion, but I have observed it in *Chlamydosaurus, Otocryptis, Sitana, Salvator, Tupinambis, Elgaria, Gerrhonotus* and *Lanthanotus*, in addition to most varanids.

Iguanian and agamid lizards typically shift into bipedality as a result of running and achieve the characteristic 30–45-degree horizontal angle as leg movement shifts solely to the long hind limbs. In almost every lizard capable of this locomotion, the hind limbs tend to be considerably longer and more muscular than the forelimbs. Van Wassenbergh and Aerts (2013) studied the agamid *Ctenophorus cristatus*, which is morphologically similar to many of the bipedal iguanians (including agamids). How, they asked, does the lizard maintain a constant speed while running? The lizard's first requirement is lifting the anterior part of the body and moving the center of mass toward the hip. In so doing, the forelimbs are lifted and supplanted by the reach of the hindlimbs and the raised height at the hips. This same posture may be seen in young of the larger species of *Varanus*. Aerts et al. (2003) suggested that static bipedality may not be adaptive, having little to do with giving the lizard a particular advantage. Instead, they propose that bipedal locomotion is a spandrel, a feature that evolved as a byproduct that may have no particular advantage to the organism (Gould and Lewontin 1979). However, they also suggest that by shifting the body weight to the hip while running the lizard might more easily and responsively manage obstacles; having only the hips, but not the hips and shoulders, in use during high-speed running could indeed be advantageous (Olberding et al. 2012). Having a single center of balance likely enhances stability and allows lizards to compensate for different substrates (Druelle et al. 2019). Monitors, though, tend to be erect, their backs nearly perpendicular to the ground, as they run. This could be an adaptation to the larger mass of bipedal varanid species; given their body mass, the more that can be shifted to the hips, the easier it could be to maintain balance. Like the horizontal runners, varanids keep the tail straight and horizontal to the ground when running. Again, the comparatively greater mass of the tail might help counterbalance the monitor enough to stay upright. I have seen only two varanids (*Varanus panoptes*) with one-third or more of the tail missing, and these could not, apparently, get to an upright posture to the degree of lizards with intact tails. It has also been proposed that legs in the positions taken during dynamic bipedality are more able to adjust for changes in terrain. According to Druelle et al. (2019), a more elevated head becomes more stable and would facilitate both visual and vestibular perception.

Bipedal running has been reported for several species of *Varanus*, and it seems confined to the larger species (Schuett et al. 2009). Across all varanids bipedal tripoding has been observed in perhaps the majority of species, including the dwarves, such as the saxicolous *Varanus storri*, middle-sized arboreal species such as *V. prasinus*

(Fig. 7.3E), and large generalist forest species like *V. indicus* (Fig. 7.3A), and most of the largest species including *V. gouldii* and *V. spenceri* (Fig. 7.3C, D). At least as many have been observed to engage in upright wrestling bouts (Attum et al. 2010, Auffenberg 1981, Bennett 1998, Deraniyagala 1958, Gaulke 1989, King and Green 1999, Philippen and Eidenmüller 2007, Thompson et al. 1992, Tsellarius and Tsellarius 1997), where the upright posture may be an intimidation behavior.

7.4 FORAGING AND FEEDING

Most phylogenetic analyses place the Varanidae in a position indicating that they are highly derived squamates, so it is not surprising that their hunting and prey capture behaviors are not limited to a single mode (Figs. 7.4–7.6). Their alleged mammal-like physiology has been reviewed and found to be misleading (Jessop et al. 2020), but important analogs remain. Like mammals, varanids are frequently in motion and when active spend considerable time foraging. The (probably) unique respiratory system, with its independently operating lungs, is important for the respiration and cooling functions that allow monitors to run longer than other lizards without becoming fatigued. Their activity periods are also longer than that seen in many other squamates and put a high demand on frequent feedings. For monitors that feed on lizards the stamina to run for a longer time than the prey is highly beneficial.

All varanids approach a food item and bite it (Fig. 7.4A–C), a simple jaw-prominent feeding mechanism (I shall return to that subject below). Some species have developed additional approaches for the capture and consumption of food and may be far from universal. For example, *V. mertensi* (Hermes 1981) and *V. niloticus* (Keith and Ginsberg 2010) have been observed to capture small fishes within a coil of their tails, then scoop the prey toward the mouth. In Sri Lanka, *V. salvator* has been seen to excavate small pools and trap fishes within (Wickramasinghe et al. 2010). These three species, plus *V. bengalensis* and *V. dumerilii*, include hard-shelled clams, snails, crabs, crayfish, and turtles as prey that, if the lizard cannot swallow whole, may be eaten in bits. In particular, *V. dumerilii* dismembers crabs

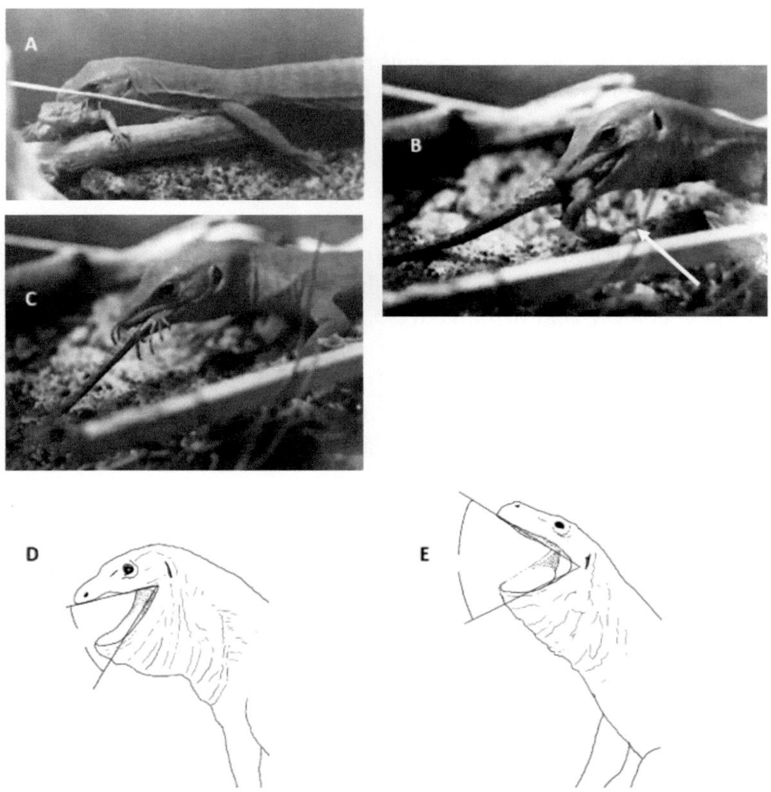

FIGURE 7.4 Upper set: young adult female *Varanus tristis* consuming a lizard (*Sceloporus*) in captivity. (A) Lizard is seized midbody by monitor, then vigorously shaken until it reduces resistance to capture. (B) Lizard, except for tail, is now in mouth and pharynx of monitor; note that the lower jaw is slightly bowed outward (arrow), accommodating the large prey. The upper jaw remains fixed in normal position. (C) Lizard is largely consumed, at which point the monitor begins the gulping sequence to complete the swallow (see text). (D) and (E) : diagrammatic representation of two stages in swallowing a very large prey animal (omitted) by *Varanus komodoensis*, showing the maximum gape of the varanid that takes the largest prey. (A) Prey in widely opened mouth (44°), its entire head in pharynx as lizard's jaws used to grip until (B) the head is raised, and gape widens to near maximum (59.5°) as lizard uses gravity to assist in swallowing. Note the greatly enlarged gular region as it accommodates the large prey. As in *tristis*, the lower jaw remains fixed and the lower jaw bulges laterally at the intramandibular hinge. Compare the gape of *komodoensis* with the varanids in Fig. 7.5. Drawn from photographs.

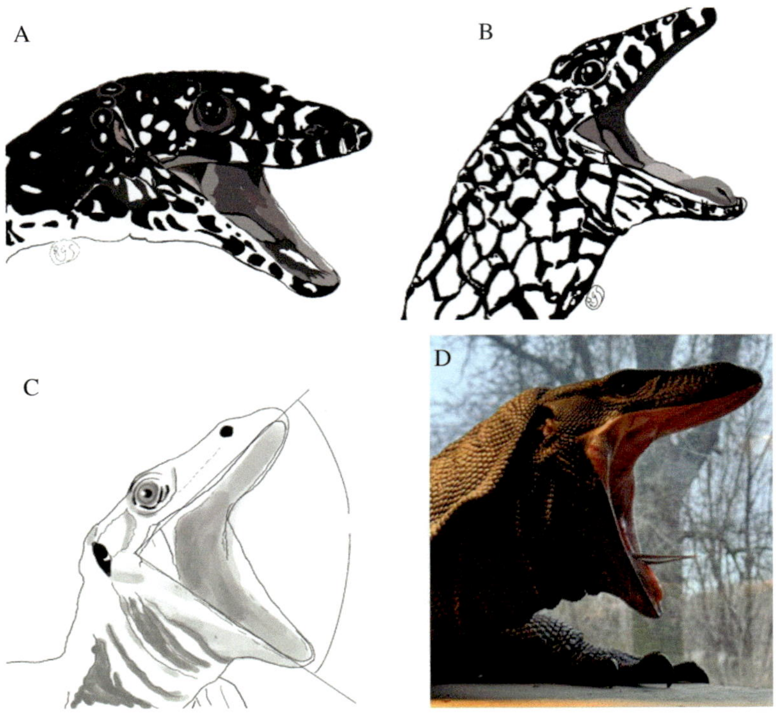

FIGURE 7.5 The gape of a varanid is greater than that of many lizards but does not approach that of caenophidian snakes. These illustrations show the maximum angle possible for a monitor to open its mouth. (A) *Varanus doreanus*, angle 77°. (B) *Varanus giganteus*, angle 64°. (C) *Varanus komaini*, angle 74°, Drawn from specimen in (D). (D) *Varanus komaini* showing gape and passive extension of tongue when doing so. Drawing from live specimens. (D) Courtesy of Gin Black.

and eats them one claw at a time. Examination of the guts of these and several other large monitors often includes hard-shelled beetles. The large, arboreal, and predominantly frugivorous *V. olivaceus* takes hard-shelled tree snails which it crushes with its array of thick, conical, and blunt teeth, while in *V. niloticus* and *V. bengalensis* there is an ontogenetic shift in morphology of the posterior teeth into something resembling molars and used for the hard prey taken by adults (but not juveniles). A similar change is seen in the African *V. albigularis*, though from a younger age, and is used to crush small tortoise shells. There is no such change in tooth shape in either *V. dumerilii* or *V. mertensi*. Hard-bodied prey is rarely shaken prior to consumption.

In contrast to the way hard-bodied prey are handled, most other prey species are captured by a direct seizure by the monitor's jaws. Small animals, such as insects, frogs, and lizards, may be consumed quickly by a few lunges and gulp movements of the head and neck (Fig. 7.3A–C)

FIGURE 7.6 The anguid lizards *Shinisaurus crocodilurus* (left) and *Xenosaurus grandis* (right). Both have long been considered part of the diploglossan line of the Anguimorph, but some phylogenetic studies have placed *Shinisaurus* on the varanoid line. Both taxa employ the tongue for prey detection, facilitating capture and intrabuccal manipulation.

as described above. Larger, struggling, or more formidable animals are generally seized and shaken vigorously until the prey's neck is broken and resistance ends. This action is almost identical to what many predators do with prey, including mammalian carnivores such as dogs and wolves. Smaller, insectivorous lizards also often shake prey violently; *Crotaphytus* and *Pogona*, for example, engage in such action with even small prey (pers. obs.), but in their cases it is to reorient that prey into a position that more easily accommodates swallowing or may serve to disorient the insect so that handling is easier. The same holds true in small *Varanus*, such as *V. storri* and *Varanus primordius*. When larger animals are prey, the shaking stops it from struggling or being able to escape. Unlike caenophidian snakes, varanids cannot open the mouth to exceptional proportions by means of disarticulating cranial bones. The hyoid apparatus can be lowered and retracted, thus allowing the pharynx to expand, and the quadrate does shift forward to some degree but does not disarticulate from the skull. Varanids can, however, open their mouths wider than most lizards in the vertical plane (Figs. 7.4D, E and 7.5) and can accommodate large prey because of the mobile intramandibular joint (see Chapter 5). The mandibular symphysis is fixed, but the intramandibular joint allows the mid-jaw to bow laterally about 5–7 degrees horizontally (Fig. 7.4D, E). The absolute range of flexibility of the intramandibular joint has barely been studied, but it may be possible that in some *Varanus* it can flex more than 10 degrees (pers. obs.).

Lizards are generally divided into two groups based on prey capture (Jessop et al. 2020): either tongue assist (Fig. 7.6) or bite to grab. Tongue-assisted lizards tend to be highly visually oriented, little or not olfactory guided, and use the tongue to make initial contact with food, whether animal or plant, and to help move it into the mouth (Fig. 7.6). In many cases, food in the buccal cavity may be further manipulated by the tongue, which plays a part in swallowing. Lizards in this group include the Iguania, Scincomorpha, Gekkota, and other non-anguid taxa. The Anguimorpha, which includes the Varanidae, instead are largely visually oriented, highly olfaction guided, and use the tongue as a sensor to detect and follow the scent of prey. Some anguids, such as *Gerrhonotus*, *Shinisaurus*, and *Xenosaurus,* combine modes, using the tongue for prey detection via olfaction, and to help move food into the mouth (Fig. 7.6). Varanids do sometimes use the tongue to reposition food already in the mouth and do so by moving the tongue between the buccal side of a mandible and the food, then pressing the tongue against the prey and retracting it, bringing the prey rearward.

Because varanids have been observed in the field and captivity for well over a century, it came as a surprise to the herpetological community when it was discovered that these lizards have venom systems (Fry et al. 2006, 2009, Fry and Scheib 2007, Mcnaughtan and Norman 2009, Vidal and Hedges 2005, but see Sweet 2016), an attribute among lizards previously attributed only to the Helodermatidae. The venom inhibits blood clotting in humans, and likely other vertebrates, and toxicity varies among taxa, being greatest in *Odatria* and the *prasinus*-group (Dobson et al. 2019). Arbuckle (2009) investigated the use of venom by varanids and postulated that it could be used for defense, digestion, prey capture, and as an oral antimicrobial agent, thereby filling the same roles that venom plays for venomous snakes. Dobson et al. (2019) favored the prey capture hypothesis. Arboreal varanids have the most active venoms, which probably evolved to permit capture of energetic arboreal prey, which could easily escape by flying, falling, or disappearing into the foliage. Those authors consider the need to quickly subdue swift prey by some varanids to equate with similar venom in the arboreal African snakes *Dendroaspis angusticeps* (green mamba: Elapidae), *Dispholidus typus* (boomslang: "Colubridae"), and *Thelotornis mossambicanus* (twig snake: "Colubridae"). Coincidentally, the two colubrids had been believed to be non-venomous until they fatally bit eminent herpetologists Karl P. Schmidt (*Dispholidus*) and Robert Mertens (*Thelotornis*). The usefulness of venom to varanids is an ongoing subject of research, questioned in part by the differences in immunity of the lizards themselves (Chandrasekara et al. 2024). If venom is primarily for defense to at least some *Varanus*, uses put forward by Arbuckle (2009) are still possible.

7.5 THERMOREGULATION AND ULTRAVIOLET RADIATION REGULATION

This section deals with behavioral aspects of regulating a monitor's body temperature and the closely related influx of ultraviolet radiation. Hormonal/neuronal aspects of the subject are briefly covered in Chapter 8. Living things, no matter their taxonomic classification, are largely assemblies of peptides, polypeptides, and proteins (plus water). These components have functions from structure to behavioral regulation and, like all chemicals that undergo reactions, are regulated by temperature. If the organism is too cold, digestive enzymes might not function; too hot, and many chemical bonds break and lead to various degrees of damage, including death. Homeotherms, the mammals and avian reptiles (birds), convert considerable body heat via metabolism of their food (fuel), particularly carbohydrates and fats. The body's thermal homeostasis is controlled by alternately burning more fuel or venting excess heat through panting or sweating. The thermostat-regulated metabolism requires considerable caloric energy to operate, up to 80 percent of a homeotherm's calories used in heat regulation. This accounts for why homeotherms, including humans, must eat so frequently.

Ectotherms are those creatures that do not expend much metabolic effort into producing heat, but instead use basking in the sunlight to get warm, and then find shade to keep from overheating. The activity expended in physically moving to and from heat sources is calorically much less costly than producing one's own heat, as evidenced by the poikilotherm's need for fewer meals. Heatwole and Taylor (1987) divide ectotherms into three sub-classes. They reserve the term poikilotherm for animals in which their body and ambient temperatures are nearly equal. Heliotherms derive

TABLE 7.3
Three Examples of Temperature (°C) Variations Taken Simultaneously for Three Organisms in the Same Habitat, Within 50 m of Each Other at Ambient Temperature of 36°C

Organism	Body Temperature (External)	Basking Site Temperature	Retreat Temperature	Substrate
Dog (black fur)	56.1	36.1	–	Grass
Dog (gray fur)	53.9	36.1	–	Grass
Dog (white fur)	40.0	36.1	–	Grass
Chelydra serpentina ♀	25.0	25.2	–	Soil
Plestiodon fasciatus	43.3	42.7	33.7	Concrete

Note: My dog (♂) is a Pyrenees Mountain dog mix with three different fur colors. The *Chelydra* was laying eggs in soil around the base of a tree. Color and basking site temperature vary considerably within a short distance.

their heat from solar radiation via basking and being diurnal. The third sub-class contains the thigmotherms which obtain heat largely from the substrate. By these criteria, varanids are primarily heliothermic, though all three environmental sources contribute to providing their heat (Table 7.3), and they straighten the limbs, raising the ventral surface off the substrate and cooling via conduction. How do monitors know when they are overheating, and when it is time to seek a cool retreat? However sensitive their skins may be, and we know little about that, it is a special cranial organ that acts as a UVR (ultraviolet radiation) dosimeter. This is the parietal organ, also known as the parietal eye, third eye, and (less accurately) pineal eye (Fig. 7.7). Most terrestrial vertebrates have a subdermal version of the organ, but it is only visible dorsally on the heads of many diurnal lizards, including most of the Iguania (excepting Chamaeleonidae), Xantusiidae, and the Tuatara (Rhynchocephalia), and it has long been of interest to researchers (de Graf 1886, Dendy 1899, Leydig 1890, Locy 1894, and see Eakin 1973 and references therein). The term "third eye" is perhaps most accurate because (a) outside, the varanids, the eye may border or be within the frontal(s), and not in the parietal, and (b) the structure has the embryological and morphological equivalents of the other eyes, though it is limited in light recognition and forms no retinal image (see Chapter 6 for morphology). "Parietal organ" at present has become the preferred term among many zoologists. The parietal organ in monitors is visible in all neonates and young lizards, and though it generally remains visible in adults (Fig. 7.7), it often becomes occluded and indistinct in old specimens.

Functions of the parietal eye have remained unclear and include sky orienteering (Beltrami et al. 2010, Foá et al. 2018), following circadian rhythms (Janik and Menaker 1990), mate acquisition (Olsson et al. 2011), magnetoreception (Nishimura 2020), ultraviolet exposure measurement (Jenison and Nolte 1980), ultraviolet vision (Fleishman et al. 1993, 2011, Pérez i de Lanuza and Font 2014), hormone regulation (Firth and Turner 1982), inducing agnostic behaviors (Phillips and Howes 1988, Rossi et al. 2023, Whiting et al. 2006), and thermal sensitivity (Avery 1979, Conley and Lattanzio 2022) among them. Experiments were carried out on a variety of lizards, including the genera *Anolis* (Grace and Beshare 1994, Hutchinson 1974), *Crotaphytus, Lacerta* (Crisp et al. 1979, Olsson et al. 2011), *Leiocephalus* (Phillips and Howes 1988), *Platysaurus* (Whiting et al. 2006), *Podarcis* (Tosini and Avery 1996), *Sceloporus* (Grace and Beshare 1994), *Tropidurus* (now *Microlophus;* Rossi et al. 2023), and *Xantusia* (Bezy 2019), a limited sampling heavily favoring the iguanines and including no varanids. One can, however, extrapolate because of the amount of overlap

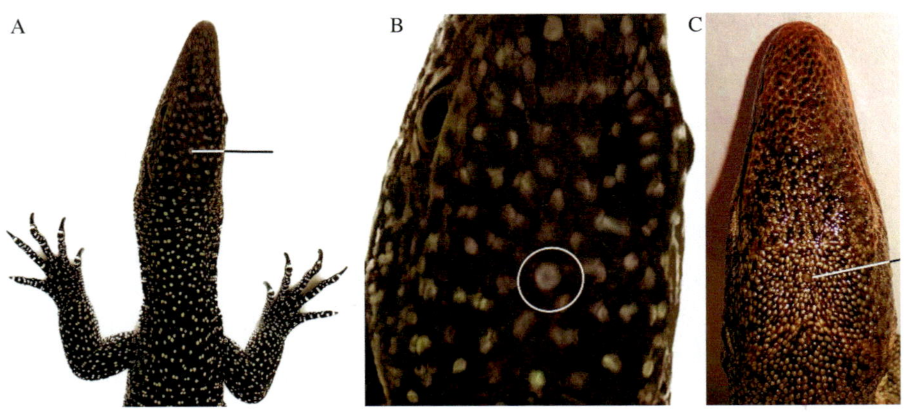

FIGURE 7.7 (A, B) External appearance of the parietal eye in the central parietal area in an adult *Varanus indicus.* (C) The same in an adult *Varanus orientalis* in which the eye is indistinct, its surface the same color as nearby scales. The parietal organs of old varanids are often obscured by thickening of the outer layer.

in results from the research on lizards as well as other squamates. First, there is now little doubt that the parietal eye acts as a UVR dosimeter and, by measuring ultraviolet exposure, helps the lizard know when to move out of the sunlight, keeping the body from overheating. More recently, experiments have shown that there is an important link between the third eye and compass orientation (Beltrami et al. 2010, Foá et al. 2018). Lizards trained in experiments to orient in fixed directions in units exposed to sunlight failed to orient themselves when the parietal eye had been painted over. The mechanics of how the parietal eye-orientation operates are still not understood but the parietal organs' sensitivity to light, especially ultraviolet wavelengths, has been confirmed. It thus acts not only as a UVR dosimeter but also as a thermostat and light intensity meter (Avery 1979, Engbretson and Hutchinson 1976, Irwin et al. 2018, Tosini and Avery 1996). There are likely several hormonal pathways regulated by the many wavelengths of light. In several lizards it has been shown that the reproductive organs are strongly regulated by parietal eye-mediated hormones, as are circadian rhythms generally (Cothran et al. 1979, Firth et al. 1988, Innocenti et al. 1994, Menaker and Wisner 1983, Table 7.3). Though UVR and thermoregulation are often considered two aspects of a rigidly linked phenomenon, studies on yet another iguanine – *Sceloporus undulatus* – have shown how the two needs are distinct, and that the lizards will allow either UVR needs, or the temperature needs to override the other (Conley and Lattanzio 2022). The pineal organ has some control over agonistic behavior (Phillips and Howes 1988, Rossi et al. 2023, Whiting et al. 2006), very likely by the control of hormones. Though several of the cited research papers use the phrase "of lizards" when discussing just one species, the cumulative work points to the inclusion of the most basic results, enumerated above, as being likely in varanids.

The need of captive reptiles to have regular exposure to quality UVB radiation has been justified by numerous studies (supr. cit.). The reptile care industry offers options for providing that light. Research has shown that the popular "conventional" UV lights and LED emitters are not created equally, and often do not perform to advertised standards (Bergmanson et al. 2011, Wunderlich et al. 2024), and may even be harmful to human caretakers. The researchers strongly advise purchasers to ascertain that they are getting the product they need even though the specific needs have yet to be quantified (Fig. 7.8).

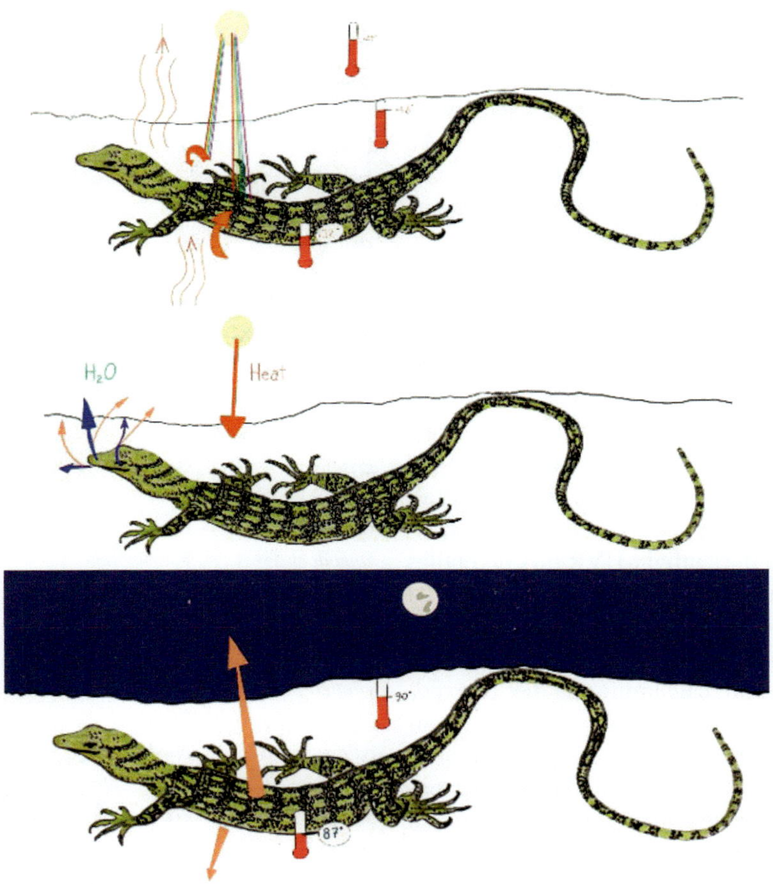

FIGURE 7.8 Absorption and dissipation of heat by a monitor lizard. Upper; full-spectrum solar radiation, including ultraviolet types A and B are absorbed directly by the lizard, plus secondary absorption from heat stored in the substrate and reflected off nearby objects. Middle; lizards exposed to sunlight may cool actively by panting, but the method is physiologically expensive: the lizard cools slightly and slowly by expelling water from the nostrils, mouth, and especially the eyes. At this point, a healthy lizard would retreat to a shaded/cooler place to escape further light absorption and to cool passively. Lower; further heat loss via convection cooling occurs at night.

See text for details.

The belief that UVR is important for good reptile care probably started with Ditmars (1956) relating some incidents at the Bronx Zoo, where he was the curator of reptiles. The zoo had received three water monitors, *V. salvator*, and displayed them in an indoor enclosure. Ditmars related that after some time the monitors seemed to know their keeper and would crawl onto his body and give him an "embrace." When the lizards were moved outdoors, they became aggressive and not only kept their distance but aggressively rushed the keeper. Ditmars wrote that "the peculiar effect of fresh air and sunlight" was somehow responsible for the change. Over the following years other herpetologists reported similar drastic behavioral change when captive lizards were exposed to outdoors, specifically to sunlight. Medical researchers had later learned of the importance for humans in getting sufficient UVR to allow the body to convert cholesterol precursors into vitamin D3, essential for use of calcium in any of its many crucial chemical roles. Eventually the knowledge of vitamin D3 synthesis and alterations in lizard behavior were (indirectly) linked, resulting in the belief that the physiological process was universal in amniotes (it is not). Extrapolate that concept to the time when private reptile keeping "boomed" as a commercially viable industry and the perceived need to supply reptiles with bulb-generated UVR began. But the question remains: must captive reptiles be exposed to UVR in order to be healthy? I have known successful keepers and breeders who have taken one side or the other of the question, based on their long-term ability to keep healthy reptiles and get them to breed into several generations. I suggest that the question is more complicated than it appears.

The research into the needs, amounts, and functions of UVR has been thoroughly studied for decades, and during any given period, there has been consensus until the next set of studies. The literature on the subject is considerable and focuses on the minutiae, such as what atoms must be changed to get vitamin D activated, to the large scale, such as how long should people in senior care centers get sunlight each day. But the limiting factor here is that all this research has been done on and for humans. The published research reports on the sunlight-UVR link in reptiles are much less voluminous than for humans, but the cited sources in each are predominantly from human studies. It is true that experiments testing blood free-calcium levels in lizards have shown that more calcium in the diet or exposure to UVR for some time each day do, in fact, alter calcium levels. What has not been addressed is precisely how much calcium is actually required to keep a particular reptile healthy? It is possible, though untested, that UVR might change an otherwise healthy lizard into one that is hypercalcemic. I have been examining this relationship for many years and cannot find a coherent model that not only links UVR with calcium activation (it does) but also addresses the question of how much UVR, if any, do reptiles in captivity need in order to thrive. The subject has far-reaching potential in fields as broad as animal thermoregulation and human mental health, particularly regarding depression. Until quality reptile-focused research is done that does not depend on data for humans, the evidence for each position remains largely anecdotal.

7.6 AGONISTIC BEHAVIOR

I find the definition of agonistic behavior given by Lillywhite (2008) to be more succinct than any I could coin: "reference to social interactions between members of the same species, involving aggression, threat, or combative behavior in response to social conflict" (Lillywhite 2008:9) (Figs. 6.5, 6.10, and 6.11 and Table 7.4). Following Lillywhite's definition, agonistic behavior is intraspecific competition, where animals of the same kind are prepared to fight over resources needed by both. Agonistic behavior has been one of the most well-studied aspects of living (vs. preserved) monitors (Table 7.4). The large size of at least half the

TABLE 7.4

Data for *Varanus* for Which Intraspecific Combat and the Resources That Were the Cause of That Combat Have Been Published.

Species ♂ ♀	Resource	Reference
albigularis ♂		Branch (1991)
*bengalensis**?		Ali (1944)
bengalensis ♂ ♀		Auffenberg (1981a)
*bengalensis** ♂		Deraniyagala (1958)
*caudolineatus** ♂		Thompson et al. (1992)
cumingi ♂		Horn (1994)
dumerilii ♂		Davis et al. (1986)
exanthematicus ♂ ♀	Female, food, territory	Attum et al. (2000)
exanthematicus ♂		Bayless (1994)
gilleni ♂	Territory	Carpenter et al. (1976)
gilleni ♂	Female	Murphy and Mitchell (1974)
*gouldii** ♂		Thompson et al. (1992)
*griseus** ♂	Status, territory	Tsellarius and Tsellarius (1997)
*indicus** ♂	Female	McCoid and Hensley (1991)
*komodoensis** ♂	Female, food	Auffenberg (1981a)
komodoensis ♂		Horn (1995)
*mertensi** ♂	Female	Horn et al. (1994)
niloticus *?		Clements (1968)
niloticus * ♂	Female	Hagen et al. (1995)
olivaceus ♂	Female	Auffenberg (1988)
*palawanensis**?	Food	Koch et al. (2010)
Salvator ♂ ♀	Food, status	Daltry (1991)
*Salvator** ♂	Food, status, territory	Vogel (1979)
*semiremex** ♂		Horn (1985)
*similis** ♂		Horn (1985)
*timorensis** ♂		Horn (1995)
*varius**?		Horn et al. (1994)

Note: An asterisk (*) denotes observations made in the field.? Means species identity uncertain.

species no doubt attracted human observers, the more so because of the conspicuous actions, such as wrestling, that they perform. Few large works about varanids fail to mention their agonistic behaviors, either in the field (Auffenberg 1979, 1981a, b, 1994, Bennett 1998, Deraniyagala 1958, Earley et al. 2002, Eidenmüller 1997, 2007, 2016, Gaulke 1989, Green and King 1993, Neugebauer 1976, Pfeffer 1965, Sprackland 1980, Thompson et al. 1992, Tsellarius and Tsellarius 1997) or in captivity (Attum et al. 2000, Brown 2012, Carpenter et al. 1976, Delean 1981, Murphy and Mitchell 1974).

There are several ways in which *Varanus* may respond to a perceived threat. These are the only options available to most animals, in order of energy expenditure: fight, flee, or freeze, the latter a matter of staying rigidly still. All three may lead to fatality. In choosing to stand and fight, there are usually three possible outcomes. First, the defender prevails and repels or kills the attacker, which is dependent upon several factors. Is the defender in good health? Is it large/powerful enough to pose a threat, or perceived threat, to the attacker? Is the defender facing just one or two or more attackers? Has the defender the stored energy to maintain its defense longer than the attacker has for attacking? Will the defense terminate early (e.g., bluff)? Second, the monitor could flee. This works if the lizard is faster or more maneuverable than its attacker. It can be successful if the monitor can escape to a retreat which the attacker cannot breach (e.g., a deep burrow, narrow rock crevice, up a tree). This is probably the defense option taken by most squamates. Finally, the monitor could simply freeze and depend on camouflage or its lack of movement to escape motion detection by the attacker. The freeze response leaves the lizard quite vulnerable, but similar inactivity is common among reptiles; snakes of the genus *Heterodon* are well known for their ability to play dead (thanatosis) to avoid predation. Similar behavior has been reported in numerous lizards, including such diverse genera as the limbless *Anelytropsis* (Torres-Cervants et al. 2004), conspicuously colored *Crotaphytus* (Gluesing 1983), and a South American sand dune resident, *Liolaemus* (Santos et al. 2010), and new cases are showing the behavior to be more common than previously known (Costa-Campos and Costa-Anaissi 2020). Anecdotal reports claim that thanatosis is exercised in *V. albigularis, V. bengalensis, V. exanthematicus* (Schmidt 1919), and *V. macromaculatus*, but it is not common within the family.

As regards the first option of standing its ground and preparing to fight, if there is any universal agnostic behavior in *Varanus*, it must be the stereotypical posture it takes that brings nearly all its defenses to bear on the antagonist. The stereotypical action taken by the lizard can approximately be broken down into four steps:

1. Taking a position in front of the intruder and turning broadside, i.e., standing perpendicular to its target (Fig. 7.9). When the lizard moves into position, it also moves as far from the target as possible; if there is ample space to run away, it usually does.

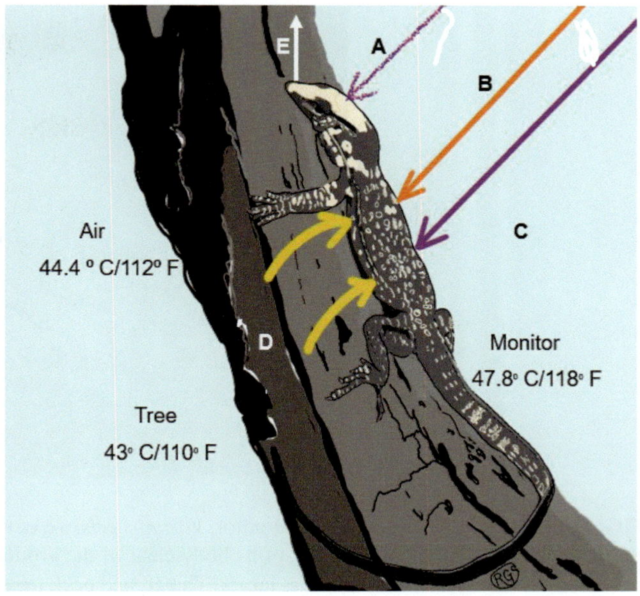

FIGURE 7.9 Absorption and dissipation of heat by a Clouded Monitor lizard (*Varanus nebulosus*) above the ground where air can circulate under the basking site. (A) Solar radiation includes ultraviolet light, predominantly UVB, which is perceived largely by the parietal organ. (B) Full-spectrum solar radiation contains visible light and (C) heat-inducing infrared wavelengths and is also secondarily delivered by conduction from heat stored and reflected off nearby objects. The surface upon which the monitor basks (D) may be warmer than the lizard and conduct heat to it, or it can be cooler, in which case conduction transfers heat from the lizards to the substrate. (E) Evaporation, primarily via the nostrils and eyes, is a passive mechanism for helping lower the body temperature but also leads to dehydration. Lizards exposed to sunlight may cool actively by panting, but the method is physiologically expensive because it leads to increased dehydration. At this point a healthy lizard would retreat to a shaded/cooler place to escape further light absorption and to cool passively. Note that in ecological work there is no such thing as "the" temperature because temperatures may vary considerably within a few centimeters of each other, as shown by the air, tree, and lizard body temperatures.

   Standing its ground is more likely if (a) the target is about the same size or smaller than the monitor, and (b) if the monitor has been cornered, with no practicable way to flee. It is generally at this stage that the monitor will forcibly expel air from its lungs and produce a prolonged hiss. At this stage, too, the monitor opens its mouth slightly.
2. Once in position, the monitor immediately rises on stiffened legs, making it look taller. Simultaneously, the lizard arches both the nape and body upward. The characteristic posture of the neck is sometimes called a roach (Fig. 7.10), with the head angled downward making the nape appear the highest anterior part of the lizard. The body often does a similar posturing, in which the mid-dorsum is the highest central part of the body. These reactions may occur simultaneously or in sequence.

FIGURE 7.10 An Algerian Desert Monitor, *Varanus griseus*, in threat display. This is an early stage of the display, in which the monitor has only pushed the anterior part of its body clear of the ground. The body has been inflated with air and tilted slightly toward the observed, to whom the lizard appears larger. The arched neck (roach) is also made to look larger by inflating the neck and lowering the snout. The tail has been pulled away from the observer, a prelude to whipping out offensively.

Terrestrial, especially large, monitors will also often raise the tail into an arched posture, while smaller and arboreal species are more likely to draw the tail into a characteristic coil that resembles the coiled line on a boat.

3. If the threat moves nearer, the monitor will inflate its body with air and then angle the torso to present a broad oval aspect toward the threat. If a threat is perceived to be serious enough, this step is performed simultaneously with those in step (2) above. During this tilted-torso display, the action of breathing alters the shape of the body, its apparent enlarging and shrinking reminiscent of the display of Frilled Dragons (*Chlamydosaurus*). The drastic changes in the monitor's aspect may confuse or frighten potential enemies; indeed, more than once I have seen a meter-long monitor frighten an adult human away. For the monitor, however, the display may or may not be a bluff.
4. At this stage there are two likely actions performed by the lizard, and they do not have to occur in a particular sequence. It is during this stage that the monitor will make its loudest and most threatening hissing. The accompanying actions are (a) snapping the tail laterally as a whipping weapon, or (b) lunging to bite the transgressor. These actions require the monitor to advance on the target, that is not possible to do them simultaneously. Being whipped by the tail is surprising to most recipients, and though generally harmless to humans, large lizards may remove skin or sting with some intensity. Tail lashing by large monitors has been reported, but not confirmed, to be able to knock a human off its feet, and that a Perentie's lash can break the legs of a dog (King and Green 1999), but I have yet to find confirmation in the scientific literature. Even a relatively small (SVL >120 mm) monitor can lash with enough force to sting; larger monitors can cause the skin to tear. Reports of varanids breaking a person's legs from tail lashing are unverified and apocryphal.

On the other hand, if the monitor lunges and bites, it can cause a serious wound, depending on what was bitten. Against small mammals and other large lizards, the bite could turn into a potentially fatal encounter. The degree of severity for humans has become a new line for inquiry since the discovery that monitors possess venom glands (Fry et al. 2006). Certainly, it has long been known that large monitors can inflict serious wounds, from severe bleeding at the bite site to excising tissue. Most of these reports have been anecdotal or presented on the Internet, and there is ample evidence to believe the reported details. With the discovery (Fry et al. 2006, but see Sweet 2016) that Anguimorphs, including *Varanus*, have both venom glands and toxins that can be quite dangerous to other animals, there is an explanation for the many reported incidents of varanid bites as being similar to envenomation by snakes. Accounts of symptoms associated with monitors bites include extreme swelling and inflammation, tissue discoloration, and extreme pain. Accounts of such envenomation have been posted for *V. acanthurus*, *V. indicus*, and *V. komodoensis*. But even lacking a venomous arsenal, bites from most medium to large monitors (>260mm SVL) can still be painful and cause bleeding, because such lizards have long, sharp teeth, and the larger the lizard, the larger the teeth (excepting *V. rudicollis*). Bites by Komodo Dragons can be life-threatening (Auffenberg 1981b). The teeth of adult *V. salvadorii* are usually longer than those found in a Komodo Dragon with a skull of equal size. Few accounts of the seriousness of the bites have been published, both in India and probably from *V. bengalensis* (Kushwaha et al. 2015, Vikrant and Verma 2014), but I have seen photographs of the injuries this species has caused. I have been "nipped" on a thumb by a *V. spinulosus* (skull 33mm long) that I barely felt. Nevertheless, the miniscule tooth marks soon produced a copious flow of blood that did not clot for over an hour.

Monitors are also well known for their intraspecific combat, which resembles wrestling (Auffenberg 1981a, Carpenter et al. 1976, Davis et al. 1976, Deraniyagal 1958a, Horn et al. 1994, McCoid and Hensley 1991, Murphy and Mitchell 1974, Thompson et al. 1992, Table 7.5). Though both sexes may practice this agonistic behavior, it is most commonly performed by males. Bouts may be preceded by body posturing (see steps 1–3 above), hissing, and tail lashing, and are soon followed by either horizontal or vertical mutual grasping. Bouts may last for seconds or for hours, and though the fights may draw blood, both animals tend to be able to walk away intact, the loser more quickly. Though combat bouts are performed by both sexes and across the range of species, from small (*V. gilleni*) to large (*V. komodoensis*), males predominate in participation and most species for which combat is commonly reported tend to be large (SVL > 180mm) as adults (Auffenberg 1981a, Earley et al. 2002, Horn et al. 1994). From literature accounts, about 60 percent of the reports about varanids wrestling are based on field observations. All encounters involved at least one male, while 22 percent involved both sexes and none were between females. The reasons for the combat were given for 46 percent of the encounters, and of those equal numbers (17.8 percent) were over territory, food, and two or more resources, while 10.7 percent were over social status.

Among larger varanid species, such as *V. bengalensis, V. gouldii, V. komodoensis, V. spenceri*, and *V. panoptes*, the tripod posture (see above) is sometimes used as a threat display (Figs. 7.3B–D and 7.11). The posture significantly changes the target aspect of the lizard from a horizontal object of indeterminate size to a tall and obviously defensive lizard. This tactic may, in function, parallel the rising and hood display of cobras and the perception of them as a danger signal and is a reaction duplicated in many nonvenomous species of snakes. Throughout the range of tripoding varanids that are not sympatric with any cobras, other venomous snakes that perform a cobra-like defensive display are often present (e.g., *Notechis, Pseudechis, Pseudonaja*). If this is the case, then varanid tripoding may be an interesting case of behavioral convergence.

When under attack, a tripoding monitor may shift from standing upright to going through the four steps of horizontal defense (see above), performing rapid tail lashing and loud hissing. The effectiveness of this display probably varies depending on the species and number of transgressing animals. It may, for example, work poorly against adult dingoes, yet fare much better against young and immature individuals. This is an intriguing subject in need of focused research. Different species of *Varanus* in captivity have developed reputations along behavioral lines from tractable ("tame") and readily accept handling, to wildly aggressive (Table 7.5). While such designations are applicable for individuals, they have not been demonstrated as consistent within a species. *V. niloticus, V. salvadorii*, and *V. yuwonoi* are generally considered to be aggressive, even dangerously so, but I have encountered many individuals that were as calm and handleable as their less contentious relatives. Study of this phenomenon, along with intelligence, is a field that has effectively been unstudied despite the wealth of examples demonstrating the likelihood of important discoveries.

TABLE 7.5
Comparison of Aggressive to Less (Non)Aggressive Species of *Varanus* Based on Wild and Captive Conditions

Aggressive	Non-Aggressive	Reference
	Varanus bivittatus	Stefani; pers. obs.
	Varanus gilleni	Stefani; Payne
	Varanus kingorum	Stefani
	Varanus pilbarensis	Stefani
Varanus niloticus		Brown; Pers. obs.
Varanus jobiensis		Stefani; pers. obs.
Varanus togianus		Stefani
Varanus similis		Stefani
	Varanus primordius	Pers. obs.
	Varanus macromaculatus	Sprackland (1992, 2009)
	Varanus varius	Pers. obs.
	Varanus albigularis	De Hart, Payne
	Varanus exanthematicus	De Hart, Payne
Varanus indicus		De Hart
	Varanus acanthurus	
	Varanus dumerilii	Templeton; Payne; pers. obs.
Varanus panoptes		Templeton
	Varanus salvator	Payne; pers. obs.
Varanus tristis		Payne
Varanus doreanus		Payne
Varanus salvadorii		Crutchfield; Payne
	Varanus rudicollis	Per. obs.
Varanus yuwonoi		Brown; Cole; pers. obs.
	Varanus prasinus	Brown; per. obs.

Note: Most of the species that are initially aggressive in captivity can acclimate and become less aggressive through careful handling, becoming more comfortable with their surroundings, or both. References without citation dates are personal communications from Kathryn Brown, Michael Cole, Tom Crutchfield, Christopher De Hart, Roy Payne, Michael Stefani, Joshua Templeton.

7.7 PATTERNS OF BEHAVIOR

Monitor lizards display a range of behaviors that are common among other squamates but often not all in the same non-varanid taxon. For example, monitor problem-solving skills, bipedality, and complex combat, when combined, create a nearly unique set of behaviors. Individually, some

FIGURE 7.11 Threat positions that may be taken by defensive *Varanus indicus*. (Upper) The neck, back, and tail form conspicuous arches that are generally accompanied by hissing and tail lashing. Some species will lunge to bite an aggressor. If the threat remains, the monitor may alternate between this and the posture shown in Fig. 7.10. (Lower) Tripod position as used in threat display. In this posture tail lashing is not employed, but hissing continues, often with gaping jaws, and lunging to bite.

of their behaviors are taken to more advanced levels than seen in other reptiles, some being uniquely stereotypic.

- The ability of *Varanus* to learn is considerable, including abilities to remember and utilize those memories in new ways. They have been observed keeping count of food items, figuring out simple puzzles, exploring enclosures for escape possibilities, recognizing individual humans, responding specifically to their keepers by name, and engaging in play. However, anecdotal reports greatly outnumber carefully crafted experiments and research into monitoring intelligence remains an embryonic field of study.
- Varanid neurobiology has been largely limited to anatomical and simple physiological studies. The relationship between brain neuron numbers and perceived intelligence is limited to four species and shows no correlation between those variables, e.g., an "intelligent" species may have far fewer neurons than a "less intelligent" species. Neither is neuron count linked to absolute size of the monitor.
- Of the numerous stereotyped behaviors – bipedality, wrestling, threat displays, thanatosis – none is unique to monitors, though the specific actions often are. Bipedal static threat display is seen in few other lizards (e.g., *Chlamydosaurus*; bipedal threat, lunging, but no tail lashing), while lateral display with body tilt with hissing and tail lashing is present in some other species (e.g., *Iguana*; no bipedal threat, body compresses but does not depress in display) with minor differences.
- Bipedality in monitors includes static and dynamic modes. In static mode lizards may rise up into threat posture or to periscope, seeing over tall objects. Nearly all of the monitors with longer hind limbs than forelimbs can run bipedally for considerable distances, a behavior that possibly improves their vision by raising the head and improves maneuverability and ability to clear obstacles in their paths.
- Monitors are predominantly foragers but many are sit-and-wait predators. Prey recognition and tracking are largely olfactory but include an important visual component. Their huge gape, a result of a laterally hinged mandibular joint and highly mobile quadrate, allows varanids to consume almost anything they can overpower. When that

fails, they effectively use the claws to tear food into edible pieces. In the latter case, they perform largely as tongue-assisted consumers.

- Varanids employ the suite of ectothermic regulatory modes. As poikilotherms they thermoregulate to keep their bodies within a set temperature range, using both thigmothermic and heliothermic behaviors to obtain the heat and shuttling to cooler places to prevent overheating. Heliothermic temperature regulation is likely a function of the parietal eye and its related brain region.
- Though the specific amount of ultraviolet light exposure needed by monitors is a hotly debated question, the behavioral effects of UVR are confirmed. Agonistic, reproductive, and hormonal cycles are all demonstrably affected by UVR and the vitamin D3 pathway. Some behavioral effects of UVR observed in varanids show correlates in human mental health, specifically regarding vitamin D3 activation, bone density, and depression, all areas that can benefit from further research into the "simpler" monitor lizard brain.
- Agonistic behaviors include the common trio of progressively lower energy-expending fight, flight, and freeze. Specific behaviors include the energy-costly responses to intraspecific competition (mating displays, threatening/fighting rivals, patrolling territory) and low-cost thanatosis, the latter known in four species.

8 Monitor Lizard Ecology

8.1 INTRODUCTION

Monitor lizards have been referred to, in print (Green and King 1993, Pianka 1986, Pianka and King 2004) and in conversation, as "the mammals of the reptiles," making them, according to some, the true mammal-like reptiles, a title to which the early synapsids may no longer lay claim. Is there justification for such an analogy? They are, arguably, among the most active of lizards, most species moving constantly when foraging or exploring, only a few being sit-and-wait hunters. Other lizards intersperse short periods of activity with nearly equal or longer periods of rest, neither being longer than a few minutes. But by use of the process of, for example, gular fluttering, varanids are able to stay active longer, cool quicker, and deal with extreme heat better than many other lizards. Their blood-based delivery of oxygen to tissues exceeds that of most other reptiles and even some mammals, due to the unusual pulmonary system that allows them to use one lung independently of the other (Wood et al 1978). Though they use chemoreception to find mates and prey, monitors, like many mammals, are unquestionably visual creatures. Unlike most mammals, though, monitors have excellent color vision, accommodated by a rich mixture of three types of cone cells, almost exclusively for diurnal vision. Portions of the digestive tract and coprodaeum can retain and reabsorb considerable amounts of water in the body despite having unspecialized squamate kidneys. These divergences from typical lizard biology increase with body size, the largest being most mammal-like, while the small species are more lizard-like, partly explaining the historically greater interest in the former than the latter. However, Jessop et al. (2020) do not consider Komodo Dragons to be mammal-like because mammalian predators control prey species' population sizes, but the monitors do not. The same applies to the other species of *Varanus*, effectively limiting the argument for at least some of their mammalian affinities.

Morphology provides organisms with their equipment for living; structure, proportions, and chemical composition. Ecology details the individual's day-to-day business of living. Physiology is the use of chemical composition and energy, and the functions they drive to effectively use the morphology. Though the parts may seem distinct, they are, like the combined Yin and Yang forming the Tao, an inseparable entity. This chapter examines what makes the working morphology do what is necessary to be alive, and the relationships between functioning morphology and survival that begin with the application of physiology to the environment, and its highly complex interactions between the biotic and abiotic components, ecology. Physiology is the chemical ability to survive or fail, for which reason its role in ecology is reviewed next.

8.1.1 Physiological Considerations

Physiology is a complex science rooted firmly in physics and chemistry. With relatively few exceptions, there are five broad principles underlying vertebrate physiological processes:

1. ***Almost all systems are one-way*:** Blood flow is unidirectional and irreversible in its vessels, going away from the heart in arteries, and back to the heart in veins (exception: pulmonary artery and vein transport blood in the opposite manner); nerves only send impulses from neuron body to axon to synapse, hence the need for both afferent (toward the central nervous system) and efferent (away from the CNS) nervous pathways; muscles can only contract, so one set can close a joint, but an opposing set must then contract to open that joint; food goes in one end (mouth) and comes out the other (barring gastric upset). Being biology, there are always exceptions. One system that is not one-way is the lungs, where air comes in (nostrils) and exits (nostrils) via the same system of tubes (bronchi and trachea).
2. ***All physiology follows circuits*:** Because one-way paths do not double back on themselves, it follows that physiological processes must travel in circuits (excepting lungs). Conditions that alter those circuits, such as embolisms or hematomas, can have minor to grave consequences for the organism.
3. ***All physiology is based on and directed by gradients***, the components of which follow basic chemical principles to attain homeostasis. Gradients allow the passive transport of substances from regions of greater concentration to those of lower concentration without using the body's energy (ATP). To move substances against a gradient, the body must engage active transport, which uses the body's ATP (passive transport: letting a rock roll downhill; active transport, pushing it back up), which in turn requires more fuel (food) than passive transport.
4. ***All physiological activity – the actual achievement of the biological goals – is done at the atomic (ionic) and molecular levels*:** O_2–CO_2 transfer in alveoli and at the individual cells is accomplished by agglomerate actions of molecular exchanges; filtration by the kidneys removes

DOI: 10.1201/9781003463528-11

wastes at the ionic and molecular levels; formation of polypeptides/proteins is directed by portions of DNA molecules and construction (synthesis) guided by RNA molecules moving amino acid molecules.

5. ***All physiological activity leads to one or more of three possible outcomes***: Make something (anabolism), break something (catabolism), or move something (active transport). The first two of these are metabolism, and the latter is transport. Anabolism involves energy use for synthesis, such as to repair injured tissues, replicate DNA, or make proteins. Digestion is an energy-using form of catabolism, reducing food particles into molecular components that can diffuse from the gut into the bloodstream. Active transport is the metabolic use of energy to move a substance against a gradient (see point 2).

To understand a particular physiological process, then, it is useful to understand (a) what the gradient is, (b) what the cycle is, and (c) where the energy is used. Analysis of a physiological process involves three questions familiar to followers of detective stories: means, motive, and opportunity. Means = does the material/energy exist in this place to do X, and is that material available? Motive = is there a reason why the means would be engaged? And Opportunity = can the means and motive actually come together to accomplish X? As an example, consider a monitor basking on a hot rock. It has become heated enough to feel the stress to get away from the heat. Does it have a way to move out of the heat (limbs), and sufficient energy to do so (from stored energy as ATP)? If so, it has the means to become cooler. Is there a reason to use the stored energy to move the limbs to the shade? If the monitor's body temperature goes much higher, the lizard might be permanently injured or die, so a motive exists. Can the monitor use the stored energy to move the limbs to carry it away to shade? If so, it probably goes into the shade.

The physiological constraints of any animal are contingent on body temperature, obtaining adequate oxygen and eliminating carbon dioxide, obtaining an adequate diet and converting it to energy, and for repairing/replenishing internal structures, finding a mate, and having a safe place to lay eggs (or give live birth). Large carnivorous lizards must achieve these objectives differently than endotherms and smaller ectotherms, and though not all varanids qualify as giant lizards, more than half the known species grow to or in excess of 300-mm SVL. That means their bodies will take longer to absorb adequate heat, they will need highly efficient respiratory systems that enhance their ability to successfully capture and subdue prey, their metabolisms will process and utilize food faster than herbivores, and they must be in good health if they are going to successfully mate and bear viable eggs. All of these processes are determined by the large set of environmental parameters.

8.2 HABITAT AND ENVIRONMENT

The single most important thing that faces every living creature is its environment, which is important because it represents a highly complex mixture of innumerable biotic and abiotic factors in which the organism is or is not able to survive. Most, if not all, elements of the environment will have some effect on its living components, each of which will have its "Goldilocks" range of acceptability: good, bad, or neutral, and if the conditions change sufficiently, the environment leaves the living organisms facing the "choice" to adapt (evolve), leave (migration), or die (extinction). Consider a "simple" environment, a desert, and you are a varanid lizard. Can you survive there? Many factors can singly or jointly make that desert a suitable home or not. Start from the ground up, literally. How often and how much rainfall can you expect? What kind(s) of soil are present? Is the predominant soil porous, quickly absorbing any moisture, making rain disappear into the ground quickly, or impermeable, allowing for at least temporary ponds or billabongs? Or is the temperature so intense that even water collected on non-porous soil will quickly evaporate? Is the soil compact and moist enough to allow excavation for a clutch of eggs, even if the excavation must reach down several meters, or will some alternative, such as a hollow tree or gaps between rocks, be acceptable? Will humidity fall within safe parameters for your lungs or not? Will you have large basking rocks that will heat well above ambient air temperatures, or have to simply lie out on the substrate, perhaps using burrows (borrowed or the result of your efforts) to offer escape from the glaring sun? Those are just the most conspicuous abiotic considerations. The other living things in the environment will present different challenges. Are there adequate plants or other animals to sustain populations of prey for the lizard and its expected offspring? Will that prey be relatively consistent, or will it change seasonally? Will that potential prey have adequate nutritional value, both in terms of proteins and energy-rich tissues necessary for survival? Is there sufficient food to support prey species? What plants are present, and will they be needed for shade, as perches to watch over your territory, or just for prey to consume? Will plants provide adequate shade to shelter you from the sun or the eyes of aerial predators, or are they the type that bear foliage briefly after bursts of rain that may come only once or twice per decade? Are you monitoring arboreal, terrestrial, or both? How much competition for resources will there be from conspecifics and other residents of this environment?

Predators are an important part of the biotic environment, and no species, despite its size, is exempt from them. The most frequently reported animals that prey upon varanids are mammals, birds, notably raptors, other reptiles, including crocodiles, snakes, and other lizards, and other monitors (Angelici and Luiselli 1999, Auffenberg 1981, 1988, 1994, Auffenberg et al. 1988, 1989, Cowles 1930, Gaulke 1989, Green and King 1993, Jessop et al. 2015, King and Pianka 2007).

It is perhaps in the choice of habitats that most reveals the non-uniformity of varanids, despite their "conservative" body plan. The morphological variety extends far beyond the degrees of difference in size and mass. The environment in which an organism lives has profound effects on its natural history and may lead not only to interspecific differences (Shine 1986) but intraspecific differences at the population level (Tinkle 1972). In Shine's (1986) study, he compared two pairs of similar and sympatric species of *Varanus* living in Australia's northern Alligator River area. The morphological similarity of *Varanus panoptes* and *Varanus gouldii* is such that the former was long considered a subspecies of the latter. Shine observed *V. panoptes* to inhabit areas near water, where it could obtain its preferred aquatic prey. *V. gouldii* inhabited woodland and more open areas and consumed terrestrial invertebrates as prey. Though the two species of near equal size are sympatric, they partition their resources in a way that allows both to successfully exploit the region. Shine also compared the ecologies of two semi-aquatic monitors. The semi-aquatic/arboreal *Varanus mitchelli* followed the changing distribution of freshwater creeks and accepted the changing variety of prey species that became available, but the larger *Varanus mertensi* was a resident of permanent or semi-permanent water holes where it had a consistent diet of crabs (Shine 1986). On the Indonesian island of Batanta, four species of *Varanus* are sympatric, but differences in size and microhabitat preference provide another example of resource partitioning. The small *Varanus macraei* generally forages on tree trunks, but quickly climbs to the thick canopy when alarmed. *Varanus doreanus* is similar in distribution and is larger, seeking larger prey. Mangrove Monitors, *Varanus chlorostigma*, are closely linked to a largely terrestrial, semi-aquatic ecology, living near permanent bodies of water from streams to the coast of the island. The fourth species is the large *Varanus salvadorii*, which forages on or near the ground and is otherwise in the canopy. It is also rarely seen on Batanta (Chris Applin, pers. comm.).

Species preferences may differ in different localities. In numerous observations, many varanids tend to stay in a particular habitat or linked habitats, e.g., the ground and trees. Two similar sympatric species of Mangrove Monitors in Papua New Guinea, like the *V. gouldii/V. panoptes* in Australia, also demonstrate preferences for particular habitats. Along the Bensbach River in southwester Papua New Guinea, the blue-tailed *V. doreanus* is highly arboreal and rarely seen on the ground or low on tree trunks, while *V. chlorostigma* was highly terrestrial, partially aquatic, and tended to climb trees only to escape potential predators (Sprackland 2009; pers. obs.). Other species, however, are much less specific in their needs. *Varanus rosenbergi*, a close relative of *V. gouldii* and/*V. panoptes* occurs on Kangaroo Island in a range of habitats that include temperate rainforest and semiarid grassland (Rismiller et al. 2010), taking on the preferences of both *V. gouldii* and *V. panoptes*.

8.3 CONTEMPORARY BIOGEOGRAPHY

The physical distribution of monitors will be presented as two related but distinct topics. The origin and long-term distribution of the varanids from their pre-*Varanus* ancestors is covered in Chapters 10 and 11. The subject of this section is the contemporary distribution of the monitors and the environmental factors that act as parameters for that distribution. Extant monitors are restricted to the southern continents of the Old World–Africa, Asia, and Australia, plus western Oceania – but inhabit lakes, rivers, savannahs, forests, deserts, beaches, and mountains. They can climb, dig, and usually swim with remarkable proficiency. Their current distribution has few habitat parameters, but they are inextricably positively linked to high temperatures, and most species live within tropical latitudes. For vertebrates, varanids might be considered extremophiles, for not only are they most numerous in hot environments, but they are also found in some of the hottest environments on the planet. Among varanid distributions, the hotspots and their highest recorded temperatures are Australia (Alice Springs 35°C to 40°C), Pakistan (47°C), Ethiopia (49°C), Tunisia (50°C), and Yemen (54°C). Monitors are also known to bask on rocks with temperatures far above that of the ambient air and in captivity have been recorded basking on objects that measured in excess of 60°C. Four large species of *Varanus* occupy regions below 31 degrees south latitude (Table 8.1), making them the most cold-tolerant species in

TABLE 8.1
Country, Southernmost Latitude, Average Annual High and Low Temperatures for the Three Most Cold-Tolerant Species of *Varanus*

Species	Country	Latitude	Annual High	Annual Low
niloticus	South Africa	31.5 S	27[a]	17
niloticus	United States	30 N	33[b]	10
albigularis	South Africa	34 S	27	17
rosenbergi	Australia	36 S	31	7.5
varius	Australia	38 S	28[c]	4.5

Note: Species listed from north to south distribution. Temperatures are in degrees Celsius. Data included for introduced *niloticus* in the United States for comparison with South African natives. See text for details.

[a] Data for Cape Town, South Africa from www.metoffice.gov.uk/weather/travel/holiday-weather/africa/south-africa/cape-town.

[b] Data for Orlando, Florida, USA, Melbourne, Victoria, and Kangaroo Island, South Australia from https://www.timeanddate.com/weather/@2161899/climate.

[c] Temperatures for Melbourne.

TABLE 8.2
Calcium-to-Phosphorus Ratios of Several Prey Items of Captive Monitors

Prey	Ca:P	Prey	Ca:P
Cockroaches	1:2.5	Earthworms	1.8:1
Adult crickets	1:2.5	Shrimp	1:3.3
Gut-loaded crickets	1:1	Crawfish	1:11.2
Chicken chicks	1:1.4	Adult mice	1.4:1
Chicken eggs	1:25	Adult rats	1:1.4
Salmon	(Atlantic)	1:40.8	

Sources: Brown (2012), Nijboer (2020), and U.S.D.A. (2018).

the family. These are (a) *Varanus niloticus*, which ranges to 31.5 degrees south latitude, (b) *Varanus albigularis*, which ranges to 34 degrees, both species native to Africa, and two Australian species, (c) *V. varius rosenbergi* (Fig. 8.3), which ranges to 36 degrees, and (d) *V. varius*, which ranges *to 38 degrees south, all of which go into inactive brumation periods during the coldest months. There are no small varanids near these latitudes. The escape of many V. niloticus* into southern Florida gives us a rare chance to see a "what if" experiment in action; what if Nile Monitors were in North American environments similar enough to those in Africa that survival would not be impossible because of abiotic factors? Weather data for a known locality of the lizards, Orlando, Florida, show average annual highs and lows well outside the South African temperatures (Table 8.1) by 5°C and 7°C, respectively. Rainfall, humidity, and habitats in Florida approximate those for *V. niloticus* in southern Africa, and given these conditions, there seems to be no reason that the invasive animals will not remain and become firmly established, and possibly expand their range to the Florida-Georgia border, wanted or not (Table 8.2).

Varanid ancestors left a fair fossil record, allowing us to trace monitor history from late Jurassic "proto-varanoids" through late Cretaceous species that are barely distinguishable from living monitors. In prehistoric times, they roamed North America and Europe, but at times when North and South America were still widely separated, and Central America was little more than a line of small, isolated islands. This explains why there is no indication that monitors ever lived in South America.

8.4 METABOLISM AND ENERGY

Just as one cannot discuss what happens in a fireplace without discussing logs and fire, metabolism cannot be understood without its association with energy. Metabolism comprises the use of energy for the complementary processes of breaking down materials (catabolism) and building new ones (anabolism or synthesis). Vertebrates derive most of their energy by cellular respiration, which converts glucose into adenosine triphosphate (ATP). To put the importance of ATP in perspective, humans produce an amount equivalent to their body weight (twice in the athletic types) each day. It has a short half-life, depending on its chemical environment, of less than one second to about 20 minutes. It is therefore not surprising that monitors have the highest metabolic rate recorded for squamates (Table 8.3). Metabolic energy use and efficiency have been frequently studied aspects of the biology of *Varanus*, in part because of the lizards' alleged approach to the mammalian condition (Auffenberg 1981, Bartholomew and Tucker 1964, Benedict 1932, Bickler and Anderson 1986, Christian et al. 1994, Dryden et al. 1992, Gleeson 1981, Green et al. 1986, 1991a, b, Horn 2004, Secor and Phillips 1997, Thompson et al. 1992, 1995, Wood et al. 1978). A group feeding off a shared carcass by *Varanus komodoensis* or *Varanus bengalensis* certainly resembles the feeding situation of some of the large felids, but so do corvid, accipitrid, and cathartid birds. Neither lizards nor birds share any similarities with their mammalian counterparts because of phylogenetic affinities, but because of their convergent feeding behavior.

A body's energy expenditure is measured as metabolic rate, and is, for varanids, most often measured indirectly as the rate at which the lizard uses oxygen. Oxygen consumption testing is done while animals are at a specific temperature and at rest, and the rate of consumption is expressed as VO_2 (volume of available oxygen, where VO^2_{max} is the maximum volume of available oxygen (O_2) that is consumed per minute per kilogram of mass at maximum activity) and is referred to as the standard metabolic rate (Table 8.3). The rate increases during activity, considerably so after feeding; e.g., the activation of the gut and the digestive process increases oxygen demands. In their study of two dwarf species of Australian goannas, Thompson and Withers (1994) discovered that in the laboratory, long-term captive specimens of *Varanus caudolineatus* from the same locality showed no notable difference in relative mass-specific standard metabolic rate (measured as VO_2 [ml g^{-1} h^{-1}]) but differed significantly within groups taken from different locations. The pattern of these differences illustrates the possible taxonomic questions that still need resolution (Fig. 8.1). The mass of the *V. caudolineatus* specimens in Thompson and Withers' study ranged from 13 to 19.5 g, and the *Varanus acanthurus* 19.6–83 g, among the lightest of varanids. In contrast, the rate of metabolism in the much more massive African *V. albigularis* (mass 6.2–9.2 kg) had its post-feeding VO_2 increase seven to ten times above pre-feeding levels (Secor and Phillips 1997). It is apparent that the degree of postprandial VO_2 increase is positively correlated with prey mass, with small, active foraging species, including *V. caudolineatus*, *V. acanthurus* (Thompson and Withers 1994), and *Varanus gilleni* (Bickler and Anderson 1986), having a much lower proportional VO_2 increase than in large sit-and-wait species such as *V. rosenbergi* (Christian and Weavers 1994), *Varanus exanthematicus* (Wood et al. 1978), and *Varanus giganteus* (Thompson et al. 21992). Thus, while active foraging might be less energy efficient than sitting and waiting,

TABLE 8.3
Standard Metabolic Rates for Several Species of *Varanus* Tested at about 35°C

Species	n	Mean Mass (g)	VO_2	References
brevicauda	11	17.4 ±		
caudolineatus (Ajana)	s	8.3	0.238	1
caudolineatus (Atley)	s	15.5	0.156	1
gilleni	s	30.0	0.169	2
acanthurus	s	47.2	0.116	1
gouldii	s	154	0.123	3
salvato	r	458	0.093	4
salvator	r	505	0.089	5
gouldii/rosenbergi	r	674	0.066	6
Australian varanids	r	714	0.102	7
albigularis	r	963	0.096	8
exanthematicus	r	1040	0.118	9
panoptes	s	1427	0.120	3

1. Thompson and Withers (1994)
2. Bickler and Anderson (1986)
3. Thompson and Withers (1992)
4. Gleeson and Bennett (1982)
5. Gleeson (1981)
6. Bennett 1972
7. Bartholomew and Tucker 1964
8. Louw et al. (1976)
9. Mitchell et al. (1981)

Note: Based on VO_2 = ml g^{-1} h^{-1} where VO_2 stands for the volume of oxygen consumed, ml is milliliters, g^{-1} per gram of body weight, and h^{-1} is per hour.

the energy required to move a small, light body requires less energy per unit of body mass. In a summary paper, Thompson and Withers (1997) claim that the maximal metabolic rate of 6.36 mL O_2 g^{-1} h^{-1} was the highest recorded in the Squamata, and that the highest rates of metabolism occurred in *V. caudolineatus, V. gilleni*, and *Varanus tristis*, all arboreal species, while the ground-dwelling species, *Varanus brevicauda, Varanus eremius, V. acanthurus, V. gouldii, V. rosenbergi*, and *V. panoptes*, had the lowest rates among varanids. Thompson and Withers conducted a follow-on study of *V. komodoensis* and *V. giganteus* (Thompson and Withers 1995), which put the Perentie in the group of other terrestrial monitors. If this pattern holds true across the Varanidae, we can expect high rates of metabolism in arboreal species, such as the *Varanus prasinus* group, and possibly in *V. salvadorii*, which, despite its large size, is also both active and arboreal. Earll (1983) measured VO_2 in another large terrestrial species, *V. bengalensis*, which was within the ranges recorded for other large varanids (e.g., *V. varius*, Bartholomew 1982).

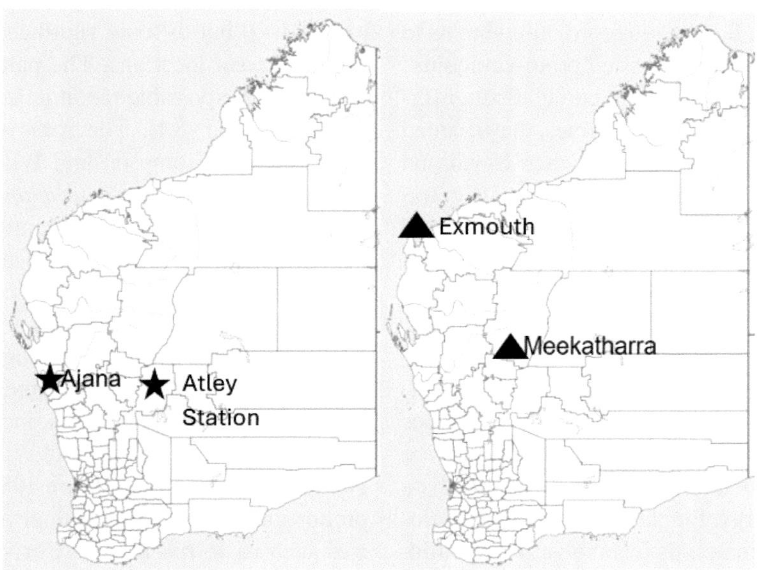

FIGURE 8.1 The sources of the specimens used by Thompson and Withers (1994) in their comparative studies of small monitor metabolism. See text for details.

8.5 RESPIRATION

To operate and sustain an efficient oxygen delivery system, the lungs and circulatory system must be able to stay coordinated during times of strenuous activity. Mammals facilitate such needs through a diaphragmatic pulmonary mechanism and the ability to completely segregate oxygenated and deoxygenated blood in the heart. To address the first problem, lizards lack a diaphragm. Monitors, however, are probably unique among lizards in being able to both breathe and run simultaneously by drawing air into the greatly distensible gular pouch, then contracting the throat muscles to force that air back and into the lungs while the lungs are squeezed to push air toward the mouth (Owerkowitz et al. 1999; Wang et al. 1998). The lizards thereby establish a rhythm of circular breathing like that employed by humans skilled in playing the Australian didgeridoo. This breathing mode frees varanids from the dash-pause-dash activity performed by most lizards, which are unable to run and breathe simultaneously. In so doing, monitors can outrun prey or predators for longer durations, enhancing their survival, and making them equal or superior to mammalian predators of similar size.

Monitors have a second way of breathing while running. The sigmoid movement of the trunk causes the lung on the same side as the flexion to exhale. Simultaneously, the opposite lung, where the torso is expanded, can inhale (Owerkowitz et al. 1999). The constant pulmonary air flow possessed by birds is essential for their abilities to fly exceptionally long distances and at altitudes where the air is too thin to sustain most other animals.

The efficiency of blood to carry oxygen is further advanced relative to other lizards thanks to the four-chambered heart (see Chapter 6) (Gregorovicova et al. 2022, Hanemaaijer et al. 2019a, b, Mather 1944, Rajani et al. 2021). Numerous studies over several decades have documented the respiratory efficiency, if not superiority, of varanids over other reptiles (Bartholomew and Tucker 1964, Bennett 1972, Christian and Conley 1994, Christian and Weavers 1994, Earll 1982, Ishimatsu 1988, Wood et al. 1978). Respiratory efficiency varies among varanid species. Frappell et al. (2002) showed that *V. mertensi* remained stable whether at rest or running a treadmill, and "pulmonary diffusion of carbon monoxide" was twice that reported from similar studies of *Varanus exanthematicus.*

8.6 WATER DYNAMICS

All reptiles require water to survive. Indeed, about 30 percent of a reptile's mass comprises water. Water serves innumerable essential physiological functions: gas exchange in the lungs must occur via a water barrier; kidneys require water to process and eliminate nitrogenous wastes; cellular metabolism is dependent upon the hydrogen and hydroxyl ions (H^+, OH^-) resulting from hydrolysis ($H_2O \rightarrow H^+ + OH^-$), the ions resulting from hydrolysis make an aqueous environment acidic (high H^+ concentration), neutral ($H^+ = OH^-$), or basic/alkaline (high OH^- concentration). The failure of any of these processes, if not corrected, is fatal. As in all animals, this involves the opposite functions of obtaining and losing water, and the ability to retain a sufficient quantity to sustain life (Benedict 1932, Braysher 1986, Seshadri 1959). The range of metabolic tools available to varanids is closely related to habitat and specific physiological needs. Many species live in hot habitats where ambient temperature often exceeds 38°C even in shade. Desert species may encounter substrates 50°C, prolonged exposure to which may be lethal. Unlike some mammals, monitors cannot perspire, but they do thermoregulate in large part by controlling the dynamics of their body water (Table 7.2). Panting, the commonest lacertilian mode of cooling when encountering excessive heat, is effective but results in rapid dehydration (Braysher 1971). The excretion of ammonia products via the urinary tract also takes a high cost in metabolic water, especially in large-bodied animals. In squamates, which have metanephric kidneys, there is no Loop of Henle, a tubular structure in mammalian kidneys that allows for the formation of hypertonic waste. Instead, squamates save water by extensive resorption in the stomach, cloaca (Seshadri 1959), and sometimes the large intestine. The business of water dynamics – that is, use of water for basic metabolism, cooling, and waste elimination – is therefore central to the ability of lizards to survive in their often harsh habitats.

The varanid urinary system consists of paired, but not always symmetrical, kidneys and ureters, located dorsally between the ilia and above the digestive tract (see Chapter 6). Lizards' metanephric kidneys contain nephrons, collecting ducts, and a protective capsule. Nephrons are the main functional component of kidney physiology and are clustered in large groups within glomeruli. Glomeruli in the kidneys are the centers of blood filtration, and their waste nitrogenous products form uric acid ($C_5H_4N_4O_3$), a form of nitrogen excretion that uses far less water than what is required to form liquid urine, thereby allowing lizards to retain considerably more water than they could if they produced urea. Uric acid is secreted from the posterior portion of long ureters and into a coprodaeum, which functions, in part, as a bladder. Unlike a true bladder, the coprodaeum in most varanids is a common preanal chamber that receives urine, fecal matter, and eggs, and is the anterior chamber to the cloaca (Baysher 1976). They can shift material in the coprodaeum to the posterior portion of the intestine, which in turn can resorb water into the circulatory system. Regulation and use of metabolic water has been studied for many varanids (Christian et al. 1996, Dryden et al. 1992, Green and Christian 2007, Thompson 1999, Thompson et al. 1995, Thompson and Withers 1992, 1994).

Sodium ion (Na^+) regulation is critical in many physiological processes, and the kidneys play a major part in that regulation. It is the major cation in extracellular fluid and is essential in controlling physiological water balance. As the complement to Ca^{++}, Na^+ is also involved in creating and propagating nerve impulses, muscle contraction, and nutrient absorption. Hypernatremia (excess Na^+ concentration) leads to rapid dehydration by drawing water out of cells and into the bloodstream, an increase that leads to an inability of the kidneys to produce filtrate. Overabundance of Na^+ will affect the amount of water in the brain, which in turn can lead to seizures and death. Fine control of bodily Na^+ levels is therefore crucial (Green et al. 1986). Many habitats used by varanids – deserts, grasslands, mangroves – are inhabited by prey species that tend to have high levels of salt in their systems, and imbalances may have drastic consequences. Facilitating ion control in many monitors are nasal glands that collect and extract excess salt (Dunson 1974). Located adjacent to the nasal capsules, between the external naris and the anterior orbit, the glands comprise columnar cells that line a series of meshing microtubules and are analogous to the tear glands of mammals. The tubules of varanids collect salt and the very small quantities of water needed to secrete salt from the nostrils. Because of the low water content, the secretions tend to dry quickly, leaving crusty white salt deposits on the skin around the nostrils. Because the glands do not operate via a passive osmotic gradient, as in the mammalian loop of Henle, the columnar cells of the tubules must employ active transport, which requires large numbers of mitochondria and ATP, making nasal salt gland physiology costly to species in which it occurs. Monitors that possess such glands dwell in many habitats. Those that frequent salt water, and include *Varanus indicus, V. komodoensis*, *Varanus salvator*, *Varanus semiremex* (Dunson 1974), and *Varanus flavescens*, all of which have comparatively short nasal tubes. Residents of dry, sandy regions also use salt glands to inhibit quick dehydration, and include *V. bengalensis*, *V. griseus*, and *V. tristis*.

Though lizards lack the complex mammalian infolded nasal structure called the ethmoid labyrinth, the varanid nasal cavity performs a similar function by creating a counter-current exchange where warm internal air is cooled by passing inspired air. Though not as efficient in water retention as the mammalian version, it remains an important inhibitor of respiratory water loss during exhalation. In many varanids, the internal nasal chamber is large compared to other lizards, and research into the effect of the chamber on both thermoregulation and water retention seems appropriate.

8.7 FOOD AND ENERGY EXPENDITURE

Food acquisition may be the single most important activity that varanids perform on a daily basis; it is the expenditure of energy to replenish energy. Though many monitors overlap in prey preferences and availability, they sometimes partition resources along environmental lines (Table 8.4). Losos and King (1992) described dietary differences among Australian monitors in the field, and this was confirmed by Losos and Greene (1988) and Sprackland (1995), who analyzed stomach contents from museum specimens. Other analyses of varanid diets include Arbuckle (2009), Böhme and Ziegler (2007), Cowles (1930), Gaulke et al. (2007), Greene (1986), and McCoid and Witteman (1993). A common observation from all the stomach content studies is that varanids are largely insectivores, secondarily consuming numerous other invertebrates, the composition of which depends on habitat. Vertebrate prey was rare, though eggs of turtles, squamates, and birds and their eggs are a staple in the majority of *Varanus* species' diets. (I can confirm the

TABLE 8.4
Most Important Prey Items for Adults of Various Species of *Varanus*

Desert Species	Primary Prey	Secondary Prey	Tertiary Prey
acanthurus	Orthopterans	Beetles	Lizards
brevicauda	Reptile eggs	Orthopterans	Beetles
caudolineatus	Roaches	Orthopterans	Lizards
eremius	Lizards	Orthopterans	–
giganteus	Lizards	Mammals	–
gilleni	Lizards	Orthopterans	Spiders
glauerti	Lizards	Frogs	Orthopterans
glebopalma	Lizards	Orthopterans	Spiders
gouldii	Lizards	Reptile eggs	Orthopterans
*griseus**	Lizards	Snakes	Mammals
kingorum	Orthopterans	Roaches	–
panoptes	Orthopterans	Ants	Frogs
pilbarensis	Orthopterans	Lizards	–
primordius	Ants	Orthopterans	–
*scalaris**	Orthopterans	–	–
spenceri	Orthopterans	Mammals	–
storri	Orthopterans	Roaches	Beetles
tristis	Lizards	Orthopterans	Lizards
Non-desert terrestrial species			
bengalensis	Lizards	Mammals	Crayfish
*exanthematicus**	Frogs	Orthopterans	Eggs
komodoensis	Deer	Boar	Rodents
mertensi	Crabs	Beetles	Frogs
mitchelli	Fish	Orthopterans	Spiders
niloticus	Molluscs	Frogs	Orthopterans
rosenbergi	Roaches	Orthopterans	Spiders
varius	Bird eggs	Rabbits	Birds
Forest species			
dumerilii	Crabs	Insect larvae	Ants
*indicus**	Lizards	Orthopterans	Fishes
olivaceus	Fruits	Snails	Bird eggs
prasinus	Orthopterans	Beetles	Roaches
rudicollis	Insects	Frogs	Misc. arthropods
salvadorii	Lizards	Birds	Frogs
*salvator**	Orthopterans	Frogs	Lizards
semiremex	Geckos	Crabs	Fish

Sources: Data from Auffenberg (1981), Dalhuijsen et al. (2015), Gaulke et al. (2007), Irwin (1994), Losos and Greene (1988), McCoid and Witteman (1993), Philipp et al. (2007), and Sprackland (1989, 1995).

Note: Orthopterans overwhelmingly dominate the diets of all species, certainly as young lizards. This chart gives data for sub-adult and adult specimens. Habitat types include peripheral environments. species marked with an (*) represent taxa for which the data sources probably represent different species because of taxonomic revisions.

latter: the very first monitor I ever collected in the field was an adult Lace Monitor that promptly vomited what had to have been the contents of an emu egg.)

Diets of varanids are known primarily from two sources: lizards in captivity and field observations. Both are supplemented by analysis of a limited number of museum specimens with food in their digestive tracts. The natural diets of relatively few species of monitor have been published, though data from the stomach contents of lizards in the field and museum specimens show that insects, particularly orthopterans, form the major prey items (Table 8.4). Insects are included in the diets of all juvenile monitors, including the fruit-eaters. Other common prey includes other invertebrates, reptiles, small mammals, eggs, birds, frogs, and fishes (Auffenberg 1978, 1981b, 1988, 1994, Auffenberg et al. 1988, 1989, Auliya 2006, Dalhuijsen et al. 2015, Gaulke et al. 2007, Greene 1986, Irwin 1994, James et al. 1992, Losos and Greene 1988, McCoid and Witteman 1993). A review of the preferred prey items of varanids shows an unsurprising pattern; within each habitat group, there are relatively few prey taxa that make up the bulk of the diets. Desert varanids feed heavily on the most abundant sympatric prey, which includes orthopterans (grasshoppers, katydids, crickets, and kin), other lizards, and a variety of arthropods, including several orders of insects (Coleoptera, Lepidoptera, Hymenoptera). Non-desert terrestrial monitors, those that live in habitats that have grass or shrubs as a major part of the landscape, tend to consume mammals, aquatic invertebrates (mollusks, crustaceans), and orthopterans. The species near water favor the aquatic invertebrates, insects, frogs, and fishes in their diet, while those from drier habitats take predominantly mammals, other reptiles, and eggs. Forest varanids follow the same pattern: those that spend time near water or are semi-aquatic take fishes, frogs, crabs, and snails, while their more arboreal counterparts feed mostly upon insects (again mostly orthopterans), eggs, and other lizards. From this information, it is clear that varanids are not particular specialist predators, but rather predators that consume the most numerous individuals of the species in their habitats (Table 8.3). It is true, of course, that some monitors have specialized feeding habits. *V. dumerilii* have a way of eating crabs that is possibly unique (Krebs 1979). The lizard stands upon the crab and then uses its jaws to reach down and pull off the crab's limbs one at a time. Once a limb is detached, the monitor swallows it until only the carapace remains. This is consumed entirely, albeit in a lengthier process. Many other varanids also feed on crabs, but in the conventional manner of positioning the crustacean and swallowing it whole. It is not the diet that is specialized, but the manner of consumption. A similar situation has been reported for a specimen of *V. prasinus*, in which the prickly limbs of a stick insect were not found in the lizard's gut with the rest of the insect. And even though few varanids will attempt to dismember or tear prey apart, this is a common feeding behavior of *V. bengalensis, V. niloticus, V. prasinus*, and *V. komodoensis.* Lizards in trees are highly likely to encounter birds' eggs and other lizards, so that is what they eat. Desert monitors live in particularly harsh environments where other lizards, insects, scorpions, and small mammals are overwhelmingly the most commonly encountered potential prey. And from examination of the data, it is clear that monitors in their natural habitats prefer and consume small prey, though the young of some species prefer proportionally larger items than the adult monitors.

Cooper (1989) demonstrated prey recognition from odor in *V. exanthematicus,* and Sprackland (1989d) noted food discrimination based on odor and texture. Despite what seems to be a preference for smaller prey, Varanids can and do consume larger food items than most other lizards, thanks to their wide gape (see Chapter 6) and modifications of the gular region and hyoid apparatus (Montuelle et al. 2011, 2012, Rieppel 1981, Schaerlaeken et al. 2012, Smith 1986, 2011). Animals secure their food in numerous ways, and squamates employ one of two: lingual assistance and bite and grab (Fig. 7.6). These mechanisms are phylogenetically linked and provide evidence for a hypothesis of where the root for squamates lies (Townsend et al. 2004). If lingual feeding arose once in squamate evolution, it should be considered an ancestral trait, making Iguania + Agamidae the most ancestral group. An alternative, though, shows a possible reversal much further up the tree, putting Iguania + Agamidae in the position of most derived. The former fits with most prior (and subsequent) phylogenies, but in both cases, the Varanidae is placed at or near the top of the tree, affirming their status as highly derived squamates.

Few species of *Varanus* are known to be sit-and-wait predators and include *V. acanthurus* and *V. brevicauda* (Thompson and Withers 1997a; see below) and, unless they are basking, seem to be always in motion. Even the frugivorous species (*V. bitatawa*, *V. mabitang*, and *V. olivaceus*) consume animal matter, mostly in the form of snails, in their diets, and it is worth investigating to see if the lack of the need to chase prey has also reduced the metabolic rates of these species. Being large and plant-eating, they may represent an analogue to mammalian herbivores.

How do varanid digestion and body temperature compare to those of the mostly herbivorous species? Clark et al. (2005) tested the hysteresis (the lag time behind the change in a property, such as temperature, and the change affecting the property) for body temperature and the effects of digestion in *V. rosenbergi.* They exposed the lizards to temperatures between 19°C and 35°C, comparing VO_2 after fasting for eight or more days, and again a day after eating. In all cases, they found no hysteresis, but VO_2 was significantly higher in lizards after they were fed. They also noted that the heart rate of postprandial lizards remained constant, though these animals also took longer to cool below 23° C. Clark et al. (2005) hypothesized that, though heart rate remained stable, blood flow varied, with the digestive tract being prioritized, leaving reduced blood flow to carry warm blood near the dermal areas. That, combined with the heat produced by digestion, would lower body temperature more slowly than heating it, an advantage for a monitor that lives in some of the coolest environments inhabited by the family. Given that dark colored material absorbs heat faster than lighter shades, the predominantly black integument contributes to the rapid heating of *V. rosenbergi*. According to Christian et al. (1996), the dark skin of this species allows greater absorption than the skin of any other varanid.

Thompson (1999) summarized reports of standard, maximal, and field metabolic rates of several predominantly Australian varanids, plus the Indonesian *V. komodoensis* and Asian *V. salvator*. Thompson and Withers (1997a) divided ten Australian endemics into groups that actively and widely forage (*V. caudolineatus, V. gilleni, V. tristis, V. eremius, V. gouldii, V. rosenbergi, V. panoptes*, and *V. giganteus*) and those that are largely sit-and-wait (ambush) predators (*V. acanthurus* and *V. brevicauda*) and found the latter had lower standard metabolic rates (Thompson and Withers 1994). Thompson (1999) showed that small varanids had metabolisms that were about equal to those of other similarly sized lizards. However, the metabolic rate of large varanids exceeds that of other similarly sized lizards (Thompson 1999, Thompson and Withers 1992, 1997a, Thompson et al. 1998). The three arboreal species studied (*V. caudolineatus, V. gilleni, V. tristis*) at 35°C had "significantly higher maximal metabolic" rates than their terrestrial varanid counterparts (Thompson and Withers 1997a).

Dermatophagy involving sloughed skin is found across numerous families of lizards but occurs only rarely in varanids. Consumption of a shed may recycle useful nutrients, such as proteins, or may help replenish energy used in the shedding process. Likewise, there is almost no information on the energy expenditure involved in ecdysis, and published data are limited to the energy content of the sloughed skin (Blem and Zimmerman 1986), but given the time and energy the process entails, it is likely to generate a sizeable expenditure of energy.

As carnivorous lizards, varanids possess a simple digestive tract. The stomach is highly distensible, which allows for the containment of often large prey, though when empty, it is oval or ovoid and compacted into a small area of the ventral thoracic cavity. The intestines are divided into a thin, short intestine and a much thicker, long intestine, both of which have a rugose inner surface. The projections of the large intestine act in concert with the kidneys in the resorption of water. Digestion of high-protein prey is rapid, with the time between ingestion and excretion being 18–20 hours. in healthy monitors. The rapidity of the digestive process allows varanids to capture and utilize energy from food at a rate that sustains their high level of activity; conversely, the activity level corresponds with the need for frequent meals.

8.8 ULTRAVIOLET RADIATION AND VITAMIN D

Among the earliest reports about the possible influence of ultraviolet light exposure on lizards were those from Ditmars (1910, 1956; see also Wood et al. 2023) and Bogert and Del Campo (1956), who noted the conspicuous and dramatic differences in the behavior of Water Monitors and Gila Monsters, respectively, depending on whether they were indoors under electric lights or outdoors and exposed to direct sunlight. The Bronx Zoo had received three water monitors, *V. salvator*, and displayed them in an indoor enclosure. Ditmars related that after some time, the monitors seemed to know their keeper and would crawl onto his body and give him an "embrace," a behavior that is commonly displayed by contemporary keepers and their lizards. When the lizards were moved outdoors, they became aggressive and rushed the keeper, hissing and lashing their tails. Ditmars wrote that "the peculiar effect of fresh air and sunlight" was somehow responsible for the change. Over the following years, other herpetologists reported similar drastic behavioral changes when captive lizards were exposed to the outdoors, specifically to sunlight. In the subsequent decades, medical researchers demonstrated that the ultraviolet portion of solar radiation – sunlight – has a profound and important role in the synthesis and use of essential physiological chemicals (Ferguson et al. 2009, Olsson et al. 2011). Eventually, the knowledge of vitamin D3 synthesis and alterations in lizard behavior were (indirectly) linked, resulting in the belief that UVR was necessary for vitamin D to be activated and the specific physiological process was universal in amniotes (it is not). Extrapolate that concept to the time when private reptile keeping "boomed" as a commercially viable industry and the perceived need to supply reptiles with bulb-generated UVR began. The chemical pathway from UVR and precursors such as 7-dehydrocholesterol to activated vitamin D3 has been well established (Fig. 8.2). But the question remains: must captive reptiles be exposed to UVR to be healthy? I have known successful keepers (me included) and breeders who have taken one side or the other of the question, and based on their long-term ability to keep healthy reptiles and get them to breed into several generations, and I claim that the question is more complicated than it appears.

The amount of research into the need for, amounts of, and functions of UVR has been thoroughly studied for decades, and during any given period, there has been consensus until the next set of studies. The literature on the subject is considerable and focuses on the minutiae, like which atoms must be changed to get vitamin D activated, to the large scale, such as how long people in senior care centers should get exposure to sunlight each day. But the limiting factor here is that all the published research reports on the sunlight-UVR link are based overwhelmingly on humans. For reptiles, it is much less voluminous and with the cited sources in each predominantly from human studies. It is true that experiments testing plasma calcium levels in the blood in lizards have shown that more calcium in the diet or exposure to UVR for some time each day does, in fact, raise calcium levels (Diehl et al. 2018, Ferguson et al. 2009, Oonincx and Van Leeuwen 2010, 2017, McWilliams 2005). What has not been addressed is precisely how much calcium is actually required to keep a particular reptile healthy (Baines et al. 2016). It is possible, though untested, that

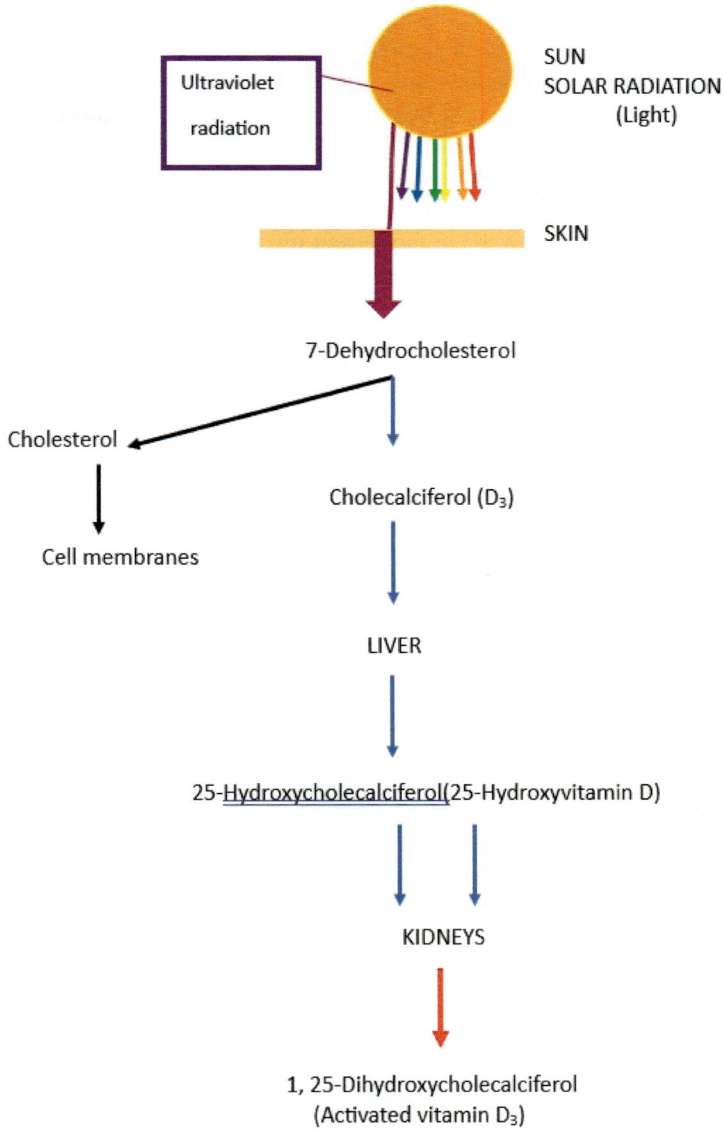

FIGURE 8.2 The chain of reactions that uses ultraviolet light and precursers in the body to produced metabolically useful vitamin D3. The pathway is well corroborated in humans, but not in reptiles.

UVR might change an otherwise healthy lizard into one that is hypercalcemic, or it could be unnecessary for captive monitors. In their critiques of earlier published experiments, two teams (Drivers and Camus 2020, McWilliams 2005) disparaged the works because they had fallen "short of the standardized requirements necessary to establish true reference ranges, and some fail to take into account the significant effects of gender, age, season, reproductive status, diet, and feeding schedule" (Drivers and Camus 2020). More rigorous studies have documented both the interconnectedness of the ions/UVR/vitamin D components and their necessity for healthy varanids, but most work has been done with smaller species from other families (see below).

Herpetolocultural and veterinary recommendations for the need to provide ultraviolet light B wavelength exposure (UVB) have been accepted practice since at least the 1980s, though experimental justification has been questionable. In testing such need, Oonincx et al. (2010) exposed 84 neonate Bearded Dragons (*Pogona vitticeps*) to a range of either oral doses of D_3 or UVB for 2–12 hours daily. After 6 months, lizards that had received oral D_3 had slightly increased plasma D_3 levels, depending on dosages received, while UVB-exposed lizards had 18 times higher D_3 levels. They arrived at two conclusions. First, only 2 hours of daily exposure was enough to produce sufficient D_3 levels for physiological needs. Surprisingly, they also concluded that oral doses of D_3 do not raise D_3 levels to anywhere near those levels. Diel et al. (2018) also conducted controlled experiments to test the effects of different types of ultraviolet generating light bulbs of UVB wavelength levels known to facilitate cutaneous vitamin D synthesis, to enhance vitamin D_3 in *Pogona vitticeps* for 120 days. They exposed five groups of seven to different types of UVB lamps for 2 hours per day, and a control group that received no such exposure. After 120 days, all lizards exposed to the UVB lamps showed lower plasma 25(OH)D_3 (calciferol) than other studies with the same species and with free-living adult Bearded Dragons. They concluded: "None of these lamps, under this regime, appeared to have provided a sufficient UVb dose to enable synthesis of plasma 25(OH) D_3 levels similar to those of free-living bearded dragons in their native habitat" (Diel et al. 2018:308). The studies experimentally tested each of the components in one species of *Varanus* and found that changes to any of the factors (Ca, P, UVR, cholecalciferol) detrimentally altered the physiological functions performed by them all.

I have been examining this relationship for many years and cannot find a coherent, objective model that demonstrably links UVR with how much UVR, if any, reptiles in captivity need in order to thrive. Publications on varanid husbandry have near universally stated that UV lighting is a requirement for captive lizards (Bennett 1998, Brown 2012, Eidenmüller 2007, Lillywhite 2023, Sprackland 1992). The efficacy of providing artificially produced UVR would depend on the ability of the bulb to do so at required UV-B wavelengths, a subject for which parameters and product quality are variable (Ferguson et al. 2009, 2010, McWilliams 2005). The subject has far-reaching potential in fields as broad as thermoregulation and mental health. Until quality reptile-focused research that does not depend on data for humans, the evidence for each position remains to be explored further.

Calcium (Ca^{++}) is often physiologically associated with phosphorus (P) in the diet (Table 8.4). Insects and, to a lesser degree, other invertebrates are a staple in the diets of all varanids, at least at some point in their lives. Under captive conditions, there is the probability that such food items will lack sufficient dietary calcium sources (Table 8.2). Vertebrate prey offers sufficient calcium, coming mostly as calcium phosphate. Until recently it was believed that bone is composed of hydroxyapatite ($Ca_3[PO_4]_2)3Ca[OH]_2$) but has been reanalyzed and found to be hydrated amorphous calcium phosphate ($Ca_{7.5}[PO_4]_{2.8}[HPO_4]_{2.6}[CO_3]_{0.6}[OH]_{0.2}$) (Von Euw et al. 2019), which provides both calcium and essential phosphate ions (HPO_4^{2-}). Insects tend to contain a low calcium to phosphorus composition (0.003 Ca:0.008 P, Table 8.2) (Nijboer 2020); commonly offered feed insects have Ca:P ratios ranging from 1:2.5 for cockroaches to 1:7 for adult crickets (Table 8.4). Supplementing the insect's food with calcium carbonate brings the ratios closer to equality, such that a "gut-loaded" adult cricket may have a Ca:P ratio of 1:1 (Brown 2012). Other prey, such as earthworms, may have balanced ion ratios (Ca:P ratio of 1.8:1). However, requirements for a proper ratio of the ions vary in the literature, from 2:1 in veterinary sources (Corcoran 2023) to 1:1 in a text on varanid husbandry (Brown 2012). Corcoran (2023) states that as the Ca:P ratio approaches 1:1 or the P levels elevate above 1, kidney damage may occur. McWilliams (2005 and references therein) provides an overview of published research with a focus on *Iguana iguana* and *V. komodoensis* (Wood et al. 2023) that provides support for the need not only for proper Ca:P ratios, but also ultraviolet radiation exposure, and dietary needs for proper calcium supplementation.

Excessive UVR exposure can be dangerous to varanids, but UV and heat (infrared) are not always linked; a cold lizard could continue basking in order to raise body temperature long after it has absorbed dangerous levels of UVR. We know that temperature induces color change in some lizards (Crisp et al. 1979, Dawson and Templeton 1963, Fitch and Turner 1982, Foá 1991), and because temperature and ultraviolet radiation typically are united experiments for UVR needs must be carefully controlled to account for temperature.

8.9 COMPETITION

All animals face competition for necessary resources from sympatric organisms, but the most intense competition comes from their own species. Each member of the species has precisely the same needs for the same resources from prey to shelter to territory and mates, with the needs and adaptability to the environment governed largely by genes. A change in gene frequencies within a population is a basic definition of evolution, a process that does not require the production of new species. As one consequence of this competition, no population stays, if it ever acquires, the textbook steady state as presented by the Hardy-Weinberg Equilibrium Equation, a formula taught to high school and undergraduate college students to show what equilibrium is supposed to look like. The variables are gene frequencies, represented by p and q, respectively, and exist in equilibrium when:

$$p^2 + 2pq + q^2 = 0.$$

Any deviation indicates the change of one gene's frequency relative to another. Deviation can be the result of several factors, including a change in availability or types of food present, the number of conspecific competitors, the number of non-conspecific predators, other predators, changes in climate, the size of the population, the availability of mates, and local pathogens. If any factor or factors allow one genotype to succeed at the expense of the other, the result is genetic disequilibrium. Sometimes an organism can adapt to significant changes in its environment through phenotypic plasticity. In the case of Darwin's Finches, the phenotype was able to modify bills to accommodate larger, harder seeds (Grant and Grant 2014), while in some *Anolis* lizards, there was major coloration and pattern change across altitudes and rainfall on a single island (Malhotra and Thorpe 1991a, b, Thorpe and Brown 1989). Competition with mammals and other monitors shaped varanid evolution (Brennan et al. 2021); the tendency for cannibalism, however unusual it might be, has been documented for *Varanus bengalensis, V. nebulosus, V. salvator, V. macromaculatus, V. komodoensis, V. gouldii*, and *V. timorensis* (Arida et al. 2023, Sirimanna et al. 2020, Zdunek et al. 2024) and likely occurs elsewhere within the genus.

Though competition is always present, even in allegedly "empty niches," it is particularly conspicuous on islands, where both taxon diversity and population size are often in greater flux than occurs in large mainland environments.

8.10 ISLAND VARIANTS

Island populations arguably live in the most competitive environments. An island, as used here, is any physical environment that is surrounded by an environment that is inhospitable to the island species and thus forms a significant barrier to dispersal. This makes mountains, coral reefs, and water-surrounded land examples of islands. Despite the formidable isolating features, islands are populated and repopulated by immigrants that arrive by chance (storms, winds, flight) or intentional human introduction. Mainland species, upon colonizing small islands, face environmental challenges that lead them to display greater dimorphism or polymorphism. In general, the further an island is from a large landmass such as a continent, the fewer other species from that continent are going to be encountered (Brown and Lomolino 1998, Losos and Rickefs 2010, MacArthur and Wilson 1967). That means every founding population from the landmass will be facing different numbers and kinds of predators, prey species, and species that already fill the niche preferred by the immigrants. The vegetation and abiotic features that include terrain, rainfall, availability of fresh water, and thermal range mean the difference between success and failure for founders. Distance alone does not dictate environmental variation. For example, the California town of Montara is comparatively cool and misty year-round, while Half Moon Bay can be intensely sunny and 30°F warmer on many summer days. Terrain, then, is a considerable variable in ecological problems because it has a strong influence on so many other abiotic features. Consider that every immigrating population is an invasive species, and reptiles have an impressive record of success even in the presence of situations that would impede them in their original homeland. The invasive Brown Tree Snake (*Boiga irregularis*) has been an extraordinary success on the remote island of Guam since it was accidentally introduced during World War II. In the decades since, it has become a dominant predator that has pushed some endemic birds into extinction and put several others on the edge. North America's Red-eared Slider (*Trachemys scripta elegans*) has managed to establish decades-old colonies in Western Europe, Japan, and Australia. The United States has several invasive species that have been there for many decades. The Caribbean Brown Anole (*Anolis sagrei*) was first established in Florida (probably the most accommodating landing port for reptiles) and has gradually been supplanting the native Green Anole (*A. carolinensis*) ever since. Additionally, *A. sagrei* has been evolving to a degree that some molecular biologists are suggesting it may soon need to be made a new species, and different populations have successfully been producing *A. sagrei* X *A. carolinensis* hybrids. Florida has more recently been settled by one of the largest monitor lizards, *V. niloticus*, a special concern because their natural diet includes crocodilian eggs, and because there have been several incidents of them harming or killing household pets and native wildlife.

Islands are recognized as biologically very active laboratories on which to study diversification and evolution because the total environment will be different in at least some ways, and that leaves room for genotypic and phenotypic divergence from the founders. In such cases, there tends to be selection for morphologies not seen on continents, such as small- to medium-sized flightless birds, large to very large lizards, significant changes in diet, more selective mate choice, or novel coloration and melanistic forms. Numerous studies on island species, especially endemics, had largely been taxonomic in nature (Barbour 1912, Mertens 1934, 1959, Wiles et al. 1989), and some have included brief ecological data that included *Varanus* as well (Mertens 1959, Rodda et al. 1991, Sprackland 2001a, Sprackland et al. 2004). Arguably the most thorough study of evolution in action on islands was conducted by Peter and Rosemary Grant (2014), who studied Darwin's Finches on the Galapagos island Daphne Major over 40 years and found correlation among the weather, variations in the foods (seeds), and millimeter changes in bill structure, documenting how even small phenotypic changes can have significant survival value. Selection pressure for convergent phenotypes may lead to the original population evolving into two or more distinct morphologies that may or may not resemble the founders (Grant and Grant 2014, Malhotra and Thorpe 1991a, b, Thorpe and Brown 1989). These examples of watching natural selection in real time provide objective support for the hypothesis about how a single group of mainland species likely evolved into such different forms on different islands (Zhue et al. 2020). It represents a view of the departure point for forming new species.

Geno- and phenotypic experimentation has occurred on numerous islands, especially in the Indo-Australian archipelago, and has proven to contain numerous varanid (and other) species that had remained undescribed until the 21st century, largely because the many small islands and remote mainland locales had been given only the most minuscule of herpetological attention. Several factors have caused this, but the number of new species being described at present is relatively consistent, and the large majority are island endemics. In addition, island varanids tend toward melanism (Mertens 1934), even if, as with *V. macraei* and *V. jobiensis*, there exist numerous, small light markings; predominantly black examples of the latter are known, and some may represent undescribed species. Juveniles often bear more conspicuous markings or colors than adults, and that has kept some species cryptic to taxonomists. In hatchling *V. beccarii*, the black body has numerous tiny green flecks that will disappear with the lizard's growth. The green likely represents a remnant of the species' derivation from a green mainland species, e.g., *V. prasinus* or *V. kordensis*).

Varanids also tend toward gigantism coupled to melanism, if even only slightly, after becoming successful on islands (Case and Schwaner 1993, Cogger and Heatwole

TABLE 8.5
Known Island Endemic Species of *Varanus* and Their Distribution

Species	Distribution
*Varanus spinulosus**	Isabel Island. Eastern Solomons
*Varanus komodoensis**	Komodo, Flores, and nearby Indonesian islands
*Varanus beccarii**	Aru Islands, Indonesia
*Varanus bogerti**	D'Entrecasteaux Islands, Papua New Guinea
*Varanus telenesetes**	Rossell Island, Papua New Guinea
*Varanus palawanensis**	Palawan Island, Philippines
Varanus tsukamotoi	Saipan Island
Varanus tanimbar	Tanimbar Islands, Maluku, Indonesia
Varanus yuwonoi	Halmahera Island, Indonesia
Varanus macraei	Batanta Island, Indonesia
Varanus melinus	Obi Island, Moluccas, Indonesia
Varanus timorensis §	Timor Island
Varanus auffenbergi §	Roti, Timor Island group
Varanus rainerguentheri §	Timor Island
Varanus louisiadensis	Louisiade Archipelago, Papua New Guinea
*Varanus juxtindicus**	Rennell Island, Solomons
*Varanus bennetti**	Palau
*Varanus togianus**	Togi Island, Sulawesi, Indonesia

Note: An asterisk (*) indicates a melanistic adult form. Of the 18 species listed, 9 are melanistic. Dwarfism (§) is unusual among endemic insular *Varanus*.

1981, Meiri 2007, Price et al. 2015). Of the 19 known island endemic monitors (Table 8.5), the most familiar example is *V. komodoensis*, the largest (in mass) of living lizards, which has dark skin and has evolved to exist in a harsh environment. It is unquestionably the top predator on its islands but also lives with large prey species (buffalo, pigs), crocodiles, and venomous snakes. It resides on several small islands in Indonesia and parts of western coastal Flores to the east, giving the biggest monitor one of the smallest distributions. It also tends toward melanism, at least into the gray and dark brown, and is the species at the top of the region's food web. Smaller Indonesian islands are homes to *V. kordensis*, the largest of the *prasinus* group species, Papua New Guinea islands for *V. bogerti*, *V. andamanensis*, *V. mabitang,* and *V. komaini* from the salvator group, populations that are presently still classified as *V. jobiensis* around Irian Jaya, and some populations of the *indicus* group. Elsewhere, varanids tend to be small enough that attacking and subduing large mammals is neither possible nor required, most having an SVL under 380 mm. The largest member of the *V. prasinus* group is *V. beccarii*, a melanistic endemic of forests on the Aru Islands south of West Papua, while another endemic that is nearly as large is the poorly known *V. bogerti* from the D'Entrecasteaux chain of islands in eastern Papua New Guinea. In contrast to these species, another *V. prasinus* group member, *V. macraei*, is smaller and black with numerous large bright blue markings. Unusually for small island endemics, *V. macraei* is not the only varanid on its island, the 60 km by 8 km Batanta Island, Indonesia. Sympatric species include the large *V. chlorostigma, V. doreanus*, and very large *V. salvadorii*, each favoring different habitats, and the latter is uncommon. A study of the ecology and interactions of just the varanids on a small island would offer a laboratory to see how different aspects of island biogeographic models play out.

Australia's large *V. rosenbergi* (Fig. 8.3) is distributed along the far west, lower central, and extreme southeast, including some Bass Strait islands, and it presents an interesting question about islands and varanid ecology. This is possibly Australia's most studied monitor (Case and Schwaner 1993, Christian and Weavers 1994, Green et al. 1991, Green and King 1993, Rismiller et al. 2010), particularly populations in SE Australia and Kangaroo Island. There have not, however, been comparative studies of those with populations in South Australia and the Perth area. Each of these populations is separated from the others by huge expanses of desert that are uninhabitable to the species, in effect making each an island population. Such barriers to dispersal, such as deserts, deep bodies of water, and mountains, have been shown to be important factors in speciation in Melanesian birds (Mayr and Diamond 2001) and African monitors (Bayless and Sprackland 2000a, b). I have examined museum specimens of *V. rosenbergi*, but there were too few in both number and for any locality, and though I failed to obtain data to substantiate the idea that there were different taxa involved, I did find features to prevent the idea of being ruled out. Like *V. komaini, V. exanthematicus*, and *V. albigularis*, its insularity is based on land features.

The situation with eastern Indonesian monitors resembles that of Darwin's Finches in several ways. First, in both locations, the islands are relatively close. The Galapagos' Isabela and Santiago, for example, are 18 km and 56 km from Fernandina, with each island having its own endemic finch species (Grant and Grant 2014). Salawati and Waigeo (with numerous islets) islands are only 2 km and 20 km, respectively, from Batanta, and each has its own endemic varanids (and other organisms, e.g., mammals and plants). The large island of Misool, at 88.5 km from Batanta, is well within the range of distances between the Galapagos Islands. Given the proximity of the small islands to each other and the western New Guinea mainland, there have been greater chances of island invasions by other species. Waigeo's endemic black and yellow *Varanus boehmei* is similar to Batanta's undescribed endemic and *Varanus* from Salawati. There exist numerous other island groups that host more than one varanid species and are close enough to other islands and large landmasses (Sulawesi, New Guinea, Borneo) that molecular study might show patterns and timing of different arrivals and, because migration can return to the original source, departures.

Islands are known for their extraordinary tendency to host endemic species, and excepting perhaps Manhattan in New York, there are few islands that do not have unique

FIGURE 8.3 Varanus rosenbergi is a large melanistic monitor with at least three widely disjunct populations in southern Australia.

biota. Multitudes of examples present a great diversity of extremes in size, coloration, habits, and reproductive modes that, despite their greater sizes, are not often seen among continental faunas. Species that are particularly removed from other landmasses and live in an environment devoid of significant predation, competition, or both may evolve to become so divergent from their founders that they become particularly ill-suited to survive when new immigrants do eventually arrive, the Dodo of Mauritius being the iconic example. Does any of this apply to *Varanus*? The once-largest monitor, †*Varanus prisca*, became extinct after humans became an invasive species in Australia, and there is some debate about how much that may have been due to hunting the lizards. The monitors were large enough to have been a prized food item for the hunters (and vice versa), and the ability for humans elsewhere to take down large predators, such as Cave Bears, demonstrates that it was possible. *V. prisca* had been the apex predator far longer than people had been in Australia, making it specialized, perhaps in a way that resulted in the loss of fear, as for the Dodo and Marine Iguanas. Phylogenetic evidence links *V. prisca* to *V. komodoensis*, implying that the species at some point had sympatric distribution with a recent common ancestor and, for some time, with each other. Humans and other predators can account for the demise of the Komodo Dragon and may explain why it is not still present on other islands, except for Flores, in the archipelago. By the time of those events, Australia was cut off from nearby landmasses except, perhaps, New Guinea across the Torres Strait. Does this imply that *V. prisca* didn't survive in New Guinea because that predator-devoid island was the wrong habitat? Was the species a resident of deserts instead of seasonally wet forests? Whatever happened, the extirpation of *V. prisca* is the largest-scale example of an island varanid going extinct (see Chapters 10 and 11).

Within the prehistoric ranges of †*V. prisca* and *V. komodoensis* lived the fossil †*V. hooijeri* on Flores (see Chapter 10). This was also a large monitor, and its skull shows many features close to those in the other two species, yet with a unique, extremely broad parietal table, the implication of which is that it was also a top predator. Not much can be said based on such meager material. But Flores was, and remains, home to another large monitor, *V. bivitattus*, or Banded Water Monitor. Excepting *V. hooijeri*, for which data have not been compared, the other three species have all been shown to be closely related and have been combined into a common subgenus.

Another putative extinct endemic varanid may have lived in New Caledonia (Balouet 1984, 1991, Bayless and Naclerio 1993). This is covered in Chapter 10.

9 Mating and Reproduction

9.1 INTRODUCTION

> It IS MORE THAN strange that we still do not have sufficient information about the reproductive history of monitor lizards... As far as I know, only Theobald gives an extraordinary account of a species of monitor in India, the Yellow Monitor (*Varanus flavescens*). 'The monitor lizards,' he wrote, 'lay their eggs in the soils [and] sometimes they use the nests of white ants...'

So wrote A. G. Brehm (1883) in his classic *Brehm's Thierleben* with the irony that this mode of egg laying wasn't reliably confirmed until Cowles' (1930) report on the life history of *Varanus niloticus* in South Africa. Using ant nests or termitaria as incubation chambers for their eggs has now been verified in several of the large varanids of Africa, Asia, and Australia. We have learned a lot since 1883, but "we still do not have sufficient information about the reproductive history of monitor lizards." Here is a review of that information.

9.1.1 Reproductive Modes

Two universal drives in living things are eating and reproducing. All other activities are collateral to achieving these two objectives. Reproduction in monitors involves several steps: (1) finding and securing a mate, (2) mating (Fig. 9.1), (3) egg development within the female, (4) laying eggs in a suitable space (may entail nest construction), (5) embryonic development post parturition, (6) hatching, and sometimes (7) help from the mother to escape from the nest. Reptiles in general have long been known to reproduce sexually, where a male must introduce sperm to a female, and the two parental sets of DNA are joined in an egg cell to form a zygote, thence embryo. Sexual reproduction lies at the base of the Biological Species Concept in which species are defined as entities in a population that can enter into such a gamete mixture and produce fertile offspring (Mayr 1963, Mayr and Ashlock 1991, Wiley 1980); that is, any two healthy mature individuals of each sex should be able to produce fertile offspring.

For some monitors, there is the possibility of facultative parthenogenetic reproduction (Table 9.1). The discovery of parthenogenetic Caucasian Rock Lizards (Lacertidae, *Darevskia*) in Armenia (Darevsky 1958) revealed a radically new mode of reproduction to our understanding of terrestrial vertebrate physiology. Subsequently, parthenogenesis has been found in other lizard families, including Agamidae, Anguidae, Gekkonidae, Gymnophthalmidae, Scincidae, Teiidae, Xantusiidae, and Varanidae (Lenk et al. 2005), though in the latter, at least, the mode seems facultative. In true parthenogenesis, females produce same sex clutches, all young being female and viable. In varanids there are accounts of females of some species producing both sexes or only one sex. It is quite possible that facultative parthenogenesis is employed rarely by wild lizards, where potential mates of both sexes are present, though there are no data available on the subject. It would seem to be useful in establishing a founder group when inhabiting new territories, such as islands (Jacobson 1909, Rawlinson et al. 1990).

Parthenogenesis, sex-change, and hybrid speciation are related topics and have been studied extensively in many animal groups. Among vertebrates only fish can change sexes within their lifetime (Casas and Saborido-Rey 2021), though the environmental pressures that select for the ability might also be involved in developing parthenogenetic populations of other groups. In most studied cases, parthenogenesis is the

FIGURE 9.1 Mating monitors. Upper, *Varanus macraei* (courtesy Julian Firth); bottom, *Varanus acanthurus*. In each case, the male is on top of the female and rotates his pelvic region to allow one hemipenis to access the female's cloaca.

DOI: 10.1201/9781003463528-12

TABLE 9.1
***Varanus* Species Known to Have Reproduced Parthenogenetically in Captivity as of 2023**

Species	Reference(s)	Species	Reference(s)
albigularis	Branch (1991)	*komodoensis*	Watts et al. (2006)
bengalensis	Auffenberg (1981b)	*mertensi*	Horn et al. (1994)
	Deraniyagala (1958)	*niloticus*	Cowles (1930)
brevicauda	Standen (2008)	*olivaceus*	Auffenberg (1988)
caudolineatus	Thompson et al. (1992)	*panoptes*	Lenk et al. (2005)
cumingi	Horn (1994)	*rainerguentheri*	Grabbe and Koch (2014)
dumer ilii	Davis et al. (1966)	*salvator*	Gaulke (1989)
exanthematicus	Attum et al. (2000)		Vogel (1979)
gilleni	Carpenter et al. (1976)	*semiremex*	Horn (1985)
	Murphy and Mitchell (1974)	*similis*	Horn (1985)
glauerti	Hörenberg (2013)	*timorensis*	Horn (1985)
gouldii	Thompson et al. (1992)	*tristis*	Harris (2011)
griseus	Tsellarius and Tsellarius (1997)	*varius*	Horn et al. (1994)
indicus	McCoid and Hensley (1991)		

result of hybridization between two closely related, genetically compatible species (Brunes et al. 2019). Wiechmann (2012) reported on parthenogenetic *Varanus* that were the result of hybridization in captivity between two alleged subspecies of *Varanus panoptes*. Sprackland (2009) united the subspecies, so Wiechmann may have been working with conspecific lizards from different populations. The parental composition for other parthenogenetic varanids is unknown, though all cases involved females that had been kept for at least a year without access to a male. Though interspecific hybrids are produced in captivity, there is no evidence of such cases in the wild. For the parthenogenetic lizards developed in captivity, there is no evidence that the mother was such a product of hybridization, leaving the question of the mechanism involved unresolved.

Parthenogenetic ability might also partially explain the vagility of the genus and its success at colonization, particularly when it comes to spreading among southwestern Pacific islands. There is a caveat, though, in the absence of parthenogenetic monitors anywhere except in captivity. So far, though, most information about parthenogenetic varanids has come from private breeders and zoos (Table 9.1), and to date, there is no clear evidence of any varanid being the product of interspecies hybridization. Extensive overviews of varanid reproduction in captivity have been published by Aller and Manago (2009), Brown (2012), Eidenmüller (2021), Horn and Visser (1977, 1989), Köhler (2005), Mendyk (2016, 2017), and Sprackland (1989). For overviews of reptile eggs and incubation generally, see Belinsky et al. (2004), Birchard (2004), Booth (2004), Deeming (2004), Köhler (2005), and Zimmermann (1986). Long-term field studies on the topic involve relatively few species (Andrews et al. 2017, Auffenberg 1981, 1988, 1994, Auffenberg et al. 1988, 1989, Auliya 2006, Biswas and Kar 1981, Carter 1999, Cowles 1930, Doody et al. 2014, James et al. 1992, Rismiller et al. 2010, Thompson and Pianka 2001). Most of the data we have on clutch sizes, incubation times, and temperatures, and hatchling emergence had not been discovered until the 1990s (Table 9.2); these reproduction data for the comparatively well-known *Varanus indicus* were unknown as recently as the field study of the species on Guam by Wikramanayake and Dryden (1988). However, the great majority of reports on varanid reproduction come from specimens kept in captivity (Table 9.2), including the prolific *V. indicus* (Brown 2012, Gregorovicova et al. 2012, Kok 1993, Kukal 1993, McCoid 1993, Wesiak 1993a, b).

9.1.2 Sexual Dimorphism

Distinguishing the sex of a monitor is possible but involves rather subtle traits of different levels of reliability. The most reliable is the examination for the presence of ZZ or ZW chromosomes. There can be a sex-linked difference in SVL of adults, but if the lizards are the same size but of different ages, the trait is not useful. The width of the head of male lizards is usually comparatively broader, with larger temporal muscles than females. This holds true in monitors, but this character is most reliable if there are both sexes at hand to compare. In the species that have preanal pores, the structures are larger and much more distinct in males than they are in females (Figs. 4.16 and 9.2A, B). Males of several of the Australian *Odatria* have lateral spiny tufts at the base of the tail (Figs. 4.25 and 9.2C, D), but females may have a few slightly enlarged, but not spiny, scales in the same region. Only two species, *Varanus pilbarensis* and *Varanus flavescens*, are known to display sexual color dimorphism, in each case the males having a much more intense coloring than females.

9.2 MATING

The broad subject of varanid sexual reproduction has been reviewed by Auffenberg (1981, 1988), Böhme (1988), Brown (2012), Eidenmüller (2016, 2021), Horn and Visser

TABLE 9.2
Representative Reproduction Data of *Varanus* Based on Published Reports

Species	Reference(s)	Species	Reference(s)
acanthurus	Doody et al. (2017)—w	*mitchelli*	Gaikhorst et al. (2009)—c
	Thissen (1992)—c		Shine (1986)—w
albigularis	Poder Le (2007)	*niloticus*	Cowles (1930)—w
auffenbergi	Wesiak (2008)—c	*panoptes*	Doody et al. (2014)—w
baritji *	Bedford and Husband (2003)—c		Shine (1986)—w
			Wolf (pers. comm.)—c
bengalensis	Thakur (2020)—w		
exanthematicus	Wesiak (2006)—c	*prasinus*	Eidenmüller (2016)—c
giganteus	Bredl (1987)—c	*rainerguentheri*	Grabbe and Koch (2014)—c
gouldii	Barnett (1979)—c	*rosenbergi*	Andrews et al. (2017)—c
	Doles and Card (1995)—c		Ehmann et al. (1991)—w
	Shine (1986)—w		King and Green (1979)—w
indicus *	Kok (1993)—c		Rismiller et al. (2010)—w
	McCoid (1993)—w	*salvadorii*	Camina et al. (2014)—c
	Wikramanayake and Dryden (1988)—w		Mays (2007)—c
juxtindicus *	Wesiak and Koch (2009)—c	*salvator* *	Biswas and Kar (1981)—w
macraei	Dedelmar (2007)—c	*similis*	Moldovan (2007)—c
	Eidenmüller (2016)—c	*storri*	Bartlett (1981, 1982)—c
mertensi	Shine (1986)—w	*varius*	Bredl and Schwaner (1983)—c
	Rhind et al. (2016)—w		Carter (1999)—w

Note: For some species there is a considerable published record. Several other species are bred in the private sector (e.g., *dumerilii*, *glauerti*, *primordius*) but reports are few or have not been published. Symbols: c = captive; w = wild; * = taxonomic identity uncertain because of recent revisions.

(1987, 1989), Murphy (1972), and Sprackland (1989). The number of species for which published data are available for wild specimens has grown steadily (Ariefiandy 2015, Biswas and Kar 1981, Carter 1999a, b, Cowles 1930, Doody et al. 2014, 2015, 2018, Ehmann et al. 1991, Rismiller et al. 2010, Thakur 2020, Thissen 1992), but it is usually based on just a few specimens of any given species. Zoos and private keepers have presented a wealth of such information, largely on the Internet, and have frequently included data on breeding parameters and hatchings. The recent statement, for example, that "there are no reports of breeding this species [*V. rudicollis*] in captivity" (Eidenmüller 2021),

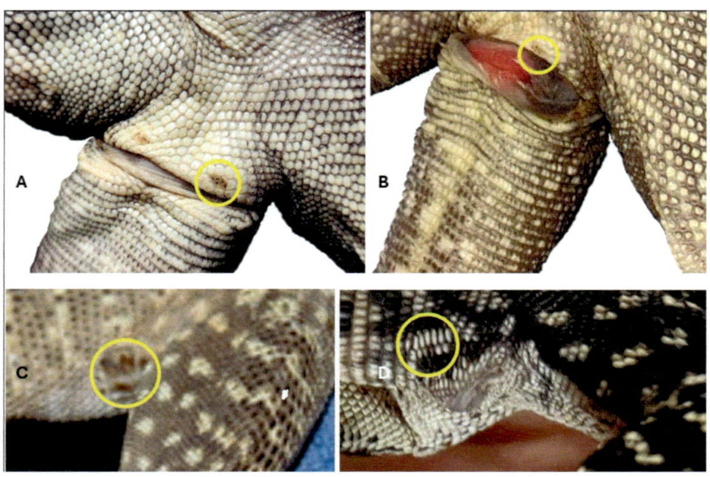

FIGURE 9.2 Male sexually dimorphic characters in adult *Varanus* with characters circled. (A) *Varanus togianus*, a putative member of the *Varanus salvator* group, adult male showing preanal pores that, though present, are less developed than in other members of the group and of other species (see Fig. 4.16). (B) Companion female for breeding, showing a scale with three small black spots that are not raised or show any further feature of a preanal pore. (C) *Varanus tristis*, adult male with five enlarged and pointed scales compared with (D) A female adult of the closely related *Varanus similis* showing several vertically elongated but not keeled or spiny scales. The right hemipenial bulge is visible in the male, and the cloaca is visible in the female. Photos A and B courtesy Mike Stefani.

applied to most varanids, there have been numerous verified reports of such presented online since 2021. Both the number of species and the number of F2 generation and beyond produced by private breeders have been considerable (Aller and Manago 2009, Bedford and Husband 2003, Bredl 1997, Camina and Koch 2017, Eidenmüller 1989, 1994, 1997, 1999, 2016, Engelmann and Horn 2003, Fischer 2012, Gorman 1993, Mitra 2001) and include numerous breedings of the rare Blue Tree monitor (*Varanus macraei*) (Dedlmar 2007, Rahmanto et al. 2022)

Though varanids have delimited territories, they do not engage in the routines found in many iguanians of showing bright color flashes, performing pushups, or bobbing their heads. However, male monitors of several, possibly all, species engage in wrestling combat (Earley et al. 2022, Horn et al. 1994, King and Green 1999, McCoid and Hensley 1991, Murphy and Mitchell 1974, Thompson et al. 1992), the winner then approaching the local female and, if she is willing, mates. Mating and reproduction have been well studied in numerous species in the wild (Auliya 2006, Carter 1999, Cowles 1930, Doody et al. 2014, 2018, James et al. 1992, King and Green 1979, McCoid 1993, Rismiller et al. 2010) and in captivity (Bartlett 1981, 1982, Bedford and Husband 2003, Bredl 1987, Bredl and Schwaner 1983, Brown 2012, Camina et al. 2014, Eidenmüller 2016, 2021 and references therein, Gaikhorst et al. 2009, Horn and Visser 1989, 1997, Kok 1993, Mayes 2007, Moldovan 2007, Thissen 1992, Wesiak 1993a, b, Wesiak and Koch 2009). Like all squamates, male monitors possess paired copulatory organs called hemipenes, the structure of which is species-specific and of systematic importance (Böhme 1988, 1991, Böhme and Ziegler 2007, Branch 1982, Brown 2012, Card and Kluge 1995). The paired hemipenes lie caudad to the cloaca on each side of the base of the tail in the corpora cavernosa. In this position, the organs are thin, elongated, and "inside out" (depending on which surface is considered inside or outside). When employed, the organs exit the cloaca by a combination of lateral muscles, the m. hemipenis repulsors, and enlargement of the organ. The now everted organs take on their characteristic shape, revealing the structures that serve to allow entrance and stability in the female. The organs are brought back into the corpora cavernosae by the m. retractor penis muscles. The different structures of the hemipenes vary in size, shape, and degree of ornamentation among species, making them both phylogenetically informative and taxon diagnostic. In most squamates, each hemipenis is Y- or V-shaped. One side may be subequal, larger, or longer than the other, and terminates in a slender tip, the hemibaculum, which is composed of fibrous and hyaline cartilage tissue (Al-ma'ruf et al. 2021, Card and Kluge 1995). Proximal to the hemibaculum is a broad and intricately wrinkled surface, the paraphasmata, which provides a pair of sulcal lips for the distal portion of the sulcus, the channel for sperm transference. Most proximal and located between the cloaca and extended hemipenes is a smooth columnar region, or truncus, that bears a distinct groove from the cloaca to the distal part of the paraphasmata, the sulcus proper. Unlike the situation in the mammalian penis, in which the sperm-carrying channel is completely enclosed, the groove in lizards is open. During copulation the male tends to lie astride the female and twists his pelvis so that his cloaca and one hemipenis can reach and enter the female's cloaca (Figs. 9.1 and 9.2). Böhme (1988) observed that the males may have both hemipenes, inserting one to each side of the female, and that in this way both of the oviducts become fertilized. In cases where the male only inseminates one side of the female, she may lay fertile eggs from only one oviduct and thus produce half a normal clutch (Böhme, pers. comm.). Böhme (1995) described hemipenes-like structures in female varanids that he called hemiclitorises, which are morphologically similar to the males' hemipenis in having hemibaculi and epidermal flounces, and, as in males, can be quite long when everted. The hemiclitoris has subsequently been discovered in other lizards (Böhme 1995, Folwell et al. 2022b, Oliveira 2021, Valdecantos and Lobo 2015) and more recently in snakes (Folwell et al. 2022a, b). Unlike hemipenises, hemiclitorises leave no conspicuous bulge in the ventral base of the tail.

9.2.1 Egg Development

Following successful mating, females go through a period of three weeks (e.g., for *Varanus mertensi*) to eight weeks before laying the eggs (Eidenmüller 2021). Initially, the females may continue feeding as usual, even up to the day of laying, but more often cease several days before. Eggs can usually be seen through the ventral surface of the female and can be felt by gentle palpation. Once eggs have been laid, the female may rest for several more days before she resumes feeding and her normal routine, and she is often thinner to the point of seeming emaciated. Post laying, the female's torso may look deflated and flaccid, with the skin folds much more pronounced than usual.

Gravid female monitors grow abdominally as eggs develop in the oviducts (Fig. 6.25) and complications can occur that inhibit the oviposition of the eggs, a condition known as egg binding or dystocia (Hedley 2016), though the latter term is generally used in reference to difficulty in birthing live young. The condition is common in captive reptiles, usually linked to incorrect husbandry, with most of the veterinary literature reporting on cases in Bearded Dragons (*Pogona vittatus*) and colubrid snakes, which are among the most commonly kept reptiles. Most accounts of egg retention in varanids are anecdotal, much on Internet forums, but some published case accounts are available (e.g., Divers and Williams 1993). Gardner and Barrows (2010) present necropsy results for an adult *Varanus albigularis* that had advanced egg binding and died shortly after being received at the veterinary hospital. The number of symptoms, both pre- and postmortem, was diverse, from lethargy and incomplete shedding to severe yolk coelomitis, inflammation, and "adhesions between the liver, gastrointestinal tract, oviducts, and coelomic membrane" (Gardner and Barrows 2010:121). A young *Varanus salvator* was presented (Govendan et al. 2020) that had been anorexic and with a swollen abdomen. Fluid treatment and improved husbandry were begun, and the lizard laid eggs four days post-treatment (Govendan et al. 2020). The authors suggested that egg retention may be more likely in young lizards. Metabolically, dystocia can be caused by a diet poor in calcium and/or vitamin D_3, leading to hypocalcemia, as can infections and physical injuries that may damage the reproductive tract. Lizards in captivity may also become dystocic from obesity, and the accumulated fat bodies crowd the eggs, making deposition difficult or impossible. Untreated dystocia is usually fatal, but veterinary surgery has been successful in removing eggs, some of which may be viable. Symptoms of egg binding include abdominal swelling, anorexia, and pronounced lethargy, e.g., limited movement not linked to basking or sleeping. In early stages of gravidity, lethargy is more likely to present, and an increased abdominal mass may first be confused with "simple" obesity. In late stages, the lizard can discharge fluid, including blood, from the cloaca.

Illness due to egg binding is sometimes difficult to correctly diagnose in the early stages because symptoms are general enough to have many possible causes. Anorexia is a common symptom of infection, but it can also be caused by insufficient heat, hydration, or nutrition. In my experience, refusing water has often been linked to infectious stomatitis if the environmental temperature is too low. Humidity levels have also been linked to numerous illnesses in captive monitors. Poor nutrition can be blamed for many illnesses in captive animals, such as insufficient or excess calcium, potassium, vitamin D_3, or excess protein. Surgical removal of bound eggs has become a standard and usually successful procedure.

9.3 EGGS AND INCUBATION

9.3.1 Nest Environments

The specifics regarding number, size, nest site, incubation time, temperature, and humidity vary, but share two features. Eggs are almost always buried in the ground (Fig. 9.3A) or at ground level (e.g., in termitaria) in an environment where the temperature (and presumably humidity) remains stable, and incubation is continuous, without a pause, and takes longer than almost all other reptiles. Monitor eggs are leathery, symmetrically oblong, and white (Fig. 6.25), the size and number of which are linked to species, size, and age of the mother. Large species tend to lay a greater number of comparatively large eggs, while smaller species lay fewer and smaller eggs (Fig. 9.4 and Table 9.2). Also, large females will lay more eggs than smaller individuals of the same species; however, data regarding female to clutch size can vary considerably between captive and wild conditions (Fig. 9.4). Clutch incubation time also varies by species and ranges from about 80 days for small taxa up to 296 days for the largest (Table 9.3). Incubation time for a given species is not necessarily linked to individual egg size plus neonate SVL and mass, the allometric relationship being negative between egg size and mass (Thompson and Pianka 2001). Very little is known about in situ nesting and development among varanids, most monitor but eggs tend to be laid deep

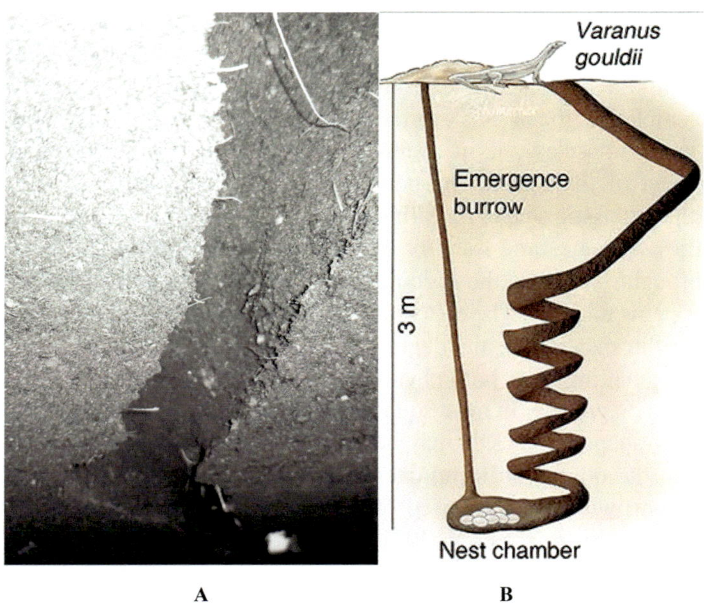

FIGURE 9.3 (A) Even arboreal species such as *Varanus macraei* may excavate nests in the ground, such as this one dug in a terrarium (arrow points to an egg). (B) Discovery of the spiral burrows of Australian *Varanus panoptes* and *Varanus gouldii* came as a complete surprise to researchers, and may be unique to these species. The spiral burrow is the access for females, who deposit their clutch, often communally, in the lower chamber (arrow). The straight tunnel in C is excavated by hatchlings, providing their own direct escape route from the nest. See text for details. (A courtesy Julian Firth.)

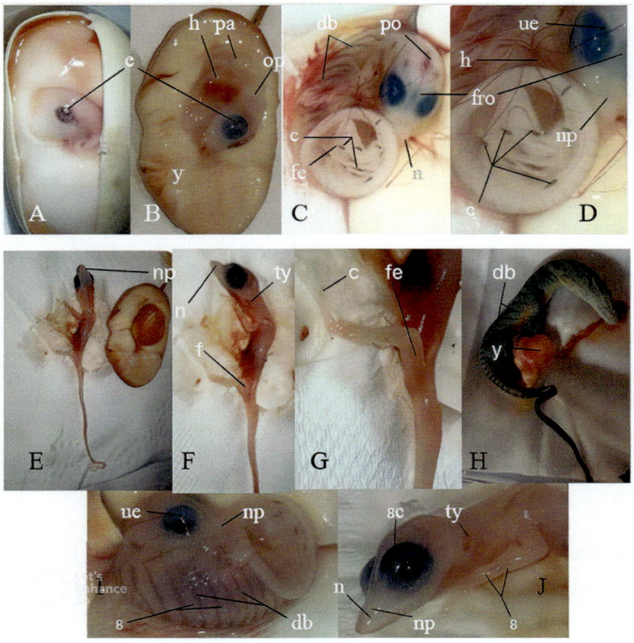

FIGURE 9.4 Developmental stages of embryonic *Varanus macraei*, an arboreal species from New Guinea. (A) DPO 32 and (B) DPO 50. Parts (C and D) are near the hatching stage, approximately DPO 100, and clearly show an amniotic sac, snout elongation, dorsal pattern bands, and as yet unfused parietal elements surrounding the parietal eye. (E and F) DPO 120 late stage embryo lacking bands, showing well-developed eyes, and limbs and tail in near terminal hatchling proportions. (G) The same specimen showing the late development of the tarsal-metatarsals and phalanges. (H) A stillborn lizard that was nearly ready to hatch. (I) Dorsolateral view of embryo in C and D, clearly showing the dark bands on the nape, body, and tail, representing a component of the melanin-rich pattern as distinct from the later coloration. See text for explanation. (J) Close-up of head of same specimen; note the early development of the nasal hump characteristic of many varanid species, and the pronounced elongation and attenuation of the snout; sclera are faintly visible around the cornea, and claws have yet to appear. Claws (c), dorsal bands (db), forelimb (f), frontals (fro), eye (e), femur (fe), heart (h), nasal prominence (np), nostril (n), otic placode (op), pharyngeal arch (pa), parietal eye (po), scales (s), rudimentary scleral ossicles (sc), tympanic opening (ty), upper eyelid (ue), yolk sac (y). Days post oviposition (DPO) are approximated based on Gregorovicova et al. (2012). Photos courtesy of Julian Firth.

enough that the chances of them being moved is limited; unlike bird's eggs that require periodic turning, such movement may be fatal to squamate eggs, though this does not seem to be universally true (Aubret et al. 2015, Birchard 2004, Deeming 2004, Eidenmüller and Wicker 1998). Some species are known to lay in communal nests (Fig. 9.3), e.g., *Varanus gouldii* (Doody et al. 2020, Whitford 1998), *V. panoptes* (Doody et al. 2014, 2015, 2018), and *Varanus acanthurus* (Doody et al. 2017, 2020, pers. obs.). In almost all cases, there is no recorded parental involvement with the eggs once they have been laid and the nest covered. However, the species that lay eggs in termitaria perform the unusual and characteristic task of returning to the nest at the time the eggs are due to hatch and again assault the hard mound and excavate to make it possible for the neonates to emerge. Presumably, the young lizards lack the strength to get through the hard material themselves, thus requiring adult assistance. It has often been assumed that the female who laid the eggs is the one who returns to the nest to free the young, and it is certainly a parsimonious suggestion, but so far, there has been no published verification that it is, indeed, the mother who performs that task.

Varanids have three known ways of depositing their eggs for incubation: putting them in termitaria (Cowles 1930, Ehmann et al. 1991, King and Green 1979), putting them in deep spiral tunnels (Doody et al. 2018, 2014, 2015), or simply burying them or laying in a safe place in trees. In picking a site for deposition of the eggs, there are three qualities that make a site suitable. Incubation of varanids requires an environment that can maintain a steady temperature and do so over the long time that is taken in embryonic development. From both field-collected and captivity data, ideal nest temperatures must stay within a range of 28°C to 33°C. Given the length of incubation, the nest site must be able to maintain that temperature across the span of two or three seasons with its temperature and humidity variations. It has long been a mystery how the termites' nests can have thermal and moisture equilibrium when they are located in harsh environments where temperatures may become excessive (>45°C) for some species, and where the period of incubation may include a season of low temperatures, flooding, and harsh drought. (An Australian folksong precisely captures the Queensland climate in one line: "where the creeks run dry or are ten foot high," which is decidedly not an exaggeration.) Research by physicists and engineers recently found out how the temperature control system works (American Institute of Physics 2014). An important feature of the mounds is their specific configuration, being both tall (some up to 11 m) and a design that lets the broad sides face east-west and the thinnest faces toward the arc of the sun (usually north where monitors live). What has been known for some time is that the nests are thereby exposed to full sun in the morning, the least during the hottest part of the day, and again full sun that allows the nest to absorb enough heat to last through even a chilly night. The recent discovery demonstrates how the nests contain long tunnels, with the vertical shafts moving hot air up to an exit at the top, which provides enough of a draft to cause temperature-stabilizing flow in the many horizontal chambers. The materials used during construction include soils and decomposing vegetative matter that, once sealed within the hard walls, will hold and maintain water and keep humidity within narrow limits. Van Damme et al. (1992) looked at the effects of various incubation temperatures in *Podarcis muralis*, a European species with activity and behavior patterns similar to small monitors, and noted a profound influence on both developmental and post-hatching lizards. Clutches of varanid eggs maintained in captivity have been shown to have high to complete failure to hatch if the temperature and humidity fall outside

TABLE 9.3
Clutch Size, Incubation Time, and Incubation Temperature for Several Species of *Varanus*

Species	Clutch Size	Incubation Time (Days)	Temp. (C)	Annual Clutches	Reference
acanthurus	5–12	(120)	27–30		Eidenmüller (2021)
	8–28	95–134	29		Paul Burrows, Jr. (pers. comm.)
	Up to 14	112–120	28		Kenneth Jensen (pers. comm.)
	4–6	120	29.4	4	Michael Stefani (pers. comm.)
beccarii	3–5	172–205	27–29.5		Eidenmüller and Wicker (1992)
bengalensis	8–32	240–270	–		Auffenberg (1994), Thakur (2020)
Bangladesh	20–30	–	–		
India	8–30	240–270	–		Khan (1987)
bivittatus	13	180–200	29.4	4	Daniel (2002)
brevicauda	2–3	70–84	25		Michael Stefani (pers. comm.)
caudolineatus	3–5				Greer (1989)
cumingi	12–14	160–200	29.4	4	
dumerilii	8–15	199–207	28.8		Greer (1989)
flavirufus	7	180–190	29.4		Michael Stefani (pers. comm.)
flavescens		155	30		Aiden Bradley (pers. comm.)
giganteus	4–9	200–220			Aiden Bradley (pers. comm.)
gillent	4–7	235–285	28–30		Visser (1985)
	3–6	99–121	30–32		Kenneth Jensen (pers. comm.)
	4–6	103–120	29		Eidenmüller (2021)
glauerti	4–6	100–120	28		Paul Burrows, Jr. (pers. comm.)
glebopalma	7	100–120	29.4		Kenneth Jensen (pers. comm.)
griseus	18–24	121	29.4	4	Michael Stefani (pers. comm.)
indicus †	4	158	30		Michael Stefani (pers. comm.)
	2–7	181–191	31.2		Greer (1989)
komodoensis	24–26	237–279	29		Dujsebayeva (1995)
mertensi	12–16	262–296	25.5–29		Roy Payne (pers. comm.)
mitchelli	10–12		29.4	4	Paul Burrows, Jr. (pers. comm.)
olivaceus	3–10	219			Walsch et al. (1993)
orientalis	3–9	109–138	27–29.5		Michael Stefani (pers. comm.)
panoptes horni *	12–15	200 days	29.4–30		Greer (1989)
prasinus	3–5	157–160	31.4		Card (1993)
	1–5	145–165	29.4		Eidenmüller (1989)
rosenbergi	10–19	180			Rowan Wolf (pers. comm.)
rudicollis	7–10				Codey Kopp (pers. comm.)
scalaris	3–7	128–155	27–29.5		Kathryn Brown (pers. comm.)
spenceri	11–35	122–129	29.5		Card (1993) AAZPA Comm April 16
storri	3–7	114–120	28		Guy Marsters (pers. comm.)
	2–7	95–130	26–30		Eidenmüller and Wicker (1991)
*tristis**	3–6	(116)	28		Peters (1971) Bull.Zoo.Mgmt.3(2)
	6	(107)	29–29.4		Kenneth Jensen (pers. comm.)
					Bartlett (1982)
					Eidenmüller (1994)
					Kenneth Jensen (pers. comm.)
					Rowan Wolf (pers. comm.)

Note: Numbers in parentheses () are approximate. References marked (pers. comm.) represent breeders and the private community, otherwise unpublished data from Sprackland. † = parthenogenetic breeding. Relationship between clutch and adult sizes. This graph depicts the pattern for lizards that are in their second or later year of laying. Actual clutch numbers are also dependent upon prey availability and maternal health.
* means specific identity uncertain.

the prescribed parameters even slightly, which highlights the importance of females to choose very specific types of places for their eggs. Finally, the eggs need reliable long-term physical protection from both the environment and predators. The termitaria are such places, almost fortresses, the strength of some so great that elephants have been observed to use them as scratching posts, yet causing no harm to the mound. Many animal species prey on termites outside the termitaria; very few can break into one. In the event that the mound is somehow breached, large numbers of its residents work quickly to effectively repair the damage. *Varanus rosenbergi* has a large range in southern Australia, with three large allopatric ranges: one in Western Australia, and one around Adelaide, the other in southeastern Australia. Prevailing cold-season temperatures of around 15°C would be lethal for varanid eggs, but *V. rosenbergi*, by laying eggs in termitaria where temperature ranges between 27°C and 37°C, circumvents the problem. This evolutionary adaptation, a symbiosis between insect and lizard, allows *V. rosenbergi* to survive in environments too cool for any other *Varanus* (Ehmann et al. 1991, King and Green 1999, Rismiller et al. 2010). Accounts of *Varanus* nesting and egg laying in captivity include many species: *V. albigularis* (Markland and Brown 2009), *Varanus beccarii* (Eidenmüller 2016, Fischer 2012), *Varanus dumerilii* (Aller and Manago 2009), *V. indicus* (Kukal 1993, McCoid 1993, Wikramanayake and Dryden), *V. gouldii* (Doody et al. 2018), *Varanus juxtindicus* (Wesiak and Koch 2009), *Varanus komodoensis* (Ariefiandy et al. 2017, Auffenberg 1981, Darevsky and Kadarsan 1964, Murphy et al. 2002), *V. mertensi* (Mayes et al. 2007), *Varanus mitchelli* (Gaikhorst et al. 2009), *V. panoptes* (Doody et al. 2014, 2015, 2018), *Varanus prasinus*-group (Eidenmüller 2016), *V. salvator* (Rodriguez 2009), and many Australian species (Brown 2012, Eidenmüller 2007, Shine 1986). Many other species have been successfully bred in captivity, but not all have been reported in peer-reviewed literature.

Cowles (1930) may have been the first to report on a varanid, *V. niloticus*, using a termite mound as an incubator for its eggs. Because the mounds can be extremely hard externally, the adult female monitor must use her claws to breach the material. Once that has been done, she carefully excavates a considerable distance into the mound and eventually lays her eggs on soft material deep inside. For eggs that have such long incubation times (>150 days), the fortified, climate-controlled nest almost certainly decreases embryo mortality. Not all varanids can use termitaria, so it remains a tool for the largest and strongest species, including *V. albigularis* (Gillet in Branch 1991), *Varanus bengalensis* (Thakur 2020), *V. rosenbergi* (Ehmann et al. 1991), *V. salvator* (Biswas and Kar 1981), and *Varanus varius* (Carter 1999); Greene (1986) has published an excerpt from field notes by A.H. Miller regarding two juvenile *V. prasinus* observed by R. G. Allen to "hatch in termite nests and feed there later to emerge as one of the long-tailed species." The statement is ambiguous; did the lizards hatch in termite nests and feed there, then later emerge (unlikely as the implication of considerable time lapsing accompanies that interpretation). The fact that the lizards were seen to hatch is questionable, given that eggs buried in termitaria are not visible during hatching. The specimens involved (MVZ 74904 and 74905) are hatchling size and still show umbilical scars (Greene 1986; pers. obs.), making the most likely event the witnessing of hatchlings emerge from a termitarium and possibly consuming the insects outside.

Arguably the most spectacular of all squamate nesting is found in at least two closely related Australian species, *V. gouldii* and the larger *V. panoptes* (Fig. 9.3). These lizards have been discovered to excavate extremely large, elaborate, and deep communal nesting burrows (Doody et al. 2014, 2015, 2018, 2021) that are unlike any nesting sites known. The first such burrows found were large descending spirals that became narrower with depth. Reaching depths of up to 4 m, they are the deepest vertebrate nests known (Doody et al. 2021). It was later learned that several lizards could cooperate and dig multiple burrows that eventually terminate in an incubation area. Eidenmüller (2021: 183) wrote, in reference to captive breeding *panoptes*, "Sand or soil are ideal substrates, the deeper the better as these lizards are very fond of burrowing." Definitely true and expressed as an enormous understatement. A typical clutch will contain 6–8 eggs; the communal nests sometimes contain >100 eggs, the combined output of 12–16 females. Additional research showed that the data for *panoptes* also applied to *gouldii* (Doody et al. 2021). It is possible that other closely related species, such as *giganteus* and *spenceri*, have similar nest-building strategies. In captivity, access to termitaria or adequate substrate to excavate 3+ m nests is lacking. Nevertheless, even the species that normally employ such extreme nesting behavior have been successfully and repeatedly bred in zoos and private collections (Barnett 1979, Bredl and Schwaner 1983, Doles and Card 1995, Poder Le 2007, Wesiak 2006). In these examples, lizards will excavate shallow nests if substrate allows, or deposit eggs in a nesting box, often with soil or moss as a substrate. The depth and structure of the exceptional spiral burrows are probably a response to the harsh environmental conditions of the Outback that are absent in captivity.

Smaller species (e.g., the *Odatria*, *prasinus*, and *indicus* groups) do not typically engage in using termitaria (but see above; Greene 1986) and instead lay their eggs in tree hollows or nests dug by the females, and a single nesting site may simultaneously be used by multiple females (Doody et al. 2017).

9.3.2 Embryo Development

Within days of oviposition, embryos will begin attaching to the inner membrane of their shells. The embryo may be visible as a small circle that, as the attachment to the shell lining advances, becomes white and gradually spreads. In studies on the fate of eggs that have been moved during incubation, it was noted that embryos cannot survive if they are on the ventral side of the egg (Booth 2004), but the data comes from observing crocodilian and turtle eggs. The suggestion that egg rotation after the white patch spreads across the entire shell will not harm the embryo (Booth 2004) because the embryos have "become quite robust" may not hold true for varanids or other squamates. Indeed, it does not hold true for a European water snake, *Natrix maura* (Aubret et al. 2015). They specifically tested the hypothesis that reptile eggs should not be turned during incubation and found that, out of 388 eggs, 59 percent of embryos developed dorsally, 28 percent ventrally, and

14 percent laterally prior to any movement. Unsurprisingly, large females laid larger clutches than smaller conspecifics, and their eggs had a higher percentage of hatchings (92.2 vs. 81.8 percent), and more significantly, a higher neonate survival rate for turned (37.5 percent) than unturned (4.5 percent). Eggs from *Calotes versicolor* (Pandha and Thapliyal 1967), *Eublepharis macularius*, and *Pantherophis guttatus* (Marcellini and Davis 1982) showed no difference between turned and unturned eggs, but such studies remain rare (Aubret et al. 2015, Birchard 2004, Deeming 2004, Eidenmüller and Wicker 1998), and have involved few species.

Embryological development has only been described for three species (*V. panoptes, V. indicus*, and *V. rosenbergi*, discussed below), and herein a fourth, *Varanus macraei*, is depicted (Fig. 9.4). Once laid, varanid eggs proceed through development, being like most other squamates (excepting chameleons, Andrews and Donoghue 2004) in having no period of diapause (Angelici and Luiselli 1999, Gorgorovicova et al. 2022). Unverified reports of captive varanids needing a diapause period (e.g., sometimes described as hibernation) are posted on some websites, but the information contained also allows for the alleged diapause to be related to improper husbandry due to low temperatures. In chameleons the length of incubation time is considerably longer than that of most other lizards, attributable to the diapause.

If varanid eggs have uninterrupted development from parturition, why do their eggs take so long to hatch? The prevailing hypothesis posits length of development being positively correlated with brain size and development (Andrews et al. 2017, Birchard 2004, Deeming 2004, Gregorovicova et al. 2012, Thompson and Pianka 2001, Warneburg et al. 2015). There are good reasons to support this hypothesis. First, the elongation of the varanid skull takes place rather late in embryonic development. Most of the skull elements are present with a morphology not unlike that seen in other late-stage lizard embryos (Vidal Maldonado and Antonieta 2008). It is shortly prior to the stage where the parietal surrounds the parietal eye and ossifies (Fig. 9.5C) that the preorbital region begins the extension and widening that leads to a characteristic varanid snout. The late parietal closing may be due to the continuing growth of the brain; the comparative morphology of the postocular region in varanid

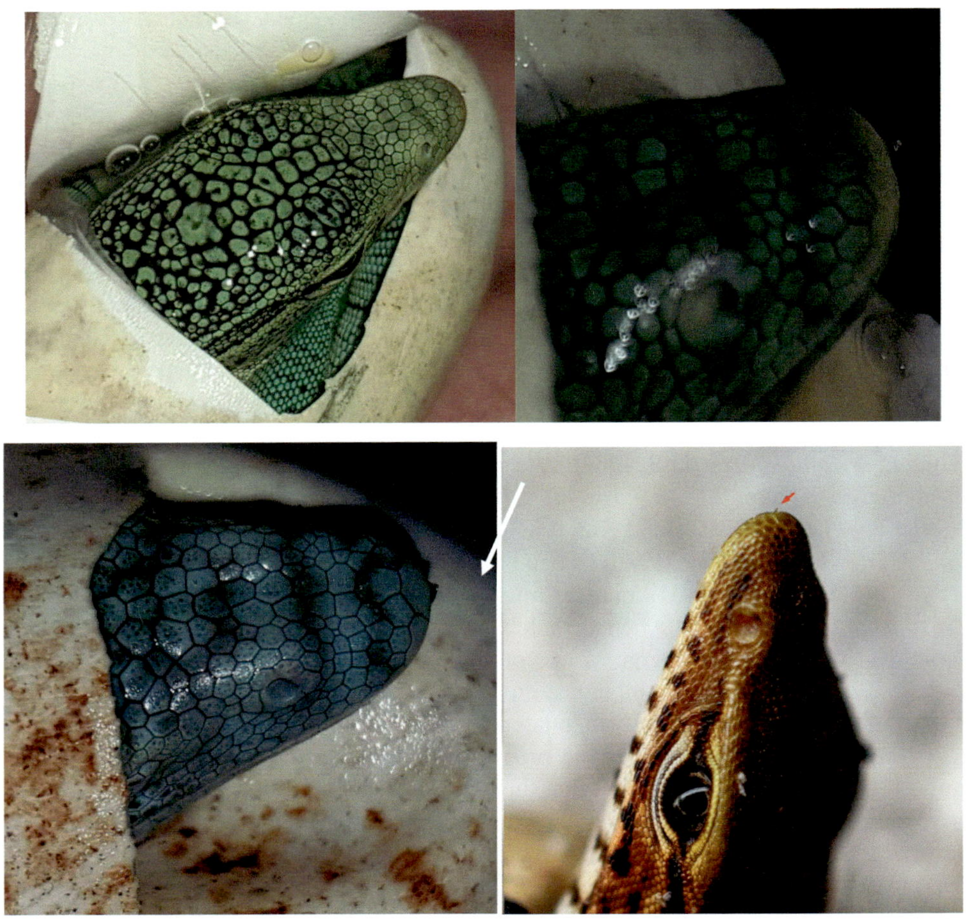

FIGURE 9.5 *Varanus* eggs and hatching neonates. (A) *Varanus prasinus* in the process of hatching from egg; (B) Closeup of *V. prasinus* showing the egg tooth, which is usually lost before hatching is completed; (C) *Varanus macraei* emerging from egg, still bearing its egg tooth; (D) *Varanus tristis* also showing the minuscule and temporary egg tooth. Arrows indicate egg teeth. (A–C courtesy Julian Firth; D courtesy Rowan Wolf).

embryos is comparatively larger than that of other species I have examined (*Anolis carolinensis*, *Coleonyx variegatus*, *Tupinambis teguixin*, *Tarentola mauritanica*) or in *Liolaemus gravenhorsti* (Lemus 1967, Vidal Maldonado and Antonieta 2008). Mammal brains, including humans, undergo a growth spurt during the late developmental stages (third trimester), accompanied by growth in the cerebellum.

Several commercial breeders keep incubating eggs in containers where they are exposed to light, reliably successful. However, monitors do not leave their eggs in the light but bury them or lay them in dark places such as inside tree hollows. Does light exposure have any effect on developing lizard eggs? In a study to evaluate the effects of illumination on lizard embryo development, Zhang et al. (2016) found that light intensity positively correlated with more rapid development, negatively affected the size of neonate skinks, *Plestiodon chinensis*, but had no effect on *Takydromus septentrionalis*, *Gekko japonicus*, and *Hemidactylus bowringii*. All these small species, however, typically lay eggs in places that are shaded or receive partial sunlight, unlike well-concealed varanid eggs. Their eggs also have a much shorter period between laying and hatching (20–40 days), which is also very unlike varanids. Though no varanids were studied, Zhang et al. (2016) suggest further research into possible correlations between shell density and egg-laying behaviors. Eggshell integrity is an important and limiting factor, but given the elasticity of most squamate eggs, I suspect that the strength and flexibility of the shell material matter more than shell thickness. It is the flexibility of the eggs that allows females to carry as many as they do; a more rigid early shell would either reduce clutch size or require larger females to maintain the numbers.

9.3.3 Hatching Dynamics

Reptiles tend to lay their eggs at times that allow hatching to take place when the environment is least hostile to neonates, and prey is readily available. Thus, the timing of laying is correlated with the length of incubation. Among the varanids for which there is considerable field data are several of the Australian species (Brown 2012, King and Green 1979, 1989, Mayes et al. 2007, Rismiller et al. 2010). Thompson and Pianka (2001) note that the smallest Australian varanids (= *Odatria* as used here) deposit eggs in early spring and hatch near the end of summer, while the larger species studied (*Varanus giganteus, V. varius*, and the *V. gouldii*-group), with their much longer incubation times (King and Green 1979, Mays et al. 2007, Rismiller et al. 2010), lay their eggs late in summer. The eggs overwinter in the environmentally stable nests and hatch again when prey is abundant, in late spring or early summer the next year. Langerwerf (2009) has suggested that reptile eggs may be biologically programmed to hatch simultaneously, a phenomenon known in many turtles (pers. obs.). Langerwerf (2009) proposed that the hatching coincides with late morning temperatures, when hatchlings will warm quickly, but not emerge into the day's most intense heat. He reports having seen the phenomenon in Australian agamids (*Intellagama*) and Argentine tegus (*Salvator*). Another advantage would be to allow hatchlings to leave the area of the nest quickly enough in a short span of time to avoid scent-driven predators. The species that consume reptile eggs also tend to feed upon hatchlings and other small animals.

Embryonic development in varanids is poorly known. The first of the few studies (Gregorovicova et al. 2012) used 14 post-oviposition embryos of *V. indicus* obtained from the Prague Zoological Garden. The eggs took 180 days to hatch when incubated at 28°C, near the upper limit known for large monitors (Eidenmüller 2016), and the authors commented on the extremely long incubation required by the lizards. Development of the skull of *V. panoptes* was the primary focus of Werneburg et al. (2015), based on 15 embryos obtained from the Australian Reptile Breeding Facility, and Andrews et al. (2017) studied 21 embryos taken from nests of *V. rosenbergi* shortly after oviposition. Our published knowledge of varanid embryonic development is therefore based on only 51 embryos of three different species. Development among the species varies, but major stages coincide. When eggs are laid, forelimb buds have already developed. By 15 days, post-oviposition (DPO, Andrews et al. 2017), embryos have three pharyngeal clefts, and by day 29, the upper eyelid has formed. Developed claws and body scales are present by day 58, and the egg tooth (Fig. 9.5) is present by day 130. The overall development of varanid embryos is not in itself exceptional among squamates, but an interesting distinction is that once a varanid embryo is nearly fully developed, further development slows down. Some researchers have suggested that the final growth period is to allow for the cranium and brain, the latter perhaps the proportionally largest among lizards.

The leathery egg presents a formidable shield around the embryo, a barrier to keep the embryo's environment intact and to keep the outside world out. In order to exit from the egg, embryos develop a small and temporary egg tooth which appears at 65 DPO in *V. panoptes* (Wernenburg et al. 2015), 130 DPO in *V. indicus* embryos (Gregorovicova et al. 2012), and 138 DPO in *V. rosenbergi* (Andrews et al. 2017). Full incubation time for the three species in the studies is 186, 161, and 179 days, respectively. The egg tooth is a medial anterior extension of the premaxilla common among reptiles and is shed at or shortly after the hatchling escapes its eggshell (Fig. 9.5).

In captivity, varanids are frequently able to produce two or more clutches (= multiclutching) annually, particularly among the smaller species (Table 9.2). Both the size of the female at maturity and incubation time are positively correlated (Table 9.2). Small species, such as *V. acanthurus, Varanus gilleni*, *Varanus primordius*, and *Varanus storri*, may reach reproductive state late in the first-year post-hatching, and incubation time is less than 135 days, commonly less than 120 days (Table 9.2). Small monitors typically lay fewer eggs than larger species, but there are exceptions (Fig. 9.6);

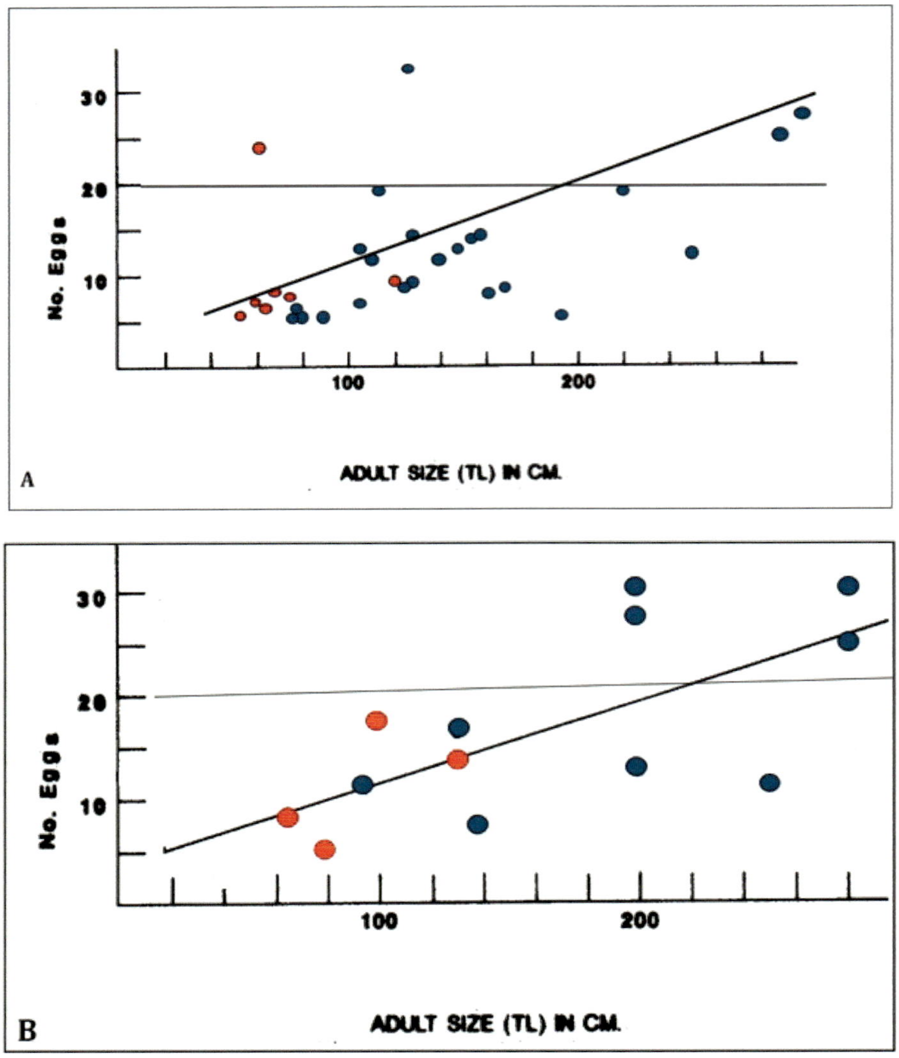

FIGURE 9.6 Comparative clutch to adult female size in (A) wild varanids and (B) lizards in captivity. Red circles indicate the smaller species (<40 cm SVL), blue the larger. Though absolute clutch and lizard size show positive correlation, there are several larger species that cluster among the smaller in laying fewer eggs per individual. *Varanus acanthurus* (uppermost red dot on both charts) consistently lays more eggs than is average for other species near the same size. (Data from Tables 9.2 and 9.3 plus pers. obs.)

V. acanthurus may produce clutches of 14–28 eggs, while *V. gilleni* and *V. storri* only 3–7. Large species (SVL > 300 mm) lay larger and more eggs; *V. dumerilii*, *V. panoptes*, and *Varanus rudicollis* lay seven or more eggs per clutch. The range of incubation periods is 80–290 days.

Proper temperature and humidity are essential for the successful incubation of eggs, and in varanids, the parameters for the incubation environment are narrow (Köhler 2005). Humidity requirements vary depending on the environment in which the species lives; eggs laid in deserts may require lower humidity than those laid in forests, but data on clutches in captivity indicate that all eggs need humidity in the 70–90> percent range (Packard and Phillips 1994, Rahmanto et al. 2022). The eggshell is selectively porous to water, oxygen, and carbon dioxide, the latter two needing a moist exchange surface (Ackerman 1991, Ackerman et al. 2002, Krebs 1999, Packard and DeMarco 1991, Packard and Phillips 1994, Packard et al. 1982, Wesiak and Koch 2009). Developing eggs will need a slow, long-term influx of water, provided by humidity, as the embryos develop, but if water absorption is too high, the embryo will drown. In *V. albigularis*, Packard and Phillips (1994) found that absorption of water, incubation length, and size of hatchlings were positively correlated with the moisture content of the incubation medium and negatively correlated with ambient temperature. Many of the well-documented accounts of humidity levels required for successful incubation are based on a few specimens or incubation events (>5) (Packard and Phillips 1994). Reports on moisture/humidity parameters for non-captive monitors have been covered in considerable detail (Dryden et al. 1990, Green and Christian 2007, Green et al. 1986, 1991, Packard et al. 1982).

Several physiological and developmental questions remain unanswered for lizard eggs and varanid eggs in particular. The normal water content of the eggs of species that lay parchment-shelled eggs is around 66 percent, versus some 78 percent for turtles, crocodilians, and snakes (Belinsky et al. 2004). But does a minor change in the amount of water affect the embryo? Hatchlings of the snake *Spalerosophis cliffordi* showed no variation due to the amount of water gained during incubation (Dmi'el 1967). Water gained has not been ruled out as an important variable for some turtles (Gutzke and Packard 1986), but there are no data available for most lizards and any varanids. Is adult lizard size linked to the quantity of water in eggs right after oviposition? Data for eggs of the large agamid, *Uromastyx aegypticus*, which like varanids are parchment-shelled, have a 70 percent water content (Thompson and Russell 1998), only slightly higher than that for most other tested lizards. And is there a limit to the rate at which eggs gain or lose water? Presumably, there would be variation in gas exchange rate and pressure (as P_{O2} and P_{CO2}) as internal and environmental water content changed, but beyond a catastrophic gain (drowning) or loss (dehydration) of water, would there be positive or negative effects? Certainly, we see effects of temperature change that either increase or decrease the rate of embryonic growth. Though there has been a vast increase in our knowledge of varanid reproduction since the 1990s, including a burgeoning ability to successfully induce breeding and rearing in captivity, there has not been a comparative growth in our understanding of embryology and its physiology in varanids.

Once laid, varanid eggs can face environmentally induced damage, particularly a tear in the eggshell, for even though the leathery structure gives the egg some flexibility, it is still a delicate structure (Packard and DeMarco 1991, Packard et al. 1982). Nothing is known about egg survival after minor injuries in wild varanids. In captivity, several different approaches have been used to successfully maintain the egg until hatching (Ackerman et al. 2002, Adragna and Madden 2009, Dawson 2018, Packard and Phillips 1994).

9.4 HATCHLINGS

Hatchlings of the species incubated in termitaria require release from the nest by an adult that can excavate an escape route in the hard mound. All other species' young are independent and must escape the nest on their own. In the cases of the large *V. gouldii* and *V. panoptes*, whose eggs are laid in 3–4 m deep spiral communal nests, the young dig their way out by themselves, excavating a straight vertical route (Doody et al. 2021). There has been some speculation that the neonates of termitarium layers feed on the termites until they can escape the nest, but that seems unlikely. For one thing, the termites are possibly too small even for young monitors to notice as prey. Second, the monitors are in complete darkness, so their instinct to attack moving objects is inoperable. Third, as is the case for all hatchlings, their primary instinctive action is to escape the nest. Except for the aid given by adult females in opening the termitaria that allows hatchlings to escape, there is no further known interaction between young and adults aside from incidental interactions with the larger population. There is no brooding of the eggs during incubation as seen in *Plestiodon* (Somma and Fawcett 1989, Vitt and Cooper 1989), nor nest guarding as seen in *Ophiophagus* (Oliver 1956), and post-hatching parental care has so far been recorded only for crocodilians and viviparous lizards (e.g., *Corucia* [Sprackland 1993], *Shinisaurus* [Sprackland 1989], *Tiliqua* [Main and Bull 1996]). Quite the opposite is likely, if rarely, the case, with adult monitors consuming hatchlings.

Most hatchlings resemble adults in most aspects of morphology, but there is considerable variation across taxa. In the young of the largest species (e.g., *V. albigularis, Varanus exanthematicus, V. komodoensis*), the ratio of tail length to SVL is usually greater than in adults, with *Varanus salvadorii* and *Varanus glebopalma* being exceptions (Horn and Visser 1991, Sprackland pers. obs.). However, it is unusual for those proportions to become equivalent to equally sized *Varanus* of a different species; that is, a meter-long (total length) specimen of a species that grows larger is not necessarily going to have equal body proportions to a species that attains a maximum length of one meter. The considerable variation in bodily proportions is a factor that makes determining the adult size of fossil varanids, such as †*V. prisca*, so difficult to ascertain with confidence, as easily seen by estimates for the species ranging from 5 to 8 meters.

The physiology of neonates differs to varying degrees from adults (Thompson 1999, Thompson and Withers 1998) but remains largely in need of study. In particular, to what degree does physiology change in species that require more effort that others to emerge from their nests (*V. gouldii, V. panoptes, V. varius*) between hatching and freedom, and how do metabolisms of neonates of large species compare with neonates and adults of small taxa (Christian and Conley 1994, Christian and Weavers 1994, Thompson and Withers 1994, 1997a, 1998, Wood et al. 1978)?

To varying degrees, the snouts of monitors are comparatively shorter as neonates than as adults. Perhaps the least amount of snout elongation with age is seen in *V. exanthematicus*, while it is much more pronounced in *V. rudicollis*. The teeth of neonate monitors are consistent among species, being slightly compressed, curved, and with an acute tip. Changes in tooth morphology seen in some species (e.g., *V. albigularis, V. bengalensis, V. olivaceus*) occur much later in life, and their ontogeny is covered in Chapter 5. Coincident with snout elongation is nostril position and shape. All hatchlings have round or slightly oval nostrils, but these may change greatly with growth. For example, in neonate *V. rudicollis*, the nostril is a small oval about midway between the tip of the snout and the anterior border of the orbit, but in adults, it becomes a thin slit more than half the length of the snout. The nostrils of young *V. albigularis* are oval and only slightly nearer the orbit than the snout tip, but as adults, the nostrils almost encroach on the eye. Many adult monitors' nostrils are enclosed in a slight but distinct hump (e.g., *V. dumerilii, V. prasinus, V. tristis*), but the rise is minimal or absent in hatchlings.

Sexual dimorphism, a phenotypic distinctiveness between the sexes in adults, is usually not apparent in hatchlings and young varanids. However, the sometimes-considerable differences between young and adult varanids may appear as distinct as what is expected between different species. The pericloacal tuft of scales possessed by males of several of the odatrians (Fig. 9.2) does not develop until the lizard nears maturity. Characteristic hemipenal bulges are also a male trait, appear earlier, but may be difficult to distinguish in very young monitors. As the lizards grow, they develop a pair of oval prominences on either side of the median postcloacal septum, and these become larger and more conspicuous with age. Females may develop similar bulges, but to a lesser degree and at a later stage than males; without having both sexes for comparison, visual sex determination can be difficult.

The limbs of varanids are also subject to change between hatchlings and adults. For example, the adpressed limbs of many hatchlings (e.g., *V. primordius*) may touch or overlap but fail to do so in adults. Across the genus the proportional distance between fore and hind limbs increases ontogenetically. The relative tail length for any species of *Varanus* varies from being about equal to SVL (*V. exanthematicus, V. komodoensis*) to two or more times SVL (*V. beccarii, V. glebopalma, V. prasinus, V. salvadorii*), with a positive correlation between short-tailed ground-favoring types and long-tailed arboreal monitors.

Hatchlings of varanids generally have a similar coloring and pattern as adults, often with the colors more intense and distinct (Table 9.4). As the lizards age, the colors tend to fade and blend, while the pattern becomes less distinct to the degree that no clear pattern is visible. In some species, though, the variation of color and color intensity differs dramatically between young and adult, and among the most conspicuous examples are the species *V. beccarii, V. flavescens, V. dumerilii, V. komodoensis*, and *V. pilbarensis* (Fig. 9.7 and Table 9.4). In *V. dumerilii* the neonates are black with orange to pale yellow bands that become progressively more caudal. Much of the head and first crossband, however, is intensely orange, leading to the suggestion that neonates are mimicking one of the red-headed venomous snakes with which they are sympatric, *Bungarus flaviceps* and *Calliophis bivirgata flaveceps* (Sprackland 1992; see also Cota 2016). The distinct light and dark banding resembles the young of King Cobras (*Ophiophagus* sp.). *V. dumerilii* remains largely unknown from in situ studies, so much of what is known comes from watching captive lizards. The question of whether neonate *V. dumerilii* is a mimic in a Batesian mimicry model has been partially tested with captive animals by use of a logical argument, e.g., comparing several potential predators and the likelihood they would be deterred by the orange coloration. Cota and Krebs (2016) discuss potential scenarios concerning the snakes already mentioned, other monitors, predatory birds, and several allopatric mammals, with no likely predator-prey association being confirmed. Adding to the question of whether the orange head is an aposematic color is the fact that it is completely lost within the first four to six weeks after hatching. Young *V. komodoensis* have bright cream or yellow markings on the head, body, and tail, all on a black background, but the pattern disappears on the head and body as the lizards grow, and sometimes faint light bands remain on the tail. As adults, the dragons become icons for the island species' tendency to grow large and become melanistic. India and Nepal's Yellow Monitor, *V. flavescens*, hatches with bright yellow transverse markings sometimes accompanied by smaller bands of small red spots (H. Petzold, per. com.). The red is known to intensify considerably in some specimens, but there is so little field data for the species that any details about how widespread, intense, or seasonal the red is unknown.

9.5 SEX DETERMINATION

Temperature stability is of major importance for egg incubation, and beyond set parameters can have detrimental or fatal consequences (Burger 1989, Van Damme et al. 1992). So far as known, all varanid sex is determined by the sex chromosomes (Holmes et al. 1975, King and King 1975), and, as in birds, are called W and Z (in contrast to X and Y in mammals), but unlike XY sex determination where the heterozygote (XY) is male, it is the female that is heterozygotic (WZ) in varanids (Iannucci 2019, King and King 1975). In ZW chromosome types, it is the W that is "highly differentiated from the Z" (Matsubara et al. 2014).

Switching from one system to the other (XY to ZW) has occurred in several squamate lineages, and the process that converts the chromosomes may be a result of inhibition of recombination of the genes on a chromosome coupled with adding repetitive DNA sequences (Matsubara et al. 2014). Though the chromosome coupling is conserved within *Varanus*, Matsubara et al. (2014) found an (AAT)n microsatellite repeat motif mapped by fluorescence *in situ* hybridization on the W chromosome of *V. acanthurus*, but a (CGG)n motif mapped onto the W chromosomes of *V. gouldii* and *V. rosenbergi*.

> Furthermore, the W chromosome probe for *V. acanthurus* produced hybridization signals only on the centromeric regions of W chromosomes of the other two species. These results suggest that the W chromosome sequences were not conserved between *gouldii* and *acanthurus* clades and that these repetitive sequences have been amplified rapidly and independently on the W chromosome of the two clades after their divergence.
>
> (Matsubara et al. 2014)

Though the chromosomes of varanids have been well documented, their genes have not. Rovatsos et al. (2019) have compared the ZZ/ZW gene copy numbers found affinities between varanids, *Heloderma*, and some anguids. They determined that ZZ/ZW sex chromosomes are not homologous to the XX/XY chromosomes of other amniotes. They also concluded that these differentiated sex chromosomes were present in early squamates around 115–180 million years ago.

TABLE 9.4
Varanids that Undergo Extreme Color Change from Neonate to Adult

Species	Color Feature of Young	Age at Change (mos)	Appearance of Young
albigularis	Light markings brighter	~18	
beccarii	Black with very small, symmetrically arranged green dorsal spots	~6–8	
doreanus	Blue tail markings much brighter	Starts at ~14	
dumerilii	Head and nape bright orange; dorsal bands black, light bands pale tan or yellowish	2–3	
griseus	Dark and light bands pronounced; tail banding distinct black and light	~12–18	
komodoensis	Light cream/yellow markings, dark areas brown or dark gray; face mottled.	~20–24	
pilbarensis	Head and most of body orange with pale rosette pattern; tail bands more distinct than in adults	–	
rosenbergi	Bright pink/orange lateral bands from nostril to mid-body, thickest on neck	–	

FIGURE 9.7 Examples of interspecific crosses in captive varanids. (A) *Varanus glebopalma* X *Varanus glauerti*; (B) *Varanus kordensis* X *Varanus macraei*; (C) *Varanus pilbarensis* X *Varanus glauerti*. (D) *Varanus cumingi* X *Varanus samarensis*. In each case the lizard is very close in color and pattern to one of the parental taxa, e.g., *glauerti, macraei, pilbarensis*, and *samarensis*, respectively. (Photos courtesy (A–C) courtesy Reptile Forums U.K.; D courtesy John Adragna.)

Many reptiles have been found to have temperature-dependent sex determination (TDS), where incubation just slightly above or below a specific temperature will skew the hatchlings to either all females or all males (Valenzuela 2004). The number of species discovered to be affected by TDS has grown since its discovery in the 1980s but has yet to be confirmed for *Varanus*.

9.6 HYBRIDS AND PARTHENOGENESIS

The reality and diversity of hybrids among reptiles is well established (Cole et al. 1993, Harrison and Larson 2014, Joly et al. 2009, Mallet 2005, Stebbins 1959), as has intraspecific breeding to discover hidden phenotypic variation (Zweifel 1952, 1981). Interest in hybrids seems divided into two major lines of inquiry: how does hybridization fit into natural evolutionary speciation models (Joly et al. 2009, Mallet 2005, Mayr 1963, Meier et al. 2017), and is there any advantage, aside from commercial, to developing hybrid monitors (Abruzzi 2021, Pavon-Vazquez et al. 2021, Taylor and Larson 2019)? Zoologists are interested in hybrid speciation and its larger role in speciation and diversity (Arnold and Kunte 2017, Brunes et al. 2019, Fujita and Moritz 2009, Harrison and Larson 2014, Mallet 2005, Mayr and Diamond 2001, Meier et al. 2007, Neaves and Baumann 2011, Oliver et al. 2014, Stebbins 1959, Wilson 1992). To classify something as a hybrid species, the organisms must consistently produce viable offspring, meaning the F1 can produce F2 and so on, and the resulting new population must remain distinct, e.g., not backcross to the founding parental group. Direct observation of phenotype will not always be useful in recognizing a hybrid (Revell 2009, Wilson 1992), and numerous molecular approaches are employed to do so (Castiglione et al. 2018, Goolsby et al. 2017, Harrison and Larson 2014, Joly et al. 2009, Mallet 2005, Mallet et al. 2016, Pavon-Vasquez et al. 2021, Wilson 1992).

Parthenogenetic organisms are the interspecific products of hybrid speciation (Brunes et al. 2019, Cole et al. 1993, Darevsky 1958, Fujita and Moritz 2009, Grabbe and Koch 2014, Neaves and Baumann 2011, Watts et al. 2006), and true parthenogenesis in vertebrates seems to be unique to the Squamata (Brunes et al. 2019, Freitas et al. 2016, Fujita and Moritz 2009). The only reported case of parthenogenesis arising spontaneously and not from hybrid speciation is found in the lizard family Xantusiidae (Bezy 2019, Sinclair et al. 2010).

Assuming that hybrids are produced, are the F1 sterile, such as mules, and incapable of reproduction? Or are the young able to reproduce, perhaps leading to a new lineage of the hybrid, or leading to a triploid (or polyploid) new species? *Aspidoscelis* (Teiidae; Manríquez-Morán et al. 2014), *Darevskia* (Lacertidae; Darevsky 1958, Freitas et al. 2016), *Leiolepis* (Agamidae; Grismer et al. 2014), *Lepidophyma* (Xantusiidae; Bezy 2019, Noonan et al. 2013) and *Gymnophthalmus* (Gymnophthalmidae; Cole et al. 1993, Pellegrino et al. 2011) are genera with parthenogenetic species that are the products of known parent species and whose offspring can reproduce well beyond the F3 generation. Varanids provide an example of parthenogenesis arising spontaneously in captivity.

The possibility of hybrid fitness was developed as early as the 1920s (Dobzhansky 1937, Mayr 1963, Mayr and Ashlock 1991), and it was discovered that the fertility of a hybrid was dependent upon whether the individual was male or female. Haldane's Rule is based on the observation that in males of species whose sex is determined by the XX/XY chromosomes, males are much more likely to be sterile. However, in species with the ZW/ZZ chromosome sets, including varanids (Iannucci et al. 2019, King and King 1985), the females tend to be sterile. In each case, it is the sex with two different chromosomes – XY or ZW – that may be infertile, making Haldane's Rule equate the mechanism for hybrid sterility to be a recessive, sex-linked gene where the genes for sex and viability are linked to a common chromosome. In varanids the products of parthenogenesis are largely male (ZZ) with some exceptions. Two Komodo Dragons parthenogenetically produced at the National Zoo in Washington, D.C., represented each sex, but the male's growth and activity were stunted compared to its sister (pers. obs.). There has still been no discovery of a natural parthenogenetic *Varanus*, but the evidence supports hybridization fitting into natural evolutionary speciation models.

The second question is, is there any advantage to developing hybrid monitors? Concern generally comes from the expectation by some people that captive-bred animals will eventually be released into the wild to replenish or reestablish decimated species (see below) and thereby affect the natural gene pools detrimentally. Such animals could very well be vectors for diseases absent from a natural population, for which reason wildlife managers rarely release animals into the wild that have been kept with humans; some biologists think that the worldwide amphibian disease, chytridiomycosis, *Batrachochytrium dendrobatidis*, is the result of amphibian–human contact in the field. Bird flu, swine flu, and Ebola are all examples of zoonoses where animals have infected humans; there are also examples of diseases passed on by humans to animals. Thus, the proposition that someone's captive-bred animals, no matter how rare, will be introduced into the natural habitat as part of a repopulation program for those species is essentially nil.

There is, sadly, a long history of escaped and intentionally released non-native animals into environments where they thrive and negatively impact native species. The state of Florida has become infamous for the number of established foreign species that have become well established, including *Agama, Basiliscus, Leiocephalus, Phelsuma, Trioceros, Salvator*, and *V. niloticus*, among lizards. The latter pair are particularly troublesome because of their size, diet, and ability to survive in a multitude of habitats. The intentional release of hybrid or modified native species seems absent – so far. Florida is not alone in having several successful invasive species; parts of California, Hawai'i, Texas, and Louisiana, for example, also have several non-native species of tropical lizards, while Kansas and New York's Long Island have colonies of the European Lacertidae that have remained viable for over half a century. In the absence of their natural predators, diseases, and other factors, these populations have thrived, but only the species in Florida seem to have expanded their ranges significantly. Of all these states, Florida has the only established large, predatory lizards (*Salvator* and *Varanus*). Their presence over several decades, plus the discovery of young lizards, confirms that, barring some effective removal process, they will stay, and their populations grow. Unlike the smaller insectivorous invasives and the herbivorous *Iguana iguana*, *Salvator* and *Varanus* present the problems of feeding on larger native animals, such as rodents and birds (plus their eggs), and their bites can inflict serious wounds on domestic pets and humans.

Hybrids may not be common in nature. Even closely related species, such as lions and tigers, that can hybridize in captivity would never have the chance in the wild, as their ranges are much further apart than their genetic distances. What of sympatric species of *Varanus*, especially between taxa that are phylogenetically closely related? The possibility exists in the case of *V. bengalensis* and *V. nebulosus*, very similar species that were for many years considered to be subspecies of the former. Auffenberg (1994) tried to determine whether one or two species were involved and posited that character states for each could be found in the other in some regions. Unlike hybrids of some other lizards (e.g., in *Crotaphytus, Aspidoscelis*) in which hybrids are distinct enough and hybrid zones effectively mapped, hybrids of other lizards may not be readily distinguishable (Arnold and Kunte 2017, Castiglione et al. 2018, Hagen et al. 2012, Joly et al. 2009). The likelihood that their phenotype will be a visible mix of traits from both parents is not particularly high. The most basic Mendelian genetics states that, generally, three out of four offspring will resemble one parent, and one will resemble the other (Mayr 1993), but none will appear mixed (excepting mutation, incomplete, and co-dominance). Therefore, there are at least three difficulties in identifying many hybrids. First, genetic outcomes cannot be predicted with precision but are expressed as probabilities. Second, a hybrid might not express a mixed genotype, having an invisible recessive trait, such as an allele for albinism. The animal would not be lighter in color than the non-albino, just as a cross between a black mouse and a white one will not yield gray offspring. Genotype and phenotype do very different things, and the principles apply to lizard as much as human genetics. And third, the model omits the possibility of mutation, which, rare though it may be, can produce a large variation from the ancestral lizards.

However, hybrid speciation is an actual phenomenon that may have resulted in the arising of more species than we once thought, via reticulate evolution (Mallet 2005, Pavon-Vazquez et al. 2021, Pellegrino et al. 2011, Sinclair et al. 2010). An important hypothesis in current biological thinking is that speciation is most typically a result of allopatry, when a contiguous species is divided by some barrier, such as a mountain range or river, into two or more populations that become isolated from each other and, without gene exchange with each other, develop novel new genotypes that lead to creation of a new species (Mayr 1963, Mayr and Ashlock 1991, Mayr and Diamond 2001, Welton et al. 2013). The degree to which members of the isolated populations may encounter each other and produce viable offspring is central to the concept of subspecies, which will be covered further in Chapter 10. It is important to understand that the ability to hybridize, whether in nature or captivity, is an indication of taxon-relatedness. Hybrids are evidence of a close relationship, implying a recent common ancestor, but not necessarily evidence of the parents being of the same species. If natural hybridization and parthenogenetic populations of *Varanus* exist, they will most probably be discovered through molecular analysis.

9.6.1 What of Hybridization in Captive *Varanus*?

One private monitor breeder attempted to get the relatively closely related *V. salvator* ♀ and *V. salvadorii* ♂ to produce hybrids (Abruzzi 2021). Though mating was attempted, and eggs were laid, the genetic data indicated that the *V. salvator* had reproduced parthenogenetically, with no input from the *V. salvadorii* (Abruzzi 2021). According to Abruzzi, there was "too much genetic distance between *V. salvator* and *V. salvadorii* in order to produce a viable offspring" (referred to as "*Varanus salvatorii*"). Based on the evidence, it appears that there is "too much genetic distance between" most species of *Varanus* to be able to produce a viable offspring. This interbreeding attempt is not particularly informative because of the lack of controls, and it involved a single event. Many private breeders have attempted interspecific crossing of *Varanus*, and many have been successful at the F1 level, but there is little reliable information available for the F2 and beyond. The majority of the reports of such hybrids are found on the Internet, usually posted by commercial breeders, and include representatives of several species groups, e.g., *glebopalma* X *glauerti* (Fig. 9.7A), *gouldii* X *panoptes*, *kordensis* X *macraei* (Fig. 9.7B), *caudolineatus* X *gilleni*, *pilbarensis* X *glauerti* (Fig. 9.7C), *exanthematicus* X *albigularis*, *prasinus* X *beccarii* (pers. obs., but unable to confirm genetics), and *cumingi* X *samarensis* (Fig. 9.7D). The validity of such crosses is not in dispute, but the viability of their offspring often is. A feature common to all these crosses is the phylogenetic proximity of the taxa involved in each pairing. Unlike the lion (*Leo*) X tiger (*Panthera*) crossing, in which each species is deemed to be in a different genus, all varanids are (so far) contained within a single genus. Furthermore, each pairing involves species that have generally been placed within the same subgenus (vide Mertens 1942; see Chapters 11 and 12), e.g., *Odatria* (*glebopalma* X *glauerti, caudolineatus* X *gilleni, pilbarensis* X *glauerti*), *Hapturosaurus* (*kordensis* X *macraei, prasinus* X *beccarii*), and *Soterosaurus* (*cumingi* X *samarensis*), *Polydaedalus* (*exanthematicus* X *albigularis*), and *Varanus* (*gouldii* X *panoptes*), as was the failed *V. salvator* and *V. salvadorii* cross. I have found no incident of an inter-subgenus level breeding attempt. Hybridization between more distantly related taxa, such as the attempted *salvator/salvadorii* cross mentioned above, seems to have all been failures, and is based on shared information from breeders. There are no reliable claims for crosses such as *exanthematicus/bengalensis* or *griseus/gouldii*, for example, taxa that are widely separated by both geography and phylogenetic distances.

The commercial sector is largely involved in creating a greater variety of animals that will be worth more on the market. Three of the most popular pet reptiles, the leopard gecko (*Eublepharis macularius*), Bearded Dragon (*Pogona vitticeps*), and Ball or Royal Python (*Python regius*), are commercially available almost exclusively as "morphs," to the extent that it has become extremely difficult to find a wild-type animal in the trade. In this regard, breeders of the reptiles are doing precisely what has been done to create the innumerable breeds of dogs, cats, chickens, goldfish (Pullan and Smith 1987), guppies (Ramsay 2014, Vila-Pouca et al. 2022), and many other species. The scientific value of such animals for experimentation is also real, as evidenced by the considerable breeding and genetic research done with them, but they are worthless for systematic studies. There is a substantial market for unusual phenotypes, or morphs, particularly for albino, melanistic, or unusual patterns, and albino monitors have sold for tens of thousands of dollars. These animals are not bred to be introduced into the wild or establish "pure" breeding lines, but to be novelty pets. To reliably and repeatedly produce such morphs, breeders keep and track genetic lineages with some care, potentially useful for testing genetic hypotheses. One example of being able to track the genes across hybrids occurs with members of the *V. salvator* group, where there is both high demand and high prices for albino and other color modifications, and these specimens have been successfully bred into the F3 and beyond generations (Internet sources).

I have mentioned the well-known interspecific hybrids that sometimes occur in zoos, such as lion/tiger crosses, two species that do not hybridize in nature because their ranges are allopatric, and by a great distance. Would they hybridize if somewhere their ranges overlapped? What selection pressures would favor the species to mate? This aspect of hybridization is being investigated in *Anolis* lizards (Bock et al. 2021), where invasive *A. sagrei*, when introduced into Florida, became highly hybridized with native lizards with which they had been allopatric. The native lizards showed little evidence of hybridization after 15 years, while the introduced anoles hybridized among themselves regularly. As the authors state, "Ecological changes during biological invasions can affect hybridization frequency and stability, which can in turn drive the success of invasive taxa"

(Bock et al. 2021). Would presently allopatric varanids successfully hybridize if they suddenly became sympatric? Currently, the only evidence that they would come from varanid breeders in many countries, and to date, such successes have been between closely related species within the same subgenus.

Initial discovery of parthenogenesis in *Varanus* was in captive lizards (Lenk et al. 2005, Watts et al. 2006). Because of the large number of private herpetoculturists breeding *Varanus*, the number of species discovered to be parthenogenetic has been growing (Grabbe and Koch 2014, Hörenberg 2013, Lenk et al. 2005, Murphy et al. 2002, Watts et al. 2006, Yadaz and Rana 1988). As of this writing (2024), 24 varanid species, about one-quarter of known species, have been reported to reproduce parthenogenetically (Table 9.1), most of those in private collections. Whether or not breeders should create hybrid reptiles is contested, and not a question to be settled here, but there are already precedents established for creating interspecies hybrids within several other animal groups.

Part IV

Systematics and Evolution

Where did monitors come from, and how have the great global upheavals of the past 70 million years affected their descent, modification, and distribution? The history presented is based on fossils, evolution models, and comparison of morphologies of extinct and extant taxa. The methods and reasoning behind the presented systematic hypothesis are covered in detail in Chapter 10, and the conclusions based on those methods are presented in Chapter 11.

DOI: 10.1201/9781003463528-13

10 Origin, Evolution, and Prehistoric Biogeography

10.1 INTRODUCTION

Herpetologists are fortunate that the varanid lizards have left what seems to be a traceable lineage in the fossil record. Fossils, however, tend to be incomplete or missing, making confident hypotheses about phylogeny difficult to produce, and naturally not all fossils represent direct ancestry. There are numerous non-*Varanus* fossils that represent different levels of platynotan diversity, and these have been placed in and out of the Varanidae depending on the methods used to assess relationships. Consequently, the best-supported fossil trail extends from the late Upper Cretaceous (80-71 mya; Fig. 10.1) to the present. Varanids are particularly well represented in the Cretaceous, Miocene, and Paleocene, with the greatest variety of genera and species known from the Cretaceous and Miocene. It is not always (if ever) possible to examine the evolution of a group from its true origin, the ancestor that was the beginning of a unique lineage to its present living descendants. The Anguimorphan lizards (Platynota plus Diploglossa) have not revealed their starting point, but within the Varanoidea (see Chapter 11) we are confident that the group of lizards that were directly on the ancestor-to-*Varanus* lineage have been recognized. Consequently, we also have excellent evidence to show that Platynota, Varanoidea, and Varanidae are monophyletic groups.

This chapter and the next present the evolutionary prehistory of the Varanidae on the geological and evolutionary scales, to be followed by examination of the genera, contemporary taxa, and their taxonomic problems and resolutions. This chapter reviews the probable evolutionary pathways taken by the Varanidae based largely on morphological evidence and compares possible branching, reticulate, and chronospecies evolution. We have more genera and species now than were available for much of the 20th century, along with re-evaluations of some of the fossil taxa as being in or out of the lineage. The roadmap we have is incomplete, but new post-Cretaceous taxa have allowed us to fill gaps and provide evolutionary signposts showing if we are on the right route. The putative relative from the Triassic, †*Cryptovaranoides microlanius* (Whiteside et al. 2022), is almost certainly a squamate and some 35 million years earlier than other fossil lizards. Its status as a lizard was questioned but convincingly defended (Whiteside et al. 2024), though it does not appear to be a varanoid. The Triassic was a time that yielded frustratingly few squamate fossils, but in the Upper Cretaceous the roadmap of varanid evolution seems paved with much better materials. The earliest definite varanoids have been found in central Mongolia and migrated east to North America, west into Europe, North Africa, and southern Asia, and later crossed the seas at Wallace's Line, "where worlds collide," establishing themselves broadly across the Indonesian-New Guinean islands and down into Australia. How they did all of this is the subject of this chapter.

One caveat must precede this section. Determining origins for species, it is easy to assume that the location with the most taxa is also the center of origin, whether the subjects are fossil or living organisms. Though Central Asia has provided the majority of fossil varanoid taxa, it by no means that the question of origin is settled. The Gobi

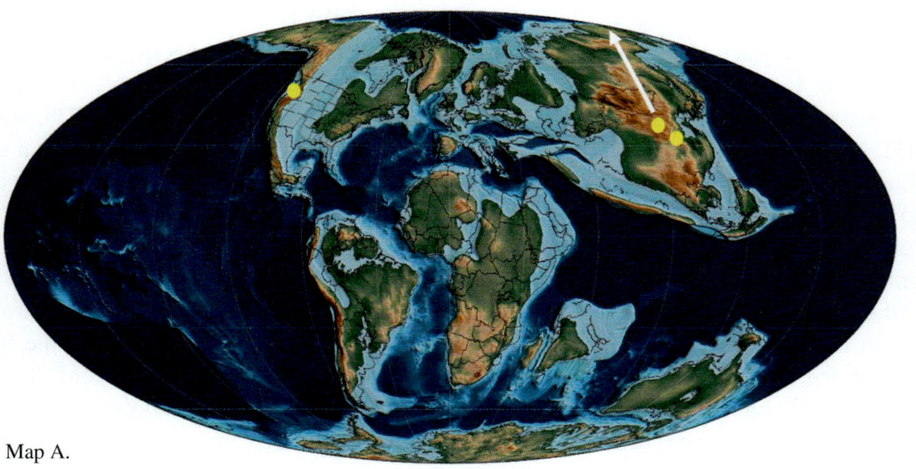

Map A.

FIGURE 10.1 Map A, configuration of the continents during the Upper Cretaceous (85mya) showing locations from which the earliest confirmed varanoid fossils have been collected (yellow dot) around 70 mya. The white arrow shows the probable path – and only known land connection – for the Asian lizards to migrate into western North America. (See Fig. 10.4 for map sources.)

DOI: 10.1201/9781003463528-14

provided an extraordinarily good combination of materials to make such fossils, and such places are rare. What the fossils have not been able to tell us is what happened to the descendants of the Gobi fauna over the next 70 million years. Monitors of some sort obviously persisted though those years and the great range of environmental changes, especially temperatures, between the Cretaceous and 2025. The fourth dimension, time, is often left out of models, sometimes to the detriment of our models. For example, were there more varanids in the Late Cretaceous of eastern Asia? How many lineages migrated and to where? If we could see back a few million more years, would we find a greater diversity of varanoids somewhere else in Laurasia? Did we find so many fossil varanoids in the Gobi because it was a preferred habitat or because possible forest species dwelt in places less conducive for making fossils?

In the sections to follow I have discussed a hypothetical scenario for varanid origins and distribution based on different lines of evidence. The original model came from one of my early analyses of the phylogenetics of living *Varanus*, primarily focused on morphology and zoogeography, and several parts have subsequently gotten considerable support. In fact, many of the species covered have been found in numerous localities, sometimes across a thousand or more miles. The paucity of fossils for any species does not, by itself, indicate rarity or delicacy. Rather, it is the inevitable result of millions of years of environmental changes that produced scorching heat, huge glaciers, temperate forests, and an increasing number of highly competent competitors. It amazes me that these lizards have not only survived but thrived through those challenges.

10.2 ORIGINS

The Anguimorpha is perhaps the oldest of the extant squamates, with potential roots reaching into the Late Triassic. Varanids have received attention from numerous herpetologists, at least partly because of the large size of many species, a factor making them rather uncomplicated bodies for dissection. The species that have been particularly well-studied morphologically have been the some of the largest members of the genus *Varanus*: *bengalensis, griseus, komodoensis, niloticus*, and *salvator* (Bahl 1937, Bellairs 1949, Benedict 1932, Muzammil, 1949, Pathak et al. 2015, Reinholz 1923). Equivalent works on the smaller species have all tended to be recent and limited to specific anatomical regions, predominantly the skull and hemipenes (Branch 1982, Brennan et al. 2021, Openshaw et al. 2016, Thompson and Withers 1977). Herpetologists and paleontologists each have an interest in the group that was very likely ignited by Duméril and Bibron (1836), who put forth the hypothesis that monitors were akin to serpents. Oddly, they did not consider limblessness or reduced limbs as relevant to finding ancestral snakes, nor did most later researchers. Nevertheless, the suite of chosen characters was convincing because monitors and snakes share several conspicuous morphological characters: (1) elongated body, (2) long, forked tongue, (3) incomplete postorbital arch, (4) loss of autotomy, (5) physical kinesis in the mandible, (6) alternate tooth replacement, (7) failure of nasal and maxilla to meet, and (8) a complete or nearly complete bony braincase. Seeking affinities among the groups of extinct marine reptiles and their living relatives, Cope (1869) compared the mosasaur †*Clidastes* to snakes, lizards, and turtles and noted,

> In its lacertilian characters it approaches nearest the Varanidae, which themselves, offer some approximations to the Ophidia. The elongation of the proötic anterior to the internal ear is a character of all the slender-tongued lizards, and the long superior nostrils and lack of malar arch belong only to the Varani.

Cope then allied Lacertilia with mosasaurids into a new group, Pythonomorpha, "On account of the ophidian part of their affinities…" The varanid-snake link held for over 100 years, with the search for snake ancestry becoming one of herpetology's (and evolution's) great questions; from what ancestors did snakes evolve? With Pythonomorpha formally recognized as being at least closely related to snakes, the hunt for those ancestors sought connections between mosasaurs and monitors with snakes. Snakes, however, have left a paltry fossil record (Caldwell 2007), and what has been found usually lacks all-important cranial elements, leaving many of the phylogenetic hypotheses based on speculation and often weak evidence (for a review see Caldwell 2021). The model for deriving snakes from lizards, specifically varanid lizards, seemed solved after the detailed comparison of anguimorph lizards to (among other groups) snakes when McDowell and Bogert (1954) presented the argument for Lanthanotus, as a platynotan sister group to *Varanus*, being the closest living relative to snakes. Naturally, this slightly shifted later research focus from monitors to the earless monitor, a term introduced by McDowell and Bogert reflecting the absence of an ear opening (the literal translation from Greek of *Lanthanotus*), a key ophidian characteristic, and it's extremely close relationship with the Varanidae. For the next 60 years, the search intensified to nail down the specific lacertilian ancestor of ophidians (Caldwell 2007, Estes et al. 1988, Lee 1997, 2005, Lee and Scanlon 2001, McDowell 1972b, Waltering 2012), but opinion was shifting on morphological (McDowell 1972a, b) and molecular (Conrad et al. 2008, Pyron 2017) grounds. Snake origins are not within the scope of this book, leaving that presentation to Caldwell (2021). I shall leave the subject with the note that as of the 2000s most studies have pulled snakes off the platynotan tree, and they are most recently considered a basal sister group to the Iguania and anguimorph clades (Streicher and Wiens 2017) and at best only very distantly related to monitors. We shall see.

The supposed role as ophidian ancestors coupled to their fair representation as fossils has kept varanoids an active area of paleontological research, and the collection of putative early species has grown considerably since 1980. The earliest history of the Platynota may predate the Cretaceous back to an origin during the Upper Triassic (Evans 1994, 1996), but fossils from that era are too incomplete to give adequate support to such a systematic position. Because Cretaceous fossils show so many varanid traits, implying that there was a long pre-Cretaceous evolution of squamates, including the Platynota, a Triassic point of origin on the geological timeline should not be ruled out (Evans 2007).

Numerous phylogenies that include fossil taxa have been put forth (Borsuk-Białynicka 1984, Cope 1869, Dong et al. 2022, Fejérváry 1918, 1935, Holmes et al. 2010, Lee 1997, McDowell and Bogert 1954, Nopcsa 1903a, b, Norell and Gao 1997, Norell et al. 1992, 2007, Pregill et al. 1986) and though the given phylogeny of each study differs in generic placements, the genera comprising Varanidae usually include †*Telmasaurus*, †*Aiolosaurus*, †*Saniwides*, †*Saniwa*, †*Proplatynotia*, *Lanthanotus*, and *Varanus* (Borsuk-Białynicka 1984, Conrad et al. 2008, Dong et al. 2022). At least eight anguimorph, probable platynotan species have been identified from Late Cretaceous fossils in Mongolia (Fig. 10.1; Dong et al. 2022), supporting the idea of that being the center of origin for the Varanidae (Sprackland 1991a, Vidal et al. 2012), though contrary views have been presented (Georgalis et al. 2020, Holmes et al. 2010). The fossils present a range of characters sufficient for them to have been placed in separate and, as far as we know, monotypic genera. However, as different or similar the resulting phylogenies may be, it is important to note that each study had different parameters, i.e., different taxa, different characters, different numbers of taxa, taxa with missing data, and different phylogenetic programs and analyses (see Conrad et al. 2008, Estes 1983b, Lee 1997, Sprackland 1991a, 1995b). Considering how many earlier (pre-computer) studies linked varanoids, mosasaurs, and snakes together, it is odd that not all subsequent studies have included mosasaurs or snakes.

The oldest of the probable varanoid fossils are †*Telmasaurus,* †*Saniwides*, †*Ovoo*, †*Aiolosaurus*, †*Proplatynotia*, and †*Palaeosaniwa,* seven monotypic genera from the Upper Cretaceous of Mongolia (Gilmore 1943), and †*Paravaranus* from the Upper Cretaceous of Canada. These eight genera come from the same period, which suggests that varanoid origins lie much earlier in time. In addition to the characters that tentatively unite these lizards, each genus has features not seen in *Varanus.* A well-preserved parietal of †*Telmasaurus* that closely resembles the morphology of some living *Varanus* was interpreted by Gilmore as the dorsal surface complete with a parietal foramen (Gilmore 1943). It was later demonstrated to be the ventral surface, and the supposed foramen turned out to be the inferior "pit for the parietal ligament" (Borsuk-Białynicka 1994) that, unlike a foramen, did not penetrate to the dorsal side. The lack of a parietal foramen is also typical in the Helodermatidae and Lanthanotidae, often accepted as sister groups of *Varanus*, and have had their position within Platynota repeatedly confirmed, the former as a basal group. The presence of a parietal foramen and attendant parietal eye is autapomorphic for *Varanus* within Varanoidea and Platynota. The Late Oligocene (tentatively) to Late Eocene genus †*Saniwa* was placed in sister taxon position to *Varanus* and has maintained that status for over a century. It is morphologically very close to *Varanus*, but with some characters not seen in any true monitor.

10.3 EVOLUTION AND SPECIATION

I feel that I must provide my chosen working definitions for the terms most important regarding organisms to which they relate: the species. For well over a century biologists have wrestled with defining species (de Queiroz 2007, Hennig 1966, Mayr 1963, Mayr and Diamond 2001) and debates continue. While classes, orders, families, and genera are categories used to sort organisms into retrievable library of life data, they are nonetheless mere names. One does not go out to collect or observe a genus but one does study a species or group of species in that genus. The only "tangible" entity in taxonomy is the species, and no matter how humans may group them, species are ultimately the only "real" taxonomic unit. A genus or family without species is non-existent, while a single member of a population is a member of a species.

The literature tackling species definitions or criteria is extensive and, as of this date, still being produced at a prodigious pace, but there is considerable agreement on two points. First, species are real and not simply human constructs (de Queiroz 2007, McGuire et al. 2023, Singhal 2023, Wiens and Servedio 2000). There *are* groups of organisms that by degree of genetic compatibility, ecology, and behavior form natural groups in nature. Note the requirement that species are based on genotype, not phenotype. Intrapopulation variation, sexual and ontogenetic di- (or poly-) morphism, and polyploidy can and have led to different phenotypes being classified as different species. (In fishes the errors have been staggering, such as classifying male and female, for example anglerfishes, as different orders!) Certainly, there are phenotypes that become fixed but are also affected by the environment, yielding various size or color morphs.

Second, given the almost unimaginable number of organisms and the numerous ways of passing along genes (reproduction by whatever means), no single universal system of determining species has been accepted. The descent–ancestor-based phylogenetic species concept is probably the most widely accepted model because if lineage can be demonstrated phylogeny is established. The older biological species concept is founded on reproductive reciprocity (all members of a population could mate with counterparts from the same population), but biologists are faced with organisms that reproduce sexually, asexually, and unisexually, and that's just for vertebrates. Furthermore, it has been repeatedly shown that even species, and this includes monitors, that normally require two sexes to reproduce may become parthenogenetic (unisexual). For these and other reasons, the biological species concept, though often useful, is not a primary model in vertebrate taxonomy and species delineation. Most definitions of species are taxon-specific, and because lizards are animals, the definition I use is the phylogenetic species concept: species are organisms that are the descendants of an ancestor and have a unique evolutionary lineage that excludes organisms with different ancestral lineages. In other words, species are irreducible populations of organisms that are diagnostically identifiable from other clusters and have a unique parent–descendant lineage (Cracraft 1989).

10.3.1 Speciation as Branching

The living varanids encompass at least 100 distinct species spread out unevenly from Senegal to the Solomon Islands. From their origin in Central Asia, they have followed two major and distinct dispersals. The Late Cretaceous migration of the ancestor †*Paravaranus* to the thin landmass that is now western North America was soon cut off, geologically speaking, from retreating as the Bering Land Bridge sank into the widening strait. Eastward dispersal was blocked by a north to south ocean over 3600 km wide. *Paravaranus* and its later descendants such as †*Saniwa*, were to be isolated from the rest of the varanids to the day the last representative became extinct. Meanwhile, the enormous early Asian continent was en route to its collision with India, Africa and, as the scattered pieces came together, Western Europe, while Australia would be accessible via archipelagos sometime later. Thus, the monitors would disperse, the ones in Asia expanding and then further diversifying. If we look at the big picture we see successful colonization and speciation that may well have been derived through the combination of branching and reticulate evolution, and perhaps by chronospecies.

From the publication of Darwin's *On the Origin of Species* in 1859, the evolution of organisms has become one of the most intensely studied aspects of biological research. Both evolution and speciation are accepted facts by scientists, but the terms are not synonymous, and clarification of this point is essential to the understanding of an evolutionary history of any group. Evolution is the change organisms acquire at the genetic level, or as many textbooks phrase it, "a change in gene frequency." If the change is "minor" it could represent the cause of variation within a population that is not fixed and does not lead to a new species. Significant change, where a trait or traits become more prevalent in a population, leads to new species formation. That is, there is no going back from the genetic change; return to the previous genotype involves changes in other genes (a reversal). Evolution leads to speciation, but speciation does not lead to evolution; one is cause, the other effect. The evolution of species involves phyletic branching where an ancestral species diverges into the ancestor plus a daughter or additional distinct daughter species (Mayr 1963). This is the familiar branching pattern that has long stood as **the** model for the evolution of new species and includes speciation by allopatry, sympatry, hybridization, and polyploidy. Polyploidy is unknown in varanids. Based on their biogeographic patterns, varanids appear to have diverged primarily via allopatric modes. Peripatric speciation occurs when an area not previously inhabited by a species is successfully colonized in that territory and may be cut off from its main population. In other words, a group of animals finds itself somehow transported to a location that is separated from the source by a barrier that usually prevents migrations from one point to the other. The newcomers become a founder population that will experience new conditions that may well lead to them evolving into a new species. This is common among island animals that can be transplanted by winds, rafting, or human introductions, and it is a feasible explanation for the numerous species in the Indonesia–New Guinea region. The thousands of islands have yielded several endemic species. Both the mangrove and tree monitors (*Varanus indicus* and *Varanus prasinus* groups) have proliferated from Sulawesi to the Solomon Islands and represent the fastest-growing lists of varanid taxa. Most were not discovered by zoologists until the late 20th century for the simple reason that few of the islands had been herpetologically explored. Subsequent exploration, in most cases based on what hunters for the live animal trade brought to market, has been involved in the descriptions of more than half the "new" species. Transplantation, intentional or not, and the relative proximity of many islands have allowed *Varanus* to be a successful vicariant genus with numerous cases of parapatric speciation. Other species where this may have happened – or still be happening – include two more Australian goannas, the Merten's Monitor, *Varanus mertensi*, and the Heath Monitor, *Varanus rosenbergi*. The former allegedly differs in average adult size and color in separated populations across northern Australia. The colors are different based on photographs I have seen (courtesy B. Eidenmüller), but I lack additional data. Because this species is poorly represented in museum collections, it will require fieldwork to determine if the differences are significant enough to warrant taxonomic revision. The other species has a peculiar distribution in SE Australia, the central south coast, and SW Australia, each separated by hundreds of kilometers. Again, there were not sufficient museum specimens when I was looking into their character states for another project, but photographs again show noticeable variations in color and pattern in each population, as do the regional environments. Hypothetically they represent the remnant populations of a once more widespread species, now undergoing speciation guided by environmental factors and isolation.

Speciation by allopatry is probably nowhere near as rapid as by parapatry. The parapatry model starts with an instant transfer of individuals to a locality devoid of that species, letting genetic and behavioral isolation begin quickly. Allopatry involves a large population that stays in place and reproductively comingles in an identifiable hybrid zone. Eventually, some natural physical change takes place, such as mountain or river formation, that effectively separates the species into two (or more) isolated populations. Because mountains and rivers do not manifest quickly (in real, not geological time) isolation is also gradual and may occur in stages, and this mode of species formation is probably more applicable to prehistoric taxa than contemporary or comparatively recent events. In Chapter 6 I suggested that Australia's Lace Goanna, *Varanus varius*, may represent three species instead of one. I based this on the geographically restricted tongue color of each population and the barriers of the Great Dividing Range and different habitats between northern New South Wales and mid Queensland. If these barriers did result in allopatric speciation, it did not happen in the recent past but over a long time as Australia's environment changed during (at least) tens of thousands of years. Smaller but effective barriers were possibly what divided so many of the small species of varanid on that continent, which could help explain the great diversity of *Odatria*.

Sympatric speciation is very much like allopatry in that members of a population become isolated from each other long enough for them to become different species. Unlike allopatry, sympatric speciation does not require any individuals to go anywhere but occurs while the evolving portion of a population begins changing in situ; that is, without physical barriers. The movers of the process could involve gene mutation, change in mate choice, or regional change in the environment.

In all cases speciation occurs at the population and genetic levels; there cannot be one without the other. The taxonomist must also consider that genetic changes can, but do not always, result in discernable phenotypic alteration. Similarly, specimens that are close phenotypically may differ more than two species very different in appearance (convergent evolution). This is a significant reason why taxonomists cannot infer species relatedness by seeking close resemblance alone (= typological species concept). But an important point is that the ability for a single population to diverge into several species is possible, though admittedly a long-time process. Given the terrestrial islands present in Australia and New Guinea, this may be an important factor in the great number of species there. The locations have suitable requirements for parapatric, allopatric, and sympatric speciation.

10.3.2 Reticulate Speciation

Reticulate speciation is evolution involving exchange of genetic material between different species or lineages, forming a network-like pattern of relationships rather than simple linear branching tree. It is a modification of branching speciation in that its new branches may cross over, away from the evolutionary line that appears more likely. Reticulate speciation occurs when horizontal gene transfer, hybrid speciation, or recombination forms specific links among organisms, and it is one of the mechanisms that produces triploid parthenogenetic species. Reticulate evolution is not accurately represented by traditional phylogenetic trees, which oversimplify the mechanisms of evolution (Thangadurai and Sangeetha 2015). Such gene exchanges are common among microorganisms but have only recently been confirmed in animals. In multicellular animals, hybridization and introgression among related species may be both common and important enough to explain why strictly tree-like phylogenies do not always follow an apparent ancestor–descendant path (Mallet et al. 2015). Reticulate evolution is unknown in varanids but should not be ruled out as possible, because varanids include many allopatric species that could be able to hybridize and perpetuate a new lineage.

10.3.2.1 Chronospecies

Not all evolutionary pathways lead to speciation branching points but instead follow a continuous route that may lead to significant genetic and phenotypic change. By the time that a species has gone from A to Z, the nodes may appear as different from each other as would members of two distinct families. Consider an analogy, where you compare what you looked like when you were six years old with what you look like now. Your phenotype has changed considerably, but you remain the same person. The object of such a non-branching sequential evolutionary process is called a chronospecies. First, the pathway from ancestor to daughter must produce no branching events, leaving only a single direct lineage. Technically, there is no daughter because it is the same species, so I shall use the term daughter variant. As each daughter variant arises its immediate predecessor, its phenotype becomes extinct. Second, the phenotypic variants must be separated by time, thus negating the possibility of interbreeding completely. In cases where the lineage involves only fossil (extinct) taxa the correct term is paleospecies. In both cases the defining feature is that the ancestor, final variant, and all intermediate variations represent a single species.

10.4 DISTRIBUTION OVER TIME

The routes taken by varanids in their geographical distribution over time are somewhat clear; it is the direction and timing of traversing those routes that remain an ongoing source of inquiry, and both are linked to changes in the physical environment during that time. Contemporary monitors can inhabit hot deserts, lush and dry forests, and urban settings, which means our one common thread is temperature. Today, few species of monitors live in temperate regions, and those that do barely penetrate those places.

10.4.1 Historical Biogeography of the Varanids

To understand biogeography, no matter what the time frame, we must be working with specimens and populations that can be identified with confidence. The literature is filled with paleontological reports of a species that, later, is recognized as being a quite distantly related organism, such as when the earliest papers about †*Varanus prisca* included data that turned out to pertain to fossil turtles. Pinning down the specific original ancestor for any group of organisms is still largely beyond the abilities of science. True, fossils and DNA studies have given us evidence of the **fact** that evolution has occurred and follows a general pattern of descent, from ancestral to more derived organisms over time (with notable exceptions), but only rarely are we able to claim a particular fossil is **the** ancestor of a living lineage. This inability to ascertain ancestor–descendant relationship results because the fossil may represent a branch it took away from the common ancestor it shared with the living species. The fossil may more closely resemble the ancestor than the extant animal, but we cannot yet conclusively say that it was or, for that matter, was not, the actual stem species. Fortunately, we can look above the species level and find fossils that are most likely ancestors to a known lineage, giving us access to a most recent common ancestor (MRCA). This process has been fruitful for monitors.

Monitor lizards are nowhere common in known fossil sites, but they are consistent in the fossil record and show remarkable similarity to the modern species. There are numerous uncatalogued specimens in museums and private collections that are presently unexamined by systematists or paleontologists. Sometimes the only chance a paleontologist has to see these specimens is online at a fossil seller's website (Fig. 10.2). In time, examination of these

FIGURE 10.2 A beautifully preserved fossil that resembles and may be a varanid as shown on a fossil seller's website in 2021. Most specimens like this command huge prices and, once purchased, are only rarely made available for scientific examination.

collections may fill many blanks regarding varanid evolution and distribution. (Late addition: as I wrote this section, an *Archaeopteryx* fossil long held by a private collector was made available to a museum for the first time.)

Many fossil lizards have been mistaken for monitors but even to the unpracticed eye monitor fossils can hardly be mistaken for anything but monitors. A continuing problem when dealing with fossil taxonomy is distinguishing similar species and genera from each other, a problem not unknown to zoologists dealing with populations of well-known, extant animals (Donoghue et al. 1989). One stumbling block is that most fossil monitors are known primarily or totally from vertebral elements. Vertebrae differ structurally depending upon their position as cervical, lumbar, sacral, or caudal, and that some differences are related to age and degree of ossification. Consequently, the sacral vertebra of one species may be given a different name than a set of caudal and cervical vertebrae of the same species found in the same geological horizon (see Estes 1983). So far, however, there are few published studies of intraspecific variation of vertebrae within the Varanidae (Auffenberg 1994, Böhme et al. 1994, Camp 1923, Cieri 2018, Delfino et al. 2013, Zamudo et al. 2016).

Among the reviews of fossil varanids, the most important are those by Dong et al. (2022), Fejérváry (1918, 1935), McDowell and Bogert (1954), Estes (1983a), Borsuk-Białynicka (1984), Gilmore (1943), and Pregill et al. (1986). Much of Fejérváry's (1918, 1935) work is confusing because of grammatical difficulties, though greater criticism must be targeted at his use of insufficient samples of extant material when comparing these to fossil varanids, leading to many erroneous assumptions and conclusions. In reviewing the fossils, Estes (1983) synonymizes many species in the genus †*Saniwa* primarily because most taxa were described from very few bone fragments and may represent minor variation, weathering effects, or differences in age or sex classes. I follow Estes in placing the Tertiary species, including *Saniwa* and the Australian Miocene †*Megalania* into *Varanus*. A review of fossil material follows.

10.4.1.1 Cretaceous Varanids

The net result of the paleontological research has been the impressive discovery of several varanid species in fossil deposits dating back to the Late Cretaceous (Fig. 10.1, 70 mya). The numerous identifiable fossil genera of varanoids are only slightly different from living forms (Balsai 2001, Borsuk-Białynicka 1984, Dong et al. 2022, Evans 1994, Gao and Norell 2000, Gilmore 1943, Nopcsa 1903a, 1912, Norell et al. 2007). Based on the paleontology and ecology, I believe the varanids arose in Central Asia (Arida and Böhme 2010, Dong et al. 2022, Fejérváry 1935, Mertens 1942a, Sprackland 1991a, Vidal et al. 2012) and radiated southwest into Africa (Branch 1982, Clos 1995), west into Europe, and eastward into Indonesia and Australia/New Guinea, the last on at least three occasions (Cogger and Heatwole 1981, King et al. 1991, Sprackland 1991b). This is covered in detail later in this chapter.

Upper Cretaceous monitors are represented by seven species in monotypic genera. The first of these fossils was uncovered during the famous Central Asiatic Expedition into Mongolia (1921–1931), organized by the American Museum of Natural History and lead by its colorful mammalogist Roy Chapman Andrews. Andrews had become enchanted with the evolutionary ideas of paleontologists Samuel Williston and museum president Henry Osborn, who believed the earliest fossils of humans were to be found, not in Africa, but in Central Asia. A previous exploratory effort had been a lackluster Soviet foray, which found only a fossil rhinoceros tooth. Undaunted by this paltry showing and fueled by a chauvinistic belief that Osborn's racist conclusion must be true (that the "higher" civilizations were centered in Central Asia, not Africa), Andrews pitched the idea of taking a motorcade into the Gobi. Osborn agreed, Andrews raised the funds, and one of the last of the great expeditions of discovery was launched.

If a mission be judged on attaining its primary goal, the Central Asiatic Expedition was a failure.

The only human remains were all nearly contemporary, and the Asian origins hypothesis went down with a whimper as other scientists unearthed important African fossils. But exploration is a probe into the unknown, and Andrews *et alia* uncovered a true paleontological buried treasure of, at the time, unimaginable materials. They stocked museum shelves with fossil reptiles, fishes, dinosaurs, anthropological specimens, film, and notebooks that are still being studied. The real success came from the efforts of paleontologist Walter Granger and his staff, for they uncovered dinosaur fossils, showing that Mongolia once had a different environment and a very diverse and important flora and fauna. It was the Andrews team who first unearthed dinosaur eggs, the only fossilized remains of two dinosaurs in mortal combat, and the earliest known monitor lizard. That site and, later, other localities in central Mongolia would yield some of the best and most complete fossils of vertebrates, including well-preserved early platynotans. Paleontologists are still uncovering precious fossils from those sites (Borsuk-Białynicka 1984, Gao and Norell 2000, Norell and Gao 1997, Norell et al. 1992, 2007).

Granger was to become busy describing the mammal and dinosaur specimens from the expedition, so the job of studying the other reptile fossils fell to colleague Charles Gilmore of the Smithsonian Institution (Gilmore 1943). Gilmore noted the similarities between the fragment of upper skull of one fossil and the homologous region in living monitors, and included this new find, which he named †*Telmasaurus grangeri*, in the family Varanidae. As a leading student of both lizard osteology in general, and fossil lizards in particular, it is ironic that Granger thought he was illustrating the dorsal aspect of the parietal region, when in fact the fossil is embedded with the interior surface (e.g., braincase) exposed. This error of inversion was revealed by Richard Estes (1983) and Magdalena Borsuk-Białynicka (1984) after the matrix was removed from the dorsal surface. However, the dorsal surface of a specimen collected years later did reveal a mid-parietal foramen (Borsuk-Białynicka 1984, *contra* Estes 1983), typical of varanids generally, and a unique mid-dorsal ridge. *Telmasaurus* also shares the unique varanoid character of possessing paired lacrimal foramina. The preorbital skull is missing, so the condition of the nasals, premaxilla, and most of the maxilla remains unknown. Gilmore illustrated a short segment bearing the only two teeth ascribed to *Telmasaurus* (Gilmore 1943, his fig. 18). They are pointed, curved posteriorly, broadest basally, and are compressed, features of, but not unique to varanids. The available fossils show no indication of dental infolding nor serrated edges. The postfrontal and postorbital bones form a characteristic X-shaped structure, their common border nearly sutured but still distinct, in contrast to extant varanids, in which these elements usually coalesce into a modified postorbital–postfrontal bone.

Also unearthed was the Upper Cretaceous †*Saniwides mongoliensis* (Borsuk-Białynicka 1984). It closely resembles the Tertiary †*Saniwa* but has fewer maxillary teeth (16 vs. 24). The skull is much more complete than †*Telmasaurus*, including the entire preorbital region, in which the nasals are paired, as in most of the Australian genus *Odatria*. There is a parietal foramen, the palate is toothless (symplesiomorphies for varanids), and the postorbital and postfrontals are fused (a synapomorphy with later varanids, including *Varanus*).

In 1990, a joint expedition between the Mongolian Academy of Sciences and the American Museum of Natural History continued the history of finding and describing numerous lizard fossils in the Gobi and contributed unprecedented information about paleoecology of the reptiles. The team uncovered another Upper Cretaceous platynotan, †*Estesia mongoliensis* (Fig. 10.3; Norell et al. 1992), which at first seemed more closely allied to living varanids than any known fossil. The excellent state of preservation of practically the entire skull looked very much like living monitors. However, †*Estesia* has only a single lacrimal foramen (living varanids have two), and the paired nasals extend further forward towards the premaxilla than in varanids. The postorbital and postfrontal bones are united into a single structure, as seen in extant monitors, but there is no parietal foramen. The teeth are varanid, being recurved, slightly compressed, with enamel-infolded bases (plicidentine). Of particular interest are the grooves along each tooth, suggesting that †*Estesia* could conduct venom in the manner still used by beaded lizards of the genus *Heloderma*. When seen from above the anterior half of the skull of †*Estesia* looks like a monitor and the rear half like a Gila Monster. Thus, †*Estesia* shows an interesting combination of ancestral and derived character states, while serving to solidify hypotheses of relationship to *Heloderma* and not *Varanus*. Phylogenetic analysis after more fossils were found has repeatedly confirmed †*Estesia* as a sister to *Heloderma*, and likely a member of the Helodermatidae.

The Mongolian fossil beds also yielded †*Ovoo gurvel*, a species whose systematic position has yet to be clarified (Dong et al. 2022, Norell et al. 2007). Norell et al. show

FIGURE 10.3 The Cretaceous fossil †*Estesia mongoliensis*. The original specimen (upper) presents a very *Varanus*-like appearance with its long preorbital area, retracted nares, and "hockey-stick" upper temporal arch. It was later reclassified among the Monstersaurs and Helodermatids based on important plesiomorphies including absence of intramandibular joint, complete and well-developed postorbital arch, single lacrimal foramen, and cranial indications of attached osteoderms.

†*Ovoo* as sister group to †*Saniwa* + *Varanus,* both nested in the subfamily Varaninae. Dong et al. found †*Ovoo* to be only distantly related and basal to other varanids. Granted that all the Cretaceous genera – *Aiolosaurus, Ovoo, Paravaranus, Proplatynotia, Saniwides,* and *Telmasaurus* – have numerous character states that ally them with *Varanus*, *Ovoo* shares a synapomorphy, with the extant genus. *Ovoo*, *Saniwa*, *Telmasaurus,* and *Varanus* are the only squamates known to possess two lacrimal foramina in the anterior orbit and, unlike in *Lanthanotus*, the dorsal foramen is surrounded by lacrimal and prefrontal bones (Fig. 5.10), not just the lacrimal. In *Varanus* the upper foramen is considerably larger than the ventral; in *Ovoo* the case is reversed. From this unique trait it seems that *Ovoo* should be closer to *Varanus* than shown in Dong et al.'s cladogram. It is also possible that *Ovoo* is a direct ancestor to *Varanus* and possibly the genus that introduced paired lacrimal foramina.

The Late Cretaceous yielded another varanid-like fossil, †*Palaeosaniwa canadensis* (Gilmore 1928), from Alberta, Canada, and known only from vertebrae and teeth. The inclusion of this taxon in the Varanidae has been controversial (Estes 1983) but was provisionally retained by Estes based on the decidedly varanoid tooth structure. It possesses anguid-like dermal scutes on the head, and the vertebrae do not exhibit the distinctly tapered shape characteristic of varanids, leading Estes (1983) to suggest that †*Palaeosaniwa* may represent a more ancestral (i.e., highly plesiomorphic) varanoid. Subsequently, and like †*Estesia*, †*Palaeosaniwa* is now considered to be a Cretaceous helodermatid (Balsai, pers. comm.) and thus excluded from the Varanidae.

†*Paravaranus angustifrons* and †*Proplatynotia longirostris* are systematic problems. Borsuk-Białynicka (1984) noted that †*Paravaranus* has characters that are a combination of Scincomorpha and Anguimorpha, but favored association with the latter based on its varanid mode of tooth replacement. She cited several character states that are inconsistent with Platynota, including lack of an alar process of the proötic, lack of subolfactory processes of

the frontals, and position of parietal foramen on the frontal-parietal suture, but considered that they "can best be regarded as plesiomorphic." The implication being that *Paravaranus* is at most basal in its position on the cladogram. From the materials available (Borsuk-Białynicka 1984) I add the following character states that separate *Paravaranus* from the Platynota: anterior of frontal barely broader than interorbital portion; single lacrimal foramen; parietal not X-shaped, lacking anterolateral processes; left and right sides of parietal-frontal suture rounded with deep medial cleft going back to parietal foramen; parietal table (anterior to paraoccipital processes) narrower than postorbital part of frontal; mandible narrow near intramandibular joint. For these reasons, I concur with Borsuk-Białynicka and do not consider *Paravaranus* a platynotan.

The phylogenetic position for †*P. longirostris* is similar to that for †*Paravaranus,* having "good" platynotan character states along with several that obfuscate its proper assignation. Features that differentiate †*Proplatynota* from *Varanus* include comparatively short premaxillary process that extends only half the rostral-orbital distance; maxilla and frontal not separated by nasal; single lacrimal foramen; anterior articular process does not flare laterally beyond the parietal-frontal suture; postfrontal and postorbital not fused; posterior processes of pterygoids slightly broader than space between them; pterygoids lack central medial process; posteroventral process of dentary does not extend caudad further than upper edge. Only two phylogenetic studies have included *Proplatynotia, but both put it in a basal position removed from the Varanoidea and more recent varanids (*†*Saniwa*, †*Archaeovaranus*, and *Varanus*).

The Cretaceous varanoid †*Aiolosaurus oriens* is known from one nearly complete skull, most of it still in matrix. It differs from extant varanids in having a single lacrimal foramen. Gao and Norell (2000) follow Bahl (1937) in accepting *Varanus* has having premaxillary foramina, which †*Aiolosaurus* lacks. In fact, many species of *Varanus* have a smooth rostral region of the premaxilla (see Chapter 5, Fig. 5.4). Bahl's data are based on a single species with pronounced premaxillary foramina. Other character states given that distinguish †*Aiolosaurus* from other varanids are incorrect when one looks at the breadth of *Varanus* morphology. Paired nasals are not universally united in the extant genus, for example in *V. acanthurus, V. caudolineatus, V. bennetti, V. ocellatus,* and some *V. marmoratus*, and *V. olivaceus* (Figs. 5.4 and 5.5), making the character state of dubious value in distinguishing genera or subgenera. The presence of an aperture between the premaxilla and maxilla is common in *Varanus*, including *V. bengalensis, V. dumerilii, V. giganteus, V. griseus* (North Africa), *V. indicus* (= *V. indicus* and *V. bennetti* of Ambon and Palau, respectively), *V. ornatus, and V. rudicollis*. The anteromedial process in some varanids separates the septomaxilla and maxilla (e.g. *V. indicus* group), and a reduced medial septomaxillary foramen is seen in *V. caudolineatus, V. dumerilii*, and *V. giganteus*, amongst others. The only character state listed by Gao and Norell (2000) that I have not noticed in *Varanus* is the placement of the anterior surangular foramen in a distinct groove. Based on these features, I see only one character to consider †*Aiolosaurus* as a distinct genus, the single lacrimal foramen. From the published description and figures of Gao and Norell, these character states are shared between †*Aiolosaurus* and *Varanus*: external naris oblong, with broad lateral orientation and slightly longer than the distance between rostrum of premaxilla and anterior rim of orbit; tooth structure; tooth row not extending beneath orbit; posterior process of maxillary with a steep, but not a right angle and long gradual posterolateral slope (cf. *V. bengalensis, V. tristis*); supratemporal with "hocky stick" shape, recurved downwards to meet quadrate; distinct rugosity at mandibular symphysis; deep anteroventral intrusion of angular beneath dentary; and strong longitudinal curvature of the post-dentary portion of mandible. It would be useful to have better illustrations of the septomaxilla, an extremely variable bone for which I have recorded 31 distinct morphologies (Chapter 5), in some cases useful as a species diagnostic. Based on the known character states, I consider †*Aiolosaurus* as part of the clade containing *Varanus* and †*Saniwa*, with retention of its generic name.

Following the suggestions of Donoghue et al. (1989), we may make character polarities based on the Cretaceous fossils. Paired nasals are ancestral in varanids, a hypothesis supported by the presence of paired nasals in †*Saniwides*, and the basal platynotans †*Estesia* and its sister taxon *Heloderma*. In living varanids, paired nasals are found in western Pacific *Varanus bennetti, eastern African V. ocellatus* (Bayless and Sprackland 2000a, b), Australia's *V. mertensi*, and several of the Australian species in the subgenus *Odatria*. In the latter case, the condition may represent a secondary reversal, being related to small size and rapid post-hatching ossification in the subgenus.

A parietal foramen is common to all living *Varanus*, plus, †*Saniwides* and †*Saniwa*, but the foramen is absent in †*Estesia, Heloderma,* and *Lanthanotus* (the sister taxon of *Varanus*). Thus, the presence of a parietal foramen as a synapomorphy for living varanids is strongly supported. Parietal foramina are common in anguids and xenosaurids, which together form the sister group Diploglossa to the Platynota, making secondary loss of the aperture in †*Estesia*, *Heloderma*, and *Lanthanotus* a parsimonious explanation for this feature.

Lanthanotus, a derived varanoid, is known today from several locations in Borneo. The putative fossil lanthanotid, †*Cherminotus longifrons*, was described from the Gobi (Borsuk-Białynicka 1984). †*Cherminotus* differs from *Lanthanotus* in having a longer snout (frontals more than twice the length of the orbits, while they are about as long as the orbits in *Lanthanotus*), paired nasals, and a parietal foramen. For these and other reasons I exclude †*Cherminotus* from the Varanidae (see Chapter 11).

The apparent diversity seen among Cretaceous varanids presents an enigma. Each taxon is too well evolved as a varanid to sit as the probable basal ancestor to the

Varanidae, leaving us to seek an earlier, yet unknown, ancestral taxon. The grooved teeth and solidly roofed parietal region of †*Estesia* support arguments of its close affinity with the Helodermatidae, though its strong upper temporal arch, complete postorbital arch, parietal and septomaxilla structure, and Y-shaped pterygoid argue for affinity with the Varanoidea. While we have no fossil evidence of pre-Cretaceous varanids, there are examples of putative earlier varanoids. *Lanthanotus* is a highly specialized species, with equally fossorial and aquatic habits (Sprackland 1970, 1972a, pers. obs.). Fossorial lizard species often have solidly roofed, short, broad skulls and lack an upper temporal arch (McDowell and Bogert 1954). Meanwhile, the generic diversity seen in Cretaceous monitors suggests that varanid origins and diversification were already well underway, both in Central Asia and North America, well before the end of the Mesozoic. My resolution to the systematics of the Varanidae is given in Chapter 11.

10.5 TERTIARY VARANIDS

Tertiary varanid fossils have been found across the Old World from Spain to Australia. Fossils from Europe have been illustrated and described (Augé 2005, Dollo 1923, Dong et al. 2022, Falconer 1868, Georgalis et al. 2017, Hoffstetter 1969, Ivanov et al. 2017, Roger 1898, Stritzke 1983, Villa et al. 2018), as have those from Africa (Clos 1995, Holmes et al. 2010), Asia (Bochaton 2019, Čerňanskŷ et al. 2021), Indonesia (Brongersma 1948, 1958), and Australia (Molnar 1990, Rich and Hall 1979); for general information on the period see Brygoo (1987), Estes (1983), and Molnar (2004).

Tertiary varanids are known from all continents except Antarctica and South America, though they are not commonly found in any locality. The literature on these taxa is largely limited to species descriptions. Estes (1983) suggested placing all Tertiary varanids into a single genus, *Varanus*, based on remarkable similarity, shared derived characters, and the lack of characters that warrant generic recognition. Alternately, he also pointed out that if †*Megalania* was to remain a genus, all the other varanid subgenera should similarly be elevated; it has since been reclassified as †*V. prisca*.

Some species that migrated further east as continents moved and climates altered with each progressive epoch caused each move eastward as species from prior epochs become absent in the west. From this observation three patterns can be posited. First, Sprackland (1991a, b) has hypothesized that following a Central Asiatic origin in the Cretaceous, varanids moved east across North America but not into Western Europe. †*Saniwa* lived in North America during the Eocene until the Miocene. European varanids likely came from the east, possibly including secondary migrations from Africa or across the Euro-African gap south of Spain. At least one Asian branch emigrated southwest into the Caucasian/Kenyan regions in the Miocene. By the Pliocene varanids had much of their present Afro-Asian distribution, reaching the Indonesian Archipelago and Australia during the Pleistocene.

Second, varanid diversity likely varied considerably with time as the northern Eurasian continent became cooler during successive ice ages. This could account for the dearth of fossil varanid variety in Europe, certainly in areas that never regained the thermal levels required by the family. The reduction of genera from the seven recognized in the Cretaceous to the one at present is suggestive that the family encountered evolutionary bottlenecks more than once in the Cenozoic, cumulatively catastrophic enough that only *Varanus* survived.

Third, the present distribution of African monitors establishes their presence across the entire northern coast of the continent, so it is probable that founding populations emigrated from present-day Morocco to Spain. Extant varanid species in Africa are most numerous from the upper Nile to South Africa and include the Miocene fossils from Kenya, providing a hypothetical second route for emigration to Europe and western Asia, and subsequently, east to SE Asia. In early times, when Africa had less desert, varanids very likely inhabited the entire continent. It is noteworthy that there are few species of *Varanus* in Africa, all are large species, and all are phylogenetically closely related, all of which suggest a single migration from West Asian ancestors. Based on the morphologies of fossil and extant material, it seems unlikely that any Indo-Australian species have African ancestors.

No fossil varanids are known from the Paleocene (Estes 1983). This is curious because the Paleocene was a time of humidity and much warmer temperatures than at present, conditions that should favor varanids. They were no doubt present, and fossils should eventually be discovered. Estes noted that the condyle of the single known vertebra of a Paleocene "*Saniwa*, aff. *S. ensidens*" was not that of a varanid but a probable necrosaur. Varanids have left a large temporal gap between the Cretaceous when, like dinosaurs, they were at a biological zenith, until the Eocene.

Eastern Asia and North America were united by land until the early Miocene, giving Cretaceous varanids a way to get into America (Figs. 10.1 and 10.4, Maps A–D) while at least one genus remained in Asia. Numerous varanid fossil taxa have been described from the early Eocene (Fig. 10.4, Map B). †*Archaeovaranus lii* from Hubei Province, China, a locality about 1500 km south–southeast of the Cretaceous fossil sites in Mongolia. Morphological analysis places †*Archaeovaranus* within the Varanidae and as the sister genus to *Varanus* (Dong et al. 2022). It appeared during the Ypresian (56 mya), and in geological terms, shortly after the Cretaceous extinction (65 mya). Granted, other varanoid taxa likely survived the extinction, but given the phylogenetic relationship between †*Archaeovaranus* and *Varanus*, the genus makes the former a possible ancestral taxon. The Eocene also saw the arrival of †*Saniwa*, the best-preserved varanid fossil known in terms of number and quality of the fossils, and from many locations in North America. The Eocene may have been the time of the first major radiation

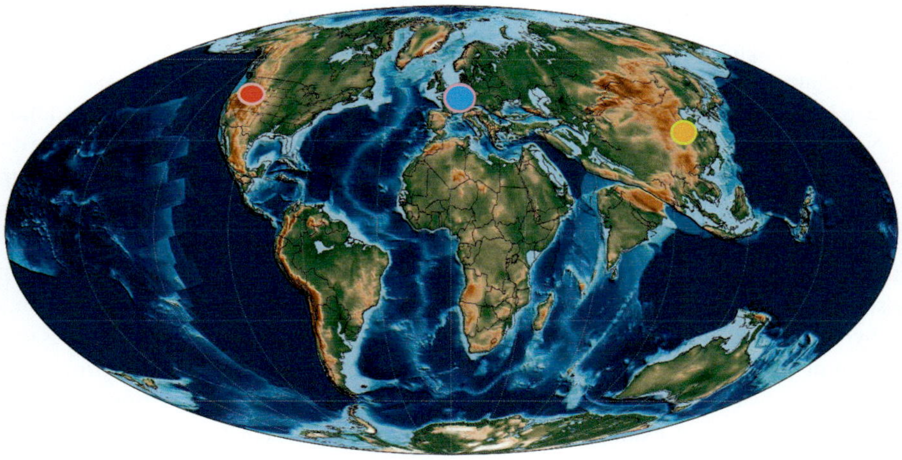

Map B.

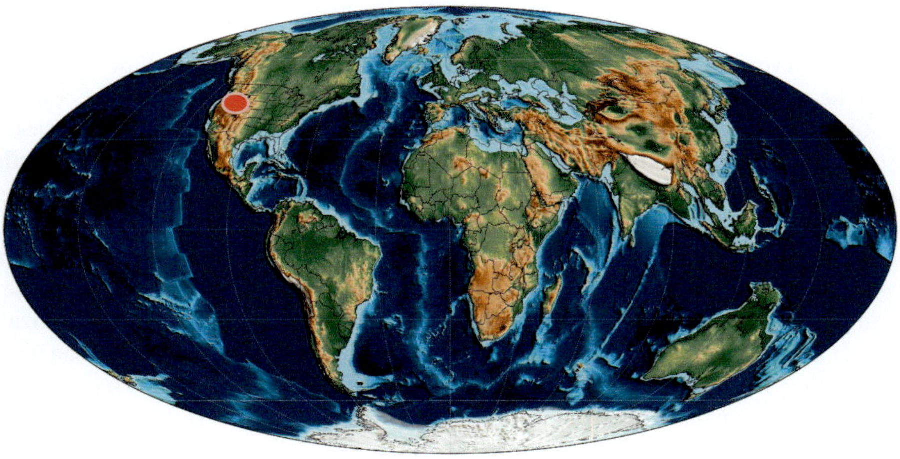

Map C.

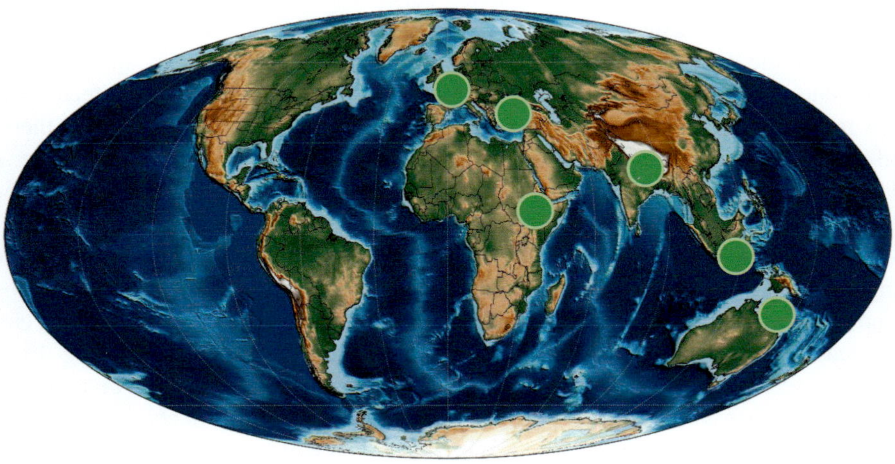

Map D.

FIGURE 10.4 The known distribution of varanids from the through Miocene. Fossils not yet known for missing periods. (B) Eocene. (C) Oligoene. (D) Miocene. See also Figure 10.1 Map A. Scotese, Christopher R.; Vérard, Christian; Burgener, Landon; Elling, Reece P.; Kocsis, Ádám T. "Phanerozoic-scope supplementary material to *"The Cretaceous World: Plate Tectonics, Paleogeography, and Paleoclimate (doi:10.1144/sp544-2024-28)"* from the PALEOMAP project". doi:10.5281/zenodo.10659112 https://zenodo.org/records/10659112

of monitors, with at least nine named taxa. *Saniwa* is known from early Eocene through middle Oligocene, having the longest known presence of any varanid genus. Seven species of †*Saniwa* have been named: *S. agilis, S. brooksi, S. crassa, S. edura, S. ensidens, S. grandis*, and *S. paucidens* from western North America (from Wyoming except *S. edura* from North Dakota).

With †*Saniwa* we have the only genus with multiple species prior to *Varanus*, and with that some intriguing questions. First is the question of size and ontogeny. Estes (1983) introduced ontogeny by suggesting that †*S. paucidens* may just be a larger specimen of †*S. ensidens* and claimed that there was very little to distinguish between †*S. crassa* and †*S. paucidens*. This may be the earliest example of a fossil squamate genus with species being named based on age classes. Prior to Estes' suggestion, Gilmore (1928) considered the synonymy of the *Saniwa* species but accepted the small variations as justification to provisionally maintain them. Second, given that most of the taxa came from the same age and formation in Wyoming, the seeming species diversity may represent the common phenomenon of several similar species coexisting. This is seen in all animal groups and in many lizard genera including *Anolis, Aspidoscelis, Crotaphytus, Phrynosoma*, and *Plestiodon*. It has certainly happened in *Varanus*, especially in the Indo-Australian region. *Varanus* not only presents great species diversity it does so with many sympatric taxa that are difficult to correctly identify from only parts of their skeleton, a problem addressed by Bochaton et al. (2019). Would it be possible to distinguish a vertebrae from *V. tristis* from *V. similis*? Some *Saniwa* species may turn out to be junior synonyms of others. Alternatively, and I think more likely, is that we are greatly underestimating the number of Tertiary species for lack of fossils. It is the confusion of identification of skeletal material that largely inspired me to write this book and why the section on the skull is so comprehensive.

Third, is it possible that all the Wyoming taxa were separated in time, with only one or two being temporally similar? If so, the deposit may have recorded an evolutionary sequence that, based on the nearly imperceptible morphological variety, may be showing a tangible limited sequence of gradualism. The span of time from Eocene to Miocene – millions of years – makes it extremely unlikely that one species went unchanged and had no branching events during that time. The possibility of chronospecies evolution must be considered because such a mechanism could allow for considerable phenotypic change not recorded in fossils coupled with the minor range expansion of *Saniwa*. Average temperatures during the Eocene varied from 19°C to 25°C (66°F to 77°F), an amenable range for the heat-loving monitors that presently thrive with a global average temperature of just 15°C (59°F). In fact, the drastic and geologically rapid drop in global average temperature at the end of the Miocene with its concurrent shift of North America into the temperate zone could well be an extinction event that decimated varanids from the continent. Living varanids survived in part because most of their ranges moved into or remained in the tropics, in particular some of the hottest regions on Earth.

The land link between NE Asia and NW North America likely led to a single migration of varanids across continents shortly before the link was flooded. The disappearance of the North American mid-continental sea during the Paleocene made the western-most part of the continent accessible for the ancestors of *Saniwa*, but there is no evidence that the genus expanded past the Rocky Mountains. North America and Europe had no such land connection, at least as far back as the early Cretaceous, and this gives us an enigma. Fossils considered congeneric with *Saniwa*, as †*Saniwa orsmaelensis* have been found in Europe, but if *Saniwa* evolved in North America from Asian ancestors, how could the genus then migrate to Europe? If *Saniwa* is derived from Asian varanoids that migrated east, how did the genus get to America? They didn't. European and American *Saniwa* are distinct and convergent genera, explained in detail in Chapter 11. No "*Saniwa*" moved east to Europe, they evolved there from Mongolian ancestors without a *Saniwa* intermediate taxon. Between the European varanids and whatever Mongolian-derived species remained, there were both and eastern and western influxes that resulted in the African migration. Descendants of the genus-rich Cretaceous varanoids from Mongolia had a lineage that led to Europe and the evolution of a *Saniwa*-like varanid, but en route there was nothing to stop some populations from going south into India and Africa. Lizards were similarly unable to traverse the Atlantic Ocean which had existed as a biogeographical barrier to such overland vicariance at least from the Cretaceous (Figs. 10.1 and 10.4). European lizards could not be the ancestral stock to American *Saniwa*, nor vice versa, but could have been ancestors to varanids of Africa. French fossils that were tentatively assigned to †*S. orsmaelensis* must represent a genus different from that in North America, covered in Chapter 11. Suggestions that the continents were briefly linked during the Eocene are not supported by geological or tectonic evidence, their last connection about 200mya. Despite their similarities, there are also differences between the American and European lizards, that, when geographic distance is considered, should be sufficient to place the latter in a new genus, †***Dytikosaurus***, that I describe in Chapter 11 (Fig. 10.5B). North American *Saniwa*, however, is so like *Varanus* that their affinity is well-supported, being close sister taxa.

The Oligocene (36–25 mya; Fig. 10.4, Map C) seems to have had a great reduction in taxa, whether due to environmental or sampling effects. During the early Eocene average global temperatures were some 12–14°C above present, steadily dropping to just 3°C above present by the end of the epoch. The Eocene–Oligocene transition was a time of significant global cooling and extinctions. If monitors of the Eocene–Oligocene transition were as thermophilic as contemporary species, it is very likely that many varanids died off as a result of the Eocene thermal decline, leaving few survivors for the Oligocene. Species rebounded from the decline in the middle Oligocene when average temperature

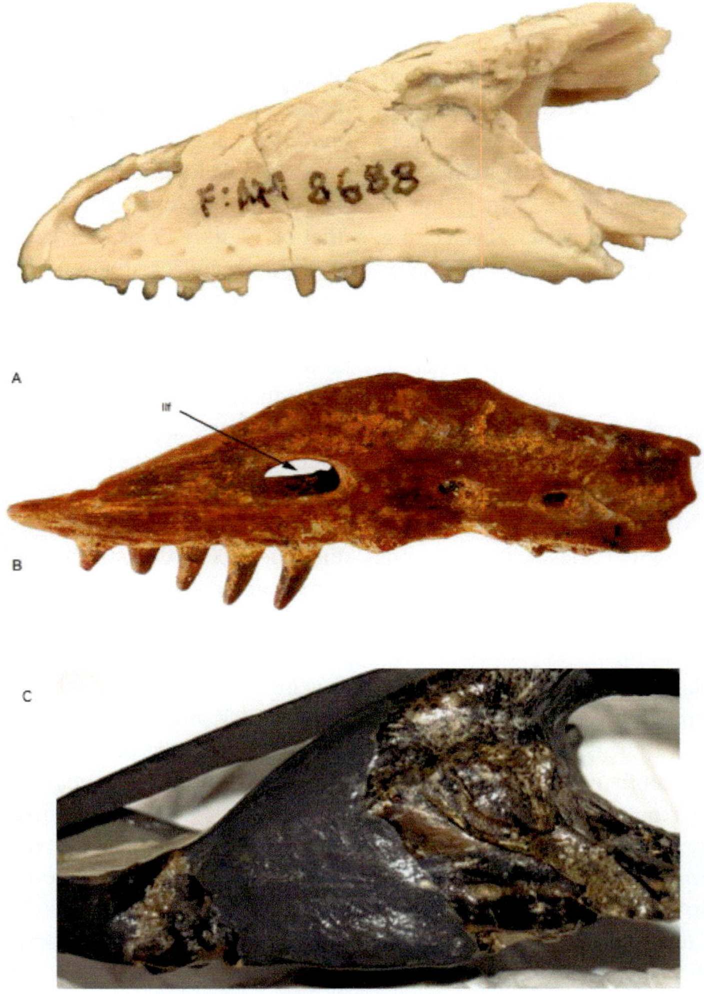

FIGURE 10.5 Comparison of the left maxillaries of *Saniwa ensidens* (A) and (C) and right maxilla of (B) †*Dytikosaurus orsmaelensis* n. gen., n. comb. In no member of the Varanidae is there a huge maxillary foramen (llf) as seen in B, nor are the teeth directs caudad. A – F: AMNH 8688. B – IRSNB R 393. C – USNM v2185, holotype.

was three to 7°C higher than at present, representing nearly 10 million years of fairly stable climate. Fossils from the Lower Oligocene of Wyoming and Nebraska, ostensibly †*Saniwa*, have not been assigned to species but Estes wrote, "It can tentatively be referred to *Saniwa*" (Estes 1983), and a second taxon "is less certainly referrable to *Saniwa*" (Estes 1983). The best that can be said about varanids during the Oligocene is that some of them existed somewhere, and they were the ancestors of the Miocene species boom.

Members of recognizably modern *Varanus* first appear in the fossil record during the Miocene (25–13 mya; Fig. 10.4, Map D). The transition from the Oligocene was accompanied by more drastic cooling than in earlier epochs and brought the formation of the great polar ice sheets. Gradually, average global temperature reached 18.4°C. Continental regions that became tropical and subtropical provided a refuge for the family and successful environments for a rapidly diversifying genus. No taxa from this time are known from North America, but the Euro-African radiation was rapid. The earliest African record for a varanid comes from the early Miocene of Kenya †*V. rusingensis*, which is similar to †*V. hofmanni* (Estes 1983) and extant *V. niloticus*. My comparison of †*V. rusingensis* and *V. niloticus* revealed no reliable feature I could use to distinguish between the species and without further comparative fossil material I leave them provisionally distinct. Recently, varanid fossils have been uncovered in Australian Middle Miocene deposits (Murray and Megirian 1992, Pledge 1990; reviewed in Lunney and Ayers 1993, Tyler 1989), but their generic identity has not yet been confirmed. Some have been assigned to the Pleistocene †*Megalania* (= *Varanus*; Molnar 1990, Murray and Megirian 1992). Until now, †*V. prisca* was the only known monitor species of the Miocene, and into the Australian Pliocene, Pleistocene, and early recent epochs. I follow Estes (1992) in synonymizing †*Megalania* with *Varanus* because it lacks unique derived characters that would justify maintaining a separate genus, and character states that are different could all be related to size. †*V. prisca*, if it reached just 5 m total length as some workers have claimed, is the largest varanid species yet discovered. The more recently found Miocene fossils, if actually varanid, could be different species, but descriptions have not yet been published. Named species of the upper Miocene include *V. bengalensis*, †*V. lungui*, †*V. tyrasiensis*, †*V. pronini*, and †*V. semjonovi* all from the trans-Caspian region, and †*V. hofmanni* (Estes 1983) from Germany. In the case of the German †*V. hofmanni*, there are some very well-preserved vertebrae from Hungary that have been depicted and strongly support that species' inclusion as *Varanus* (Venczel 2006), though cranial material is lacking. Delfino et al. (2013) have also rejected naming taxa based solely on vertebrae and have designated several European varanids as nomina dubia, a logical decision in my view (see Section 10.6). The only information available about †*V. lungui*, †*V. tyrasiensis*, †*V. pronini*, and †*V. semjonovi* are very short descriptions of vertebrae without illustrations or useful diagnostic data. Their status is therefore dubious and without further material cannot confidently be included here. I have examined the fossil cf. *V. bengalensis* material at the BMNH and based on comparisons with numerous recent specimens see no reason to doubt their identity as that species. †*Varanus mokrensis* from the Czech Republic shows numerous features in common with *Varanus*, with which it is confidently placed.

†*Iberovaranus catalaunicus* from eastern Spain was based on a few vertebrae, and its status is questionable. †*Iberovaranus* was differentiated from *Varanus* in having "a more elongate, narrower neural arch" (Estes 1983) than *Varanus*. The few fossils have the characteristic spherical condyle of *Varanus*, but in their incomplete and worn condition they also resemble vertebrae in some anguids. Resolution of the tenuous taxonomic status of †*Iberovaranus* requires further specimens to ascertain its systematic position, though I accept its tentative reclassification into the genus *Varanus* (Delfino 2012).

The discovery of two species, †*V. rusingensis* and †*V. hofmanni*, which may be the extant monitors (*V. niloticus* and *V. bengalensis*), confirms the earliest known *Varanus* in the Miocene of Africa and Europe, respectively. To these I also include the Indian †*V. sivalensis* which also shows considerable agreement with Recent *V. bengalensis* to the extent that if they are not synonymous, I suspect a chronospecies. By the Pliocene (13–1 mya) modern *Varanus* becomes the only genus, including †*V. marathonensis* of Europe and Asia Minor.

The earliest Australasian fossils are from the Pleistocene after yet another significant drop in average global temperature to about 12°C (53.6°F). Three species have been described from this period, the Javanese †*V. bolkayi*, and †*V. prisca* of much of eastern Australia (Estes 1983; Fig. 10.4, map D). †*V. emeritus* is a nomen dubium because it is not based on any type specimens, while †*V. bolkayi* is only known by its vertebrae.

Subfossils of some extant species are known from the present distribution of varanids. Apart from †*V. hooijeri* from Flores and Sumbawa islands, Indonesia, all Asian subfossils are identifiable to living taxa (Bochaton et al. 2019). †*V. hooijeri* is distinctly different from either *V. komodoensis* or *V. bimaculatus*, the two species sympatric on Flores today (Brongersma 1958, Turvey et al. 2017). Both are predators and opportunistic, and both approach 3000 mm in total length. The jaws contain barrel-like blunt, closely packed teeth, like those of the Philippine *V. olivaceus*, *V. mabitang*, and *V. bitatawa*, species that eat fruits and hard prey such as large snails. †*V. hooijeri* may have had a similar diet (Turvey et al. 2017). This dentition closely resembles that of *V. albigularis*, decidedly not a frugivorous monitor but a powerful carnivore. All these monitors have few, if any, empty tooth sockets and only *V. albigularis* has numerous replacement teeth abutting active teeth. Estes (1983a) noted that *V. hooijeri*, too, could approach "the same size as often reached by *V. niloticus* today" (which is an odd thing to say because (1) *V. niloticus* is not referred to in Estes' species account, (2) the dentition much more closely resembles at least two other extant species known at that time, and (3) why choose *V. niloticus* when several Asian species meet or exceed that size?) which may slightly exceed 2700 mm or about 1000 mm shorter than the other extant species (Sprackland 2009) and near adult size for the sympatric *V. bimaculatus*. Available fossils of †*V. hooijeri* do not include juveniles, so we cannot tell if, like *V. albigularis*, *V. olivaceus*, and *V. niloticus*, juveniles had pointed teeth later to be replaced by the blunt dentition. It's disappearance from Flores while two other nearly equally large (in length) and larger (mass) *Varanus* persisted is a matter of conjecture, but it coincides with the extinction of the fossil hominid, *Homo floresiensis* (the "Hobbit"; Turvey et al. 2017) and other endemic fossil species.

Flores has never had a land link with most other islands of the Lesser Sundas, but it is close enough (51 km) to Komodo Island that *V. komodoensis* could swim or raft the distance successfully, as could *V. bimaculatus*. Both species have been seen many miles at sea and are likely that *V. hooijeri* could have done the same. Two subfossils from Java have been named †*V. bolkayi* (Fejérváry 1935) and described as similar to those from Pleistocene "Indonesian Timor" (Hooijer 1972). They are known only as poorly preserved vertebrae, and do I consider them as established varanids here. Fejérváry and Hooijer suggested that because of the size the vertebrae †*V. bolkayi* and the Timor fossils could represent "subspecies" of Komodo Dragons.

It will take additional fossils, hopefully including skull material, to clarify the systematic positions of several platynotans, most especially in making species determinations. However, that will play out, we can safely claim that by sometime in the early Miocene the only varanids were in the genus *Varanus* and that they had more or less occupied their present world distribution.

10.6 VARANID ZOOGEOGRAPHY AND TAXONOMIC CONSIDERATIONS

10.6.1 Insularity and *Varanus*

The widespread insularity of most species of *Varanus* makes them excellent subjects for studying island biogeography and the effects of island isolation on speciation. The number of recognized taxa has grown from 27 since the first family monograph (Mertens 1942a–c) to nearly 100 as of 2025, and most of the additions have come from remote islands and isolated (by Australian standards, extremely isolated to others) regions of Australia. Investigations about varanid distribution and evolution must include an island biogeography component.

Island biogeography has always been a subject of interest to zoologists, including herpetologists (Barbour 1912, Kopstein 1926, Mertens 1934, Sternfeld 1920) but most publications almost exclusively cover species lists and taxonomic accounts. They also provide modern researchers considerable baseline data that is helpful to studies of insular evolution. The move towards quantifying insular population ecology and evolution became a major field of study with the publication of *The Theory of Island Biogeography* (MacArthur and Wilson 1967) and has been central to research ever since (Brown and Lomolino 1998, Losos and Rickefs 2010). Varanid insular zoogeography has been covered by several authors (Böhme and Ziegler 2007, Burden 1927c, Horn et al. 2007, Mertens 1934, Rodda et al. 1991, Weijola 2010). Among the features considered to be linked to island endemism include (1) dwarfism, (2) gigantism, and (3) melanism.

Regarding dwarfism Mertens (1934) reported that "entire island regions can sometimes be distinguished by the dwarfism of their reptiles," including *Varanus*. He found that the largest *V. salvator* (= *V. bivitattus*) from the eastern Indonesian islands Lombok, Sumbawa, and Flores measured 1.5 m total length, the majority around 1.07 m, all smaller than the 2.5–3 m specimens from Sumatra west (= *V. macromaculatus*). Here lies a caveat: the "species" has subsequently been divided into several newly recognized species. His smaller lizards are now recognized as *V. bimaculatus*, and the larger as *V. macromaculatus*. Members of the species group, however, do become larger in western populations (Hocknull et al. 2009), with a *V. salvator* (which is still *V. salvator*) from Sri Lanka holding a record at 3.21 m (Randow 1932), just larger than the record for *V. komodoensis* at 3.13 m. The theory of island biogeography has a rule saying counterparts to mainland small species tend towards gigantism on islands and that counterparts to large mainland species tend towards dwarfism on islands. The case for water monitors is not that clear probably because between Sri Lanka and the lesser sunder islands there is no continuous pattern of change in island size. Nevertheless, among the tree monitor lizards of the New Guinea region the largest species live on comparatively small islands while the smallest members of the group are widespread on the New Guinea mainland. Whether *V. komodoensis* qualifies as a giant or dwarf remains to be seen; if it evolved from smaller ancestors, such as a population of water monitors, it would qualify for the gigantism camp, but if its ancestor was the much larger †*V. prisca*, then its presence on the islands would suggest relative dwarfism.

The opposite pattern appears to be true among the mangrove monitor lizard group, where the largest specimens are from mainland New Guinea and the smallest from outlying islands. In this latter case there is a trend for smaller species to inhabit the small islands, and the smallest member of the group inhabits one of the smallest islands in the region. Alternately, the continent of Australia includes almost all the *Odatria* dwarf monitors (<170 mm SVL; Cogger 1992, Sprackland 2001, Storr 1980), known from the mainland and a very few islands (e.g., Groote Island). Despite repeated attempts to find an *Odatria* on the continental Fitzroy Island, Queensland, over a three-year period, the only varanid encountered was *V. panoptes* (Sprackland et al. 2004) despite the island being just 5 km from the mainland. Is *Odatria* and continental dwarfism an exception to the island rule? Australia's extant mammal species were larger prior to 40,000 years ago and have been getting smaller ever since (Flannery 1994), allegedly after the arrival of humans and their hunting methods. This time-dwarfing hypothesis fits all the continent's mammals except some wombats. Though this model may seem to be an explanation for *Odatria*, which evolved from larger varanids (Sprackland 1991), the cause does not. Admittedly, the fossil record for Australian varanids is miniscule, with only a few larger species such as †*V. prisca* and putative *V. gouldii*, making an empirical comparison, at present, impossible. If we consider that †*V. prisca* and possibly larger individuals of extant species such as *V. gouldii* and *V. panoptes* were the rule 40,000-plus years ago, then dwarfing over time could apply. Hypothetically, if any *Odatria* species founded an island population, and if the island rule holds true, we would expect those immigrants to become larger than the mainland individuals, and given the broad distribution of *Odatria, it is likely that such populations might so far have simply escaped notice.*

Monitors also provide examples of island species becoming melanistic (Fig. 10.6) (Mertens 1934). Not only are islands home for the largest of the tree monitors, but those species are also melanistic with no apparent markings except in hatchlings and juveniles. Mertens ascribed melanism to a specimen of *V. boulengeri* from Coquet Island, Queensland, that was described by Kinghorn (1923; Fig. 10.6A) and a population of *V. varius* from Kangaroo Island, Victoria (Mertens 1934). The former has been reclassified as *V. semiremex* and the specimen described and illustrated is not melanistic. The latter represents *V. rosenbergi*, a melanistic species throughout its range, not just in island populations. Mertens' only true melanistic species was *V. beccarii* (as *V. kordensis beccarii*) from the Aru Islands (Fig. 10.6C, D). Looking into patterns of insular speciation requires good taxonomy, and varanid taxonomy has been frequently altered since 1934. Another melanistic insular varanid, *V. bogerti*, inhabits islands of eastern Papua New Guinea and is only known from a few specimens.

Melanism and gigantism are present in the Komodo Dragon, *V. komodoensis*, which inhabits a few small islands and a portion of western Flores. The Togian Islands

FIGURE 10.6 (A) Figure of the specimen of *Varanus boulengeri* (= *V. Semiremex*) that Kinghorn (1923) interpreted as melanistic but clearly shows a normal pattern for the species. Species-level melanism, in which dark color and little or no pattern, is seen in (B) *V. Togianus*, and (C) and (D) *V. beccarii*. (A from Kinghorn, J. 1923. A new Varanus from Coquet Island, Queensland. *Records of the Australian Museum* 14(2):135–137.)

in Sulawesi's Teluk Tomini (Tomini Bay) are inhabited by a melanistic species (*V. togianus*) closely allied to the *V. salvator* group (Fig. 10.6B). A large, isolated population in northern Thailand has an endemic melanistic water monitor (*V. komaini*) that, even as hatchlings, are entirely jet black. The conditions that make them isolated make them a terrestrial island population, a common situation that often leads to "subspecies" recognition, as has been the case for the black monitors. Hatchlings of all the other melanistic species have some pattern with light markings (Fig. 10.7A, B). Mangrove monitors include species that have few enough light markings to be considered melanistic. Though there are few specimens of melanistic monitors in preserved collections, they are occasionally produced by breeders.

Evolutionary processes may take thousands or millions of years (gradualism), making the testing of many hypotheses rather troublesome. Islands, however, and the given distribution of varanids on terrestrial or oceanic islands, provide excellent opportunities to study more rapid aspects of microevolution, as has been amply demonstrated with *Anolis* (Losos 2011). Punctuated equilibrium (Eldredge and Gould 1972, O'Brien et al. 2024) may also operate, perhaps predominate, in such environments. Punctuated equilibrium describes a comparatively rapid evolutionary

A B

FIGURE 10.7 (A) Hatchling of *Varanus beccarii* from the Aru Islands off the southwestern coast of New Guinea. Juveniles possess numerous small green spots on a jet-black body, that disappear within the first year of life. Photo courtesy Ben Siegel. (B) Hatchling *Varanus komodoensis* showing complex pattern of light markings. These fade by early adulthood, and the body becomes a dark slatey-gray.

process, where a species may remain static for generations, but then undergo rapid changes that result in new species. It differs from chronospeciation in that the ancestral population branches into more than one descendent species. Confirming examples can be difficult (Murphy and Lovejoy 1998). Studies on other vertebrates, including finches (Grant and Grant 2014), blue-tongued skinks (Brennan et al. 2024), and anoles (Losos 2011, Tollis et al. 2018) have found examples of island populations undergoing punctuated equilibrium, but very few lizard taxa, including varanids, have been investigated for the process.

10.6.2 Environment and Distribution Patterns

The distribution of fossil varanids corresponds with development of tropical and subtropical forest and desert areas from the Cretaceous onwards, adding support to hypotheses that the family has remained linked to similar ecological factors over the past 80 million years. Sprackland's (1991) proposal of distribution since the Cretaceous accounts for known fossils and ecology of their locations, but admittedly may reflect the biases of sampling, and is largely revised in this volume.

The North America-Asian distribution is in accord with the geographic proximity of those locations during the Cretaceous. Africa was connected to Europe by the Paleocene, and it is probable that further examination of African sites will reveal more varanid material dating earlier than the Miocene. The Indian subcontinent collided with Asia during the Pliocene, at which time the fossils of †*Varanus sivalensis* occur. The distribution of varanids among the Indo-Australian landmasses similarly coincides with tectonic history (Sprackland 1991a). Numerous proposals of the evolutionary trajectory of the Varanidae have been proposed, and most are in close, if not identical, agreement (Ast 2001, Augstenová et al. 2021, Baverstock et al. 1993, Cogger and Heatwole 1981, Conrad et al. 2008, Gray et al. 2019, Hocknull et al. 2009, King and King 1975, Pianka 1994a, Pregill et al. 1986, Sprackland 1991a, Srikulnath et al. 2013, Welton et al. 2013, 2014). The most significant variable is the timing and direction of the various dispersals.

The models of fossil distribution can be used as a framework for hypotheses of intrafamilial relationships. If Cretaceous varanids carry more plesiomorphic states than later species, such as paired nasals, and these states can be corroborated through outgroup analyses with sister taxa (fossil lanthanotines and helodermatids), then certain character polarities can be assigned with high levels of confidence. An important point related to resolving varanid origins is whether the paired nasals represent a purely plesiomorphous condition, or does it arise secondarily in several species, linking most in the Australian subgenus *Odatria* (Sprackland 199a)? In embryonic specimens and juveniles, the nasals are distinct, with fusion a result of ontogeny (see Chapter 5). *Odatria* is made up of typically small (under 1 m) species. If the paired nasals in *Odatria* are related to allometry, then the character may be considered secondarily acquired, and thus an apomorphy supporting a relatively recent origin and radiation in Australia. In contrast, the fossil taxa displaying this condition are large (over 1 m) species, in which case it may be hypothesized that they are carrying the true plesiomorphy.

The availability of fossil materials provides evidence of greater (†*Saniwides*) and lesser (†*Estesia*) relatedness among Anguimorphan taxa to *Varanus*. With few exceptions, the Cretaceous fossils are more enlightening, consisting as they do of skulls plus post-cranial elements, while Tertiary material is largely restricted to post-cranial elements and rare skull fragments, excepting the several excellent *Saniwa* specimens. Nevertheless, it is apparent that, like sharks, varanids evolved into an adaptable generalist morphology that has shown little gross modification over 80 million years of development. The considerable number of small modifications, as in the case of finch bills (Grant and Grant 2014), add up to significantly enhance survival. While it might be argued that a single character such as grooved teeth, or †*Saniwa*'s vomerine teeth might warrant generic or subfamilial designation, I am not convinced that such variants are significant departures from the *Varanus* morphology. Estes (1983a) suggested combining all Tertiary varanids into a single genus, but examination of better, more complete specimens since Este's catalogue have failed to provide justification for such a taxonomy.

10.7 OTHER ANATOMICAL FEATURES

10.7.1 Notes Regarding Fossil Vertebrae

I must preface this section by stating that I am not a supporter of taxonomy based upon a single bone, especially if it is the only bone collected from a site and if it shows sufficient wear or damage to make discerning characters an exercise in ambivalence (Estes 1983a). While diagnostic features of a partial skull can be preserved enough to allow one to present a hypothesis of relationships, I have not been convinced that one or a few vertebrae are equally revealing. This bias is far from universal, and given the frequent paucity and poor quality of fossils, I understand why single-bone taxonomy has prevailed for over 150 years. Much of what we assume to know about the lives and habitats of extinct varanoids has come from vertebrae fossils. Because varanoid vertebrae are so distinctive, the information they provide can be very useful for determining prehistoric distribution and evolutionary history, signs declaring that "varanids were here." The axial skeleton of varanids is composed of numerous strong, constricted, procoelous vertebrae, each centrum ending in a large ventral condyle that articulates with a corresponding anterior condylar fossa. Characteristic of their vertebrae is the distinct pericondylar flange, anterior to which the condyle is narrower in diameter. When viewed ventrally the structure resembles an isosceles triangle. The post-axis centra are distinct with ventral Y configurations, the lower portion comprised the convex body and condyle, the upper by the diverging anterior ventral processes. Within the fork of the Y is the partially downward-facing condylar fossa; this structure is always visible in ventral view. The relative depth of indentation of the upper part of the Y shape is useful in distinguishing among genera. In †*Saniwa ensidens* the Y has a short main body, the V section is shallow, and lateral constriction of the centrum is minor. Allied *Lanthanotus borneensis* has centra that are extremely similar to those in †*Saniwa*, save the postzygapophyses in *Lanthanotus* are considerably larger and the condylar fossa is not as exposed ventrally. †*Varanus sivalensis/bengalensis* has a ventral process that is more slender, with a longer distinct central body and more laterally facing anterior processes than †*Saniwa*, a morphology very close to contemporary *V. bengalensis, V. nebulosus*, and *V. niloticus*. The highly modified *V. salvadorii* has unique centra in which the Y is actually T-shaped, having a long straight central body, and anterior processes that jut laterally at 45-degree angles from the midline. In this species, too, the condylar fossa, though visible from beneath, is less so than in any other varanid.

Varanid vertebrae are structures that have been studied (Cieri 2018, Delfino et al. 2013), but we lack a review of the morphology of the sections of the spines. Part of the need relates to fossil and subfossil varanid vertebral specimens, by far the most common bones encountered. Taxa have been named on the basis of one or a few vertebrae, and I question how valid such a designation can be. There are three major problems with vertebra-based taxonomies. First, the fossils are rarely in good enough condition to allow examination of important characters (Fig. 10.8). Even subfossil materials may show considerable wearing and weathering that make bilateral examination of a single bone impossible. Second, we do not yet have a solid data set of the morphologies of extant monitors, though we do know that their vertebrae may vary considerably among species. Alternatively, some taxa are so similar that vertebral identification is not possible. For example, †*Saniwa* has at least three synonyms – *Saniwa, Megasaurus*, and *Thinosaurus* – based on just a

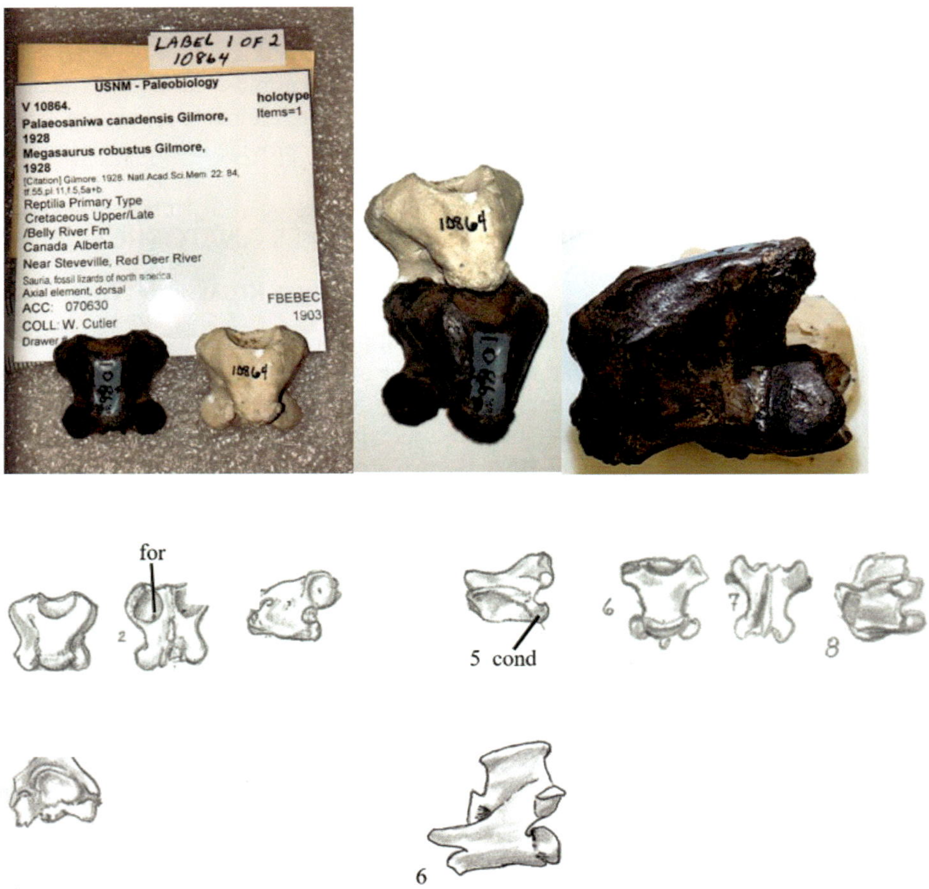

FIGURE 10.8 Comparison of vertebrae from three Platynotan taxa showing the variation that makes species level classification from a single or few spinal components impractical. Upper: holotype element (black), (dorsal view), and cast (ventral view) from †*Palaeosaniwa canadensis* Gilmore 1928, USNM v.10864, holotype. Center, same showing articulation. Right, left lateral view of lumbar of same. 1–4: diagrammatic representation of vertebrae in the photographs. †*Palaeosaniwa canadensis* Gilmore 1928, USNM v.10864 lumbar (?) in ventral, dorsal, left lateral, and anterior views, respectively. 5 – post-pelvic and 6–8 mid-lumbar (?) vertebrae of †*Palaeosaniwa canadensis* Gilmore 1928, USNM v. 2910, holotype for †*Megasaurus robustus* Gilmore 1928 (junior synonym). 6 – Cervical vertebrae of contemporary *Varanus dumerilii* (Schlegel 1839), author's collection. (Figures drawn from photos by author.)

few vertebral fossils. Either the describers failed to know of earlier descriptions or interpreted the fossils differently from each other. Third, the morphology of the vertebrae varies from region to region within a single animal. Cervical, thoracic, lumbar, sacral, and caudal vertebrae not only differ from each other, they also differ within their region (Figs. 10.8 and 5.33). It is therefore conceivable that specimen A based on a cervical, specimen B based on some sacrals, and specimen C based on a few caudals may all represent the same species. This subject is covered in more detail in Chapter 11.

Varanid vertebrae are structures in need of nomenclatural definitions, especially since few fossils are intact. Part of the need relates to fossil and subfossil varanid vertebral specimens, by far the most common bones encountered (Bochaton et al. 2019, Brongersma 1958, Fejérváry 1935, Gilmore 1928, 1943, Hooijer 1972). Cervicals tend to be complex, each notably different from the other, especially the atlas and axis. Thoracic and lumbar vertebrae are large, stout, and fairly similar in form. Caudals are small and become gradually smaller, are elongate compared to other vertebrae, possess a shallow longitudinal ventral groove, bear ventral chevrons, and have high neural crests. These morphological types also correspond well to different "species" among the fossils. For example, cervicals of *Varanus dumerilii* resemble some vertebrae from †*Saniwa*, thoracics look like †*Palaeosaniwa*, lumbar resemble †*Telmasaurus*, and so on. By the same token, the Miocene †*Varanus rusingensis* of Kenya is practically indistinguishable from contemporary *V. niloticus.*

11 Relationships
The Systematics of Monitor Lizards

11.1 SYSTEMATIC REVIEW OF THE VARANIDAE

Systematics is the scientific study of organisms and their evolutionary relationships, while taxonomy is the part of systematics involved with classifying and naming taxa. Varanid systematics has been a subject that herpetologists and paleontologists have frequently visited and nearly as frequently modified, both in relationship to the species in the family and within the greater contexts of Squamata and Reptilia. Not surprisingly, the world views of the two sciences do not always coincide, and it doesn't take a particularly careful reader to guess to which discipline the writer belongs. In both cases, studies were predominantly morphology-based until the 1970s, followed by a period of using biochemistry and computer programs to compare chromosome structure (Iannucci et al. 2019, King and King 1975), immunological distances (Baverstock et al. 1993), genes (Ast 2001, Mulcahy et al. 2012), and venoms (Arbuckle 2009, Chandrasekara et al. 2024, Fry and Scheib 2007, Fry et al. 2006). With the advent of better, faster, and less expensive DNA analysis, most of the post 1995 studies used or were wholly molecular, and by the 21st century, various "combined data set" analyses had become common. Not surprisingly, results varied between different groups using similar techniques and between DNA and morphology workers (Ast 2001, Baverstock et al. 1993, Conrad et al. 2008, Fry and Scheib 2007, Mulcahy et al. 2012, Pyron 2017, Srikulnath et al. 2013, Townsend et al. 2004, Vidal et al., Welton et al. 2013). Mertens (1942a–c) developed a taxonomic scheme that included the use of eight subgenera, to which he would later add. Though the usefulness of the subgenus allocations was questioned by some (Estes 1983, Sprackland 1991a), it was nevertheless embraced to some degree by most subsequent workers.

Of Mertens' (1942a–c) subgenera, all but three, *Varanus, Empagusia*, and *Odatria*, were monotypic and erected almost entirely on external morphological characters. For example, Mertens described the subgenus *Varanus* as having a laterally compressed tail, without annuli of equal-sized scales; round nostrils located nearer the snout tip than the eye; and fused nasals. *Odatria* has a round tail, with caudal scales forming annuli; a round nostril closer to the eye than in *Varanus*; and paired nasals. Mertens (1942c) claimed that *Varanus salvadorii* had a compressed tail, though it is subtriangular to slightly round in section, as he noted when later erecting the subgenus *Papusaurus* (Mertens, 1962). Though Mertens (1962:333) claimed he did not generally condone monotypic subgenera, he nevertheless did so when he made *Papusaurus* his ninth such taxonomic group. *Papusaurus* was diagnosed primarily on its peculiar tail shape (according to Mertens, convergent in *Varanus griseus, Varanus spinulosus*, and some *Odatria*); its extraordinary tail length of 2.3–2.6 times snout-vent length (= SVL) (convergent in *Varanus (Odatria) glebopalma, O. glebopalma*, and others); and vaulted snout (convergent in *Varanus doreanus*, of which Mertens was unaware). While the preceding characters may in fact represent unique evolutionary novelties within the Varanidae, the other diagnostic characters are certainly symplesiomorphies, and include: very large size (cf. *Varanus komodoensis, Varanus salvator, Varanus niloticus, Varanus giganteus*, etc.); nostril lateral and oval (cf. most varanids excluding *V. griseus, Varanus albigularis, Varanus bengalensis*, and several related species; see Chapter 5); supraoculars undifferentiated; caudal scales not subequal and forming whorls (vs. *V. griseus, V. spinulosus*, and subgenus *Odatria*); maxillae with long, gradually rising dorsal border (cf. most varanids); nasals paired or unpaired (covering both character states, and one that is sometimes ontogenetic, for platynotans); and tooth structure subconical, with basal fluting (= plicidentine). The undifferentiated supraoculars are interesting because their absence clearly distinguishes *V. salvadorii* from all other large (2 m +) varanids, excepting the morphologically and ecologically distinct *V. giganteus*, and all sympatric New Guinea varanids except the dwarf members of the *Varanus timorensis* group (including *Varanus tristis* and *Varanus similis*). Only the tooth structure is unique to *V. salvadorii*, being long, triangular, and with only the anterior teeth being slightly recurved (see Chapter 5). Two of Mertens' subgenera, *Odatria* and *Varanus*, contained most species, and the former has held up, with modification (Baverstock et al. 1994, Brennen et al. 2020, Holmes et al. 1975, Sprackland 1991a, b, Vidal et al. 2012). Mertens' subgenus *Varanus* was diagnosed by symplesiomorphies, and many of its taxa have since been redistributed (Baverstock et al. 1994, Böhme 1988, Dong et al. 2022, King and King 1975). Among these species is *Varanus indicus*, which Mertens considered to include three subspecies (see below). He later added *Varanus karlschmidti* (Mertens 1951), a junior synonym for *Varanus jobiensis* (see below), and considered it closely related to *V. indicus* (Mertens 1971). The type species for *Varanus* is *V. varius*, to which *V. indicus* is not particularly closely related (Ast 2001, Baverstock et al. 1994, Holmes et al. 1975, King et al. 1991). *V. indicus* was subsequently moved (Böhme 1988) to *Euprepiosaurus* Fitzinger, 1843, for which *V. indicus* is also the type species. Böhme (1988) included the New Guinea tree monitors (*Varanus beccarii,*

DOI: 10.1201/9781003463528-15

Varanus bogerti, Varanus prasinus, Varanus telenesetes, Varanus macraei, Varanus keithhorni, and others) in Euprepiosaurus. Sprackland (1991b) confirmed the close relationships between *V. indicus* and the tree monitors, but did not recommend any subgeneric assignment, pending further study of related taxa, particularly the *V. salvator* complex and New Guinea endemic *V. (Papusaurus) salvadorii*. All these taxa, excepting *V. salvator* and *V. spinulosus*, have centers of distribution in New Guinea, with only the *V. indicus*-group and *V. keithhorni* having extralimital distributions.

Mertens' other subgenera included *Polydaedalus, Indovaranus, Psammosaurus, Empagusia, Dendrovaranus, Tectovaranus*, and later *Philippinosaurus* for *Varanus grayi* (= *V. olivaceus*; Mertens 1959), and *Papusaurus* (Mertens, 1960) for *V. salvadorii* (Table 2.3). By 1960 he recognized 49 taxa (including subspecies), and relationships were based upon few characters, particularly (1) sectional shape of the tail, (2) parietal fenestration, (3) nasal bone condition, (4) height and length of the skull, (5) size and arrangement of supraoculars, (6) scale counts, and (7) position of the nostrils. Unfortunately for contemporary researchers, Mertens diagnosed his taxa based on symplesiomorphies and virtually no synapomorphies.

Mertens worked in the pre-molecular, pre-cladistics era of taxonomy, when relationships were inferred from overall perceived morphological similarities among species. Little, if any, data were taken from life history, physiology, fossils, genetics, or other biochemical sources. His subjective weighting of characters was inconsistent, and he treated many variations, notably ontogenetic ones, as species-level criteria. For example, though Mertens remarked upon the tremendous variation among mangrove monitor lizards (*V. indicus*-complex, vide Sprackland 1995), he alternately established (Mertens, 1926) and removed (Mertens 1942c, 1963) subspecific taxa from them. In essence, Mertens assigned any predominantly black monitor with tiny cream or yellow dots to *V. indicus* (see Mertens 1942c, 1951, 1959, 1971). He was also a firm believer in the subspecies concept (which is still popular with ornithologists generally, and German herpetologists in particular), naming many among the monitors. Over the ensuing decades, most of those subspecies have been either synonymized with other taxa, elevated to species status, or determined invalid. As examples, the "subspecies" of *Varanus exanthematicus* Mertens referred to as *V. e. angolensis, V. e. albigularis, V. e. exanthematicus*, and *V. e. microstictus* have been synonymized into *V. albigularis*, no subspecies (Bayless and Sprackland 2000a, b), retaining *V. exanthematicus*. The *V. salvator swarti* Mertens named based on a written Italian account turned out to be a *V. salvator* with a small-caliber bullet-hole in its snout that was misidentified as a nostril.

Fossil monitors have also suffered from taxonomic uncertainty. The skull of †*Telmasaurus* resembles the Recent *V. albigularis, V. niloticus*, and *Varanus flavescens* in having nearly parallel maxillary bones that give the anterior snout a broad aspect, broad nasal bones, a well-ossified parietal that largely roofs the supratemporal region, and a thick quadrate, and similarities in the prehistoric habitats to sub-Saharan Africa and northern India. The extant taxa also have broad dorsal premaxillary surfaces (this feature is still unknown for *Telmasaurus*), and though the single character is not sufficient to determine taxonomy, it is suggestive that the lizards shared similar ecologies and habitats. The systematic position of *Telmasaurus* has been consistent as a basal anguimorph, usually a varanoid, but variable within that clade. The same can be said of *Saniwides*, "*Saniwa*" *orsmaelensis*, the European "*Saniwa*" *feistei* (Smith and Habersetzer 2021) and *Estesia* (Norell and Gao 1997) *Saniwides* is a basal varanoid, not varanid, and "*Saniwa*" *orsmaelensis* requiring a new genus (see below).

11.2 MACRO-COMPARED WITH MICROMORPHOLOGICAL TAXONOMIES

In comparing the morphological and molecular analyses of varanid systematics, there are obvious discrepancies inherent in both. As an example, consider the taxonomic history of the subgenus *Empagusia*. Originally established by Gray (1838) to include the species now recognized as *V. flavescens* and *Varanus ocellata* (see Bayless and Sprackland, 2000a, b), Mertens resurrected *Empagusia* (1942c:241) for *V. flavescens, V. exanthematicus*, and the subspecies he allocated to the latter, designating *Eimeria flavescens* as the type species. The small suite of characters did not separate the subgenus from others. The species were diagnosed as (1) middle- to large (to 2 m), (2) with blunt, short, and high snouts, (3) enlarged nuchal scales, (4) nostrils as slits, (5) nostrils far from tip of snout and are (6) located laterally, (7) upper tail with keels, (8) supraoculars sub-equal, (9) scales of posterior half of tail not forming annuli, (10) maxilla either mesoprosopic or platyprosopic, (11) septomaxilla partly depressed with a (12) convoluted upper surface, and (13) nasal almost always single. None of these is apomorphic to *Empagusia*, characters 1, 7, and 10 being vague, while characters 2, 3, 4, 9, and 12 are too variable to carry systematic value for establishing a clade. Additionally, each of the characters is applicable to other varanids. In essence, the subgenus as defined by Mertens is based mainly on a superficial resemblance between adults of both taxa.

Sprackland, in performing a combined data set analysis of *Varanus*, obtained a tree (Fig. 11.1) in which Mertens' *Empagusia* appeared as two widely separated taxa (Sprackland 1991b:562). Instead, *V. exanthematicus* was shown as a sister group to *V. niloticus*, while *V. flavescens* was a sister to *V. olivaceus*, reflecting later taxonomic arrangements based largely on molecular evidence. Furthermore, in the absence of molecular data, Sprackland's tree arranged most extant taxa in ways concordant with later studies, including establishing the sister group arrangement of the *V. prasinus* and *V. indicus*-complexes, the close affinity between *V. komodoensis* and *V. varius*, and sister group relationship between the latter and the subgenus *Odatria*, the last elaborated in Sprackland (1991a).

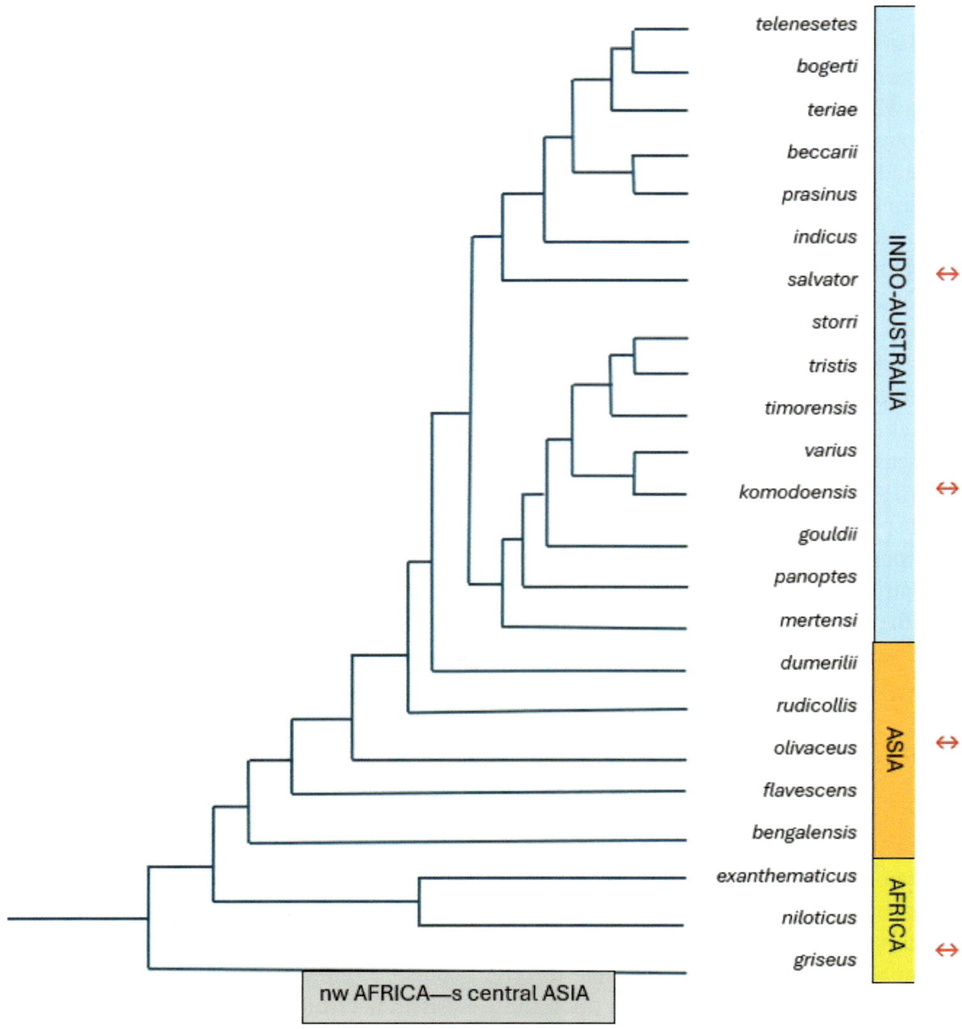

FIGURE 11.1 Combined data cladogram created by PAUP for analysis of the *Varanus prasinus* group (Sprackland 1991b) using 58 morphological, molecular, ecological, and biogeographical characters to find relationships among 23 extant species of *Varanus*. The analysis consistently found *V. griseus* to be the most basal species, congruent with an a priori hypothesis that *V. griseus* evolved in a location from which descendants migrated west across northern Africa, south into Africa, and east through Asia and to Australia and Oceania. Taxa represent eastern expansion as one moves up the cladogram. Exceptions are noted (↔) showing where a taxon diverged from the major east-west dispersal to either move north/south (*V. olivaceus*) or reverse migration westward (*V. komodoensis, V. griseus*). Note also the close relationships implied between the *V. prasinus* group (top 5 species) and *V. indicus*, possibly the earliest study implying such a close affinity.

Systematists have been arguing over the merits of molecules versus morphology since molecular studies began, and though remnants of that debate exist, there seems to be support for more integration of the approaches, provided the methods of data collection and analysis of both are rigorous (van den Ende et al. 2023). Van den Ende et al. "propose indices of molecular disparity analogous to those of morphological disparity. Despite numerous shortcomings… molecular sequence data can be obtained in a more objective, automated and scalable manner than morphological data." One benefit of molecular studies has been the inclusion of calibrated evolution timescales, or molecular clocks (Bromham 2002, 2021), and these are now possible with morphological data (Barba-Montoya et al. 2021). Finnegan et al. (2025) compared the use of individual or grouped mitochondrial genes (mtGenes) with mitochondrial genomes (mtGenome) in phylogeny building, finding that grouped mtGenes outperformed individual mtGenes, and that they "are not appropriate for use as proxies for the mtGenome when tree building." Careful use of molecular and morphological character states has provided increasingly refined results. A useful review of the pros and cons of molecular techniques is beyond the scope of this book; for reviews of the subject, see Rubinoff and Brenden (2005) and van den Ende et al. (2023) (Fig. 11.1).

11.2.1 Toxicofera and the Varanidae

A report by Fry et al. (2005) provided evidence of venom glands in many lizard taxa, including some iguanians and the Anguimorpha. Subsequently, studies have shown that the size and complexity of said glands are not nearly as developed in other lizards as they are in the Helodermatidae or venomous snakes. *Heloderma* and some of the larger monitors have venoms that can cause serious injury, possibly death, in humans. *V. komodoensis* is considered dangerously venomous to humans and the large mammals upon which it feeds (Fry and Scheib 2007, Fry et al. 2005). Though some other workers cling to the idea that the toxicity of the bite is due to the microbes in the saliva of the dragons, this has been disproven (Chandrasekara et al. 2024). Furthermore, the basis for considering many lizard taxa as venomous is not based on the presence of well-developed, venom-producing glands, but on the presence of the genes responsible for the glands (Fry et al. 2009, Srikulnath et al. 2013). Possession of allegedly well-defined dental venom grooves in the fossil †*Estesia mongoliensis*, a character previously deemed apomorphic in Helodermatidae, led Norell and Gao (1997) to remove the species from its provisional place in the Varanidae and group it with the Helodermatidae.

Fry et al. (2005, 2009) considered possession of venom glands so phylogenetically significant that they established the taxon Toxicofera (Fig. 11.2) to include all lizards bearing venom glands: the Anguimorpha, Iguania, and Scincomorpha. Their definition of "venom" is, "a secretion produced in specialized cells in one animal, delivered to a target animal through the infliction of a wound and that disrupts endophysiological or biochemical processes in the receiving animal to facilitate feeding, defense or competition by/of the producing animal." This seems to contradict their comment that venomous designation is not based on the presence of venom glands, but on the presence of genes that code for the glands. The resulting phylogenetic tree conflicts with most other studies of squamate relationships, with several tenuous purported clades. Sweet (2016) makes a strong argument against a Toxicofera clade, noting that such a phylogeny would require the reversal of numerous characters across many clades and have no synapomorphies to justify the model. In my opinion, there is no more justification for grouping all allegedly venom-producing lizards in a single clade than there is for including all the venomous snakes in such a clade. Anyone familiar with snake morphology, behavior, and physiology would be loath to unite such genera as *Hydrophis, Chrysopelia, Tantilla, Vermicella, Crotalus, Bothriechis, Cerastes*, and *Notechis* in a single, allegedly evolutionarily close group.

An interesting outcome of adding Toxicofera to phylogenies is the degree of congruence such trees produce when compared to older ones. Toxicofera has not, for example, resorted in including Varanoids to include Scincids or Teiids with monitors. Instead, it is the association of larger clades, e.g., Scincomorpha, Lacertoidea, Amphisbaenia, and such, that have been rearranged. But perhaps more to the point, character states can arise more than once without implying affinities. Consider how many times characters, including limblessness (25 times; Roscito et al. 2022, Wiens et al. 2002) and viviparity (at least 115 times; Domínguez-Guerrero et al. 2024), have independently evolved across most of the Squamata. For this reason, I do not use the taxon Toxicofera herein, largely because in the context of Varanoidea, it is not necessary.

11.2.2 Taxonomy and Phylogenies

Taxonomy, that part of systematics that recognizes, describes, and classifies species, has been prominent in varanid research since its inception. It is not my intent to review the details of the numerous classifications and species assignations, but to give a brief overview that reflects how collecting, collection bias, and taxonomic approaches have handled the varanids since the early 19th century. The years 1802, 1839, 2010, and 2014 tied for having four new species described. Except for those years, between 1762 and 1940, there was a declining number of species described, a trend that reversed in 1941 and continues through today (Table 11.1). The first group includes the species from the "Great Age of Exploration," when Europeans visited and collected specimens from places new to them. Then as now,

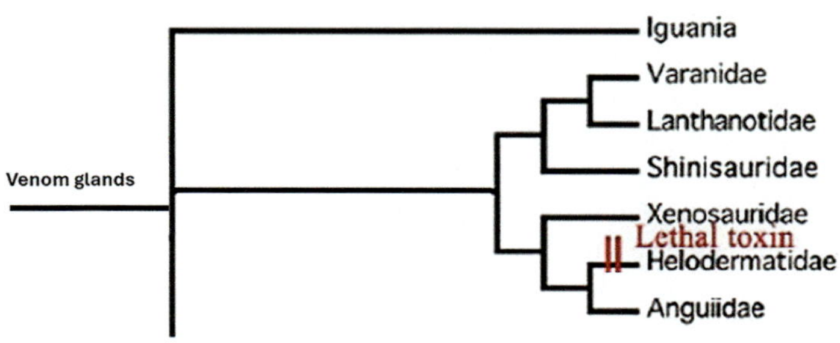

FIGURE 11.2 Cladogram showing the phylogeny of non-snake squamates within Toxicofera based on venom protein evolution (red bars). Aside from exchanging the long-held positions of the Shinsauridae and Helodermatidae, the relevance of Toxicofera to the taxonomy and relationships of Varanoidea (= Varanidae + Lanthanotidae) is uninformative. (From Fry, B., S. Wroe, W. Teeuwisse, M. J. P. van Osch, K. Moreno, J. Ingle, C. McHenry, T. Ferrara, P. Clausen, H. Scheib, K. L. Winter, L. Greisman, K. Roelants, L. van der Weerd, C. J. Clemente, E. Giannakis, W. C. Hodgson, S. Luz, P. Martelli, K. Krishnasamy, E. Kochva, H. F. Kwok, D. Scanlon, J. Karas, D. M. Citron, E. J. C. Goldstein, J. E. Mcnaughtan, and J. A. Norman. 2009. A central role for venom in predation by *Varanus komodoensis* (Komodo Dragon) and the extinct giant *Varanus (Megalania) priscus*. *Proceedings of the National Academy of Sciences* 106(22):8969–8974.)

TABLE 11.1
Synonyms for *Varanus scalaris* and *V. similis* Showing the Frequency of Specific and Subspecific Epithets Being Shuffled in Taxonomic LiteratureData accurate as of June 2025.

Varanus scalaris	Varanus similis
Varanus (Odatria) timorensis scalaris (Mertens 1941)	*Varanus (Odatria) timorensis* similis (Mertens 1958)
*Varanus timorensis scalaris (*Storr 1980)	*Varanus timorensis similis* (Branch 1982)
Varanus scalaris (Storr 1983)	*Varanus scalaris similis* (Cogger 2000)
Varanus timorensis (Cogger 1983)	*Varanus (Odatria) similis* (Berghof 2001)
Odatria kuranda (Wells and Wellington 1985)	*Varanus similis* Smith in (Pianka et al. 2004)
Odatria pengilleyi (Wells and Wellington 1985)	*Varanus (Odatria) similis* (Koch et al. 2013)
Varanus scalaris (De Lisle 1996)	
Varanus (Odatria) scalaris (Hörenberg and Koch 2013)	

specimens were not always described or even recognized as new at first, so we encounter numerous taxa collected in, say, 1870 but not described until 1940. Museum collections remain a valuable source of new materials.

Between the 1890s and 1960s, varanids were described at a rate of about one per five years and were still largely based on discoveries in museums. From the 1970s onward, however, more zoologists and collectors were investigating places that were new to herpetological exploration (though most of those same places had been well collected for shells, insects, birds, and mammals). Consequently, almost every new region and every island presented new species, and in the Indo-Australian region, the number of new monitors was greatest. As taxonomists focused more on phylogenies than naming, they found many more characters for every taxonomic level and, with the addition of genetic and biochemical features, built ever more complex cladograms. Cladistic analysis is a wonderful tool, and I embrace its use (admittedly, it took me a few years to "sign on"). But there is a caveat that I feel compelled to mention. In short, there can, in reality, be only one "correct" tree, and given the number of parsimonious, least-steps trees computers can produce, how close are the phylogenies that are chosen by the computers? My concern with computer-generated trees is based on four points. First, cladistics is based on Occam's Razor, which states that the simplest solution is likely to be the correct one. We are asking programs to find the simplest routes to the evolution of (usually) many organisms. How can we trust such a program when we can see in nature and human activity that the parsimonious path is not always followed? What percentage of the time do organisms behave or evolve in the simplest manner? That brings me to my second point, that of the literally thousands (or more) trees that a computer creates and analyses, and given my first concern, how close is our likely cladogram to actual evolutionary paths? Allowing for just one change to a program to allow a single non-parsimonious event into the process, how many more least-step trees will be generated? In other words, not only could our program-chosen trees be slightly incorrect, but they could also be completely wrong. My third concern involves software choice. In my early days of using phylogenetics, I tested identical data sets on PAUP and Hennig86 (looking to see which one was more user-friendly) and was shocked by the differences in results. I also changed the sequence of the character states, and again, there were different results. More modern programs have likely ironed out some of these pitfalls, but they should not be dismissed simply because they are unlikely. Fourth is my concern for what 1980s Silicon Valley people called GIGO: garbage in, garbage out. Programs only process what we give them, the way we tell them, so (a) how do we find an error to fix it, and (b) if we are unaware of the error, is there any way to run a diagnostic to find it? For (a) I see a solution by comparing results with those of other researchers to see if something anomalous has occurred. Many of us were doing just that after one paper published a cladogram that moved *Heloderma*, traditionally and consistently considered the sister group to Varanidae for over a century, showed up completely off the Varanoidea line, and was placed with the Diploglossa. Every prior and subsequent analysis has returned *Heloderma* to its traditional place, based on studies using different programs, characters, and other parameters, suggesting the Diploglossa result was indeed an error. The issue of higher-level classification has not yet been satisfactorily resolved. Phylogenetic analyses should also compare results between the very different methodologies of molecular and morphological processing (van den Ende et al. 2023). With possible problem (b), it is more likely to discover discrepancies by comparing data output, as in (a), but it is also possible to perform checks by changing the sequence of data input, changing outgroups, and changing the number of character states examined. In each case the resulting phylograms will either remain largely convergent, or diverge, at nodes that help track possible errors.

11.3 SYSTEMATIC REVIEW OF THE GENERA OF THE VARANIDAE

The Varanidae present a relatively good fossil representation because of the fossilization-capable environments they inhabited, and the large size of prehistoric species as contrasted with most other lizards. The Cretaceous †*Estesia* from Mongolia was initially described as "a new fossil varanoid" (Norell et al. 1992), and the similarity to *Varanus* is a striking case of convergence. The shape of the snout and shape of the facial bones do resemble some *Varanus*, but *Estesia*'s lack of a parietal foramen, single lacrimal foramen, and complete, strong postorbital arch have shown it to be more closely allied to the Helodermatidae than the Varanidae (Norell and Gao 1997). Convergence has likewise affected the placement of several alleged varanoid taxa, and a review of the taxonomy of fossil species will show their placement to be rather fluid. The composition of the Varanidae, including its fossils, is still not set in stone, or, more correctly, the matrix is still not completely removed from that stone. Guidelines for including fossil materials have improved and remained essential components for complete data analyses despite their limitations (Donoghue et al. 1989, Estes et al. 1989, Evans 1988, 1996, Lee 2005). Dong et al. (2022) recognize 11 genera in the Varanidae: †*Aiolosaurus*, †*Archaeovaranus*, †*Cherminotus*, *Lanthanotus*, †*Ovoo*, †*Paravaranus*, †*Proplatynotia*, †*Saniwa*, †*Saniwides*, †*Telmasaurus*, and *Varanus*. Based on phylogenetic reviews (Ast 2001, Brennen et al. 2020, Dong et al. 2021, Reippel and Grande 2007, Streicher and Wiens 2017, Vidal 2012, Welton et al. 2013) and my own morphological work, I accept genera as members of the Varanidoidea: *Varanus*, †*Aiolosaurus*, †*Archaeovaranus*, *Lanthanotus*, †*Ovoo*, †*Proplatynotia*, †*Saniwa*, †*Telmasaurus*, and the replacement generic name for "*Saniwa*" *orsmaelensis*, †***Dytikosaurus*** gen. n. I shall present my rationale for excluding †*Cherminotus*, †*Paravaranus*, and †*Saniwides*, from the Varanidae below.

REPTILIA

SQUAMATA Oppel, 1811

ANGUIMORPHA Fürbringer, 1900

Diagnosis – The most recent common ancestor of Platynota and Diploglossa, and all their descendants.

PLATYNOTA Camp, 1923

Diagnosis – the most recent common ancestor of Monstersauria and the Varanoidea and all their descendants.

Description – Squamates that share these character states: elongated, oval external bony naris half or more of snout length; intramandibular hinge between dentary and posterior mandibular bones; teeth broadest at base and with horizontal striations (= plicidentine); teeth do not or just barely extend beneath orbit; replacement teeth from posterior positions; teeth lack resorption pits; vomer extends caudad to near the end of maxillary teeth; lateral configuration of surangular somewhat rectangular with near vertical border at intramandibular joint; quadrupedal and pentadactyl; tail without autotomy planes; oviparous.

VARANOIDEA Camp, 1923

The most recent common ancestor of †*Telmasaurus*, *Varanus*, and all their descendants. lizards, including *Varanus*, †*Aiolosaurus*, †*Archaeovaranus*, †***Dytikosaurus*** gen. n., †*Ovoo*, †*Proplatynotia*, †*Saniwa*, and †*Telmasaurus*, *Lanthanotus*, and all lizards that are more closely related to them than to other Squamata.

Description – Squamates that share these character states: teeth with grooves related to venom delivery; venom glands present, small; two lacrimal foramina on anterior orbital rim; frontals form nearly closed ventral conduit that encloses CN I; precondylar constriction of vertebrae; vertebral condyle much wider than its base, oriented slightly ventrally; tongue at least as long as skull, bifid, and protrusible.

FAMILY VARANIDAE Hardwicke and Gray, 1824

Diagnosis – The most recent common ancestor of †*Telmasaurus grangeri* and *Varanus griseus* and all its descendants, including *Varanus*, †*Aiolosaurus*, †*Archaeovaranus*, †***Dytikosaurus*** gen. n., †*Ovoo*, †*Proplatynotia*, †*Saniwa*, †*Telmasaurus*.

Description – Squamates that share these character states based on the living species, with fossil taxa provisionally included:

Osteology – SKULL. Lacrimal with two foramina; dorsal, more medial lacrimal foramen larger of the pair; premaxilla longer than broad; posterior dorsal process of premaxilla long, more than half the length of the bony naris; nasal bone(s) small, elongate, extending as far as anterior of frontals; nasal process of maxilla rises from a posterior position on the bone; jugal well-developed and not completing postorbital arch (though there is often a cartilaginous section uniting the two), not meeting prefrontal; postorbital + postfrontal fused, at least in adults; gap between vomers, narrow anteriorly, expanding posteriorly to form triangular opening; external bony naris oval in lateral view, widest section posterior; external bony naris roughly half the snout length as measured from rostrum to most anterior border of orbit; external nostril lies along anterior edge of posterior process of maxilla; prefrontals separate maxilla from nasal(s); posterior process of prefrontal extends no more than half the upper rim of orbit, does not meet postorbital + postfrontal; descending processes of frontals vary from triangular to rectangular; well-developed subolfactory processes of the frontal, directed posteromedially and in contact or nearly so; maxilla excluded from suborbital fenestra; maxilla and frontal not in contact laterally; posterior border of the opening for the vomeronasal organ is universally closed by the maxilla and vomer, though maxilla and vomer may not actually touch; interpterygoid space broad, about as wide as anterior of parietal; parietal variable from to barely to greatly elongated, parietal table not or slightly narrower than along frontoparietal suture; parietal constricted medially, single, and pierced by parietal foramen; parietal has small anterior, and frontals posterior, flat articulation processes; parietal always has paired, long posterolateral extensions that border braincase and occiput; stapes thin, unpierced by foramen; 7–9 premaxillary teeth, up to 15 maxillary; teeth do not extend posterior to maxilla; teeth labyrinthine (plicidentine) in section, with slight posterior serrations; pterygoid Y-shaped with medial tubercle or spine (= ectopterygoid facet + pterygoid flange); palpebral present, small; frontals paired; temporal arch present, complete and enclosed between parietal and temporal arch; temporal arch composed of anterior bones and posterior squamosal; squamosal shaped like hockey stick, with pronounced downward posterior curve to where it meets quadrate; supraoccipital universally makes broad contact with the parietal bone; quadrate vertical with enlarged dorsal conch; hypoglossal foramen enlarged, containing hypoglossal and vagus nerves.

MANDIBLE – Dentary at least half the length of mandible, with curved (downward) ventral border; subdental shelf pronounced, thick; anterior dentary symphysis a small, distinct process largely confined to anterior and medial surfaces; intramandibular joint on labial side between dentary and postdentary bones; Meckelian groove present; splenial in Meckelian groove from near anterior end of dentary to the inferior center of coronoid; in lateral aspect the disarticulated surangular bone is greatly expanded posteriorly, the rear of the bone with a large vertical process; splenial posterior process square or rectangular, giving bone a hatchet-like shape; splenial with short, acute posteroventral spine; splenial pierced by anterior inferior alveolar foramen just anterior to vertical process; splenial extremely thin; splenial has restricted kinesis distinct from that of dentary; splenial and dentary form intramandibular joint, expands laterally; mandible posterior to intramandibular joint with with curved (upward) ventral edge; coronoid with two fenestrae, located far back on the bone; coronoid with thin paired anterodorsal prongs the loosely articulate with dentary; coronoid eminence varies in shape and size, but is always prominent and vertical; coronoid with medial vertical posterior descending process that encloses posterior of Meckelian grove and part of posterior splenial; angular almost entirely ventral, may extend beneath dentary; prearticular reduced, does not extend much anterior to coronoid; suture between prearticular and articular often fused or indistinct in adults; retroarticular process short, mostly dorsal, and with distinct concave surface that attaches to quadrate.

POSTAXIAL SKELETON – nine cervical vertebrae; cervical and caudal vertebrae with long peduncles; cervical and second epibranchial lost; vertebrae with greatly enlarged, widely hemispherical condyle; vertebral centra distinctly constricted; ventral surface of vertebrae with Y-shaped surface; sternum narrow, anchor-shaped; clavicle narrow; three pairs of ribs attached to sternum; osteoderms, when present, do not fuse to skull; caudal chevrons and cervical hypapophyses attached only to vertebral centra; pelvis complete; caudal vertebrae without autotomy lines.

SCUTELLATION – Upper head scutes distinct from each other, the configuration highly variable among taxa; scale shape and morphology differ on head, face, belly, dorsum, limbs, and tail; upper head scales firmly attached to underlying bones; supralabials not enlarged, with 5–7 rows beneath eye; supralabials number 22–28; scales with 0–9 cutaneous sensory structures; dorsal scales round or ovoid, surrounded by 2–4 rings of minute micro-scales; ventrals rectangular, smooth or slightly keeled, and arranged in rows and columns. Osteoderms present; those of the nuchal region are the largest. Osteoderms are free, not attached to the underlying bone, and embedded within scales, often absent in some parts of the ventral.

APPENDAGES – Quadrupedal and pentadactyl with well-developed, strong limbs and digits; claws always present, variable in shape and relative length. Tail longer than snout-vent length, proportions and shape variable; tail strong, somewhat to highly prehensile; tail cylindrical (round) in sagittal section near cloacal.

MUSCLES – Tongue long, deeply bifid, thin, protrusible for length of head or more; posterior body of tongue forms sheath adjacent to glottis; tongue surface of smooth, keratinized, stratified squamous epithelium; papillae and taste buds lost; free part of tongue bifid for >50 percent its length; jaw adductor muscles universally attach to the dorsal surface of the parietal; genioglossus lateralis muscle subdivided into separate muscles; anterior head of pseudotemporalis muscle enlarged; gular fold prominent.

OTHER – Oviparous; long egg incubation time, more than 150 days. Carnivorous, largely insectivores, though some also consume fruits; diurnal or slightly crepuscular. Lengths of head and neck nearly equal; viewed dorsally, head triangular or nearly so. Predominantly terrestrial and diurnal.

Distribution – Circum-tropical, with the earliest known fossils from the Late Cretaceous of Mongolia, followed by finds in western North America, Europe, the Middle East, eastern Africa, southern and eastern Asia, Australia, and Oceania. The largest number of species is found in Australia. Extant and fossil taxa are absent from Antarctica, Central and South America, Madagascar, and New Zealand.

VARANUS Merrem, 1820

1820 *Varanus* Merrem, *Versuch eines Systems der Amphibien*. Johann Christian Krieger, Marburg. xv + 191 pp.– Type species (by monophyly): *Varanus varius* (White, in Shaw, 1790).

Description – Varanid lizards with an adult total length greater than 700 mm; nasal bones small, elongate, extending as far as anterior of frontals; teeth present only on premaxilla, maxillae, and dentaries; 7–9 premaxillary teeth, up to 15 maxillary, usually less; teeth pleurodont, labyrinthine (plicidentine) in section, with slight posterior (and sometimes anterior) serrations; long dorsal process of premaxilla tapers, narrowest near nasals; maxilla excluded from suborbital fenestra; posterior orbital process of maxilla bears a pair of small tines, one above the other; nutrient/maxillary foramen largest and most posterior; convex depressions present (carina delimitans excavationis nasalis) lateral to cristae intranasalis of premaxilla; maxillary-frontal process of maxilla does not extend further caudad than posterior orbital process; anterior edge of septomaxilla even with or posterior to protrusion medialis alae intranasalis; lacrimal with two foramina, the upper not entirely confined to lacrimal but bordered by the prefrontal; palpebral present, small, triradiate; jugal well-developed and not completing postorbital arch, not meeting prefrontal; crista proötica extends laterally forming cover over several foramina for cranial nerves; posterior portions of exoccipital and paraoccipital process fuse in adults, obscuring suture; quadrate vertical with enlarged dorsal conch bearing large, arched dorsal head and thick capitum. Dentition pleurodont, replacement teeth emerging posterior to functional teeth. Teeth, when laterally compressed, have a fine serrated anterior, sometimes also posterior, edge. Serrated edges of teeth are reportedly iron-coated in *V. komodoensis* (LeBlanc et al. 2024), possibly present in other large taxa. Vertebrae with median constriction and enlarged posterior condyle; vertebral sockets angled somewhat downward; 9 cervical vertebrae. Pelvis complete. Clavicle slender.

Upper head scutes enlarged, the configuration highly variable among taxa; head scales often fused to underlying bones; scales contain osteoderms; supralabials not enlarged, with five to seven rows beneath the eye; moveable, complete eyelids and external ear opening present; pupil round; gular fold prominent. Dorsal scales are round or ovoid, surrounded by rings of minute scales. Ventrals rectangular, smooth or slightly keeled, and arranged in rows. Tongue long, thin, deeply bifid, and protrusible; posterior tongue forms a sheath for the bifid anterior portion.

Post-cranial anatomy

Quadrupedal and pentadactyl, with well-developed, strong limbs and digits; bipedal posture and locomotion often possible; claws always present, variable in shape and relative length. Tail longer than snout-vent length, proportions and shape highly variable; tail strong, somewhat to highly prehensile, lacking autotomy. Eyes with three color types of opsin-producing cone cells. Chromosome number is $2n = 40$; sex determination by ZZ/ZW chromosomes, the homozygote being female. Females with large paired hemiclitorises.

Distribution – Africa, Europe, Miocene; Australia, Java, Italy, Pleistocene; Australia, Tajikstan, eastern Europe, Russia, India, Pliocene; Africa, Asia, Australia, Melanesia, Micronesia, New Caledonia (possibly introduced by humans), Pliocene to present. The largest number of species is found in Australia. Absent from New World, New Zealand, Antarctica, Madagascar, and northern Asia (Fig. 11.3).

Comments – Prior to the appearance of *Varanus*, ancestral varanids were confined to Laurasia, and though fossils are known from coastal Portugal to India, most are fragments that leave their specific and sometimes generic identities unidentifiable. Two taxa of questionable affinities to *Varanus* are †*V. catalaunicus* (formerly *Iberovaranus* probably a nomen nudem; Delfino 2012) from Spain and †*V. pronini* from Russia, neither represented by cranial material, dates from the Middle Miocene. Limited and non-cranial materials likely make the following names nomina dubia: *V. tyrasiensis* (Lungu et al. 1983); *V. lungui* (Zerova and Čhkhikvadze 1983) and *V. semjonovi* (Zerova and Čhkhikvadze 1983); *V. atticus* (known from a single vertebra) and *V. amnhophilus* (Conrad et al., 2012) are likely junior synonyms of *Varanus marathonensis* (Fig. 11.4).

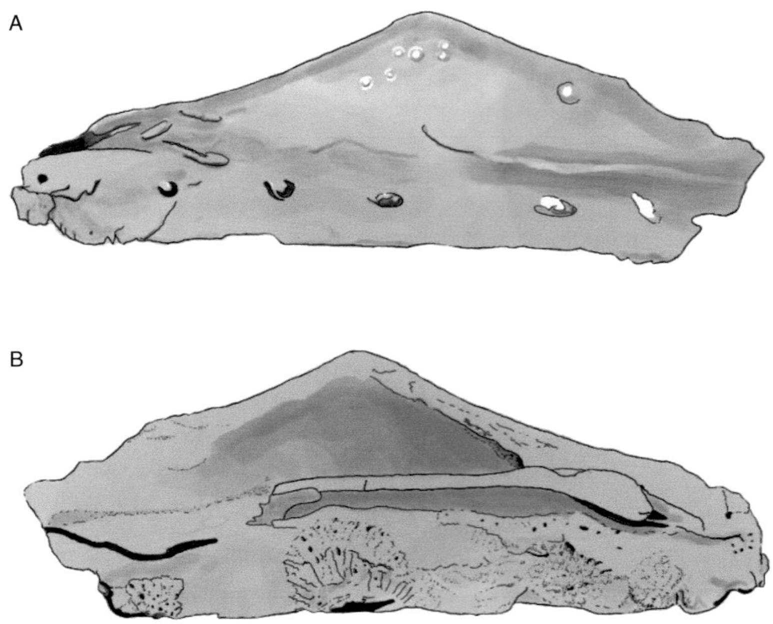

FIGURE 11.3 †*Varanus mokrensis* from the Early Miocene of the Czech Republic is known from several cranial materials that allow considerable point-by-point comparisons with extant *Varanus*. The fragmentary left maxilla, Pal. 1023 from the middle Miocene, in (A) – lateral aspect, anterior to left, and (B) – medial aspect presents varanid features, notably the tooth sockets showing imprints of plicidentine dentition, indications of a depressed nasal recess of the maxilla, broad, thick palatal shelf, thick anterior vomeral facet, enlarged medial recessus nasalis, and well-defined triangular maxillofacial/lacrimal recess. The medial morphology somewhat approaches that of *V. salvator* (see Fig. 5.8B).

Among the best preserved *Varanus* fossils are the Australian *V. prisca* (Figs. 11.5 and 11.6), the jaws (Fig. 11.5) bearing indubitable affinities with living taxa, the differences probably linked with its considerably larger size. Though an example of a Pliocene species, it provides excellent comparative material for the study of gigantism and within-genus outlier variation for monitors. Verifiable varanids became numerous during the Miocene, a time of considerable speciation among *Varanus*, though most taxa are based on material that makes species-level identification nearly impossible. †*Varanus rusingensis* from the Lower Miocene of Kenya represents the first known incursion of the genus into the southern hemisphere (Clos 1992) and possibly the oldest example of the genus, though as I note elsewhere, I find nothing to distinguish the species from extant *V. niloticus*, and several tentative species have been described from the Upper Miocene of Europe and southwestern Asia. †*Varanus hofmanni* (1898) is known from a few distinctly *Varanus*-type vertebrae from Germany.

As *Varanus* dispersed, it became the sole varanid to survive into the Pliocene, by which time it had taken on much of its present global distribution. Where skull material is lacking, three apomorphies of *Varanus* that have been found in fragments of the humerus. These are a proximal foramen on the condyle, a distinct sub-triangular fossa adjacent to the condyles, and an

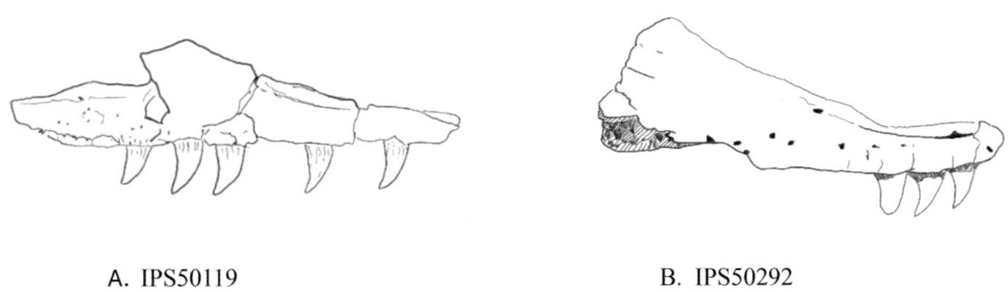

FIGURE 11.4 †*Varanus marathonensis* is represented by numerous fossils from late Miocene to early Pliocene sites across Europe. Illustrated middle Miocene right maxillae in lateral aspect, of A – IPS50119 and B – IPS50292, based on photographs in Villa et al. (2018), with plicidentine teeth clearly visible. The conformation of the dorsal surfaces of the maxillae closely resembles the morphology of *V. bengalensis*, being a long, gradual slope without a distinct, enlarged preorbital process, though without the posterior lacrimal-jugal border available, it is not possible to further refine the comparison. Drawings show slightly reconstructed interpretations based on published figures.

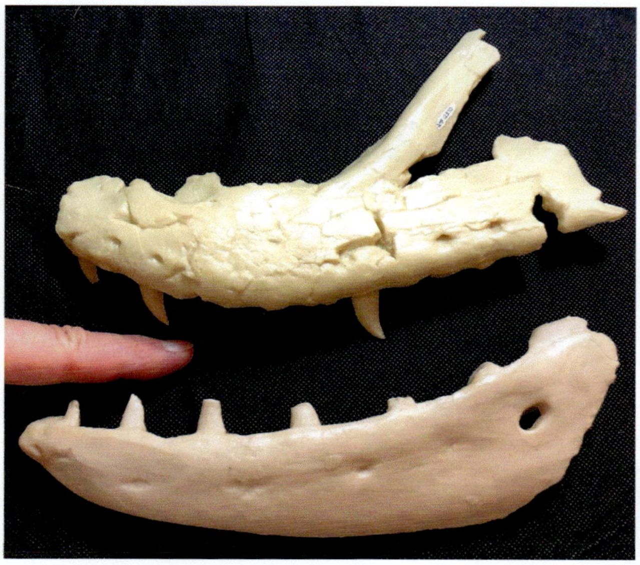

FIGURE 11.5 By far the largest known terrestrial lizard, †*Varanus prisca* (QM F6562, left dentary, and AM F2212, left maxilla, both in lateral view) from the Pleistocene of eastern Australia, including South Australia and (predominantly) Queensland. The minute anterior and posterior dental serrations common to most *Varanus* are clearly visible in the maxillary specimen. All varanids from this time forward are assignable to the genus *Varanus*. Bar = 10 mm.

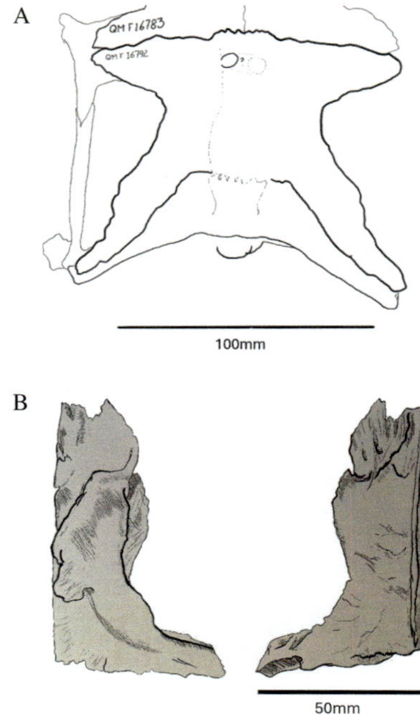

FIGURE 11.6 Skull roof components of †*Varanus prisca* from the Pleistocene of eastern Australia. A – Posterior of frontals, QM F16783 and parietal (left half plus reconstruction of left half), QM F16792, with probable morphology of left supratemporal arch. In comparison with other varanids, including the largest species, the parietal of *V. prisca* differs in having (a) rounded, not squared, edges on the anterior articular processes of the lateral anterior processus externalis, (b) drastic difference in diameters of processus externalis and post-articular dorsal body of parietal table, (c) massive nearly vertical paraoccipital process that is not an extension of raised cristae parietalii, (d) long and deep inferior fossa to paraoccipital process, (e) great thickness of bone (21 mm) at thickest position along frontoparietal suture, (f) lateral parietal flanges not broadly flared, barely beyond parietal table, buy nearly vertical. The partial parietal foramen is slightly off center to the left (solid circle) but may be broad in complete bone (to dotted circle). Note the shape of the lateral articulating surfaces of the frontal and parietal that allow for greater union with the postorbital + postfrontal bones. B – Ventral (left) and dorsal views of frontal, QM F16783. The frontal is extremely thick horizontally, 23 mm at the anterior end the just posterior to the lappet underlying the nasal articulation, 26 mm at the densest point dorsal to the descending processes. The olfactory canal is open broadly in the anterior 60 percent, but the processes unite medially for most of the posterior section. The nasal lappet has two large dorsal foramina and a smaller, anterior-facing one just anterior to the post-lappet dorsal surface.

ectepicondylar crest, often with a distal foramen. The morphologically comparable and numerous Pliocene fossils include †*V. marathonensis* (Fig. 11.4) from central Europe to Turkey, a large species (>2 m) known from western Portugal east to Greece, Georgia, and NE Africa. This species may represent the last presence of *Varanus* in Europe, and has recently been reviewed (Augé 2005, Conrad et al. 2012, Dong et al. 2022, Georgalis et al. 2017, Villa et al. 2018). Additional eastern dispersal is represented by †*Varanus sivalensis* (= cf. *V. bengalensis* or cf. *V. salvator*, Čerňanský and Patnaik 2025, Falconer 1868; Fig. 11.7), one of the largest species, from India. However, *sivalensis* is only known from a partial humerus that, though possessing varanid characters, is insufficient to confirm its inclusion in *Varanus*. Indian maxilla and dentary fossils identified as *Varanus* cf. *bengalensis* (Fig. 11.7), though similar to extant lizards, differ in being a small, possibly subadult, monitor with blunt, closely packed teeth more characteristic of large adults of *bengalensis* and adult *olivaceus*. It is further distinguished from contemporary *Varanus* in having almost all dentary teeth intact (missing only positions 1,5, and 11, with 8 teeth present), teeth in broad horizontal contact with each other, and each tooth with a distinct vertical groove on the lateral surface, each groove beginning about one-quarter the tooth's length above the dental crest and terminating in an apical gap. In each tooth the groove is nearly identical, the edges not showing signs of compression damage or natural wear, and are well defined on both dentary (all) and maxillary (posterior 3; and excepting a more typical second maxillary, only those three remain) teeth, suggestive of venom grooves, a character state common to *Varanus* and, more broadly, Varanoidea/Toxicofera (Arbuckle 2009, Ballard and Antonio 2001, Fry 2007, Fry et al. 2009). If this is the case, then *Varanus* fossils assignable to *Varanus* cf. *bengalensis* warrant description as a new species. Pleistocene fossils represent the earliest known incursions into the Sunda Island archipelago and Australia (and probably

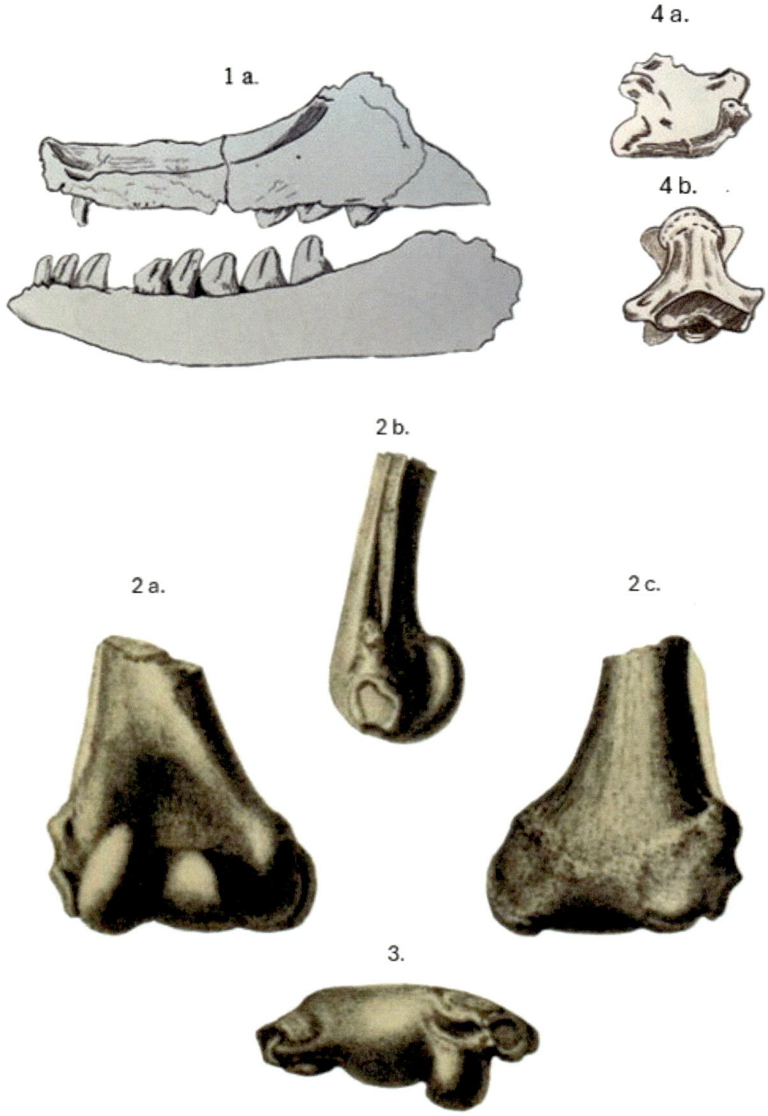

FIGURE 11.7 1. Left maxilla and dentary of *Varanus* cf. *bengalensis*, BMNH Pal. R. 719 from the Pleistocene of Madras, India. These were compared with a recent *V. bengalensis* (BMNH 1931.1.10.4) on site, and other specimens later, and though the bones closely agree, the fossil is likely from a young specimen (dentary = 25 mm), though teeth are far more blunt and closer spaced than is usual for a specimen this size. The distinct vertical slits on the teeth are not seen in any other known varanid. Given the comparative recency of the fossil plus its locality, the likelihood of ancestor–descendant relationship with extant *V. bengalensis* and *V. nebulosus* cannot be discounted. Drawings made with camera lucida at BMNH.

New Guinea). †*Varanus bolkayi* (Fejérváry 1935) of Java and †*V. hooijeri* of Flores represents the earliest Indonesian subfossil records. †*Varanus bolkayi* is only known from a few vertebrae, thus it must be included in the group of doubtful species.

2–7. The Pleistocene †*Varanus sivalensis*, BMNH Pal. R. 40819, 4–7 humerus head, and 2–3 BMNH R. 739a and R. 740, vertebrae from the type locality, the Siwalik Hills of Madras, India. Though there is too little material to make a useful identification, I include the illustration because it represents the entirety of the material for this doubtful varanoid. The vertebrae are distinctly similar to living *Varanus*, having the characteristic broad condyle with strong precondylar constriction, but lacking the slight posteroventral downward angle of *Varanus*. The humeral head is 35 mm at its widest, suggesting the animal was about the size of an adult *V. komodoensis*, in which the humeral head can exceed that width. If so, then *V. sivalensis* significantly exceeds in size the largest of extant *V. bengalensis*. Drawings from Falconer (1868) in the BMNH.

In the case of †*V. hooijeri* the fossil material contains a limited number of well-preserved skull elements. Though Brongersma did a thorough comparison of his Indonesian fossils with the extant species *V. niloticus*, *V. grayi* (= *V. olivaceus*), both illustrated, and the sympatric *V. komodoensis*, his choice of the former pair needs reconsideration because *niloticus* is African and *olivaceus* is Philippine, neither phylogenetically, geographically, or biogeographically (that is, tectonically) close to *hooijeri*. The excellent parietal of hooijeri bears very little resemblance to *komodoensis* at any age in that the living monitor has a broad parietal table, and though it tapers somewhat on the posterior half, a flat surface remains as a nearly rectangular extension all the way from the parietal foramen to the posterior crista juxtafovealis overlying the posterior parietal foramen. In *hooijeri* the anterior of the parietal, or processus externalis that spans the distance between the articular processes with the postorbital + postfrontal and forms the frontoparietal suture is extremely narrow longitudinally, forms a broad, strongly concave V-shaped mesa around and posterior to the parietal foramen, the lateral parietal cristae the joining to form a thin, blade-like table that extends to the posterior edge of the bone. In adult *komodoensis* the pattern is somewhat similar except that the table posterior to the parietal foramen does not come to a sharp single crest, but remains flat and distinctly expanded transversely. This morphology is approached, however, in *olivaceus, niloticus*, and *ornatus*, with relatively minor variations. In *olivaceus* the triangle formed around the parietal foramen is straight-edged and also forms a single blade that exetends to the most posterior edge of the bone. In *niloticus* the triangle has concave edges, as does *hooijeri*, but the single crest is shorter, flaring broadly shortly before reaching the posterior of the parietal. The closest structure is found in the eastern African *V. ornatus*, in which the processus externalis and posterior crest are nearly identical to *hooijeri*. Adult *V. albigularis*, also from Africa, have a very similar morphology and, in some skeletal material, are barely distinguishable from *V. ornatus,* while the closely related and externally similar *V. exanthematicus* has a more common, broad parietal table morphology and lacks the posterior blade. Another feature of parietal morphology shared to some degree by all the species mentioned is the great expansion and lateral downturn of the descending lateral flanges. These processes are dorsal attachment points for huge jaw muscles that in most other lizards attach ventrally. In *olivaceus, albigularis,* and *niloticus*, the processes are larger than in any other varanid, with two exceptions. Adult *V. komodoensis* has lateral flanges so broad that, in dorsal view, they are only slightly less broad than the parietal, and with the narrow posterior parietal table, these present truly huge surfaces that accommodate muscles with extraordinary crushing power. In juveniles and subadults, the flanges are not nearly so pronounced. In young *niloticus* that entire parietal has a nearly rectangular dorsal aspect in which the parietal table is only somewhat more constricted than in most varanids, but the lateral flanges are already huge, so much so that they contribute the rectangular aspect. Though the parietal of *ornatus* and adult *niloticus* very closely parallels that of *hooijeri*, in neither living monitor do the flanges approach the length, breadth, and degree of flaring as seen in *hooijeri*. The Indonesian species has truly massive lateral processes that begin to flare about even with the parietal foramen, then continue to expand to the posterior tips of the paraoccipital processes, probably a unique character state in the Varanoidea. The flanges are thicker than their counterparts in other *Varanus* and ventrally have deep oblong concavities bordered by a thick lateral bony crest. Additionally, between the concavities lies another deep process, a central fossa nearly half the length of the parietal, including paraoccipital extensions. To put these structures in context, the much larger Pleistocene *V. prisca* has a nearly "conventional" varanid parietal (Fig. 11.6), much broader than long, with a broad flat table, deeply indented slightly posterior to the parietal foramen, but instead of converging medially, flare outwards, the lateral flanges gradually descend and are proportional as compared with most *Varanus*. *V. prisca* further differs in having rounded, but not squared, terminal lateral articular processes.

The dentition of *V. hooijer*i recalls that seen in *Varanus* cf. *bengalensis* from India (see above) in being plicidentine, conical, and blunt, and at least some abutting horizontally. The fossils show significant variations, with anterior teeth being either acute or at least less blunt than posterior teeth. The most posterior teeth are barrel-shaped with nearly flat apices. Such dentition is found in adults of several extant *Varanus*, all of which share diets that largely (or seasonally) consist of hard objects such as large snails, crabs, and clams, and include, again, *olivaceus, niloticus, albigularis*, and *ornatus*, as well as *bengalensis* and old *exanthematicus*. This dentition is unlike the acute, typical varanid teeth of most species and the young of the blunt-toothed taxa. The maxilla of *hooijeri* is not well preserved regarding fine points of the anatomy, and though there is similarity to the maxilla of *Dytikosaurus orsmaelensis* n. gen. and comb. nov., this could be largely because those fossils also show signs of wear that obscure important features. The same can be said of the dentary of hooijeri, showing signs of wear that have erased traces of foramina and details of the intramandibular joint.

The right proötic of *V. hooijeri* is remarkably well-preserved compared to the maxillae and dentaries, the angled alar process showing a clear merge to form the crista proötica and the foramen ovalis and more posterior hyomandibular fossa. The opisthostic rises to form a broad, low V with the alar process. In the medial aspect, the palatal foramen for CN VII is distinct, as are the foramina for CN VIII, and the anterior vertical and horizontal semicircular canals. Brongersma was correct in asserting that *V. hooijeri* cannot be placed within any of Mertens' or subsequent subgenera, but it can be assigned to *Varanus* with confidence.

The Pleistocene was a time when Australian varanids began to diversify into the greatest number of taxa and with the most ecological and morphological diversity. †*V. prisca* (formerly *Megalania*; Figs. 11.5 and 11.6) is the largest known varanid as well as the largest known terrestrial lizard, being exceeded in size only by the distantly related marine mosasaurs. The alleged Pleistocene †*V. emeritus* (de Vis 1889) is likely a nomen dubium because the few limb fragments were not designated as types, are not helpfully diagnostic, and given their size and horizon could represent a small *V. prisca* or large remains of extant *V. gouldii* or a similar species.

The genus *Varanus* has, since 1942, gone from 27 recognized taxa (species + "subspecies") to over 100 (Table 11.1), some still designated as subspecies and about 20 undescribed taxa for which I am aware and have photographic, geographic, and some ecological information. The formal designation of subspecies (vide Mayr 1963, Mayr and Ashlock 1991, Wiley 1980) is not something I endorse (Sprackland, 1991a, b, 1992, 1993a, 1994a, 1995b, 2009). Historically, subspecies have so often been elevated to specific status with reliable frequency, for which reason I consider the category no more than the taxonomic equivalent of a question mark, indicative of insufficient information to make a confident species assignment. About this I have written elsewhere:

> "To the best of my knowledge – and I specialize in studying variation – any detailed scrutiny into so-called subspecies has resulted in either a) the recognition that more than one species is involved, b) the recognition that the degree of variation was simply unknown prior to detailed study, or c) both of these situations. …It follows, then, that if we cannot yet agree on how to define "species," the whole point of "subspecies" is a bit of an irrelevance" (Sprackland 2009:51). This approach to subspecies has been addressed in detail elsewhere (Burbrink 2001, Burbrink et al. 2000, 2022, Mohan et al. 2018, but see also Hawlitschek et al. 2012)

Of the 103 recognized species (as of June 2025), 32 of those were first described as subspecies (Table 11.1) and others represent junior synonymies or nomena dubium later raised again to species status (e.g., *V. chlorostigma, V. douarrha, V. kordensis, V. ocellatus, V. tsukamotoi*). A few excellent and impressive tomes have catalogued and described all of the known extant species of *Varanus* as of the time of their writing (Auliya and Koch 2020, Bennett 1995, 1998, Böhme 2003, Eidenmüller 2007, Mertens 1963, Pianka and King 2000), each of which became dated soon after publication because of the number of new taxa discovered. Because of the thoroughness of these works, however, I shall not contribute yet another species catalog and inventory here. The subgenera of *Varanus* and the number of assigned species as of February 2026 include the following:

Empagusia, 4	*Polydaedalus*, 9
Euprepiosaurus, 20+	*Psammosaurus*, 4
Hapturosaurus, 9	*Solomonsaurus*, 1
Odatria (see below), 28	*Soterosaurus*, 11
Papusaurus, 1	*Varanus*, 8
Philippinosaurus, 3	

When Mertens (1942a, 1942c, 1958, 1959, 1962) drew up his original subgenera, most were monotypic, but the new and reassigned groups include only 2 of 11 monotypic subgenera, six of which have at least 8 current species. *Euprepiosaurus* with 20 species and counting is often informally referenced as if it was a genus, i.e., as the Mangrove Monitors or *V. indicus* group, but as there are three fairly distinct groups within the taxon, any one or all of which could have subgeneric reassignment in future, I refrain from referring to it, as it now exists, as a genus-equivalent term. However, the subgenus *Odatria* contains numerous species that, though recently divided into two subgroups, have so many synapomorphies that for convenience's sake I have used the term *Odatria* without constant addition of the cumbersome "subgenus." Unlike other polytypic subgenera for which exceptions to most statements are common, such is rarely the case when I refer to the dwarf Austral-Indonesian goannas. Hence, I describe the group below to clearly explain and define its composition and characters as used herein; this should not be considered elevation to generic status.

SUBGENUS *ODATRIA* Gray, 1838

1838 *Odatria* Gray, Catalogue of the slender-tongued saurians, with descriptions of many new genera and species. Part 3. *Annals of Natural History*, London, 1:388–394. – Type species (by monotypy): *Odatria punctata* Gray= *Varanus tristis* (Schlegel, 1839).

> *Description* – Varanid lizards under 700 mm in total length; nasal bones small, equally long and wide; head wedge-shaped, snout acute, less than 45 percent of length of head; nostril small, round (or nearly so), lateral and situated much closer to rostrum than to eye, and below canthus rostralis; distinct, often acute, canthus rostralis; 7–8 maxillary teeth; long dorsal process of premaxilla barely tapers, is nearly straight, and has a broad (vs. blade like) dorsal surface; crista protica present, reduced; posterior portions of exoccipital and paraoccipital process do not fuse but remain distinct in adults; quadrate more nearly uniformly columnar than that in *Varanus*; largest part of quadrate head is anterior process (= cephalic head); quadrate capitum not or slightly enlarged.

Osteology – premaxilla with broad dorsal ascending process, a pair of large anterior foramina, and a smaller pair of lateral foramina; nasals short, paired, divided by premaxilla; maxilla very slightly less than 50 percent of skull length, platyprosopic, with rectangular ascending lateral process on posterior third of maxilla; ascending process nearly vertical anteriorly, forming posterior border of elongated processus nasalis maxillaria; jugal very nearly meeting postfrontal-postocular; quadrate short, with long quadrate head and very narrow base at mandibular articulation.

Scutellation – head scales small, undifferentiated, and with pebble-like texture; supraoculars not or very slightly different in size from interocular scales; dorsal pattern of linear spots forming bands, the spots becoming hollow-centred rosettes in subadults and adults; tail round (secondarily largely compressed in *Odatria semiremex*) in section, without dorsal crest; caudals subequal in sequential bands, keeled or sharply spinose, but never smooth.

Comments – To this subgenus belong the world's smallest monitors (*O. bushi*, *O. brevicauda*, and *O. caudolineatus*), as well as two species with exceptionally long tails (*O. glauerti* and *O. glebopalma*). In many species, the limbs are comparatively short, and in some, they do not meet when adpressed, a condition unique to *Odatria*. There is also a unique sexual dimorphism seen in *O. acanthurus*, *O. tristis*, and their closest relatives, in that males possess a small circular cluster of enlarged, spiny scales ventrolateral to the cloaca. Two phylogenetic analyses (Brennen et al. 2020, Vidal et al. 2012) have not only disclosed two distinct lineages, but the cladograms are in agreement: small species with keeled or spinose caudal whorls (= *acanthurus* group) and larger, thinner, largely arboreal species (= *tristis* group). For taxonomists so inclined ("splitters") these groups could eventually split, resulting in another subgenus.

Distribution – Extant; *Odatria* are distributed from Timor to eastern Australia, Papua New Guinea, and some adjacent islands. Fossils unknown.

†*AIOLOSAURUS* Gao and Norell, 2000. Fig. 11.8 1a–b

2000 †*Aiolosaurus* Gao and Norell, Taxonomic composition and systematics of Late Cretaceous lizard assemblages from Ukhaa Tolgod and adjacent localities, Mongolian Gobi Desert. *Bulletin of the American Museum of Natural History*, New York 249:97–100. – Type species (by monophyly) = †*Aiolosaurus oriens*.

Diagnosis – Varanid lizards with an anteromedial process of maxilla separate septomaxilla from premaxilla; paired nasals; dorsal septomaxillary foramen strongly reduced and close to midline; anterior surangular foramen with a groove of the coronoid process (Gao and Norell 2000); only one lacrimal foramen confirmed.

Description – As per diagnosis. Length of head of holotype ~62.5mm, about the size of many extant species such as *Varanus dumerilii* and *V. exanthematicus*. Gao and Norell cite a gap between maxilla and premaxilla as a character state, but this condition exists in many *Varanus*, as do anterodorsal foramina and a smooth anterior surface of the premaxilla (contra Bahl 1937, who used few taxa in his study).

Comments – In *Aiolosaurus* we have a fossil that is barely distinguishable from contemporary *Varanus*, much more so than other Cretaceous taxa. In the type description, Gao and Norell (2000:97–100) primarily distinguish the genus *Aiolosaurus* from *Cherminotus*, noting that combining the two in a common taxon "is unsupported by any character evidence," with which I concur.

Distribution: Upper Cretaceous, Mongolia.

†*ARCHAEOVARANUS* Dong, Wang, Zhao, Vasilyan, Wang, and Evans, 2022. Fig. 11.8 2

2022 †*Archaeovaranus* Dong, Wang, Zhao, Vasilyan, Wang, Evans, A new stem-varanid lizard (Reptilia, Squamata) from the early Eocene of China. *Philosophical Transactions of the Royal Society* B3772021004120210041 https://doi.org/10.1098/rstb.2021.0041. Type species (by monophyly) = †*Archaeovaranus lii*.

Diagnosis – This new genus and species is a medium-sized stem-varanid that resembles *S. ensidens* and *Varanus* and differs from helodermatids, *Lanthanotus*, and the Late Cretaceous stem-varanids in combining the following characters: a U-shaped palatine; a short maxillary tooth row confined to preorbital region, bearing well-spaced recurved teeth with basal plicidentine; and splenial positioned anterior to level of dorsal prominence of the coronoid. *Archaeovaranus* resembles *S. ensidens* and differs from *Varanus* in having descending processes of frontal widely separated in ventral midline; separate postfrontal and postorbital; a complete bony postorbital bar; teeth on palatine and pterygoid; posterolaterally placed basal tubera on basioccipital; a single coracoid emargination; having a longer rostrum; two pterygoid tooth rows (versus a single row), and resembles *Varanus* but

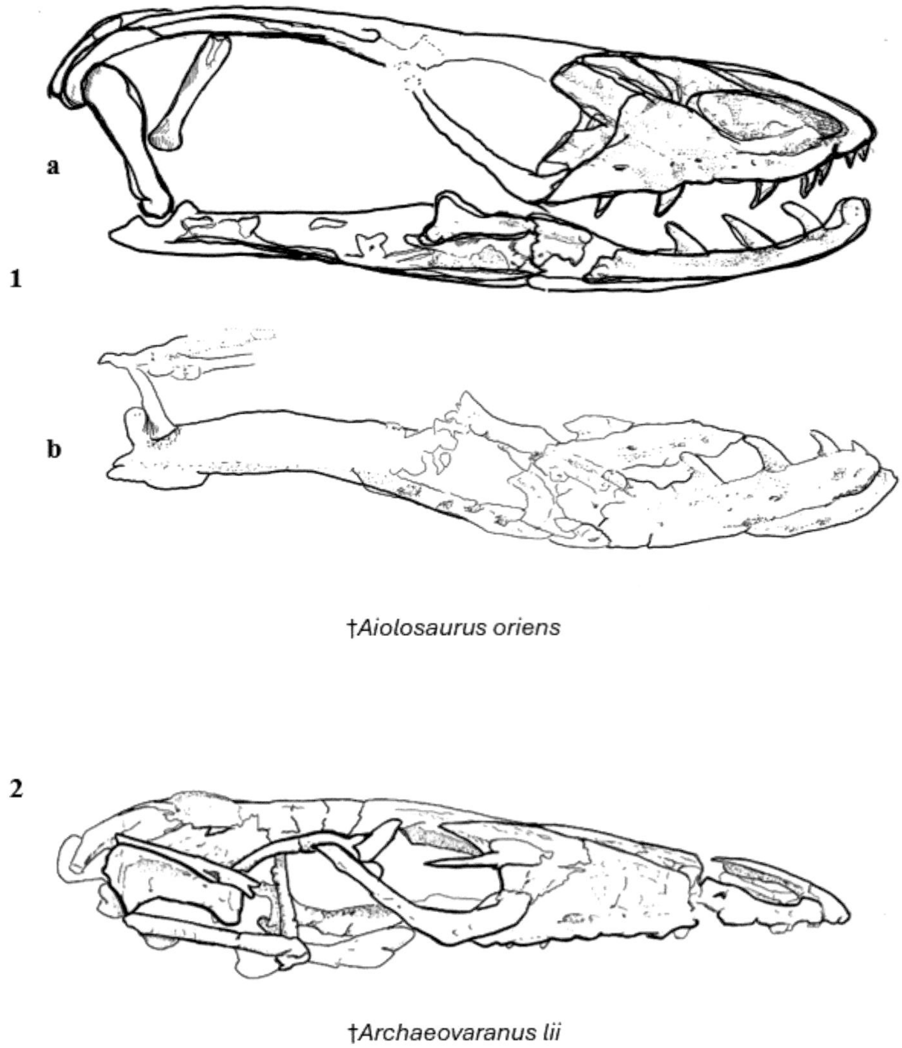

FIGURE 11.8 Fossil varanid that very closely approaches *Varanus* in morphology. 1 – †*Ailolosaurus oriens* from the Upper Cretaceous of Mongolia, (a) in partial restoration based on photographs by Gao and Norell (2000) and (b) showing mandible in ventrolateral view. In morphological features †*Ailolosaurus* shares with *Varanus* the length and proportions of the premaxillary dorsal ascendant process, maxilla with characteristic posterior triangular maxillofrontal process, maxilla with shallow depressed, horizontal nasal excavation nearly half the length of maxilla, slight medial nasal protrusion, distinct lateral cleft between premaxilla and maxilla, spaces for at least 8 maxillary teeth, and maxilla does not approach orbit. 2 – †*Archaeovaranus* from the early Eocene of China is comparatively close to †*Ailolosaurus* in time and geographically, and is also notably similar to living *Varanus*, suggesting a close relationship between the taxa. Character states shared with *Varanus* include length of ascending process of premaxilla, overall premaxillary morphology with thin medial ventral process, the length of the ascending process, maxilla with deep depression, horizontal nasal excavation about one-quarter the length of maxilla, distinct upturned medial nasal protrusion, distinct lateral cleft between premaxilla and maxilla, spaces for at least 8 maxillary teeth.

not *Saniwa* in having a dorsal lacrimal foramen enclosed by the prefrontal and lacrimal (rather than being restricted to the lacrimal). *Archaeovaranus* differs from both *S. ensidens* and known species of *Varanus* in having the forelimb and hindlimb, and the humerus and femur, of near equal length.

Description – A medium-sized stem-varanid combining the following characters: two lacrimal foramina, the ventral of which is fully enclosed by the lacrimal; elongated vomers; intramandibular joint; anterior process of coronoid long; and distinct precondylar constriction on the vertebral centra.

Comments – This species has been recently described, so additional information is not available. However, the describers (Dong et al.) provide further details of character states, morphology, location, and phylogeny. This genus, along with *Aiolosaurus* and *Saniwa*, comes closest to modern varanids in cranial configuration, and of the generic distinctions, some fall within the range of variation within *Varanus*.

Distribution – Early Eocene of China.

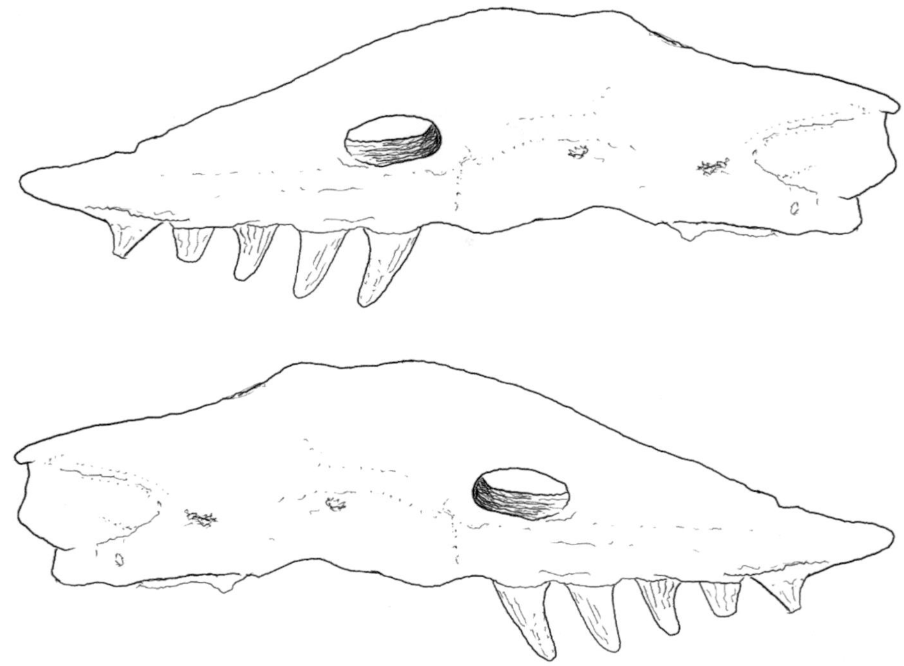

FIGURE 11.9 †*Dytikosaurus orsmaelensis* n. gen., n. comb., showing partial left maxilla in lateral view (upper, anterior at right) and with image reversed (lower) to facilitate comparison with other species. Though long considered a European member of *Saniwa*, plate tectonic-based biogeographical factors make that event nearly impossible, leaving the origins to be found from descendants of Cretaceous east Asian stock.

†*DYTIKOSAURUS* gen. n. Fig. 11.9

2025 †***Dytikosaurus*** *orsmaelensis* (Dollo, 1923), n. comb. Eocene. Type species (by monophyly) = *Saniwa orsmaelensis* Dollo 1923.

Etymology – Greek (δυτικός) for "western" plus (σαύρα) "lizard," referring to its occurrence in western Europe, far west of other known contemporary Eurasian varanids in China-Mongolia.

Diagnosis – Extremely similar to *Saniwa* from which it is distinguished by the following character states: large posterior maxillary foramen; maxillary supradental shelf behind last maxillary foramen, bearing four teeth; posterior of maxilla extends a short distance beneath orbit; lateral parietal crests not raised; lateral processes of parietal large, arched downwards; parietal foramen large, surrounded by ovoid raised ridge; parietal not as broad as in *Saniwa*; teeth with more lateral striations than in *Saniwa*.

Description – European varanid lizards that closely resemble *Saniwa* but differ as per diagnosis, plus presence of large osteodermal scutes on the dorsal and facial surfaces; dorsal surface of skull with distinct "osteodermal incrustations"; crest surrounding parietal foramen more distinct than in *Saniwa*, with irregular surface.

Comments – Varanid lizards previously assigned to the North American genus †*Saniwa*, but for biogeographical reasons could not have shared a recent common ancestor (see Chapter 10 for details; see also Augé et al. 2021) with that genus.

Distribution – The Lower Eocene of Belgium and Germany.

†*OVOO* Norell, Gao, and Conrad, 2008. Fig. 11.10

2008†*Ovoo* Norell, Gao, and Conrad, A new platynotan lizard (Diapsida: Squamata) from the Late Cretaceous Gobi Desert (*Ömnögov*), Mongolia. *American Museum Novitates* 3605, 22pp. Type species (by monophyly): †*Ovoo gurvel*.

Diagnosis – (Modified from Norell et al. 2008.) A varanid lizard distinguished from closely related varanoids by the following character states: nasals paired; presence of aperture between premaxilla and maxilla; dorsal septomaxillary foramen strongly reduced and close to midline. Sharing with *Aiolosaurus oriens* character states, including the presence of a premaxilla-maxilla aperture; dorsal septomaxillary foramen strongly reduced and close to midline. Distinguished from *Aiolosaurus oriens* and other closely related varanoids by the following derived character states: presence of paired mystery bones roofing the nasal/frontal suture; absence of dermal sculpturing on the skull bones (region unpreserved in *Aiolosaurus oriens*); premaxillary nasal

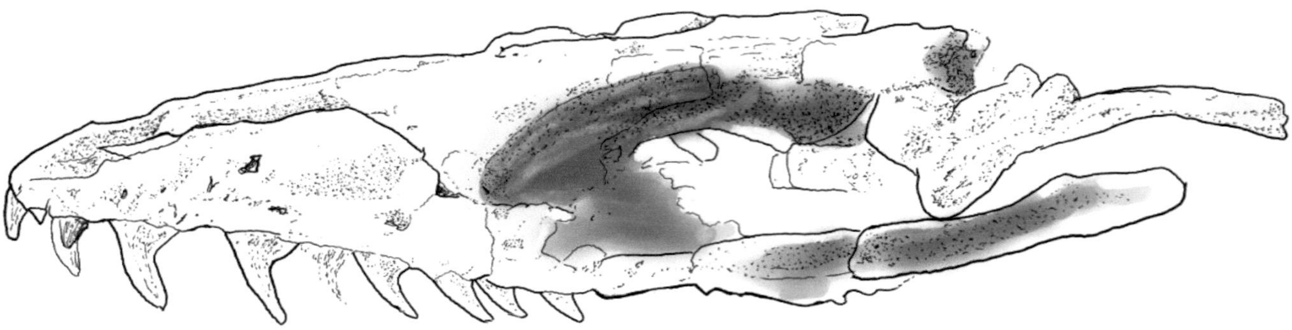

FIGURE 11.10 †*Ovoo gurvel*, a Cretaceous basal varanoid, characterized by the presence of paired "mystery bones," a unique feature unknown in any other squamate. Based on drawings and photographs in Norell, Gao, and Conrad (2008).

process broader than deep; premaxilla-maxilla aperture large and rounded; premaxillary teeth 12 (cf. 7–9 in other varanids); maxillary teeth 12–13 (7–15 in other varanids); most posterior maxillary tooth under extreme anterior of orbit; palatine and pterygoid teeth present, the latter distinctly larger.

Description – A varanid lizard with the unique character of having a pair of "mystery bones" (Norell et al. 2008), having no homolog in any other known squamate. Mystery bones overlie nasals and frontals in a way that looks inserted – that is, the posterior shape of the nasals and anterior shape of the frontals look similar to the shapes seen in other varanids, except that they are separated by a length equal to that of the mystery bones. There is a resulting short point of contact between nasals and maxillae. Dorsal process of premaxilla one-half the length of bony nares; nasals with anterior prongs that extend forward to half the length of septomaxillae; premaxillary foramina present; medial premaxillary tubercle present; vomers long, reaching to anterior rim of orbit; vomers narrow, rectangular, with a slight medial gap their entire length; choanae quite retracted; pterygoids a broad, stout Y-shape, medial process bearing teeth; parietal missing, but frontoparietal suture line of left frontal is straight, as in varanids.

Distribution – Upper Cretaceous of Mongolia.

†*PROPLATYNOTIA* Borsuk-Białynicka, 1984

1984 †*Proplatynotia* Borsuk-Białynicka, Anguimorphans and related lizards from the Late Cretaceous of the Gobi Desert, Mongolia. *Palaeontologica Polonica* 46:28–34. Type species (by monophyly) = *Proplatynotia longirostra*.

Diagnosis – A basal platynotan lizard distinguished from the Varanidae by these character states: maxillae and nasals in contact; alar process of proötics lacking or small; teeth lack basal fluting; osteodermal covering (based largely on Borsuk-Białynicka 1984).

Description – Skull pointed; snout 1.5× length of orbit; external bony nares not much retracted, barely larger than length of septomaxillae; dorsal premaxillary process less than half the snout length, expands to arrow shape at paired nasals; premaxilla with foramina; frontals paired; frontoparietal suture nearly straight, very slightly angled posteriorly from midline; parietal foramen present, oval, and surrounded by low ridge; external parietal crests slightly raised; jugals come to acute point near postorbital; postorbital and postfrontal separate; posterior process of prefrontal and postfrontal widely separated; frontal in broad contact with orbit; nasals, frontals, and prefrontals all in contact with maxilla; dorsal ascending process of maxilla almost entire length of bone; vomers parallel, somewhat rectangular and straight; vomers not in contact posterior to vomeronasal fenestrae; palatines and pterygoids fused (?), with small denticles; basisphenoid small, not as wide as parietal; posteroventral processes of basisphenoid extend to embrace much of the basioccipital; dorsal surface of skull with distinct "osteodermal incrustations."

Comments – This genus represents a suite of plesiomorphies for both Platynota and Varanoidea, yet with sufficient features to be included in the Varanidae as a sister taxon to the clade containing the more recent varanids †*Saniwa*, †*Dytikosaurus*, †*Archaeovaranus*, and *Varanus*. It is not represented by sufficient material to be more specific regarding its relationships.

Distribution – Late Cretaceous of Mongolia.

†*SANIWA* Leidy, 1870. Figs. 11.11–11.13

1870 *Saniwa* Leidy, On *Emys jeanesi, Etnys haydeni, Baena arenosa, Saniwa ensidens*, with a description of the new fossil species. *Proceedings of the Academy of*

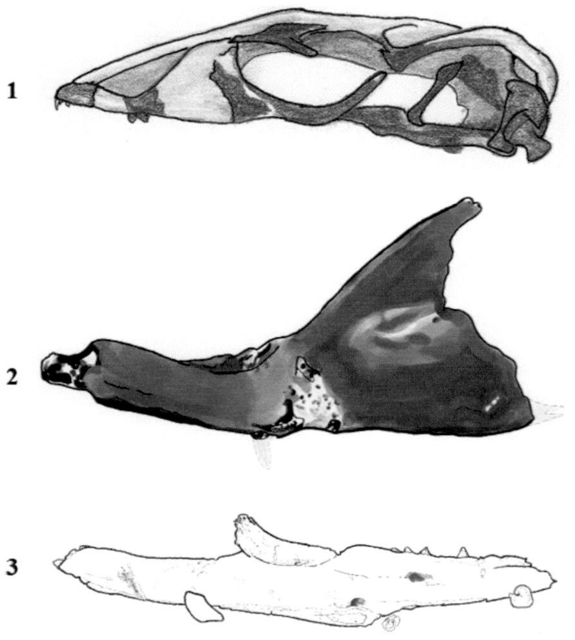

Saniwa ensidens holotype, USNM v 2185.

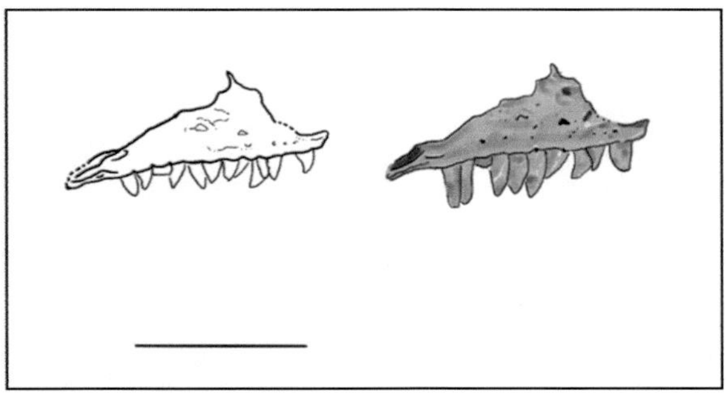

FIGURE 11.11 *Saniwa ensidens* represents a Late to Middle Eocene, perhaps Lower Oligocene varanid with considerable morphological affinities with *Varanus*, and the only confirmed genus of the Varanidae known from North America. Depicted is the holotype, USNM v 2185 in 1 – left lateral view, 2 – detail of left maxilla showing remainder of broken tooth and posterior maxillary-jugal articulation in light gray, and 3 – right dentary in lateral view.

Natural Science, pp. 123–125. Philadelphia – Type species (by monotypy): *Saniwa ensidens* Leidy, 1870.

Diagnosis – Medium-sized and terrestrial lizards that can be distinguished from *Varanus* by the following character states: premaxilla nearly rectangular in dorsal aspect, anterior face curved; ascending dorsal process of premaxilla one-third the breadth of anterior premaxillary process, with broad flat surface; paired lacrimal foramina entirely within lacrimals; palatine and pterygoid with miniscule teeth; postorbitals and postfrontals not fused.

Description – Varanid lizards that exceed 1100 mm in total length and closely resemble extant *Varanus*.

Osteology – Nasal bones small, elongate, extending as far as anterior of frontals; teeth present on premaxilla, maxillae, dentaries, pterygoids and vomers; teeth on pterygoids and vomers extremely small, conical; seven to nine premaxillary teeth, up to 15 maxillary; maxillary teeth extend posterior to anterior border of orbit; teeth labyrinthine (plicidentine) in section, with slight posterior serrations; descending processes of frontals incompletely known; maxilla excluded from suborbital fenestra; lacrimal with two sub-equal foramina that may be entirely contained within lacrimal; jugal well-developed, much more stout than in *Varanus*, and not completing postorbital arch; frontal single; parietal constricted medially and pierced by parietal foramen; temporal arch present; quadrate vertical with enlarged dorsal conch. Intramandibular joint present; splenial sutured to ventral coronoid; coronoid with two fenestrae, located far back on the mandible. Vertebrae with median constriction and enlarged posterior condyle; 9 cervical vertebrae. Pelvis complete. Clavicle slender.

Variation – Holotype (USNM v2185; Fig. 11.11, 1–3): maxilla of common *Varanus* configuration, e.g., having greatly protracted nasal portion with horizontal surface that yield to drastically rising posterior half (= hypsiprosopic incline) and gentle curve terminating in sub-rectangular maxillo-frontal process (Fig. 11.11); maxilla and nasal meet; nasals fused; anterior of nasal with pair of elongated prongs that embrace posterior of premaxillary process; supraorbital large, robust, and triradiate with acute termini; supraorbital extends half the length of orbit; jugal robust, but uncertain if postorbital arch was complete; frontals much broader posteriorly, most constricted at mid-orbits; frontals with feeble ridge along horizontal midline; extremely small (pinhole) foramen lies posteriorly to left of frontal crest (Fig. 11.12); parietal eye just slightly to right of midline; lateral parietal crests long, raised, with sharp edges; parietal arches and degree of constriction like typical *Varanus*; quadrate short, stout, upper crest twice as broad as articular process; posterior crest of quadrate strongly hooked; palatine and pterygoid with miniscule teeth (Fig. 11.12); vomers narrow, not flaring, not in contact medially; pterygoid flange narrow.

AMNH 8688 (postorbital region missing; Figs. 11.11 and 11.12) differs from the holotype in the following features: maxilla unusual with short nasal portion barely one-quarter length of bone; rising process

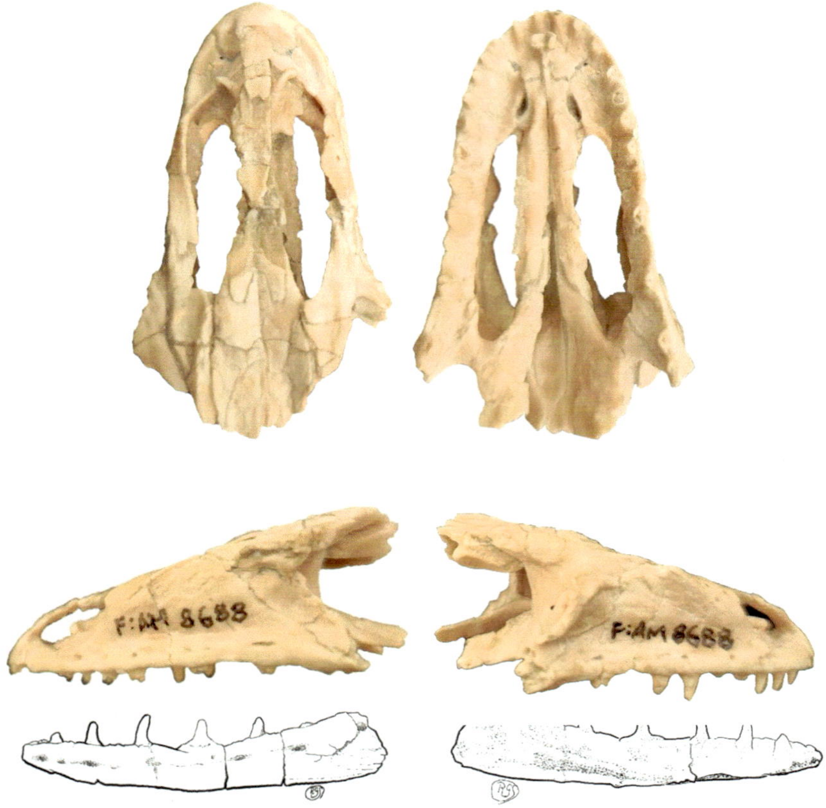

FIGURE 11.12 *Saniwa ensidens* AMNH 8688, right maxilla, lateral view. This specimen is in excellent condition and clearly displays important osteological features of the snout and dentary. Bar = 10 mm.

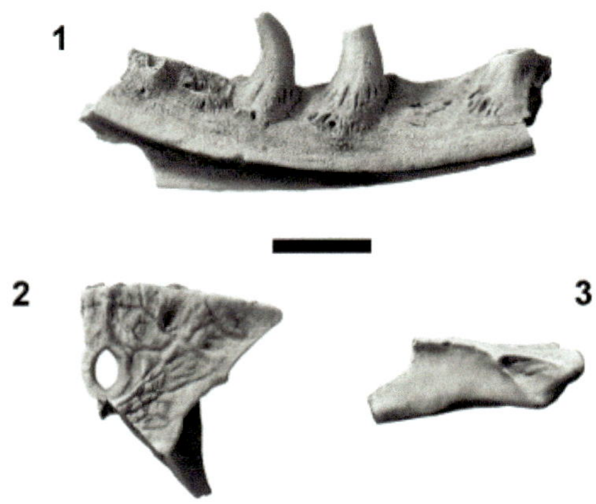

FIGURE 11.13 *Saniwa edura* holotype, PTRM 5379, from the Late Eocene of North Dakota. 1 – Partial left dentary in medial view. 2 – Right fragment of parietal in dorsal view, clearly showing a large, ovoid parietal foramen, and a low, continuous sagittal crest that meets and surrounds the parietal foramen, a character state found in varanids from the post-Cretaceous fossil sites. 3 – Same, in lateral view. This species represents the last known varanid to inhabit North America. Bar = 5 mm. (From Smith, K. and J. Habersetzer. 2021. The anatomy, phylogenetic relationships, and autecology of the carnivorous lizard "*Saniwa*" *feisti* Stritzke, 1983 from the Eocene of Messel, Germany. Pp. 441-506. In: A. Folie, E. Buffetaut, N. Bardet, A. Houssaye, A. Gheerbrant, and M. Laurin(eds). *Palaeobiology and palaeobiogeography of amphibians and reptiles: An homage to Jean-Claude Rage. Comptes Rendus Palevol* 20(23):441–506. https://doi.org/10.5852/cr-palevol2021v20a23.)

of maxilla huge, beginning above 4th maxillary tooth position at 45-degrees, then becomes somewhat horizontal to meeting with frontal and prefrontal; no maxilla-frontal process; maxilla and nasal not in contact; nasals appear divided; anterior of nasals, supraorbitals, and jugals unknown; frontals paired, of even width at least to mid-orbit; no midline crest along frontals; no indication of pinhole foramen in frontals; no visible palatal teeth, those on pterygoid barely distinct; pterygoid flange broad.

Appendages – Quadrupedal and pentadactyl, with well-developed, strong limbs and digits; claws always present, variable in shape and relative length. Tail longer than snout-vent length, proportions and shape variable; tail strong, somewhat to highly prehensile, lacking autotomy.

Comments – *Saniwa* was probably common if the number, condition, and completeness of the available materials are any indication. The prize is a complete, articulated skeleton in situ in an aquatic habitat, surrounded by numerous fishes (FMNH PR2378). The holotype (USNM v2185) is a remarkable, nearly complete, well-preserved specimen, but the teeth and came from another species, not a varanoid, but a xenosaurid, *Restes* (Caldwell 2003), that was

> collected at or near the same location as the *Saniwa*. Later *Saniwa* specimens have been collected with teeth intact on the maxillaries and dentaries. The number of shared characters between *Saniwa* and *Varanus* is numerous and supported by considerable phenotypes. In 1921–22, the holotype of *S. ensidens* was prepared by Smithsonian preparator, Norman H. Boss, to be a museum exhibit, to which the missing bones were crafted to look like varanid structures (Gilmore 1922).

An interesting feature was described by Smith et al. (2018) from partial parietals of two specimens of *Saniwa* referred to *S. ensidens* in the Yale Peabody Collection (YPM 613 and YPM 1074). Both specimens, taken from the Bridger Formation, possess the characteristic parietal foramen, but also an unusual, smaller posterior foramen. This possibly unique condition resulted in "the only known jawed vertebrate with four eyes..." one pineal, the other parapineal in origin. The fossils are about 49mya within a comparatively recent population of *Saniwa*, but are not known in earlier (52mya) lizard (CM 47930), nor the holotype (USNM v2185), also from the Bridger. I have not seen the second foramen in FMNH PR 2378 from 48mya. As I noted above, the long period during which *Saniwa* was extant makes assignation of species a questionable practice, meaning the four-eyed specimens may represent a distinct later species than *S. ensidens*.

If *Saniwa* is a near-unequivocal member of the Varanidae, the number of its species is debated. Because the genus lived for more than 30 million years, it is unlikely that only one species was involved. *S. orsmaelensis* is the name given to specimens collected in the Lower Eocene in Belgium. Estes (1983) accepted recognition of the species provisionally, noting it might be conspecific with *S. ensidens*, but I have herein placed it into †*Dytikosaurus* gen. n. Eight species have been described, all but one synonymized with *ensidens* by Estes (1983). The remaining species (*agilis, crassa, grandis*, and *paucidens*) from the Middle and Upper Eocene of Uinta County, Wyoming, were named by Marsh (1872) and placed in a new genus, *Thinosaurus* Marsh (1872). Estes (1983) considered *Thinosaurus* a junior synonym of *Saniwa*, and in the absence of anything but a small number of vertebrae, Marsh's "species" should be placed in junior synonymy of *Saniwa ensidens* or, possibly, *S. edura* (Fig. 11.13; see below) depending upon their ages. The holotype of *S. paucidens* was known from the "greater part of a skeleton... (now lost)" (Estes, 1983:187) that was only discernible from *ensidens* by its size Gilmore (1928), likely making *paucidens* a nomen nudum. Given the extreme range of sizes in varanids, it is not unlikely that the specimens of *Thinosaurus* are simply lizards at different stages of growth and may represent an example of a paleospecies. However, the presence of numerous sympatric species that would be difficult or impossible to differentiate from fossil material is a possibility; *Aspidoscelis*, *Draco*, *Plestiodon*, and *Sceloporus* are extant taxa for which such a situation would be likely. Resolution of the species-level taxonomy of American *Saniwa* requires access to more, and more complete, fossils. Smith (2006) described and illustrated *Saniwa edura* from a partial left dentary and partial parietal from the Late Eocene of North Dakota, United States (Fig. 11.13). Only two of the three *S. ensidens* I have before me can be compared with Smith's material because AMNH 8688 lacks the parietal and dentary. In both Smith's partial right parietal (PTRM 5380, Fig. 11.13 2, 3) and the holotype (USNM v 2185), there is a low but distinct sagittal crest that meets the anterior rim of the parietal foramen and diverts to produce a low circum- foraminal rim. In the complete skeleton (FMNH PR2378), there is a similar ridge that, as in the holotype, originates at least at the anterior midline of the frontals. Smith's *S. edura* has the posterior of the parietal missing from immediately behind the parietal foramen, so it cannot be seen if the crest continues. However, in both the other specimens, the parietal is flat posterior to the foramen. The holotype and *S. edura* have a slightly curved frontoparietal suture, and all three have pronounced descending supratemporal processes. The holotype has small descending lateral parietal flanges, FMNH PR2387 has large processes (Rieppel and Grande 2007), and *S. edura* seems to have an intermediate form. All three of the *S. ensidens* differ in features that, based on specimen sizes, do not seem to be ontogenetically influenced. Most notable is the variation seen in maxillary conformation (Fig. 11.13), though all share sufficient characters that, without further comparable material to determine otherwise, warrant their inclusion in *Saniwa*. Considering that the Eocene spanned about 22 my, I am inclined to think of the variations in *Saniwa* are indicative of distinct species, excepting *S. edura*, possibly the last American varanid (Smith 2006), all to be provisionally retained in a metaspecies *S. ensidens**. A metaspecies indicates a valid established species and any cryptic species it likely contains (Archibald 1994, Lee 1995, Sprackland 1995b:65).

Saniwa has long been recognized as closely related to *Varanus*, and justification for placing it in its own genus (and subfamily, Saniwinae) has been based on minor characters (Estes 1983, Gilmore 1928, Romer 1956). I present the given arguments for separation or combination and my responses to each, those in boldface representing what I consider the six valid criteria for retaining two distinct genera.

1. "The premaxilla of *Saniwa* has only six teeth, while *Varanus* has seven to nine." This is incorrect; Saniwa has seven premaxillary teeth or at least sockets for the same. The

holotype does possess only six teeth, but there is a clear and empty seventh socket present. Like the pterygoid teeth, this would be easy to miss if you weren't looking for it.

2. "*Saniwa* has separate postorbital and postfrontal bones, while these fuse to form a postorbitofrontal bone in *Varanus*." This is not true for all *Varanus*, and especially so for neonates and very young, so it cannot be used as a trait upon which to base a genus.
3. ***Saniwa* has a toothed palate**; *Varanus* does not.
4. "*Saniwa* has a parietal foramen on the frontal-parietal suture." This is incorrect and possibly slipped past editing by Gilmore (1928). The foramen is quite distinct and well inside the parietal bone.
5. "Post-cervical vertebrae are relatively longer in *Saniwa* than *Varanus*." Not always true when near-equally sized specimens are examined.
6. **Dorsal lacrimal foramen wholly enclosed in lacrimal in *Saniwa***, but within lacrimal + prefrontal in *Varanus*. This is a consistent difference in character states between the taxa.
7. "Vomers wider in *Saniwa* than in *Varanus*." Not universally true. The vomers are wide in some large monitors, including *V. indicus*, but also in many smaller species, including *V. caudolineatus*.
8. "Retroarticular process unlike *V. salvator*, but similar to *komodoensis, varius, indicus, exanthematicus*," therefore not a useful genus-level diagnostic.
9. Coracoid of *Saniwa* has only one fenestra, as is true in some *Varanus*.
10. **Maxilla of *Saniwa* extends suborbitally** and has suborbital teeth.
11. **Maxilla of *Saniwa* has a long zygomatic process**.
12. *Saniwa* has long neural spines. This and the next character state have been quoted by various authors; this statement is true, but the following is not.
13. *Saniwa* has low neural spines.
14. *Saniwa* has reduced intercentra.
15. ***Saniwa* has rudimentary zygantra and zygosphenes**.

Distribution – Late to Middle Eocene, perhaps Lower Oligocene; North America: Wyoming, North Dakota; perhaps California, Upper Eocene. These data need reexamination and confirmation.

†*TELMASAURUS* Gilmore, 1943

1943 †*Telmasaurus* Gilmore, Fossil lizards of Mongolia. *Bulletin of the American Museum of Natural History* 81:361–384. –Type species (by monotypy) = *Telmasaurus grangeri* Gilmore, 1943.

Diagnosis – Basal varanoid lizards that differ from *Varanus* in the following character states: frontal single, extending well beyond anterior orbital rims; descending processes of frontal do not approach meeting medially; postorbital arch complete (?).

Description – Extinct lizards with a frontal bearing low crest on dorsal and ventral surfaces, the dorsal extending caudad to slightly past parietal foramen; frontals with well-developed ventral olfactory conduit that is not closed ventrally; lacrimal foramen confined to lacrimal; median to small lacrimal foramen is bordered by lacrimal and prefrontal bones; prefrontal and postorbital + postfrontal do not meet; parietal crests form continuous curvature from frontal suture to end of paraoccipital process; lateral parietal flanges minute; supraorbital process of prefrontal extends about 75 percent of the orbital rim; frontal and postorbital do not meet, but approach much closer than in *Varanus*; jugal laterally compressed; alar processes of supraoccipital reduced; squamosal reaches forward to a point even with parietal foramen, contacts postorbital-postfrontal broadly; squamosal not strongly arched into hocky-stick shape; jugal and prefrontal in lateral contact on anterior orbital rim; postorbital-postfrontal sutures visible; posterior portion of pterygoid and palatal plate of pterygoid with numerous tiny conical structures; anterior of pterygoid very broad, nearly triangular.

Comments – Even with the snout elements missing, *Telmasaurus* features character states one would expect in a basal varanoid, including a simple frontoparietal suture, large orbits that exclude the maxillae, united (possibly fused) postorbitals + postfrontals, a centrally constricted frontal, large descending frontal processes, paraoccipital processes for nearly a 180-degree arc, and well-developed supratemporal arches.

Distribution – Late Cretaceous of Mongolia.

FAMILY LANTHANOTIDAE Steindachner, 1877

1877 *Lanthanotus* Steindachner, 1877 Über zwei neue Eidechsen-arten aus Süd-Amerika und Borneo. *Denkschriften der Kaiserlichen Akademie der Wissenschaften, Mathematisch-Naturwissenschaftliche Classe*, XXXVIII, Bild I, Abhteilung 4:1–5, Taf. 2, 5 Juli 1877. Preprint.

Diagnosis – Varanoid lizards readily distinguished from other varanoids by the following character states: snout broad and blunt; head widest across the orbits; body cylindrical, slightly depressed dorsoventrally; body with heterogeneous scalation; two or three longitudinal rows of enlarged, oblong scales from occiput (dorsally) and axilla (lateral) to base of tail, becoming smaller and extending caudad; tail cylindrical, scales forming distinct, equally wide whorls; head scales uniform, lacking differentiated labials, supraorbitals, or parietals; pineal eye and pineal foramen absent; external ears concealed; nostrils tiny, round, dorsally situated in low prominences; eye tiny, dorsolateral; lower eyelid transparent; no gular fold; fore- and hind limbs nearly equal in length.

Description – Varanoid lizards with an adult total length less than 500mm; paired lacrimal foramina within lacrimal bones; premaxilla broad with dorsal process no more than half the length of the bony nares; nasal single, broader than dorsal process of premaxilla and anterior half of paired frontals; jugal forms complete postorbital arch, meets postfrontal dorsally; postorbital absent; no temporal arch; squamosal rudimentary; vomers narrow anteriorly, then flare broadly to suture with palatines; palatines and pterygoid with minute teeth; no gap between vomers; in ventral view, vomers conceal nasal; external bony nares oval, broad, dorsolaterally oriented; process prefrontalis of maxilla half the length of maxilla, roughly rectangular without distinct apex; prefrontal separates maxilla from nasal; posterior process of prefrontal long, nearly length of orbit, and meets postfrontal; frontals with large triangular descending processes that do not meet ventrally; interpterygoid space broader than parietal; parietal elongated; parietal foramen lacking; stapes present; ventral surface of mandibles with single, slight convex curvature; splenial reduced and entirely anterior to the medial intramandibular suture; Meckelian groove present, reduced compared with *Varanus*; coronoid somewhat rectangular in lateral view; coronoid eminence not much differentiated from dorsal surface; phalangeal formula for manus and pes 2-3-4-4-3.

Nostrils tiny, dorsally situated within small prominences; eyes tiny, pupil round; lower eyelid clear; external ear opening lacking; excepting serial enlarged tubercules, head scales small, round, and undifferentiated; temporal regions with 3–5 (usually 4) greatly enlarged tubercules in a row; three rows of similar enlarged tubercules extend from near axilla to groin, the two dorsal rows largest; body elongated, cylindrical to slightly depressed; snout blunt and broad; limbs short, when adpressed do not meet; claws tiny; tongue bifid, protrusible; ventral scales in rows and columns, similar to *Varanus*; tail cylindrical, with annuli of large and small scales; scales with osteoderms; no gular fold.

LANTHANOTUS Steindachner, 1877. Figs. 11.14 and 11.15

1877 *Lanthanotus* Steindachner, 1877 Über zwei neue Eidechsen-arten aus Süd-Amerika und Borneo. *Denkschriften der Kaiserlichen Akademie der Wissenschaften, Mathematisch-Naturwissenschaftliche Classe*, XXXVIII, Bild I, Abhteilung 4:1-5, Taf. 2, 5 Juli 1877. Preprint. Type species (by monotypy) = *Lanthanotus borneensis.*

Diagnosis and description – As for the family.

Distribution – Extant, widely distributed across Borneo, including Sarawak, Brunei, Kalimantan, and possibly Sabah.

INSERTAE SEDIS †*CHERMINOTUS* Borsuk-Białynicka, 1984

1984 †*Cherminotus* Borsuk-Białynicka, Anguimorphans and related lizards from the Late Cretaceous of the Gobi Desert, Mongolia. *Palaeontologica Polonica* 46:60–65. Type species (by monophyly) = *Cherminotus longifrons.*

Diagnosis – "The skull morphology of *Cherminotus longifrons* is poorly understood. Consequently, published diagnoses are inadequate," (Gao and Norell 2000) remains true for the species. The condition of IGM 3/170 (MAE 24/93-26), however, displays features that were not present in the holotype material (Borsuk-Białynicka 1984). Based on this, the diagnosis distinguishes *Cherminotus* from *Lanthanotus* in having a long squamosal, extending anterior to postorbital; supratemporal arch present; parietal foramen present in the center of the parietal body; paraoccipital processes of the parietal extend posterior to the occipital condyle to near the length of the atlas. *Cherminotus* differs from *Varanus* in having a single lacrimal foramen, frontals with large ovoid descending processes that do not meet or nearly meet ventrally; jugal forms a complete postorbital arch; jugal contacts postfrontal; no gap between vomers; external bony nares roughly rectangular; process prefrontalis of maxilla absent, entire dorsal edge level; splenial reduced and extends to just anterior of coronoid eminence.

Description – Varanoid lizards probably smaller than 450 mm, with a single lacrimal foramen; premaxilla broader than long; anterior process of premaxilla broad, dorsal process thin, about half the length of the bony nares; nasals

FIGURE 11.14 Live *Lanthanotus borneensis*. (A) in lateral view while submerged. (B) submerged with transparent lower eyelid clearly visible. (C) dorsal view next to 15 cm ruler. (D) extreme close-up of head. Note (1) lack of tympanic opening, (2) clear lower eyelid with pupil visible, (3) very small nasal opening, and (4) great variety of enlarged nuchal tubercles, all features absent in known *Varanus*.

paired, not or barely wider than anterior half of frontals; teeth slightly laterally compressed and recurved; vomers much longer than broad, uniform in width their entire length; vomerine suture with palatines narrow; in ventral view vomers do not conceal nasals; external bony nares comparatively narrow compared to *Lanthanotus*, longer than broad; prefrontals broadly separates maxillae from nasals; posterior process of prefrontals long, less than length of orbit, and does not meet postfrontals by a short gap; complete postorbital arch; interpterygoid space as broad as parietal; parietal elongated, posterior of parietal table much narrower than along fronto-parietal sutures; parietal foramen present; stapes unknown; ventral surface of mandibles with slight convex curvature along dentary; minor concave curve posterior to intramandibular joint; splenial crosses intramandibular suture; Meckelian groove present, more pronounced than in *Lanthanotus*; coronoid with distinct triangular eminence in lateral view.

Comments on "lanthanotids." – *Lanthanotus* has been an enigma for most of the time since its description in 1877, compounded by the rarity of study specimens until 1961. Its alleged position as a putative ancestor to snakes (McDowell and Bogert 1954, Reippel 1980) has made it the focus of considerable attention, and it has relatively recently been reassigned to a clade far removed from serpents (e.g., Caldwell 2021, McDowell 1972b). *Cherminotus* was described as closely allied to *Lanthanotus*, though any morphological similarities are limited. Of the 29 character states I have listed in describing the Lanthanotidae (and by extension, *Lanthanotus*), fully 17 differ between the genera. Subsequent to its description, *Cherminotus* was repeatedly paired as a sister group to *Lanthanotus*. Why? The major phylogenetic studies (Conrad and Grande 2008, Dong et al. 2022) have been ambiguous about the relationship, though it has been disputed (Gao and Norell 2000). Given the incomplete nature of *Cherminotus* material and the numerous soft anatomical apomorphies describing *Lanthanotus* that cannot be compared with *Cherminotus*, I exclude the latter from the Lanthanotidae. The number and degree of

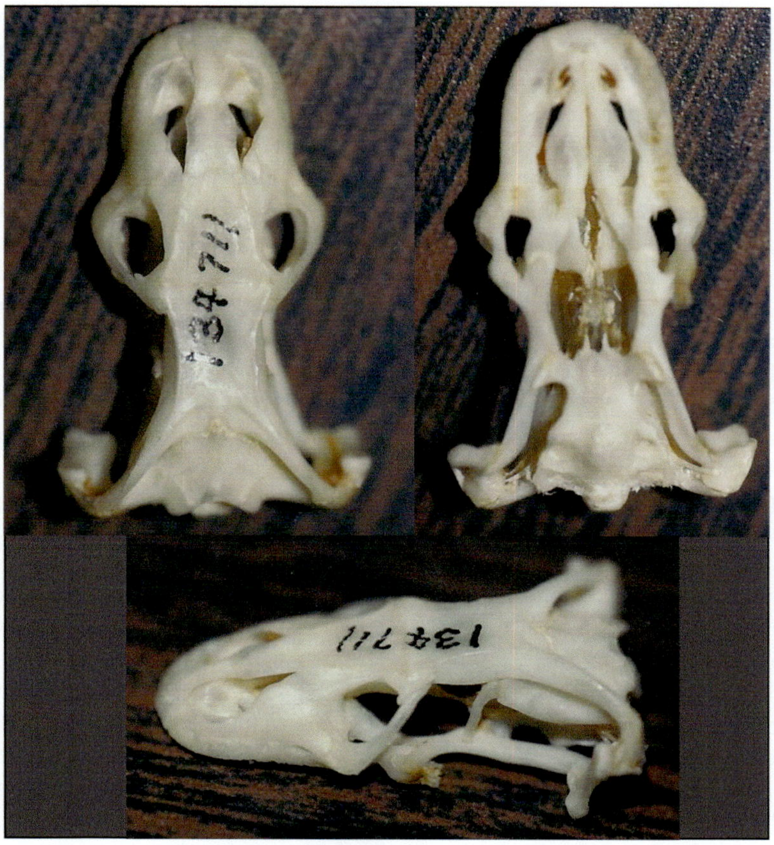

FIGURE 11.15 The skull of a juvenile *Lanthanotus borneensis* (FMNH 134711) in dorsal, ventral, and left dorsolateral views. Not the absence of supratemporal arches and the broadly expanded jugals.

morphological features that distinguish *Lanthanotus* warrant it remaining a monotypic taxon separate from the Varanidae. Comparing available cranial features of *Cherminotus* and *Varanus*, I find 14 of 46 differ, some significantly, e.g., in *Cherminotus* one versus two lacrimal foramina, complete postorbital arch, no gap between vomers, no process prefrontalis of maxilla, and reduced splenial clearly distinguish it from the Varanidae but warrants inclusion in the Varanoidea.

INSERTAE SEDIS †*PARAVARANUS* Borsuk-Białynicka, 1984

1984 †*Paravaranus* Borsuk-Białynicka, Anguimorphans and related lizards from the Late Cretaceous of the Gobi Desert, Mongolia. *Palaeontologica Polonica* 46:14–15. Type species (by monophyly) = †*Paravaranus angustifrons*.

Diagnosis – A small putative platynotan with varanid-type tooth replacement; retracted external bony nares; huge orbits. Differs from *Varanus* in the following character states: proötic alar processes lacking; unusual frontoparietal suture of paired hemispheric shapes; frontal single; frontal with greatly elongated anterior and tripartite posterior processes along suture lines; parietal small, lateral crests angled sharply medially; nasal fused, leaf-shaped; vomers strongly Y-shaped, anterior portion in contact, posterior widely divergent; maxillae without dorsal rise.

Description – A lizard with little resemblance to *Varanus* or related genera. Snout nearly as long as orbit; snout tapers to near-acute point, much narrower than width across orbits; anterior of premaxilla incomplete, but dorsal remnant varanid-like; posterior of nasal longer than broad, with posterior medial crest; nasals fused; gap between nasal and prefrontal; bony external nares almost the length of snout, more rectangular than oval; anterior half of frontal extremely narrow, barely or not wider than posterior part of nasal; space between pterygoids tear-shaped, narrow anteriorly; no alar processes on supraoccipital; paraoccipital processes of parietal form a 180-degree angle.

Comments – Borsuk-Białynicka, in describing the family, genus, and species, states that "the dorsal surface of the prefrontal is rhomboid in shape owing to the reduction of its anteromedial part caused by the extreme elongation of the external nares. This, along with the huge orbits and the large supratemporal fossae, gives the skull its varanid appearance"

(Borsuk-Białynicka 1984:16). However, those very character states make †*Paravaranus* appear extremely unlike *Varanus* based on my observations. The main point of this book is to document the breadth of morphological diversity within the Varanidae, the vast majority relating to the extant genus *Varanus*, which has allowed me to record features rarely reported in paleontological accounts of the lizards. Among these are the number and position of premaxillary foramina, variation in the ascending dorsal process of the premaxilla, shape and medial variation of maxillae, intricacies of parietal structure and proportions. Based on that information, it is clear that the overall skull shape in dorsal aspect resembles embryonic monitors and, except for the shortness of the snout, *V. griseus* itself is characterized by a wide orbital region. It is therefore safe to say that *Paravaranus* may display, at best, outlier morphology for the varanids. Despite the numerous missing skeletal elements, those present that bear no appreciable resemblance to the Varanidae include the nasal, frontal, parietal, vomers, massive jugals and pterygoids, and an abbreviated acute snout. For all these reasons, I exclude *Paravaranus* from the Varanidae and Varanoidea; its status as a platynotan remains to be demonstrated.

INSERTAE SEDIS †*SANIWIDES* Borsuk-Białynicka, 1984

1984 †*Saniwides* Borsuk-Białynicka, Anguimorphans and related lizards from the Late Cretaceous of the Gobi Desert, Mongolia. *Palaeontologica Polonica* 46:51–55. Type species (by monophyly) = *mongoliensis*.

Diagnosis – Medium-sized varanoid-like lizards with nasals and maxilla in contact; vomers not in contact medially; pterygoids and palatines with small teeth; descending processes of frontals large, do not meet medially; maxillary teeth reach beneath the anterior rim of the orbit.

Description – Premaxilla with very thin dorsal ascending process that just meets the divided nasals; external bony nares short, less than half the length of the snout; anterior of maxillae broad with wide nasal excavation; posterior of maxillae with steep triangular maxilla-frontal processes; frontals extend well anterior to orbits, each with two long anterior prongs; no median crest along frontals or parietal; postorbital + postfrontal united; straight frontoparietal suture; pineal foramen small, present; "variable number of lacrimal foramina" (Borsuk-Białynicka 1984); pterygoids stout, barely Y-shaped, medial processes with teeth; maxillary process of premaxilla rudimentary; sockets for at least ten maxillary teeth, four premaxillary; teeth barely recurved, semi-conical; braincase narrow compared to *Saniwa* and *Varanus*; alar processes present, oriented anteriorly; basisphenoid ventrolateral processes long, embrace most of basioccipital; pterotica of exoccipital wings reduced; quadrate stout with broad ventral articulation, cf. *Varanus macromaculatus*.

Comments – Though numerous morphological features can ally *Saniwides* with the Varanoidea, there are sufficient character states to include it in the Varanidae.

11.4 UNRESOLVED SPECIES-COMPLEXES

The number of recognized varanid species has increased from the time of Mertens' monograph (Mertens, 1942c) from 23 to over 85, and numerous species have been discovered but remain to be named. Thus, there are two sets of new species waiting to be studied: those being collected presently, or soon to be collected, and those residing in museums already but that have not been formally recognized. Of the latter, some are known to be new to science, but for different reasons have not yet warranted the attention of a taxonomist; the remainder are lying hidden and undetected for what they are. It is still possible to find "new" species in museums that were collected many decades prior.

Why this delay? In part because the contemporary process of naming a new species requires considerably more work than it did even as recently as the mid-20th century. Consider this taxonomic tale from the 19th century: The Green Tree Python (*Morelia viridis*) was named by Dutch zoologist Hermann Schlegel (1872:54), who, decades earlier, also named the Green Tree Monitor, *V. prasinus* (Schlegel 1839, in 1837–1841). This is the complete text that gives him credit for naming the python species:

> Wij maken nog gewag van eene andere nieuwe sort van de Aroe-eilanden, *Python viridis*, omdat zij in hare groene met witte viekjes afgebrokene kleur, de grootste overeenkomst met Boa canina [sic] van Zuid-Amerika heft.[1]

Absent from that account are numerous features that are generally required today: designation of a holotype specimen and where it is deposited, diagnosis, comparison with similar species, and a description, though he does note a type locality, the Aru Islands. Descriptions, even at that time, customarily included measurements, scale counts, number of teeth, color in life or, if unavailable, in alcohol (which is why several species were named "*azureus*" when

[1] Translation: "We mention another new species from the Aru Islands, *Python viridis*, because in its green color with broken white spots there is similarity to *Boa canina* of South America."

in life they were bright green), sex, and approximate age. Most descriptions now also include genetic data, including material deposited onto GenBank, and often a phylogenetic analysis of where the new species fits within its genus or family. Spurious taxonomic actions have occurred in the recent past, in which an author merely claimed that a specimen described (but not named as a new taxon) is to be the holotype, without the describer ever actually having seen the specimen or providing a description (Wells and Wellington 1985; but see also Sprackland et al. 1997b and ICZN 2014).

That said, there are a few notable problems in varanid taxonomy that have been known for several years but have been adequately complex to defy resolution to date. I present details here for three such groups.

11.4.1 *Odatria tristis/Odatria scalaris*-Group

The subgenus *Odatria* represents the most species-rich assemblage within the genus *Varanus*, and from the time the group was defined by Mertens (1942a, c) has largely been treated as a genus (Figs. 11.16–11.18). This is due to several considerably similar morphological and ecological traits among its species that include having somewhat blunted snouts, small, round nostrils, paired nasals, short limbs, distinctly cylindrical tails with keeled to spinose scalation, and relatively small sizes, making them dwarves among giants. None of these traits, however, except nostril size and shape, and relatively short limbs, is common to all species. Informally, the several spiny-tailed species have been dubbed the "*acanthurus* group," with several of its taxa having taken turns as subspecies or junior synonyms of *Varanus* (*Odatria*) *acanthurus*. Only relatively recently has their taxonomy reflected *Odatria* being composed of the *acanthurus* and *tristis* groups (Brennen et al. 2020, Vidal et al. 2012). The recognition is sound on ecological and morphological grounds:

a. *acanthurus* group: small, spinose and relatively short tails, primarily terrestrial, favoring rocky terrain with low-growing vegetation and low humidity. Endemic to Australia.
b. *tristis* group: medium-sized, keeled but not spinose, longer-tailed species that are significantly to highly arboreal, favoring habitats with trees, hills, and some degree of humidity. Inhabits Australia, New Guinea, nearby islands, and the Timor Island group (Figs. 11.14–11.16).

The spiny-tailed, or ridgetailed, monitors were the object of systematic research that elevated several subspecies and described a new eastern species (Pavón-Vázquez et al. 2022), one of the projects that had been waiting for researchers long after informal recognition of the groups had been made. Using a combined data set approach, the authors revealed a pattern highly consistent with

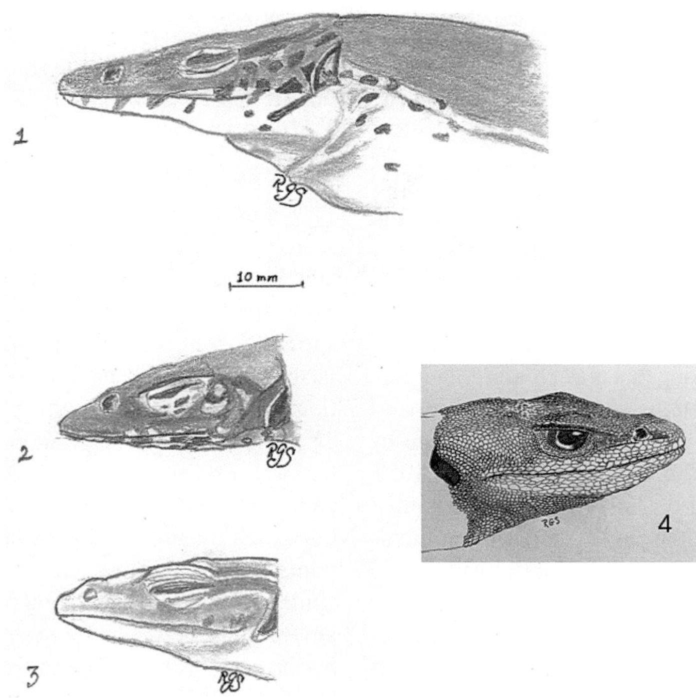

FIGURE 11.16 Diagrammatic comparisons of the taxonomically enigmatic *Varanus (Odatria) tristis* group. Specimens sketched from (1) "*Varanus timorensis orientalis*" USNM 128387 adult, Groote Eyelandt, Australia, (2) "*Varanus timorensis*" USNM 204027 adult, Moorehead, Western District, Papua New Guinea, 3) "*Varanus timorensis orientalis*" USNM 128569 half-grown, Yirrkala, Australia, and (4) *V. tristis*, NMW 12400:1 adult, Australia. The *tristis*-allied species and *V. scalaris* are characterized by an acute, distinct canthus rostralis, replaced by a rounded or indistinct canthus rostralis in the remaining *tristis*-group lizards. These specimens can be appropriately identified as (1) *V. orientalis*, (2) *V. similis*, (3) *V.* cf. *similis*, and (4) *V. tristis*.

FIGURE 11.17 A–C, *Varanus scalaris* holotype (SMF 32806), from Beagle Bay, far NW Western Australia, A and C photographed in partial natural light from a window, B using a standard lab area fluorescent desk lamp. The distinctions between rusty and gray areas are due to preservation and partially shed skin when the goannas were preserved. Rusty-orange markings are visible around the edges of the transverse dark bands, on the head, and among several of the rosettes (ocelli), the latter identifiable as such by the small dark centers of each. Note that the bands of rosettes are made of broad units that meld together, obfuscating individual rosette form. In D, the colors and pattern of a freshly preserved specimen from Queensland are apparent. Also characteristic of *V. scalaris* are the acute, not rounded, canthus rostralis ridge, and dark, unmarked posterior 50- to 60 percent of the tail.

morphology in revising the taxa, finding *V. acanthurus* to be monotypic (no subspecies), synonymizing *V. baritji* with *V. insulanicus*, and recognizing the two subspecies of *V. storri* to full species as *V. storri* and *V. ocreatus*, and naming a new species *V. citrinus*.

It is, however, the *tristis* group that is a taxonomic enigma, and much of the ambiguity stems from the numerous subspecies, species, and the mixing of them around done by Robert Mertens (Mertens 1926, 1941, 1942a, c, 1958, 1959, 1963). Coupled with unusual or "tentative" names applied to published photographs without any taxonomic comments or justification, the group requires phylogenetic analysis and taxonomic clarification. Among the many names applied to the species are *auffenbergi, centralis, karanda, ocellata, ocellatus, orientalis, pengilleyi, punctatus, scalaris, semiremex, similis, timorensis, timoriensis,* and *tristis*, along with a few nomina nudem for putative taxa never formally described. At present, only seven of those names are formally accepted.[2]

Within this group are taxa that have been, and would seem to remain, taxonomically stable. *Varanus auffenbergi* and *V. timorensis*, plus a third possible species awaiting study, are the only species within the group that seem to be entirely extralimital to Australia and endemic to the Timor Islands, and do not have "subspecies" requiring clarification, and about which nothing more needs to be said here. The name "*timoriensis*" is a lapsus. Australia and New Guinea, however, contain the remaining described species and, undoubtedly, some awaiting description. Given the huge area inhabited by these lizards, there are almost certainly species awaiting discovery. The extent of phenotypic similarity and overlap among these small goannas is unlike anything seen elsewhere in *Varanus*, with island-like distribution patterns very close to those described by Cogger and Heatwole (1981, Fig. 11.4), especially their continental subdivisions 1 through 4. Different colleagues have suggested anything from three to ten species await naming from known populations. A few herpetologists have indicated that they were about to undertake or were engaged in systematic studies of the *tristis* group, but to date, none of that work has come to light. For the present, and in sequence,

[2] While the manuscript was about to go to printing, six new taxa of Varanus were named, three of which relate to the *scalaris* group.

FIGURE 11.18 Enigmatic members of the *Varanus tristis* group potentially allied to *V. similis* and whose taxonomy and relationships need study. A – *Varanus pengilleyi*, AM R94494, holotype. B – *V. kuranda* vide Wells and Wellington (1985:21) from Worrell (1966). C – *V. similis* from interior, non-coastal region of the Northern Territory. D and E – *V. similis* showing common Northern and New Guinean patterns. F – *Varanus kuranda*, AM 68820, holotype. G – *Varanus* sp. undet., from the Atherton tableland, Queensland.

the names so far assigned, and comments on their status, are reviewed next.

1. *ocellata/-us*. First used by Heyden (in Rüppell 1830) and applied as *ocellatus* to *Varanus* from Kordofan (= Sudan), which was subsequently called *V. exanthematicus*, and resurrected for those populations by Bayless and Sprackland (2000a, b). Gray (1831) applied the name to *Monitor ocellatus* for specimens from Senegal, the terra typica of *V. exanthematicus*, that he later emended to *Empagusia ocellata* (Gray 1838). Gray (1845) later named what is now *V. acanthurus* as *Odatria ocellata*, and once the genus *Varanus* incorporated all extant monitors, *ocellata* became unavailable for the Australian *acanthurus* and members of the *tristis* group, being preoccupied for the African species.
2. *punctata/-us*. Merrem (1820) first used the combination *Varanus punctatus* for *V. bengalensis* when he established the genus *Varanus*. Merrem predated Gray's 1838 (see also Gray 1845b:pl. 1) use of the name for *Odatria punctata*, now *V. tristis*. Thus, *punctatus* remains a junior synonym for *Tupinambis* (= *Varanus*) *bengalensis* Daudin (1802) (Merrem also used *punktirter* for *bengalensis*), and is not available for the *tristis* group.
3. *semiremex* is the current name of the valid *Varanus semiremex* of eastern Australia. It was applied in error by Sternfeld (1924) and Mertens (1937) to *Varanus orientalis* (see 4).
4. *orientalis* was named as a variety, *Varanus punctatus* var. *orientalis* (Fry 1913), and retained as such by Zietz (1920). Barbour and Loveridge (1929) first used *orientalis* to design a variant of *V. tristis*, and would be redefined as a subspecies of

V. timorensis by Mertens (1942c), Worrell (1966), Storr (1980), and others. Using the same principle of populational distinctness and integrity as Pavón-Vázquez et al. (2022) for the *acanthurus* group, I recognize *Varanus orientalis* as a full species.

5. *scalaris* was also named as a subspecies of *timorensis* (Mertens 1941) and has remained in use as both a species and subspecies since, while *similis* came later (Mertens 1958) and shared a similar history. Published descriptions have been insufficient to distinguish the different populations, and the preponderance of color and pattern variety has led to a confusing array of names. To date, five names have been assigned to the goannas of this group (*centralis, kuranda, pengilleyi, scalaris, similis*), and though a few herpetologists have set out to find some resolution (e.g., Smith et al. 2004), this has yet to occur. While I was studying the huge collection of varanids in Frankfurt's Senckenberg Museum, I had, serendipitously, taken the work table near a large window that looked onto the ground's garden, which meant that I was able to look at specimens under daylight. This type of arrangement had already come in useful to me at the Natural History Museum in London, where, by briefly holding specimens in the sunlight and being able to examine the coloring as it shown through the abdominal scales along the axilla to groin sutures, was able to see that the skin of *Varanus beccarii* was not true black but extremely dark forest green (this was disputed by a later worker who restricted examination under the slightly yellowish table lamps). Given that preserved specimens rapidly change color in spirits, I also performed similar examinations with many other varanids. In many of those that appeared black, the lighting helped confirm that, or revealed the skin as being very dark brown or gray. Consequently, when I looked at the holotype of *V. scalaris* (SMF 32806) at Senckenberg Museum, a dull but distinct pattern of pale rusty-orange markings was apparent (Fig. 11.15A–C). Most of the orange skin was from bits that were being shed when the lizard was preserved, but among the intact skin markings were evident. I showed this to another varanid researcher, plus two other herpetologists, along with other specimens of *V. scalaris* and some labelled as *V. similis*. Soon after, my colleague showed me a recently published color photograph of a monitor that was depicted as "*Varanus* spec." (Bennett 1998:255), "*Varanus* cf. *similis*," (Eidenmüller and Philippen 2008:90) or *Varanus* "*pelewensis*" (King and Green 1999:pl. 3, Kirschner et al. 1996:230–31) that seemed to represent *scalaris* with the orange markings. Though the name "*pelewensis*" has persisted across several publications, almost always in quotation marks, and has become the name for a familiar color form for an Australian monitor of the *V. tristis* group, it has not been formally described as such and is probably a nomen nudum, e.g., no designated type, no formal description. Based on the several specimens I have examined, including live lizards in the area of Mareeba, Queensland, I conclude that "*pelewensis*" is a particularly conspicuous color phase of *V. scalaris*. Furthermore, if the holotype is actually that coloration, then "*pelewensis*" would certainly be a junior synonym. Considering the numerous spurious or incomplete descriptions that have appeared since the 1980s, I cannot state with confidence that at some time someone will not claim that "*pelewensis*" has, in fact, been published in accord with ICZN rules, in which case it would have to be considered a junior synonym of *Varanus scalaris*.

   The status of *scalaris* specimens still presents a complication. The holotype, which of course establishes scalaris as a properly described species and the oldest available name for any taxa within its group, was collected from Beagle Bay, a site of the far northwestern coast of Western Australia, and though it bears some small rusty-orange markings, it is distinct from the "*pelewensis*" lizards in two important ways. First, the limbs and entire nape of "*pelewensis*" goannas has the pattern characteristic of all populations referred to as *V. similis*, i.e., a "snowflake" pattern of small, solid white (not yellow or yellowish) spots that form lace-like bands on a black background, the spots and spaces equidistant from each other, and those on the nape extending laterally, sometimes as far forward as the labials. In *V. scalaris,* such white spots, when present, may be pale gray and are restricted to the limbs (compare Fig. 11.15C, D).

6. *centralis* was named by Mertens (1957), but the similarities between *centralis* and *orientalis* led him synonymize the former with the latter (Mertens 1958). The purported range of *centralis* is in many places sympatric with other members of the *tristis* group, including *orientalis*, but the lack of research on the systematics of these goannas leaves open the possibility that *centralis* could represent an evolutionary lineage distinct from *orientalis* or *tristis*. I suspect that a photograph in Worrell (1966:pl. 27) may represent the *centralis* pattern. Probably because Mertens synonymized *centralis* so soon after naming it, there are almost no specimens so labelled. It will require a fresh examination of *tristis*, *orientalis*, and many of the *scalaris* populations to determine their status, as well as that of several other populations. Based on my observations, though, one character state that seems common to *tristis* and *orientalis* (and

presumably *centralis*) is the dorsal pattern, which consists of distinct, well-formed light ocelli with dark centers, these aligned in well-defined transverse bands across the dorsum and posterior half of the nape. The clarity of this pattern is seen in the young of *V. scalaris*, and a similar pattern with less clearly defined markings is seen in some possible adult *V. scalaris*. The pattern contrasts starkly with that of *V. similis* and its potential allied species.

7. Of the remaining available names for taxa of the *tristis* group (*kuranda, pengilleyi, similis*), only *similis* has a clear and detailed taxonomic description. I have remarked elsewhere that it seemed that Mertens had taken almost any black Pacific monitor that had small light spots and added it to *Varanus indicus* (Mertens 1942a, c), and I suspect the same procedure was used with *V. similis*. Phenotypically these lizards differ most conspicuously in color and pattern as follows: *V. pengilleyi* (Wells and Wellington 1985) is black dorsally with large, mostly well-defined pale gray ocelli in transverse rows, these interspersed with broken transverse lines from axilla to hips; limbs and nape with solid white spots surrounded by narrow black borders, and around these are white flecks of single scale size. Hollow dorsal bands become smaller and solid, bands continuing onto the tail. *V. similis* (Mertens 1957) is like *pengilleyi* except pale ocelli do not form distinct transverse bands, and band markings on the tail become a salt-and-pepper mix of pale gray and black scales, not forming a distinct pattern. *V. kuranda* (Wells and Wellington 1985; Fig. 11.16F) differs from all other members of the *tristis* group in having a marbled pattern of irregular, often solid yellowish-tan dorsal markings, some of which are ocelli (with dark centers), no obvious transverse bands, no solid light spots on nape, these replaced by smaller versions of the dorsal markings. Limbs are dark with comparatively few light spots. Anterior fifth of tail with distinct pale and black rings, but these become indistinct caudad, and tail is almost entirely black in posterior half.

Of the *similis*-taxa, only *similis* per se is reported as extralimital from Australia, found in much of southern New Guinea and islands between New Guinea and Australia, e.g., in Torres Straights. The names of neither *kuranda* nor *pengilleyi*, however, have appeared in the literature apart from naming and minimal descriptions (Wells and Wellington 1995). Included are type specimens and localities, and from my examinations, these seem to be valid taxa requiring more detailed descriptions. Worrell (1966:pl. 27) presented four monochrome photographs of specimens, one of which ("A central Queensland phase of *Varanus timorensis similis*") appears to be a typically patterned *similis* (Fig. 11.16D). The live specimen he illustrated immediately above ("Spotted Tree Goanna... ...of the Far North"; Fig. 11.16B), though not clear enough to pinpoint species, shows character states of canthus rostralis, form of rosettes, and pattern (but not black tail) of the *tristis* clade, which Wells and Wellington (1985) state represents *V. kuranda* (Fig. 11.16F). The "Far North" of Australia spans more than 13,000 km; the location for that specimen, though vague, does include the northeast and center of Queensland, from which the species allegedly occurs. Worrell does not illustrate *scalaris*, notes in the text of its similarity to *similis*, "with a tendency to dark cross-bars on the dorsal surface" (Worrell 1966:81). *Varanus similis*, though, is a taxonomic enigma for several reasons. First, it concerns its broad distribution across much of Australia's top end and southern New Guinea, where it has been collected often enough that specimens are present in most of the museums I have visited. As with many specimens, locality data are often less than precise, but there are enough records to confirm a very large west-east distribution on either side of the Arafura Sea. Of the many specimens I have examined, I have noted no appreciable variation at the species level in terms of color, pattern, scale counts, morphology, or habitat preference. This is a case where molecular analyses will probably be the best approach to refining the taxonomy and understanding relationships.

Second, of the few details regarding color within this putative species, Australian specimens from some localities have yellow gular and sometimes pectoral areas. Though this has sometimes been claimed for *V. scalaris* also, the few such specimens I have been able to examine turned out to be *V. similis* (or at least not *scalaris*), in which the range of coloring is confined to black, gray, and white.

Third, though the pattern distinctions between *V. pengilleyi* (Fig. 11.16A) and *V. similis* are distinct despite the identical coloring, the pattern variation among populations of the latter is extensive (Fig. 11.16B, C). Though the differences can be subtle, the patterns – clarity of ocelli and regularity of transverse bands – are most clearly seen in Australian populations, becoming gradually more orderly in western localities, e.g., Western Australia and much of the Northern Territory (Fig. 11.16C). Eastern populations display the pattern where ocelli have irregular borders (Fig. 11.16C), many of which bleed into each other, and lack any pattern of linear banding. Further complicating this point is that specimens from New Guinea can show a range between both extremes in some localities, such as Western Province, Papua New Guinea, while Western Papuan lizards (Fig. 11.16E) have very irregular, randomly distributed ocelli and no distinct banding. Then too, there are populations that are distinguished in having a pattern and nape-head coloring typical of *similis*, a dorsal pattern of considerably merged and fragmented ocelli, and orange coloring as seen in *scalaris*, though not in any distinct pattern (Fig. 11.16G). Of course, taxonomic decisions should be based on more than color patterns, but when patterns vary to this extent, it is likely a phenotypic clue that different species are likely involved.

11.4.2 *Varanus varius* Possible Group

The Lace Monitor, *V. varius*, was among the first varanids given a post-Linnean description and name, dating to 1790. As one of the largest monitors, it has received attention from ecologists (Carter 1999a, b) and physiologists (Bartholomew and Tucker 1964, Weavers 2011), but not nearly as much as *V. giganteus, V. gouldii*, or *V. rosenbergi*, or the many *Odatria* species. It was the first Australian species discovered to use ant hills as places to deposit and incubate its eggs (Carter 1999b), after the initial reports of the behavior in Nile Monitors (Cowles 1930). Taxonomically, *V. varius* has been stable from Shaw's original designation as *Lacerta varia* in 1790, followed by different versions of *varia, varius*, and *variegatus, though* Daudin (1802) called it both *Tupinambis variegatus* and *Tupinambis bigarré*. The only significant addition was when Duméril and Bibron (1836) named a light tan and black-banded specimen Varanus belli, which was later learned to be simply a color variant of *V. varius*. Numerous captive-bred clutches producing both banded and "normal" patterned lizards have confirmed the identity. Suggestions that the "Bell's phase" was sex-linked or limited to populations west of the Blue Ridge Mountains of New South Wales have been proven false.

However, during my field studies in New South Wales and Queensland between 1999 and 2003 I noted (Sprackland 2009) that (a) Lace monitors possessed three tongue colors (pink, purple, and pale yellow), (b) the variants were confined to well-defined areas in the two states, (c) the populations were divided by significant natural barriers to dispersal, and (d) specimens did exist in museum collections and zoos but had largely gone unnoticed with the possible exception of Smissen et al. (2013). Given that species-specific tongue colors have been confirmed in the mangrove and Nile monitor groups, it is certainly worth investigating the systematics of Lace Monitors.

11.4.3 *Varanus indicus*-Group

The large group of Mangrove Monitors has moved from taxonomic near-stasis between 1941 and 1989 to considerable re-evaluation, species discovery, and systematic updates ever since (Böhme et al. 1994, Böhme and Ziegler 1997, Iyai and Pattiselanno 2006, Koch et al. 2009, Philipp et al. 1999, Sprackland 1993b, 1994b, 1995b, Weijola et al. 2016, Weijola et al. 2020, Weijola and Sweet 2010, Ziegler et al. 1999). In terms of finding and describing new species and erecting new subgenera, this is the most actively studied group within the Varanidae. I am aware of three species presently awaiting description. Mangrove and Tree Monitors were recovered as sister groups (Sprackland 1991a, b) and repeatedly confirmed as such (Ast 2001, Baverstock et al. 1993, Böhme et al. 1994, Hocknull et al. 2009, King et al. 1991), initially united in the subgenus *Euprepiosaurus*. Later, the Solomon Islands *V. spinulosus* was moved to a monotypic subgenus *Solomonsaurus* (Bucklitsch et al. 2016), and the tree monitors were placed in the subgenus *Hapturosaurus* (Bucklitsch et al. 2016). It remains to be seen if additional subgenera will be warranted as the different species-rich groups (see below) are studied and as new taxa are described.

Distribution of this subgenus is huge, from Halmahera and nearby islands of eastern Indonesia to Rennell and San Cristobal in the extreme eastern Solomon Islands (= 3,873 km), and from Guam in the Northern Marianas south to Tropical Northern Queensland and the Northern Territory, Australia (= 4,200 km), an area of some 16,266,000 km^2 or 2.1 times larger than Australia. Informal species groups include, by number of recognized species, the predominant Mangrove Monitors (*V. indicus, V. chlorostigma*, and kin), Peach-throated Monitors (*V. jobiensis* and kin), and Blue-tailed Monitors (*V. doreanus* and kin). The distribution of Pacific, presumably mangrove, monitors has long included their possible introduction by humans (Balouet 1984, Bayless and Naclerio 1993, Kishida 1929, Uchida 1966, 1967, Weijola et al. 2020). In his doctoral thesis, Balouet (1984) wrote, "Les ossements de Pindai permettent en oute d'établir la contemporaneité de l'homme et de Meiolania, du Mégapode, du Varan," but his information was relayed by a local (who ran over the lizard with his car, ate it, then disposed of the remains) as an incident from two decades earlier. Balouet (1991:1386–1388) later recorded "*Varanus* cf. *indicus*" on Pindai (100+ bones) and Gilles (1 vertebra) Islands of New Caledonia, and depicted important and well-preserved elements of the skull (parietal, right maxilla, left mandible), pectoral girdle, vertebrae, and forelimb. These all conform well with recent materials of the *V. indicus* group, notably the nearly straight lateral parietal crests that taper to about one-half the width along the frontoparietal suture, the condition seen in the *V. salvator* and *V. indicus* groups. Whether the subfossils represent an extinct native population, lizards introduced as a food source, or rafted to the islands, the possibility of a New Caledonian varanid cannot be summarily dismissed, and given the speciation across the SW Pacific by the group, the New Caledonian monitor would very likely represent a new species. Though numerous species are involved, the considerable distribution of mangrove monitors across Oceania and Australia-New Guinea is not unlike that of the Red Shiner, "This species is ubiquitous in Kansas… We have yet to find it in drinking fountains or swimming pools, but we believe that Red Shiners occur at some time in every natural body of water in Kansas" (Collins and Cross 1995:56). The multi-approach research by Weijola and colleagues (e.g., Weijola and Sweet 2010, Weijola et al. 2016, 2020), using ecology, molecular data, and examination of freshly collected specimens from numerous field localities, is precisely the type of combined data analysis needed to better understand the relationships and biogeographic history of the lizards.

Balouet (1991:1386), an ornithologist, was incorrect on one point, however, claiming that "the family Varanidae in New Caledonia today includes two genera, the fossil genus *Iberovaranus* and the modern genus *Varanus*." The inclusion of *Iberovaranus* (see above) is unlikely, and *Varanus* is not presently extant in New Caledonia. He was unable to compare the subfossils with "*Varanus karlschmidti*" (now *V. jobiensis*), though as a member of the *indicus* group, it might have been quite similar. He correctly noted the differences between his material and the "New Guinean species *Varanus grayi*" (now *V. olivaceus*), which is from the Philippines (and both taxonomic changes were made prior to 1991).

On this latter point, regarding biogeography, the *V. indicus* assemblage (I think the term "supergroup" suits, given the number of taxa and potential for additional subdivision beyond the three closely allied subgenera) is an extraordinary lacertilian model of island biogeography for large lizards that parallels the situation with West Indian *Anolis* (see Losos 2011).

These unresolved taxa do not represent the entirety of members of *Varanus* in need of additional study, just those that are most conspicuous and have been problems for over three decades. I have no doubt that the tree monitors, the *prasinus*-group, will, like the *indicus*-group, be shown to consist of numerous, largely allopatric populations in need of recognition.

11.5 RESOLVED HIGHER-LEVEL RELATIONSHIPS (HOPEFULLY)

The systematics of varanomorph lizards has been studied by numerous researchers using the breadth of taxonomic experiments, from simple morphology to molecules to complex "complete data sets." Mertens (1942a, c) provided the first major revision of extant varanids, his morphology-based system a framework of similarities that, in large part, reflected convergence. Not surprisingly, the large number of species described after his monograph provided so much new data that taxonomic changes were inevitable.

In order to find appreciable relationships among taxa that offer a variety of character states, I have employed a highly morphological investigation of the Varanids based on several sources. Despite their variable results, I have chosen to examine certain phylogenies because of their results that are good fits of temporal, biogeographic, tectonic, morphological, and complete data analyses. These are (chronologically) Evans (1984, 1988), Baverstock et al. (1993), Gao and Norell (2000), Ast (2001), Townsend et al. (2004), Lee (2005), Norell et al. (2007), Hocknull et al. (2009), Conrad (2008), Conrad et al. (2008, 2012), Delfino (2012), Gauthier et al. (2012), Vidal et al. (2012), McCurry et al. (2015), Lin and Wiens (2016), Pyron (2017), Zhue et al. (2020), Brennen et al. (2021), Simões and Pyron (2021), and Dong et al. (2022). From these sources, I sorted data into comparable sets, omitting missing characters or characters for which comparative information is absent (e.g., skull elements). The data accepted for this study are provided in the taxonomic section above, broken down into families and genera. Taxa with objective positive morphological evidence were sorted to establish hypotheses of taxonomic affinities, these results represented by solid lines in Fig. 11.17. The groups were then appraised for temporal compatibility, meaning examination to see if taxa were close in geological time and location, which closely followed the classificatory output. Finally, data were plotted on maps representing the location of landmasses during each geological period (Fig. 10.4) in order to eliminate unlikely routes of dispersal. The most conspicuous of these unlikely routes is the failure of Europe and North America to show any connection or near connection at any time from at least the middle Cretaceous, which negated the hypothesis that Eocene *Saniwa* could have been present on both continents so long after their separation. At the time *Saniwa*, or its most recent ancestor(s), allegedly migrated to North America. Not only was America isolated from Europe by the Atlantic Ocean, but by two other formidable barriers. First was the wide sea that divided North America into two large islands, a deep and formidable obstacle that the lizards either did not cross or, possibly, crossed but could not survive for whatever reason on the eastern shores. There was no alternative land link between western and eastern North America, nor was there a southern link between the still submerged Caribbean archipelago or Central-South America. The northern lands also became subject to a developing temperate climate with seasons that may have been too cold for varanid survival, especially if the *Saniwa* were physiologically and ecologically like living *Varanus*. If, despite the odds, some *Saniwa* migrated to and prospered on the eastern subcontinent, the second barrier to Europe was the Atlantic Ocean. The combined inhibitory effects of distantly separated parts of North America, wide mid-continental and intercontinental seas, and a significantly cooling climate would have made an eastward migration of *Saniwa* an extremely lucky, and unlikely, case of vicariance.

So where did "*Saniwa*" *smallness* come from? The parsimonious solution was that each genus evolved independently of and convergent relative to the other. This requires taxonomic re-evaluation of European "*Saniwa*" *orsmaelensis*, which, in its most recent review, was kept in the genus *Saniwa* (Augé et al. 2022). In view of the biogeographical near impossibility of a single genus being involved, and because there are no junior synonyms available for the European monitors, I have established ***Dytikosaurus*** gen. n. for what is now ***Dytikosaurus*** *orsmaelensis* n. comb. The phylogeny presented in Fig. 11.19 provides parsimonious and mutually congruent hypotheses to account for the distribution and relationships of Varanoidea, plus clarification of the relationships between Platynota and Varanoidea, terms that are not synonymous and are taxonomically informative given the variety of assigned genera.

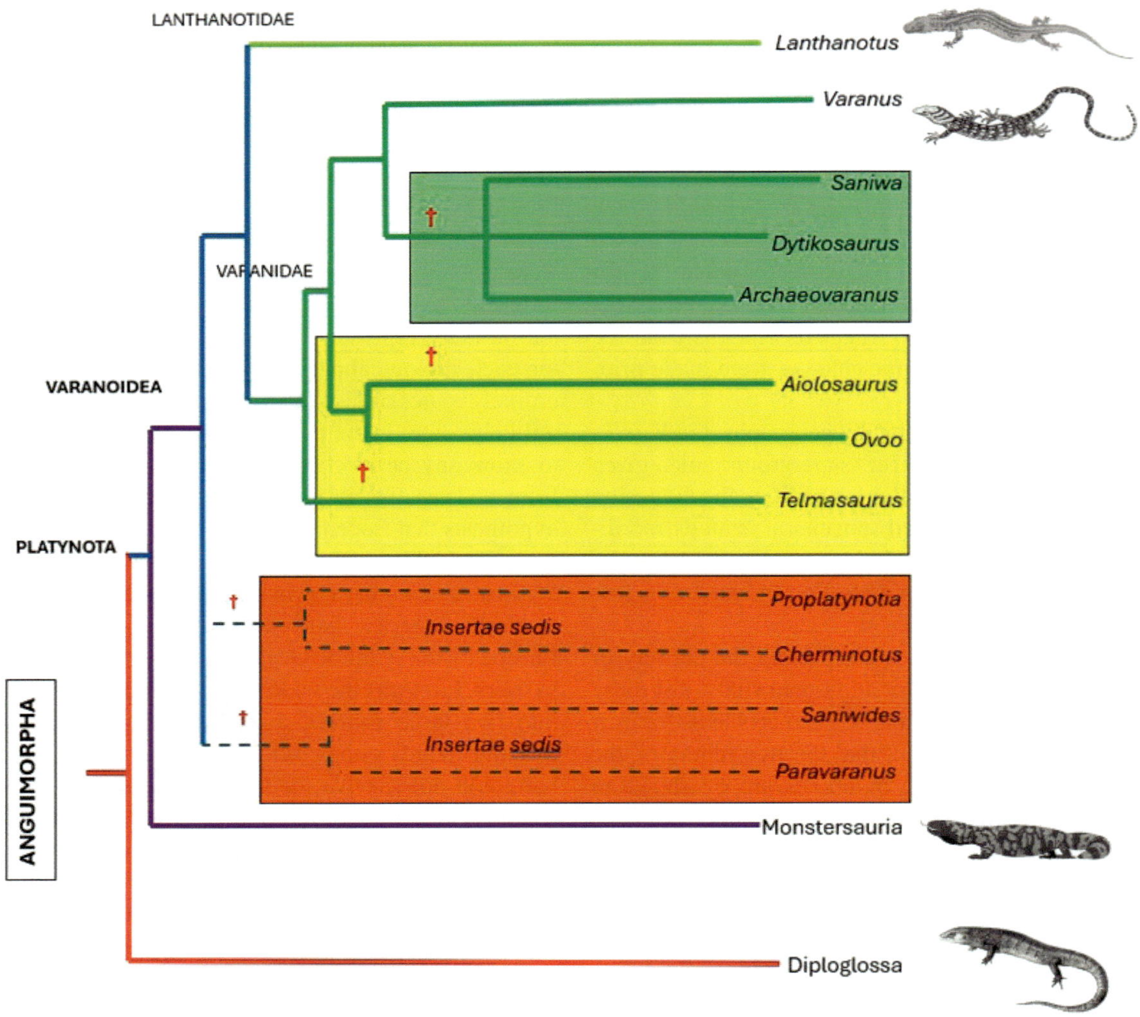

FIGURE 11.19 Phylogram of the Varanidae relative to the Anguimorpha based on the synapomorphies described in the text. Diploglossa and Platynota are sister clades within Anguimorpha; Lanthanotidae and Varanidae are sister clades within Varanoidea. Taxa in orange are unresolved largely for lack of adequate fossil materials and, where materials are comparable to extant taxa, do not contain any or sufficient synapomorphies to be accepted in Platynota except tentatively (see text). Orange and yellow boxes represent Cretaceous genera. Green represents Cenozoic (Eocene – Oligocene) genera. Uncolored taxa are extant. Dotted lines indicate taxa whose relationships with Platynota and with each other are uncertain.

11.6 PHYLOGENETIC SUMMARY

Monitors and their related taxa are among the oldest representatives of the Squamate clade; their known origin in the Late Cretaceous is likely subsequent to unverified Jurassic roots. Fossils are known from some, but not all, geological periods from the Late Cretaceous, with the Miocene arrival of *Varanus* and its rapid and profuse speciation events. The increased exploration of SE Asia and the Indo-Australia archipelago, coupled with exceptional fossil finds since the 1960s, has turned Varanoidea from being thought of as a species-poor to a species-rich assemblage. Their distribution over time spans five continents. Their taxonomic diversity is such that ten genera are included in Varanoidea, eight of which are contained in the family Varanidae. All the fossil genera are monotypic, pending additional and diagnosable finds, while the single extant genus has about 100 described species.

Appendix I

ABBREVIATIONS

a fac	anterior facet of frontal
a. pr.	alar process of prootic
a. sp.	alar process of sphenoid
add. f.	adductor foramen
aia f.	anterior inferior alveolar foramen of splenial
alp	alar process
ang	angular
ant myl f.	anterior mylohyoid foramen of coronoid
ap cor.	anterior process of coronoid
aper. max-prm	aperature maxilla-preamxillaris
aper. vom	vomer aperature
art	articular/retroarticular process
avc	anterior vidian canal
bas fo	basioccipital fossa
boc	basioccipital
bpt p.	basipterygoid process
bsp	basisphenoid/sphenoid
bsp sp	basisphenoid spine
cap sq	capitum squamosum of quadrate
ccon	cephalic condyle of quadrate
cd	lateral edge of maxillary crest
cho	choana
ci	intranasal crest of maxilla
col cen	quadrate central column
col fos	fossa columellae
cor	coronoid
cor cr	coronoid crest
cor fo	coronoid fossa
cor lap	coronoid lappet
cpfo	crista postfovealis of parietal
cr. pro.	crista prootica
crc	crista cranii
crp	septomaxillary crest
crpa	parietal crest
cs. tr.	crista transversus
dc pro	descending coronoid process
den	dentary
den. shelf	dental shelf of maxillary
e	eye
ecp	ectopterygoid
epp	epipterygoid
est	extracolumella
exo	exoccipital
f. mag.	foramen magnum
f. oval.	fenestra ovalis
f12	foramen for CN IIX, hypoglossal
f5	foramen for CN V, trigeminal, main branch
f6	foramen for CN VI, abducens
f7	foramen for CN VII, facial, main
f8	foramen for CN VIII, auditory, vesibulococlear
f9	foramen for CN IX, glossopharyngeal
fac ect	ectopterygoid facet
fac pal	palatine facet
fac. p.	facial process of maxilla
facq	facet for quadrate
fcht	foramen for CN VII, chorda tympani n.
fed	endolymphatic foramen of otic capsule
fen pal	palatal fenestra/choana
fen sub	suborbital fenestra
flh	foramen lacerum posterius
fm5	foramen for CN V3, trigeminal mandibular branch
fmX5	foramen for CN V^2, trigeminal maxillary branch
fn	alveolar foramen
fnm	maxillary foramen
fo5	foramen for CN V^3, trigeminal ophthalmic branch
fp	premaxillary process
fp7	foramen for CN VII, facial palatal ramus
fro	frontal
frs	frontal suture
hym f.	hyomandibular foramen
hymf	foramen for CN VII, hyomandibular branch of facial
hyp fo	hypophisial fossa of sphenoid
iacp	inferior anterior coronoid process
icf	foramen for internal carotid
if	inferior alveolar fossa
imj	intramandibular joint
int. carot.	interior carotid foramen
ios	interorbital septum
ip	incisura prootica, exit for CN V, trigeminal
jac	posterior maxillary process/articulation of jugal
jug	jugal
jug f.	jugular/vagus foramen
lac	lacrimal articulation of maxilla
lac	lacrimal
lac. f.	lacrimal foramen
lfl	lateral flange of parietal
lrp	lateroanterior process
lrst	lateral recessus scala tympani
max	maxilla
max. shelf	maxillary shelf
Me. f.	Meckelian fossa
Meck pr	Meckelian process
med. rec. nas.	medial recess for nasal sac

men. f.	mental foramen
mfo/lac. rec.	maxillo-frontal/lacrimal recess
mfr	maxillary nutrient foramen
mx5	maxillary branch of CN V, trigeminal
nas	nasal
nas fac	nasal facet of frontal
nf	nutrient foramen
oc con	occipital condyle
opo	opisthotic
orb. p.	orbital process of maxilla
otc	otic capsule
oto	oto-occipital
pa	prearticular
pal	palatine
pal pl	palatine plate
pal. shelf	palatine shelf
pap	palpebral
par	parietal
par f	lateral parietal fossa
par. f.	parietal foramen
pf	pineal foramen
pf fac	posterior frontal facet
pfa	prefrontal articulation of maxilla
pfo	median parietal fossa
pfr f	postfrontal facet of frontal
pjo	posterior orbital (jugal) process
plyf	perilymphatic foramen
pma	anterior process of maxilla
pmp	premaxillary process of maxilla
pn fo	posterior parietal nuchal fossa
poc	paraoccipital process
pof	postfrontal/postorbitofrontal
poo	postorbital
pop	postoptic/laterosphenoid
pp cor.	posterior process of coronoid
ppat	posterior parietal table
ppm	posterior process of maxilla
pr. as.	ascending process of supraoccipital
pra	prearticular process
prc	parietal recessus
prf	prefrontal
prf f.	prefrontal facet of frontal
prm	premaxilla
pro	prootic
proc. ext.	external process of parietal
proc. vm	processus vomer maxilla
psp	parasphenoid
ptb	parietal table
ptd	pterygoid
ptdp	pterygoid process
pter	pteroticum
pty fl	pterygoid flange
pty hd	pterygoid contact of quadrate
pvc	posterior vidian canal
qhd	quadrate head
qrg	quadrate ridge
qsf	quadrate shaft
qut	quadrate
qut fac	quadrate facet
qut proc	quadrate process
ra	retroarticular process
sacp	superior anterior coronoid process
sdn. s.	subdental shelf
sep	septomaxilla
sep fos	septomaxillary fossa
sep t	septomaxillary table
set	sella turcica
sp sep	septomaxillary spine
spf	splenoccipital foramen
sph	sphenoid
sph w	sphenoid wing (process)
spl	splenial
spl sul	splenial sulcus
squ	squamosal
ssc	semicircular canals
sta	stapes
stf	supratemporal facet
su tr proc	supra trigeminal process
suo	supraoccipital
suoc	supraoccipital crest
sur	surangular
sur pr	surangular process of cornomoid
sut p	supratemporal process of parietal
svj	sulcus for jugular vein
sym	mandibular symphysis
tb	turbinal
tr. cra.	base for trabeculae cranii
tym	tympanic membrane
tym col	tympanic column of quadrate
vf	vagus foramen for CN X
VI f.	foramen for VI nerve
VIII f.	foramen for VIII nerve
vom	vomer
XII f.	foramen for XII nerve

Appendix II

PRESERVED SPECIMENS EXAMINED

This is a list of preserved museum specimens, exclusive of the mangrove monitor group, examined for this study. Included are also the author's (RGS) collection and uncataloged specimens in museums. Only holotypes/syntypes for recognized species are separately listed; those for subspecies are noted with the specimen number. Only types for about 15 Australian taxa were unavailable for study. A further six species have no types. These include *Varanus exanthematicus*, *Varanus indicus*, and *Varanus varius* (types lost), and *Varanus flavescens*, *Varanus niloticus*, and *Varanus salvator*, which were described based on illustrations. Some species (e.g., *Varanus pilbarensis*) were only available to examine as live specimens and are therefore not included below. The only described taxa completely absent from this list are *Varanus acanthurus insulanicus* (slides of holotype were made available by H. Cogger, AM), *Varanus baritji* (for the prior two, see the recent work by Pavón-Vázquez et al. 2022), *Varanus glauerti*, *Varanus rubidus* (slides of living examples of the latter three were kindly provided by several acknowledged sources, as well as photos of an undescribed form from northern Australia), *V. salvator andamanensis*, and *V. s. nicobaricus*. Several other putative species are awaiting description. My comparative material includes data for over 1000 preserved specimens, in addition to innumerable live animals (I examined about 500 mangrove monitors). Species are listed alphabetically by genus and species. I have omitted subspecies designations. Taxa for which the holotype was examined are marked with an asterisk (*), extinct taxa with a dagger (†).

†Estesia mongoliensis * (N=1) IGM 3/196

Heloderma

horridum: CAS- 134061

suspectum: CAS- 102577

Lanthanotus borneensis * (N=28): AMNH- 87375 BMNH- 1906.2.7.1, 1909.3.4.1, 1920.4.6.1, 1976.1336-9 FMNH- 121986, 130981, 134698-9, 134711, 141973, 148589, 151714 NMSZ- 1913.136.5 NMW- 16365 (holotype), 16366 RGS- 36 RMNH- 14880 SDSNH- 56749 SMF- 57806, 59464-5, 66188, 68143, 72641

†*Palaeosaniwa canadensis* * (N=1): NMNH v10864 (holotype)

†*Saniwa ensidens* * (N=1): NMNH v2185 (holotype)

Varanus

acanthurus * (N=16): AM- R 49418, R 90873 BMNH- 1946.8.30.94 (holotype) CAS- 12629, 104407, plus 1 uncat. NMW- 12404 RGS- 18 RMNH- 23411 SMF- 11639-40, 11642, 53251-3, 73527

albigularis * (N=68): BMNH- 96.9.24.15, 98.1.8.6, 1901.6.24.23, 1902.12.1.37, 1902.12.18.37, 1905.3.7.99, 1905.11.7.11, 1905.11.10.1, 1908.5.15.16, 1908.12.28.36, 1911.1.26.1-5, 1928.10.19.58, 1932.5.2.85, 1937.12.5.379, 1928.10.19.58, 1931.7.20.366-7, 1932.5.2.85, 1936.9.4.33, 1937.12.5.377-79, 1959.1.3.61, 1975.103 MHNP- V 4/3 (holotype) NMSZ- 1859.13 NMW- 11247, 12499:1, 12499:2, 12499:3, 12501:1, 12501:2, RMNH- 3162, 14964 SMF- 11541 (holotype of *microstictus*), 11543, 21573-4, 25387, 26578, 33261-2, 60619, 69459 ZFMK- 5138, 21652 ZMB- 445, 6587, 6602, 9244, 11552-3, 11555 11868, 21932-3, 25035, 28915, 29996, 30378, 37650, 44048 ZSM- 132/1933, 603/0

beccarii * (N=19): BMNH- 1905.11.29.4, 1910.4.26.25-6 MCSNG- 28723 (syntype) MCZ- 7489 (2 spec.) NMW- 12395 RGS- 199, 201, 1990-1993 SMF- 11616, 11636, 11638 ZFMK- 7829 ZMB- 7993 (syntype), 5903 ZSM- 42/1990

bengalensis * (N=37): BMNH- PAL R 719, 1901.3.8.6, 1940.6.4.2-3 CAS- 13989, 16883, plus uncat. skull. MHNP- Va 8-13, V. 10 2179 (holotype), plus 1 unnumbered NMSZ- 1859.13.624, 1859.13.736 NMW- 615a-c, 22285, 79991 NMSZ- uncat. RMNH- 1838, 3857 SMF- 11550-1, 11555, 11652, 40160, 48236, 52897, 55443, 56445-8, 60428, 63180, 63456, 64164 ZSM- 441/1920

bogerti * (N=5): AMNH- 41639 (holotype), 41638, 76722 BMNH- 76.7.6.2, 89.7.1.8

brevicauda * (N=8): BMNH- 98.10.26.6.4 (syntype), 1946.8.30.46.4 (syntype) RMNH- 25898 SMF- 51778, 69302-4, 72331

caudolineatus * (N=8): BMNH- 64.7.22.3 (holotype) NMW- 12407 SMF- 40036, 53255-7, 69300-1

chlorostigma (N=4); BMNH 1.18a MHNP 2202(1536), Va. 8.3, 2202 (lectotype for *chlorostigma*, paralectotype for *kalabeck*) RMNH 3151

cumingi (N=4): NMNH- 305884 (skull) RGS- 3 uncat

†*dirus* * (N=2); QM F874 (holotype), F42105

doreanus * (N=56); BMNH 1913.10.31.107-9 MHNP Va. 8.5 NMW 12431:8-9, 12431:12 RMNH 5164, 7035, 21017, 21029, 21032a, 21051, 21055, 23424 SMF 33290, 33291 ZFMK 19219-21, 20326-7, 24668, 48420, 51425, 52922 (neotype) ZMA 10191, 10193, 10194a, 10195, 10197b, 10199, 12125 ZMB 790, 18838, 18841, 18856-9, 21258a-b, 21396, 22247, 22272a-c, 22400, 24866a-b, 25735, 26453, 26573 ZSM 11/1970, 14/1970, 141/1975

dumerilii * (N=57): BMNH- 48.2.11.6, 87.2.7.15.17, 92.6.3.1, 92.10.7.4, 1951.1.8.34, 1975.2159 MCSNG- 28770 NMNH- 166730, 595348 NMNH- 166730, 30974, 70828 NMW- 12463, 12465-6, 12468-9, 20819, 21255, 27288:1, 27288:2 RGS- 52-2, 203, 226 RMNH- 3168 (holotype), 3167, 3185, 4194, 6727-8, 7400, 9610a, 9610b, 23415, 23436, 23460 SMF- 11556-8, 11557-8, 34345, 52779, 52780, 69138 ZFMK- 14877 ZMB- 4750, 8424, 29684, 37396, 38888, 47654 ZSM- 1/1918a, b, e, 104/12, 130/1975, 448/1911, 817/1920, 2717/0, plus 3 uncat *eremius* (N=8): BMNH- 97.10.27.59, 1907.5.6.10, 1908.5.28.48, 1910.5.28.14, 1932.3.7.91-93 NMW- 12405 RMNH- 6426 SMF- 11648

exanthematicus (N=42): BMNH- 69.11.4.1, 99.8.23.6-7, 1906.7.25.2, 1911.7.15.20, 1913.9.16.12, 1920.1.20.3755, 1924.5.21.5, 1933.11.17.48, 1935.12.1.118-20, 1937.12.4.15 NMNH- 161744 NMW- 12495 RGS- 202 (skull), 205, 370 SMF- 11544 (holotype of *ocellatus*), 33259 ZFMK- 14884, 17528, 33632, 39872-4, 51721-2, 56399 ZMB- 381, 802, 6387, 6903, 11254, 16040-1, 16275, 24920, 27971, 30285, 37704, 38877 ZSM- 30/1991, plus one uncat

finschi * (N=28); NMW 12429:4, 12429:12 ZFMK 26347 (holotype), 26343-5, 26348 ZMB 8133a-b, 9452, 14596, 14598, 21228, 22252, 26559b, 26580a-c, 26659a-b, 28378, 283390a-b, 30123a, d, 38624, 52998, 52999, 53000

flavescens (N=23): BMNH- I.4e NMNH- 267002 NMSZ- 1859.13.734 NMW- 12460, 12461:1, 12461:2, 12462:1, 12462:2, 17059, 18841, 19994:1, 19994:2 SDSNH-66533 SMF- 11546, 53639, 54157, 56457-60, 63179 ZFMK-14879, 21653 ZSM- 2/1957

giganteus * (N=6): BMNH- I.22a (holotype) NMW-17215 SMF- 53263-5 ZFMK- 14882

gilleni (N=13): BMNH- 97.10.27.58, 1908.5.28.44-7 CAS- 2 uncat. NMW- 12406 RGS- 12-13 SMF- 11627, 53258 ZSM- 12/1917a-b

glebopalma (N=6): RGS- uncat. SMF- 69144, 69954, 70137 ZFMK- 33625, 40785, 41436

gouldii * (N=25): BMNH- 1946.9.7.61 (lectotype), I.17b, I.17i, 57.10.24.30, 68.4.3.58, 81.10.12.1, 90.1.14.7, 97.12.10.17, 1904.10.7.36, 1986.11.81 NMNH- 64083, 67734 NMW- 17230, 19150 RMNH- 3153 SMF- 11620-1, 11625, 53267-9, 53271-3, 54507, 59018 ZFMK- 14885

griseus (N=55): CAS- 23008, 99980, 135137-8 NMNH-196455, 220282, 292065 NMW- 1618, 12130 (4 spec.), 12451, 12452 (4 spec.), 12456:1, 12456:2, 12458 (6 spec.), 17227, 20338, 31651 SMF- 11533-6, 11538, 46784-6, 51074, 54153, 59637, 60623, 61334, 61784, 63169-71, 63172-8, 71694 ZFMK- 7848, 14883, 53533 ZSM- 4/1953, 161/1985

indicus (N=190); AM 3 uncat. BMNH 1.18b, 19.11.93, 49.3.2.47, 71.4.16.26, 71.5.22.4-5, 76.7.6.1, 77.2.24.37, 77.2.24.38a-b, 77.2.24.39, 82.8.29.36, 83.3.26.14, 83.6.28.24, 84.3.24.77-78, 88.1.7.1, 88.8.30.3, 90.1.14.8, 95.10.17.7, 98.3.3.9-10, 98.5.27.1, 1910.4.26.24, 1913.10.31.22, 10.31.107-8, 1932.7.2, 1935.5.10.21-27, 1936.7.7.19, 1955.1.12.96, 1961.1891, 1963.557, 1969.2907, 1986.129 CAS 10987, 17292, 18441-2, 72033, 72046, 94024, 111549-50, 114099, 121786, 126919, 126920, 135453, 152219, 153759, 154190, 154191, 154192 FMNH 13805, 41315-6, 97851 MCSNG 28747, 28760, 28765 NMW 1886, 12429:1, 12429:9 RMNH 3149, 3150, 3152, 1180, 3184, 3189, 3190, 3800, 4193, 5260, 5359, 5370, 5374, 6726, 7167, 7168, 7169, 7197, 7223, 7297a-f, 21018, 21020, 21021, 21024, 21026, 21033, 21034, 21036, 23424, 23422 SMF 11578,11579, 11580, 11581-2, 11583 (holotype for *V. i. rouxi*), 11584, 11585, 11586, 11587, 11588, 11589, 11590, 11592, 11591, 11593, 11635, 32805, 32808a-b, 32809, 32810,32850, 34437-8, 56469, 56470 ZFMK 14234-5, 19001, 19222, 20328, 24669, 26342, 26346, 26349, 45043, 47537-8, 50487, 52818, plus 2 uncat. ZMA 10192a-b, 10194b-c, 10196a-b, 10197a, 10200, 15412a-c, 15413, 15414a-c, 15415a-b, 15417 ZMB 7819, 8789, 8854, 8895, 8896-7, 10610, 14520, 14627, 17521-2, 21260a-b, 21424, 21877, 22645, 24219, 24700, 24747, 26580a-c, 29508, 29552a-b, 30123b-c, 37679, 39182 (formerly MCZ 72928) ZSM 13/1970, 129/1975, 579/1911c, 880/1920

jobiensis *(N=45); BMNH 1938.6.7.81 FMNH 14201 MHNP Va. 8. 1986-798, Va. 8/9 A NMW 12431:1, 12431:6, 12431:10 RGS 3 uncat. RMNH 7150, 7225, 21019, 21022a-b, 21023, 21028, 23406 SMF 7038, 41179, 61292, 61293, 70327-8 ZFMK 19212, 47535-6, 48421, 50090, 53347 ZMA 10212, 11602 ZMB 8833, 18841, 18856, 21382, 22222-3, 22247, 25742, 25820, 34102 (holotype) ZSM 5/1970, 41/1990

komodoensis (N=6): NMNH- 519762 SMF- 23189, 37209, 56444, 57550

melinus (N=6): RGS 6 uncat

mertensi (N=9): RGS- 1970-72 RMNH- 181b SMF-53274-5, 60583, 61291, 62684

marmoratus (N=): NMNH- 498916 (skull)

mitchelli (N=6): BMNH- 57.10.24.34 QM- J 61775 RGS-150-151 SMF- 54512, 70953 ZFMK- 54250

niloticus (N=52): BMNH- I.15g, 97.5.31.2, 1935.8.3.1, 1946.8.30.45, 1946.8.30.96, 1970.1983 MHNP- V 12/28 NMW- 605, 1678 SMF- 52820-1, 53197, 54117, 59635-6, 59646, 59852-63 ZFMK- 5821, 7847, 14888-9, 26374, 32389, 45758 ZSM- 5/1953, 18/1964, 26/1954, 55/78, 80/1953, 81/1953, 89/1926, 122/1909a-b, 123/1909, 144/1975, 155/1908, 165/1912, 182/1909, 288/1975, 614/0, 617/0, 619/0, 620/0

olivaceus (N=6): NMNH- 27776, 222400 (skull) NMW-30175 SMF- 72154-6

orientalis (N=6): NMNH- 57948, 57949, 128387, 128569, 128570, 128574 RGS- 444-447

pengilleyi * (N=1): AM R 94494

prasinus * (N=94): AMNH- 59051, 59053, 92335, 92337, 92371, 92663, 99610-11, 101071, 105877-78 BMNH-89.7.1.8, 95.4.26.14-16, 96.7.8.1, 1946.9.1.42 CAS- 10119, 126909, 126922, 135589 FMNH- 14102-3 MCSNG- 28724 MCZ- 4435, 10119, 126798, 126800-1, 137514, 137529-30, 140843, 141304, 149745-6 MVZ- 74904-5 NMW- 12038, 12390:1, 12390:2, 12391:1, 12391:2, 12392-4, 12396, 25203, 30753 QM- J 2118 RGS- 100-1 RMNH- 4812 (holotype), 5369, 5624, 6454, 7210, 7148, 7151-2, 21007-16 SMF- 11626, 59639, 62261, 69454, 69530, 69649 SJSU- uncat. USNM-195775 (3 spec.) UTACV- 6736, 6744, 6816 ZFMK- 3 uncat. ZMB- 8831 ZSM- 3/1970, 4/1970, 23/1916a-b

primordius * (N=8): QM- J 25947, J 25949 RGS- uncat. SMF- 65775, 70138-40 ZFMK- 27108-9, 39187 ZMB- 6204 (holotype)

†*prisca* (N=5): BMNH- PAL R 12007 QM- F 871, F 6562, F 16782, F 16783

rainerguentheri * (N=3): NMNH- 215907, 215908, 306185

rosenbergi (N=14): AM- 3694, 7634, 130775, 13077, 19029, 19031, 19723, NMNH- 63119, 63120 BMNH-1906.11.1.6, 1931.7.1.136-8, 1955.1.4.52 NMW- 12438 *rudicollis* * (N=21): BMNH- I.13a (holotype) CAS- 11017 NMNH- 24002, 26205, 26510, 94952, 101309, 259133 RGS- 53, 371-2, 441-443 SDSNH- 66302 SMF- 46885, 59216, 59239, 60804, 67586, 67644 ZFMK- 54253 ZSM- 8/1919, 68/1914, 366/07, plus 1 uncat

salvadorii * (N=12): BMNH- 1938.6.7.82 MCSNG- 28726 (holotype) RGS- 410, 510 RMNH- 4619, 4620 SMF- 57877-8, 58064, 67670 ZFMK- 53348, plus 1 uncat. ZSM- 2/1970

salvator (N=103): BMNH- 56.11.17.5, 68.11.14.1, 74.8.14.16, 77.10.7.4-5, 94.7.9.1, 96.4.29.18, 1946.8.31.5, 1946.9.1.1, 1980.1717 CAS- 11013, 11015, 13665, 13988, 15428, 15766, 16887, 15428, 20463-4, 61119-21, 62130, 124879, 124880, 153827 MHNP- V 17/22-23 NMNH- 24021 (skull), 28083 (skull), 60898 (skull), 68631 (skull), 216258 (skeleton), 220287 (skull), 287418 NMW- 614, 1668, 12290a RGS -90 RMNH- 3170, 3174, 3176, 3178, 3191a-b, 7222a-b, 7328, 7367, 8713 SMF- 11559-68, 11570-3, 11577, 23192-4, 33126-7, 35148, 48094, 55423, 56441-2, 56465, 65817, 69744, 73947, 74295, 74378-9 ZFMK- 14860 ZMA- 15459-61 ZMB- 7380, 7388-9 ZSM- 11/1918, 132/1909, 137/1907, 137/1909, 140/1907a-b, 340/1909, 342/1909, 609/0, 610/0, 611/0, 613/0, 2608/0, plus 3 uncat

scalaris * (N=19): QM- 10873, 11171, 11218, 28863, 45277, 45357 SMF- 32806 (holotype), 53260 ZMB- 322, 3656, 5202, 6203, 7334, 7821, 9231a-b, 10557, 30125, 30550, 31446, 37686

semiremex * (N=16): MHNP- Va 20/1 NMW- 12389, 12418, 17229 QM- 454, 2151, 2152, 26696, 37574, 38268 SMF- 53259, 57138, 71620-21 ZFMK- 54249 ZMB- 5776 (holotype) ZSM- 281/1982a-b

similis (N=44): BMNH- 64.5.13.23, 78.10.16.33-34, 81.10.12.2, 82.3.27.12, 84.9.13.18, 90.1.14.9-11, 1903.10.19.11-14, 1926.2.25.49-50, 1926.2.25.58-60 CAS- 44134 NMNH- 128481, 128482, 128483, 128484, 128485, 128487, 128489, 128490, 128491, 128492, 128493, 128574, 204026, 204027 SMF- 11634, 54506, 54511, 66723-8, 70149-50 ZMB- 5775

†*sivalensis* * (N=1):[extinct] BMNH- PAL 40819 (holotype)

spenceri (N=2): BMNH- 1906.11.2.3 SMF- 53277

spinulosus * (N=): CAS 180743 NMW 23387 (holotype) RGS 2 uncat

storri * (N=26): BMNH- 1970.2372 QM- 2144, 32236, 60998, 64475, 61776, 67867, 70673 RGS- 16, 146-48 SMF- 59011 (holotype), 53254, 60620-1, 62204, 65887, 66459, 66788, 68230, 71927-8, 71961-2, 72896-7, 73237, plus two uncat. ZFMK- 25536, 40526, 41962, 52914, 54848, 57851 ZSM- 272/1982

telenesetes * (N=1): QM- J 1190 (holotype)

keithhorni * (N=3): QM- J 31566 (holotype), J 35450-51, J 43860

timorensis * (N=42): AM- 97454 BMNH- I.5a, 51.7.17.48, 55.10.16.35, 55.10.16.270, 1926.6.21.15-16, 1926.10.30.54 CAS- 126911-18, plus 1 uncat. MHNP- V/18a, V/18b (syntypes) NMNH- 579388, 579498, 579499 (Timor Leste)] NMW- 12398-9 RGS- 15, 51, 189, plus 2 uncat. RMNH- 3145, 3147, 3181, 5902a-c, 6949, 7016, 7169, 7263, 10409-10 SMF- 35660, 54513, 69973

togianus (N=3): ZMA 11459, 15460, 15461

tristis * (N=52): AM- 51908 BMNH- I.3a, I.3d-f, I.5b, 55.10.16.43, 57.10.24.28-9, 69.5.28.8, 95.11.14.9, 97.1.20.14, 1904.10.7.37, 1905.11.6.1, 1908.2.25.4-6, 1908.2.25.23, 1910.5.28.11, 1924.1.24.6, 1924.3.3.39, 1924.10.25.16, 1932.3.7.94, 1978.882-5 CAS- 77673, 77679, plus 1 uncat. NMNH- 64080, 64081, 64082 NMW- 12400:1, 12400:2, 12401:1, 12401:2, 12402 (3 spec.), 12403, 12419 RGS- uncat. RMNH- 3858 (holotype) SMF- 11630, 53261, 54508, 68243, 70150 ZMB- 33337, 37677 ZSM- 523/1920

varius (N=12): AM- 47899, 94678, 94688, 104940, 107268, 133160 NMW- 1669 QM- J 27564, 33691 SMF- 53278, 66914 ZSM- 41/1978

yemenensis * (N=7): BMNH- 1903.6.26.9-10, 1987.855, 1987.2160 MHNG- 2453.55 ZFMK- 46500 (holotype), 511677

yuwonoi (*): RGS 3 uncat

Bibliography

Abel and Werneburg. 2021. Morphology of the temporal skull region in tetrapods: research history, functional explanations, and a new comprehensive classification scheme. *Biological Reviews* 96(5):2229–2257.

Abler, W. 1992. Serrated teeth of tyrannosaurs. *Paleobiology* 18:161–183.

Abruzzi, T. 2021. Project *Varanus salvatorii. Reptiles online April 7, 2021 at* https://reptilesmagazine.com/project-varanus-salvatorii/#more

Acher, R., J. Chauvet, and M. Chauvet. 1969. The neurohypophyseal hormones of reptiles: Comparison of the viper, cobra, and elaphe active principles. *General and Comparative Endocrinology* 13(3): 357–360.

Ackerman, R. 1991. Physical factors affecting the water exchange of buried reptile eggs. Pp. 193–211. In: D. C. Deeming and M. W. J. Ferguson (eds.) *Egg incubation: Its effects on embryonic development in birds and reptiles.* Cambridge University Press, Cambridge.

Ackerman, R. 2004. Thermal, hydric and respiratory climate of nests. Pp. 15–43. IN: D. Deeming (ed.). *Reptilian incubation: Environment, evolution and behaviour.* Nottingham University Press, Nottingham, UK.

Ackerman, R., T. Barker, G. Birchard, D. Boyer, M. Gamer, S. Hammack, et al. 2002. Egg.

Aerts, P., R. Van Damme, K. D'Aout, and B. Van Hooydonck. 2003. Bipedalism in lizards: Whole-body modelling reveals a possible spandrel. *Philosophical Transactions of the Royal Society B, Biological Sciences* 358:1437.

Ahmadzadeh, F., et al. 2012. Rapid lizard radiation lacking niche conservatism: Ecological diversification within a complex landscape. *Journal of Biogeography* 40:1807–1818.

Ahl, E. 1932. Eine neue Eidechse und zwei neue Frosche von der Insel Jobi. *Mitt. Zoolog. Museum Berlin* 17(6):892–898.

Allam, A., R. Abo-Eleneen, and S. Alothman. 2019. Microstructure of scales in selected lizard species. *Saudi Journal of Biological Sciences* 26(1):129–136. doi:10.1016/j.sjbs.2017.03.012.

Ali, S., E. Stephenson, and W. Elliot. 1982. Bile acids. LXVII. The major bile acids of *Varanus monitor. Journal of Lipid Research* 23(7):947–954.

Aller, B. and M. Manago. 2009. Breeding and reproduction of *Varanus dumerilii.* Pp. 264–265. IN: R. Sprackland (ed.). *Giant lizards.* 2nd ed. TFH Publications, Neptune City.

Al-ma'ruf, A., R. Sari, I. Mustofa, S. Utama, C. Anwar, M. Mafruchati, E. Hestianah, L. Tehupuring, D. Legowo, and B. Setiawan. 2021. Morphology and histology of paryphasmata and hemibaculum of *Varanus salvator* based on sexual maturity. *Open Veterinary Journal* 11(2):330–336.

American Institute of Physics (AIP). 2014. Climate control in termite mounds. ScienceDaily, 25 November 2014. www.sciencedaily.com/releases/2014/11/141125074646.htm.

Amiel, J., T. Lindström, and R. Shine. 2014. Egg incubation effects generate positive correlations between size, speed and learning ability in young lizards. *Animal Cognition* 17:337–347.

Amiel, J., S. Bao, and R. Shine. 2017. The effects of incubation temperature on the development of the cortical forebrain in a lizard. *Animal Cognition* 20:117–125.

Amiel, J. and T. Shine. 2012. Hotter nests produce smarter young lizards. *Biological Letters* 8:372–374. https://doi.org/10.1098/rsbl.2011.1161

Ananjeva, N., M. Dilmuchamedov, and T. Dujsebayeva. 1991. The skin sense organs of some iguanian lizards. *Journal of Herpetology* 25(2):186.

Andres, K., M. von Düring, and H.-G. Horn. 1999. Fine structure of scale glands and preanal glands of monitors Varanidae. Pp. 277–289. IN: H. Horn and W. Böhme (eds.). *Mertensiella Supplement zu Salamandra, number 11, advances in monitor research II.* DGHT, Bonn.

Andrews, R. and S. Donoghue. 2004. Effects of temperature and moisture on embryonic diapause of the veiled chameleon (*Chamaeleo calyptratus*). *Journal of Experimental Zoology* 301A:629–635.

Andrews, R., N. Pezaro, J. S. Doody, F. Guarino, and B. Green. 2017. Oviposition to hatchling: Development of *Varanus rosenbergi. Journal of Herpetology* 51(3):396–401.

Aplin, K., A. Dutton, and D. King. 2006. A new species of *Varanus* Merrem (Squamata: Varanidae) from the Pilbara region of Western Australia, with observations on sexual dimorphism in closely related species. *Zootaxa* 1313:1–38.

Arbuckle, K. 2009. Ecological function of venom in *Varanus*, with a compilation of dietary records from the literature. *Biawak* 3(2):45–56.

Archibald, J. 1994. Metataxon concepts and assessing possible ancestry using phylogenetic systematics. *Systematic Biology* 43(1):27–40. https://doi.org/10.1093/sysbio/43.1.27

Ariano-Sanchez, D., R. Mortensen, R. Wilson, P. Bjureke, S. Reinhardt, and F. Rosell. 2022. Temperature and barometric pressure affect the activity intensity and movement of an endangered thermoconforming lizard. *Ecosphere* 13(3):e3990. https://doi.org/10.1002/ecs2.3990

Arida, E. and W. Böhme. 2010. The origin of *Varanus. Biawak* 4(4):117–124.

Arida, E., A. Stubbs, U. Arifn, B. Karin, S. Hykin, S. Reilly, et al. 2023. Cannibalism in *Varanus timorensis* (Squamata: Varanidae) reveals some reproductive traits of wild population [sic] on Sabu Island, Indonesia. *Russian Journal of Herpetology* 30(2):118–123.

Ariefiandy, A., D. Purwandana, S. Nasu, M. Imansyah, Y. Benu, M. Chrismiawati, P. Kamil, C. Ciofi, and T. Jessop. 2017. *Panduan Lanagan Biawak Komodo (Komodo Dragon field guide).* Yayasan Komodo Survival Program, Denpasar.

Ariefiandy, A., D. Purwandana, S. Nasu, M. Surahman, C. Ciofi, and T. Jessop. 2015. First record of Komodo Dragon nesting activity and hatchling emergence from North Flores, Eastern Indonesia. *BIAWAK* 9(1):33–35.

Arnold, E. N. 1984. Variation in the cloacal and hemipenial muscles of lizards and its bearing on their relationships. *Symposium of the Zoological Society of London* 52:47–85.

Arnold, E. 1989. Towards a phylogeny and biogeography of the Lacertidae: Relationships within an Old-World family of lizards derived from morphology. *Bulletin of the British Museum (Natural History) Zoology* 55:209–257.

Arnold, M. and K. Kunte. 2017. Adaptive genetic exchange: A tangled history of admixture and evolutionary innovation. *Trends in Ecology and Evolution* 32:601–611.

Ast, J. 2001. Mitochondrial DNA evidence and evolution in Varanoidea. *Cladistics* 17:211–226.

Attum, O., R. Earley, M. Bayless, and P. Eason. 2000. The agonistic behaviour of Bosc's monitor (*Varanus exanthematicus* Bosc, 1792) in captivity. *Herpetological Bulletin* 73:22–26.

Aubret, F, G. Blanvillain, and P. Kok. 2015, Aug. 21. Myth busting? Effects of embryo positioning and egg turning on hatching success in the water snake *Natrix maura*. *Scientific Reports* 5:13385. https://doi.org/10.1038/srep13385. PMID: 26294250; PMCID: PMC4543940.

Auffenberg, W. 1976. First description of an adult *Varanus grayi*. *Copeia* 1976(3):586–588.

Auffenberg, W. 1979a. A monitor lizard in the Philippines. *Oryx* 15(1):38–46.

Auffenberg, W. 1979b. Intersexual differences in behavior of captive *Varanus bengalensis* (Reptilia, Lacertilia, Varanidae). *Journal of Herpetology* 13(3):313–315.

Auffenberg, W. 1981a. Combat behaviour in *Varanus bengalensis*. *Journal of the Bombay Natural History Society* 78(1):54–72.

Auffenberg, W. 1981b. *Behavioral ecology of the Komodo monitor*. University Presses of Florida, Gainesville.

Auffenberg, W. 1983. The burrows of *Varanus bengalensis*: Characteristics and use. *Records of the Zoological Survey of India* 80:375–385.

Auffenberg, W. 1988. *Gray's monitor lizard*. University Presses of Florida, Gainesville.

Auffenberg, W. 1994. *The Bengal monitor*. University Presses of Florida, Gainesville.

Auffenberg, W., H. Rahman, F. Iffat, and Z. Perveen. 1988. A study of *Varanus flavescens* (Hardwicke & Gray) (Sauria: Varanidae). *Journal of the Bombay Natural History Society* 86:286–307.

Auffenberg, W., H. Rahman, F. Iffat, and Z. Perveen. 1989. Notes on the biology of *Varanus griseus koniecznyi* Mertens (Sauria: Varanidae). *Journal of the Bombay Natural History Society* 86:26–36.

Augé, M. 2005. *Évolution des lézards du Paléogène en Europe. Mémoires du Muséum national d'Histoire naturelle*, Tome 192. 369 pp.

Augé, M., A. Folie, R. Smith, A. Phelizon, A. Phelizon, A. Gigase†, and T. Smith. 2022. Revision of the oldest varanid, *Saniwa orsmaelensis* Dollo, 1923, from the earliest Eocene of northwest Europe. *Comptes Rendus Palevol* 21:511–529. https://doi.org/10.5852/cr-palevol2022v21a25

Augstenová, B., E. Pensabene, L. Kratochvíl, and M. Rovatsos. 2021. Cytogenetic evidence for sex chromosomes and karyotype evolution in anguimorphan lizards. *Cells* 10:1612.

Auliya, M. 2006. *Taxonomy, life history, and conservation of giant reptiles in West Kalimantan*. NTV Science Books, Münster. 432pp.

Auliya, M. and A. Koch. 2020. Visual identification guide to the monitor lizard species of the world (genus *Varanus*). *BfN-Skripten* 552:1–201.

Avery, R. 1979. *Lizards – A study in thermoregulation*. Institute of Biology's Studies in Biology no. 109. Edward Arnold Publishers, London.

Avila-Perez, T. 1995. *Lizards of Brazilian Amazonia (Reptilia: Squamata). Zoologische Verhandelingen* no. 299. Backyhys, Netherlands.

Ax, P. 1987. *The phylogenetic system*. J. Wiley, New York, NY.

Bahl, K. 1937. Skull of *Varanus monitor* (Linn.). *Records of the Indian Museum* XXXIX:133–174.

Baird, I. 1970. The anatomy of the reptilian ear. Pp. 193–275. IN: C. Gans and T. Parsons (eds.). *Biology of the reptilia, vol. 2, morphology B*. Academic Press, New York, NY.

Baines, F., J. Chattell, J. Dale, D. Garrick, I. Gill, M. Goetz, T. Skelton, and M. Swatman. 2016. How much UV-B does my reptile need? The UV-tool, a guide to the selection of UV lighting for reptiles and amphibians in captivity. *Journal of Zoological and Aquarium Research* 4:42–63.

Baird, T. and J. York. 2021. A decade of sexual selection studies reveals patterns and processes of fitness variation in male collared lizards. *Animal Behaviour* 180:37–49.

Bakker, M. 1947. The seasons in the tropical rainforest. Part 6. Lizards (*Emoia*) Zoological. *Journal of the Linnean Society of London* 41:243–259.

Ballard, V. and F. Antonio. 2001. *Varanus griseus* (desert monitor): Toxicity. *Herpetological Review* 32:262.

Ballinger, R., J. Nietfeldt, J. and Krupa. 1979. An experimental analysis of the role of the tail in attaining high running speed in *Cnemidophorus sexlineatus* (Reptilia: Squamata: Lacertilia). *Herpetologica* 35:114–116.

Balouet, J. 1984. Paléontologie des vertébrés terrestres de Nouvelle Calédonie et palébiogéographie du Pacifique sud-ouest. Thèse de 3ème cycle, Université Pierre et Marie Curie - Paris 6, no. 84–39, 84 pp. (unpublished).

Balouet, J. 1991. The fossil vertebrate record of New Caledonia. Pp. 1383–1409. IN: P. Vickers-Rich, J. Monaghan, R. Baird, and T. Rich (eds.). *Vertebrate palaeontology of Australasia*. Monash University Publications Committee, Melbourne.

Balsai, M. 2001. *The general care and maintenance of popular monitors and Tegus* (The Herpetocultural Library). Advanced Vivarium Systems.

Barba-Montoya, J., Q. Tao, and S. Kumar. 2021. Molecular and morphological clocks for estimating evolutionary divergence times. *BMC Ecology and Evolution* 21, 83.

Barbour, T. 1912. A contribution to the zoogeography of the East Indian Islands. *Memoirs of the Museum of Comparative Zoölogy* XLIV(1).

Barbour, T. and A. Loveridge. 1929. Typical reptiles and amphibians. *Bulletin of the Museum of Comparative Zoology* 69(10):205–360.

Barnett, H. 1979. Incubation of Sand Goanna *Varanus gouldii* eggs. *Herpetofauna* 11(1):21–22.

Bartlett, R. 1981. Notes on captive reproduction of *Varanus storri*. *Bulletin of the Chicago Herpetological Society* 16(3):65–66.

Bartlett, R. 1982. Initial observations on the captive reproduction of *Varanus storri*. *Herpetofauna (Sydney)* 13(2):6–7.

Bartholomew, G. 1982. Physiological control of body temperature. Pp. 167–211. IN: C. Gans and F. Pough (eds.). *Biology of the reptilia*. Academic Press, New York, NY.

Bartholomew, G. and V. Tucker. 1964. Size, body, temperature, thermal conductance, oxygen consumption, and heart rate in Australian varanid lizards. *Physiological Zoology* 37:341–354.

Basson, C., O. Levy, M. Angilletta, Jr., and S. Clusella-Trullas. 2017. Lizards paid a greater opportunity cost to thermoregulate in a less heterogeneous environment. *Functional Ecology* 31:856–865.

Batabyal, A. and M. Thaker. 2019. Lizards from suburban areas learn faster to stay safe. *Biological Letters* 15:20190009. https://doi.org/10.1098/rsbl.2019.0009

Bauer, A., A. Griffing, T. Dujsebayeva, Z. Davis, and A. Russell. 2023. Morphology of cutaneous sense organs of gekkonid geckos (Squamata: Gekkonidae): Structural variation in the most diverse gekkotan family. *Breviora* 576(1):1–28. https://doi.org/10.3099/0006-9698-576.1.1

Bauer, A. and A. Russell. 1988. Morphology of gekkonid cutaneous sensilla, with comments on function and phylogeny in the Carphodactylini (Reptilia: Gekkonidae). *Canadian Journal of Zoology* 66:1583–1588.

Baum, W. 2013. What counts as behavior? The Molar Multiscale view. *Behavioral Analysis* 36(2):283–293.

Baverstock, P., D. King, M. King, and M. Krieg. 1993. The evolution of species of Varanidae: Microcomplement fixation analysis of serum albumins. *Australian Journal of Zoology* 41:621–638.

Bavya. N.D. Skeleton of *Varanus* (With Diagram) | Vertebrates | Chordata | Zoology. https://www.notesonzoology.com/vertebrates/varanus/skeleton-of-varanus-with-diagram-vertebrates-chordata-zoology/8396

Bayless, M. 1993. Reproductive notes on the black roughneck monitor lizard (*Varanus rudicollis* Gray, 1845). *Varanews* 3(2):3.

Bayless, M. 2004. The local names of Pacific Monitor Lizards (Sauria: Varanidae) of Oceania & Indo-Malaysia, excluding Australia. *Micronesica* 37(1):49–55.

Bayless, M. and B. Aller. 2004. Dumeril's monitor lizard (*Varanus dumerilii*). *Reptiles* 12(3):42–46.

Bayless, M. and G. Naclerio. 1993. A report of *Varanus* on New Caledonia. *VaraNews* 3(3):1–2.

Bayless, M. and R. Sprackland. 2000a. The taxonomy of Africa's savannah and Cape monitor lizards. Part 1. *Reptiles* 8(6):76–85.

Bayless, M. and R. Sprackland. 2000b. The taxonomy of Africa's savannah and Cape monitor lizards. Part 2. *Reptiles* 8(7):40–47.

Becker, H., W. Bohme, and S. Perry. 1989. Die Lungenmorphologie der Warane (Reptilia: Varanidae) und ihre systematisch-stammesgeschichtliche Bedeutung. *Bonner Zoologischen Beitreitag* 40(1):27–56.

Bedford, G. and K. Christian. 1996. Tail morphology related to habitat of varanid lizards and some other reptiles. *Amphibia-Reptilia* 17:131–140.

Bedford, G. and G. Husband. 2003. Captive breeding of the Lemon Throated Monitor, *Varanus baritji*. *Herpetofauna* 33.

Belinsky, A., R. Ackerman, R. Dmi'el, and A. Ar. 2004. Water in reptilian eggs and hatchlings. Pp. 125–141. IN: D. Deeming (ed.). *Reptilian incubation: Environment, evolution and behaviour.* Nottingham University Press, Nottingham, UK.

Bellairs, A. 1949. Observations on the snout of *Varanus*, and a comparison with that of other lizards and snakes. *Journal of Anatomy* 83:116–146.

Bels, V. and A. Russell (eds.). 2019. *Behavior of lizards: Evolutionary and mechanistic approaches*. CRC Press, Boca Raton, FL.

Beltrami, G., C. Bertolucci, A. Parretta, et al. 2010. A sky polarization compass in lizards: The central role of the parietal eye. *Journal of Experimental Biology* 213:2048–2054. https://doi.org/10.1242/jeb.040246

Benedict, F. 1932. The physiology of large reptiles with special reference to the heat production of snakes, tortoises, lizards, and alligators. *Carnegie Institute of Washington* 425:1–539.

Bennett, A. 1972. The effect of activity on oxygen consumption, oxygen debt, and heart rate in the lizards *Varanus gouldii* and *Sauromalus hispidus*. *Journal of Comparative Physiology* 79:259–280.

Bennett, D. 1994. Some remarks on longevity in monitors. *Southwestern Herpetological Society Journal (UK)* 2(3):27–29.

Bennett, D. 1995. *A little book of monitor lizards*. Viper Press, Aberdeen, UK.

Bennett, D. 1998. *Monitor lizards: Natural history, biology & husbandry*. Edition Chimaira, Frankfurt.

Bennett, D. 2014. The arboreal foraging behavior of the frugivorous monitor lizard *Varanus olivaceus* on Polillo Island. *Biawak* 8(1):15–18.

Bennett, D. and W. Akonnor. 1995. *A preliminary investigation into the exploitation of monitor lizard Varanus exanthematicus in the coastal savannah of Ghana*. Unpublished Report to Department of Game and Wildlife, Accra, Ghana.

Bennett, D. and L. Liat. 1995. A note on the distribution of *Varanus dumerilii* and *V. rudicollis* in peninsular Malaysia. *Malaysian Nature Journal* 49:113–116.

Berghof, H.-P. 2001. Die Pflege und Vermehrung des kleinen neuguineischen Baumwaranes *Varanus (Odatria) similis* MERTENS 1958. *Herpetofauna* 23(135):5–13.

Bergmanson, J., J. Walsh, and K. Schulle. 2011. Commercially available reptile lights – Good for animal bad for handler? *Investigative Ophthalmology & Visual Science* 52(14):1547.

Bernstein, J. M., et al. 2022. Undescribed diversity in a widespread, common group of Asian Mud Snakes (Serpentes: Homalopsidae: Hypsiscopus). *Ichthyology and Herpetology* 110:561–574.

Berry, R. 2012. Circadian rhythm sleep disorders. Chapter 26. Pp. 515–543. IN: R. Berry (ed.). *Sleep medicine*. Saunders, Philadelphia, PA.

Bezy, R. 2019. *Night lizards; field memoirs and a summary of the Xantusiidae*. ECO Herpetological Publishing and Distribution, Rodeo, NM. 220 pp.

Bezzina, C., J. Amiel, and R. Shine. 2014. Does invasion success reflect superior cognitive ability? A case study of two congeneric lizard species (*Lampropholis*, Scincidae). *PLoS One* 9:e86271. https://doi.org/10.1371/journal.pone.0086271

Bhullar, B. 2011. The power and utility of morphological characters in systematics: A fully resolved phylogeny of *Xenosaurus* and its fossil relatives (Squamata: Anguimorpha). *Bulletin of the Museum of Comparative Zoology* 160(3):65–181. https://doi.org/10.3099/0027-4100-160.3.65

Bickler, P. and R. Anderson. 1986. Ventilation, gas exchange, and aerobic scope in a small monitor lizard, *Varanus gilleni*. *Physiological Zoology* 59:76–83.

Birchard, G. 2004. Pp. 103–123. IN: D. Deeming (ed.). *Reptilian incubation: Environment, evolution and behaviour.* Nottingham University Press, Nottingham, UK.

Biswas, S. and S. Kar. 1981. Some observations on nesting habits and biology of *Varanus salvator* (Laurenti) of Bhitarkanika Sanctuary, Orissa. *Journal of the Bombay Natural History Society* 78:303–307.

Blem, C. and M. Zimmerman. 1986. The energetics of shedding: Energy content of snake skin. *Comparative Biochemistry and Physiology* 83A:66–65.

Bochaton, C., M. Ivanov, and J. Claude. 2019. Osteological criteria for the specific identification of monitor lizards (*Varanus* Merrem, 1820) remains in subfossil deposits of Sundaland and continental Southeast Asia. *Amphibia-Reptilia* 40:219–232.

Bock, D., S. Baeckens, J. Pita-Aquino, Z. Chejanovski, S. Michalides, P. Muraldihar, O. Lapiedra, S. Park, D. Menke, A. Geneva, J. Losos, and J. Kolbe. 2021. Changes in selection pressure can facilitate hybridization during biological invasion in a Cuban lizard. *Proceedings of the National Academy of Sciences* 118(42):e2108638118.

Bogert, C. and R. del Campo. 1960. The Gila Monster and its allies: The relationships, habits, and behavior of the lizards of the family Helodermatidae. *Bulletin of the American Museum of Natural History* 109(1):article 1.

Böhme, W. 1988. Zur Genitalmorphologie der Sauria: funktionale undstammesgeschichtlichte Aspekte. *Bonner Zoologischen Monographien* 27:1–176.

Böhme, W. 1991. New findings on the hemipenial morphology of monitor lizards and their systematic implications. *Mertensiella* 2:42–49.

Böhme, W. 1995. Hemiclitoris discovered, a fully differentiated erectile structure in female monitor lizards (*Varanus* ssp.) (Reptilia, Varanidae) *Journal of Zoological Systematics and Evolution Research* 33:129–132.

Böhme, W. 2003. Checklist of the living monitor lizards of the world (family Varanidae). *Zoologishes Verhandlungen Leiden* 341:4–43.

Böhme, W., H.-G. Horn, and T. Ziegler. 1994. Zur Taxonomie der Pazifikwarane (*Varanus indicus*-Komplex): Revalidierung von *Varanus doreanus* (A.B. Meyer, 1874) mit Beschreibung einer neuen Unterart. *Salamandra* 30(2):119–142.

Böhme, W., H. Jacobs, T. Koppetsch, and A. Schmitz. 2019. The Kei Islands monitor lizard (Squamata: Varanidae: *Varanus*: *Euprepiosaurus*) as a distinct morphological, taxonomic, and conservation unit. *Russian Journal of Herpetology* 26(5):272–280.

Böhme, W. and T. Ziegler. 1997. *Varanus melinus* sp. n., ein neuer Waran aus der *V. indicus*-Gruppe Von Den Molukken, Indonesien. *Herpetofauna (Germany)* 19(111):26–34.

Böhme, W. and T. Ziegler. 2007. Notes on the distribution, diet, hemipenis, morphology and systematics of *Varanus spinulosus* Mertens, 1941 *Mertensiella* 16:100–108.

Bonino, M., D. L. M. Azócar, J. A Schulte, II, and C. F. B. 2015. Climate change and lizards: Changing species' geographic ranges in Patagonia. *Regional Environmental Change* 15:1121–1132.

Booth, D. 2004. Artificial incubation. Pp. 253–270. IN: D. Deeming (ed.). *Reptilian incubation: Environment, evolution and behaviour.* Nottingham University Press, Nottingham, UK.

Borsuk-Białynicka, M. 1984. Anguimorphans and related lizards from the Late Cretaceous of the Gobi Desert, Mongolia. *Palaeontologica Polonica* 46:5–105.

Boulenger, G. 1885. *Catalogue of the lizards in the British Museum*. Trustees of the British Museum (Natural History).

Bowmaker, J. 2008. Evolution of vertebrate visual pigments. *Vision Research* 48, 2022–2041. https://doi.org/10.1016/j.visres.2008.03.025

Branch, W. 1982. Hemipenial morphology of platynotan lizards. *Journal of Herpetology* 16(1):16–38.

Brandenberg, T. 1983. *Monitors in the Indo-Australian archipelago.* Unpublished thesis, University of Leiden. 121 pp.

Brattstrom, B. 1978. Learning studies in lizards. Pp. 173–181. IN: N. Greenberg and P. MacLean (eds.). *Behavior and neurology of lizards: An interdisciplinary colloquium.* US Department of Health, Education, and Welfare Publication No. ADM 77–491, Washington, DC.

Braysher, M. 1986. The excretion of hyperosmotic urine and other aspects of the electrolyte balance of the lizard *Amphibolurus maculosus*. *Comparative Biochemistry and Physiology* 54A:341–345.

Bredl, J. 1987. First captive breeding of the perentie (*Varanus giganteus*). *Thylacinus* 12(1):2–3.

Bredl, J. and T. Schwaner. 1983. First record of captive propagation of the Lace Monitor, *Varanus varius* (Sauria: Varanidae). *Herpetofauna* 15(1):20–21.

Brehm, A. 1883. *Brehm's Thierleben. Algemeine kunde des Thierreichs. Dritte Abtheilung—Kriedjthiere, Lurdje und Fisdje. Drister Band.* Verlag des Bibliografischen Instituts, Leipzig.

Brennan, I., D. Chapple, J. Keogh, and S. Donnellan. 2024. Evolutionary bursts drive morphological novelty in the world's largest skinks. *Current Biology* 34(7):P3905–3916.E5.

Brennan, I., A. Lemmon, E. Lemmon, D. Portik, V. Weijola, L. Welton, S. Donnellan, and J. Keogh. 2021. Phylogenomics of monitor lizards and the role of competition in dictating body size disparity. *Systematic Biology* 70(1):120–132. https://doi.org/10.1093/sysbio/syaa046. PMID: 32521014.

Brewster, C. and S. Beaupre. 2019. The effect of body posture on available habitat and activity- time in a lizard: implications for thermal ecology studies. *Journal of Thermal Biology* 82:10–17. PMID:31128637.

Brogniart, A. 1800. Essai d'une classification naturelle des reptiles. *Bulletin de la Société Philomatique* 2(36):89–91.

Bromham, L. 2002. Molecular clocks in reptiles: Life history influences rate of molecular evolution. *Molecular Biology and Evolution* 19(3):302–309. https://doi.org/10.1093/oxfordjournals.molbev.a004083

Bromham, L. 2021. *Causes of variation in the rate of molecular evolution BT—The molecular evolutionary clock: Theory and practice.* Pp. 45–64. IN: S. Ho (ed.). Springer International Publishing, New York, NY.

Brongersma, L. 1948. Lizards from the island of Morotai (Moluccas). *Proceedings of the Koninklijke Nederlandse Akademie van Wetenschappen* LI(4):3–12.

Brongersma, L. 1958. On an extinct species of the genus *Varanus* (Reptilia, Sauria) from the island of Flores. *Zoologische Mededelingen* 36(7):113–125.

Brooks, D. and K. Jackson. 2007. Do crocodiles co-opt their sense of "touch" to "taste"? A possible new type of vertebrate sensory organ. *Amphibibia-Reptilia* 28:277–285. https://doi.org/10.1163/156853807780202486

Brown, C., K. Laland, and J. Krause. 2011. *Fish cognition and behavior.* Wiley, New York, NY.

Brown, D. 2012. *A guide to Australian monitors in captivity.* Reptile Publications, Australia.

Brown, J. and M. Lomolino. 1998. *Biogeography.* 2nd Ed. Sinauer Publishing, Sunderland, MA.

Brunes, T., A. da Silva, S. Marques-Souza, M. Rodrigues, and K. Pellegrino. 2019. Not always young: The first vertebrate ancient origin of true parthenogenesis found in an Amazon leaf litter lizard with evidence of mitochondrial haplotypes surfing on the wave of a range expansion. *Molecular Phylogenetics and Evolution* 135:105–101.

Brygoo, E. 1987. Les types de Varanides (Reptiles, Sauriens) du Museum National d'Histoire Naturelle Catalogue critique. *Bulletin du Museum national d'histoire naturelle* 4(9):21–28.

Bucklitsch, Y., W. Böhme, and A. Koch. 2016. Scale morphology and micro-structure of monitor lizards (Squamata: Varanidae: *Varanus* spp.) and their allies: Implications for systematics, ecology, and conservation. *Zootaxa* 4153(1):1–192.

Bullet, P. 1942. Beitrage zur Kenntnis des Gebisses von *Varanus salvator* Laur. *Vierteljahrsschrift der Naturforschenden Gesellschaft Zürich* 87:139–142.

Burbrink, F. 2001. Systematics of the Eastern Ratsnake complex (*Elaphe obsoleta*). *Herpetological Monographs* 15(15). https://doi.org/10.2307/1467037

Burbrink, F., B. Crother, C. Murray, B. Smith, S. Ruane, E. Myers, and R. Pyron. 2022. Empirical and philosophical problems with the subspecies rank. *Ecology and Evolution* 17. https://doi.org/10.1002/ece3.9069

Burbrink, F., R. Lawson, and J. Slowinski. 2000. Mitochondrial DNA phylogeography of the polytypic North American rat snake (*Elaphe obsoleta*): A critique of the subspecies concept. *Evolution* 54(6):2107–2118.

Burden, D. 1927a. The quest for the dragon of Komodo. *Natural History* XXVII: 3–18.

Burden, W. 1927b. *Dragon lizards of Komodo.* Putnam and Co., New York, NY. 206 pp.

Burden, W. 1927c. Observations on the habits and distribution of *Varanus komodoensis* Ouwens. *American Museum Novitates* 316:1–10.

Burger, J. 1989. Incubation temperature has long-term effects on behavior of young pine snakes (*Pituophis melanoleucus*). *Behavioral Ecology and Sociobiology* 24:201–207

Burghardt, G. 1978. Learning processes in reptiles: Detour problems and delayed reactions. Pp. 610–614. IN: C. Gans and D. Tinkle (eds.). *Biology of the Reptilia, vol. 7. Ecology and behaviour A.* Academic Press, London.

Burghardt, G. 2013. Environmental enrichment and cognitive complexity in reptiles and amphibians: Concepts, review, and implications for captive populations. *Applied Animal Behaviour Science* 147(3–4):286–298.

Burghardt, G., B. Ward, and R. Rosscoe. 1996. Problem of reptile play: Environmental enrichment and play behavior in a captive Nile soft-shelled turtle, *Trionyx triunguis. Zoo Biology* 15(3):223–238.

Burggren, W. and K. Johansen. 1982. Ventricular haemodynamics in the monitor lizard *Varanus exanthematicus*: Pulmonary and systemic pressure separation. *Journal of Experimental Biology* 96:343–354.

Caldwell, M. 2021. *The origin of snakes: Morphology and the fossil record.* CRC Press, Boca Raton, FL. 297 pp.

Caldwell, M. 2007. Snake phylogeny, origins, and evolution: The role, impact, and importance of fossils. Pp. 253–302. IN: J. Anderson and H.-D. Sues (eds.). *Major transitions in vertebrate evolution.* Indiana University Press, Bloomington, IN.

Caldwell, M. 2003. Holotype snout elements of *Saniwa ensidens* reassigned to cf. *Restes* sp. indet. (Xenosauridae). *Journal of Paleontology* 77(2):393–396.

Camina, A. 2019. *Varanos: Biologia, mantenimiento en captividad y clinica. Segunda Edición.* Fardatxo Ediciones, Valencia.

Camina, A. and A. Koch. 2017. First breeding report of the Pilbara Rock Monitor (*Varanus pilbarensis* Storr, 1980) in Spain with taxonomic comments on the recently described *V. hamersleyensis* Maryan et al., 2014. *BIAWAK* 11(1):19–27.

Camp, C. 1923. Classification of the lizards. *Bulletin of the American Museum of Natural History* XLVIII(11):289–481.

Campbell, H. 1969. The effects of temperature on the auditory sensitivity of lizards. *Physiological Zoology* 42:183–210.

Card, W. 1993. Significant monitor hatchings at Dallas Zoo. *AZA Comminique*, April 16.

Card, W. and A. Kluge. 1995. Hemipeneal skeleton and varanid systematics. *Journal of Herpetology* 29(2):275–280.

Carpenter, C., J. Gillingham, J. Murphy, and L. Mitchell. 1976. A further analysis of the combat ritual of the pygmy mulga monitor, *Varanus gilleni* (Reptilia: Varanidae). *Herpetologica* 32:35–40.

Carter, D. 1999a. Reproductive cycle of the lace monitor *Varanus varius.* Pp. 129–135. IN: H. Horn and W. Böhme (eds.). *Mertensiella Supplement zu Salamandra, number 11, advances in monitor research II.* DGHT, Bonn.

Carter, D. 1999b. Nesting and evidence of parental care by the Lace Monitor *Varanus varius.* Pp. 137–147. IN: H. Horn and W. Böhme (eds.). *Mertensiella Supplement zu Salamandra, number 11, advances in monitor research II.* DGHT, Bonn.

Casas, L. and F. Saborido-Rey. 2021. Environmental cues and mechanisms underpinning sex change in fish. *Sex Development* 15(1–3):108–121.

Case, T. and T. Schwaner.1993. Island/mainland body size differences in Australian varanid lizards. *Oecologica* 94:102–109.

Castiglione, S., G. Tesone, M. Piccolo, M. Melchionna, A. Mondanaro, C. Serio, M. Di Febbraro, and P. Raia. 2018. A new method for testing evolutionary rate variation and shifts in phenotypic evolution. *Methods in Ecology and Evolution* 2018:181–110.

Čerňanský, A. and R. Patnaik. 2025. The first cranial material of *Varanus* from the Miocene of India. *The Anatomical Record/Early View.* https://doi.org/10.1002/ar.25676

Čerňanský, A., N. Singh, R. Patnaik, K. Sharma, R. Tiwari, R. Sehgal, N. Singh, and D. Choudhary. 2021. The Miocene fossil lizards from Kutch (Gujarat), India: A rare window to the past diversity of this subcontinent. *Journal of Paleontology* 96:213–223.

Chandrasekara, U., M. Mancuso, L. Seneci, L. Bourke, D. Trembath, J. Sumner, C. Zdenek and B. Fry. 2024. A Russian doll of resistance: Nested gains and losses of venom immunity in varanid lizards. *International Journal of Molecular Science* 25:2628. https://doi.org/10.3390/ijms25052628

Christian, K. 2004. *Varanus panoptes.* Pp. 423–429. IN: E. R. Pianka and D. King (eds.). *Varanoid lizards of the world.* Indiana University Press, Bloomberg, IN.

Christian, A. 2007. Scaling effects in monitor lizards and consequences for the evolution of large size in terrestrial vertebrates. *Mertensiella* 16:1–9.

Christian, K. and G. Bedford. 1996. Thermoregulation by the spotted tree monitor, *Varanus scalaris*, in the seasonal tropics of Australia. *Journal of Thermal Biology* 21:67–73.

Christian, K. and K. Conley. 1994. Activity and resting metabolism of varanid lizards compared with "typical" lizards. *Australian Journal of Zoology* 42:185–193.

Christian, K., L. Corbett, B. Green, and B. Weavers. 1995. Seasonal activity and energetics of two species of varanid lizards in tropical Australia. *Oecologia* 103:349–357.

Christian, K., B. Green, G. Bedford, and K. Newgrain. 1996. Seasonal metabolism of a small, arboreal monitor lizard, *Varanus scalaris*, in tropical Australia. *Journal of Zoology* 240:383–396.

Christian, K. and B. Weavers. 1994. Analysis of the activity and energetics of the lizard *Varanus rosenbergi*. *Copeia* 1994:289–295.

Christian, K. and B. Weavers. 1996. Thermoregulation of monitor lizards in Australia: An evaluation of the methods in thermal biology. *Ecological Monographs* 66:139–157.

Christian, K., B. Weavers, B. Green, and G. Bedford. 1996. Energetics and water flux in a semiaquatic lizard, *Varanus mertensi*. *Copeia* 1996:354–362.

Cieri, R. 2018. The axial anatomy of monitor lizards (Varanidae). *Journal of Anatomy* 233:636–643.

Ciofi, C. and M. de Boer. 2004. Distribution and conservation of the Komodo monitor (*Varanus komodoensis*). *Herpetological Journal* 14(2):99–107.

Clark, B., J. Amiel, R. Shine, D. Noble, and M. Whiting. 2014. Colour discrimination and associative learning in hatchling lizards incubated at "hot" and "cold" temperatures. *Behavioral Ecology and Sociobiology* 68:239–247.

Clark, T., T. Wang, P. Butler, and P. Frappell. 2005. Factorial scopes of cardio-metabolic variables remain constant with changes in body temperature in the varanid lizard, *Varanus rosenbergi*. *American Journal of Physiology and Regular Integrative Comparative Physiology* 288:R992–R997.

Clemente, C. 2014. The evolution of bipedal running in lizards suggests a consequential origin may be exploited in later lineages. *Evolution* 68(8):2171–2183.

Clemente, C. J., G. G. Thompson, and P. C. Withers. 2009. Evolutionary relationships of sprint speed in Australian varanid lizards. *Journal of Zoology* 278(4):270–280.

Clemente, C. J., P. C. Withers, and G. G. Thompson. 2009. Metabolic rate and endurance capacity in Australian varanid lizards (Squamata: Varanidae: *Varanus*). *Biological Journal of the Linnean Society* 97(3):664–676.

Clos, L. 1995. A new species of *Varanus* (Reptilia: Sauria) from the Miocene of Kenya. *Journal of Vertebrate Paleontology* 15(2):254–267.

Cogger, H. 1959. Australian goannas. *The Australian Museum Magazine* XIII(3):71–75.

Cogger, H. 1983. *Reptiles and amphibians of Australia*. 1st Ed. Reed Pty. Ltd. & Cornell University Press, Ithaca, NY.

Cogger, H. 1992. *Reptiles and amphibians of Australia*. 5th Ed. Reed Pty. Ltd. & Cornell University Press, Ithaca, NY.

Cogger, H. 2000. *Reptiles and amphibians of Australia*. 6th Ed. Ralph Curtis Books, Sanibel Island, FL.

Cogger, H. and H. Heatwole. 1981. The Australian reptiles: Origins, biogeography, distribution patterns and island evolution. Pp. 1333–1373. IN: A. Keast (ed.). *Ecological biogeography of Australia*. Dr. W. Junk bv Publishers, The Hague.

Cole, C., H. Dessauer, and A. Markezich. 1993. Missing link found: The second ancestor of *Gymnophthalmus underwoodi* (Squamata, Teiidae), a South American unisexual lizard of hybrid origin. *American Museum Novitates* 3055:1–13.

Collins, J. and F. Cross. 1995. *Fishes in Kansas*. University of Kansas Natural History Museum Publication, Lawrence.

Condon, J. 1987. A kinematic analysis of mesokinesis in the Nile monitor (*Varanus niloticus*). *Experimental Biology* 47(2):73–87.

Conley, D. and M. Lattanzio. 2022. Active regulation of ultraviolet light exposure overrides thermal preference behaviour in eastern fence lizards. *Functional Ecology* 36:2240–2250. https://doi.org/10.1111/1365-2435.14114

Conrad, J. 2008. Phylogeny and systematics of Squamata (Reptilia) based on morphology. *Bulletin of the American Museum of Natural History* 310:1–182.

Conrad, J. 2015. Skeletons of the little-known Palawan monitor, *Varanus palawanensis* (Squamata: Varanidae). *Journal of Herpetology* 49(3):485–490.

Conrad, J., A, Balcarcel, and C. Mehling. 2012. Earliest example of a giant monitor lizard (*Varanus*, Varanidae, Squamata). *PLoS One* 7(8):e41767. https://doi.org/10.1371/journal.pone.0041767

Conrad, J., O. Rieppel, and L. Grande. 2008. Re-assessment of varanid evolution based on new data from *Saniwa ensidens* Leidy, 1870 (Squamata, Reptilia). *American Museum Novitates* 3630: 1–15.

Cooper, T., A. Liew, G. Andrle, E. Cafritz, H. Dallas, T. Niesen, E. Slater, J. Stockert, T. Young, and J. Mendelson. 2019. Latency in problem solving as evidence for learning in varanid and helodermatid lizards, with comments on foraging techniques. *Copeia* 107:78–84.

Cooper, T., C. Zabinski, E. Adams, and R. Hill, 2020. Long-term memory of a complex foraging task in monitor lizards (Reptilia: Squamata: Varanidae). *Journal of Herpetology* 54(3):378–383.

Cooper, W. 1989. Prey odor discrimination in the varanoid lizards *Heloderma suspectum* and *Varanus exanthematicus*. *Ethology* 81:250–258.

Cope, E. 1869. On the reptilian orders Pythonomorpha and Streptosauria. *Proceedings of the Boston Society of Natural History* 12:250–266.

Corcoran, M. 2023. How to evaluate a reptile chemistry panel. *DVM36* 54(4):26. online at https://www.dvm360.com/view/how-to-evaluate-a-reptile-chemistry-panel

Costa-Campos, C. and J. Costa-Anaissi. 2020. Death-feigning as defensive behaviour of eight lizard species of the Amazon rainforest. *Herpetological Bulletin* 152:26–27.

Cota, M., T. Chan-Ard, and S. Makchai. 2008. Geographical distribution, instinctive feeding behavior and report of nocturnal activity of *Varanus dumerilii* in Thailand. *Biawak* 3(4):134–143.

Cota, M., T. Chan-Ard, and S. Makchai. 2009. Geographical distribution and regional variation of *Varanus salvator macromaculatus* in Thailand. *Biawak* 3(4):134–143.

Cota, M. and U. Krebs. 2016. Do the hatchlings of Dumeril's Monitor (*Varanus dumerilii*) Schlegel, 1839 display Batesian mimicry? A conspicuous phenomenon and its presumptive evidence. Pp. 69–92. IN: Cota (ed.), *Proceedings of the Interdisciplinary Conference on Monitor Lizards convened at the Grand View Hotel Pranakhon Rajabhat University, Bangkok, Thailand 27-29 July 2015*. Institute for Research and Development, Suan Sunandha University, Bangkok.

Cousteau, J. 1972. The odyssey of the Cousteau team. Episode 19, Octopus, Octopus. Online at https://www.youtube.com/watch?v=CGcl_R5kDBU

Cowles, R. 1930. The life history of *Varanus niloticus* as observed in Natal, South Africa. *Journal of Entomology and Zoology* 22(1):1–30.

Cracraft, J. 1989. Species as entities of biological theory. IN: M. Ruse (ed.). *What the philosophy of biology is*. Nijhoff International Philosophy Series, vol 32. Springer-Verlag, Dordrecht. https://doi.org/10.1007/978-94-009-1169-7_3

Crisp, M., L. Cook, and F. Hereward. 1979. Colour and heat balance in the lizard *Lacerta dugesii*. *Copeia* 1979:250–258.

Crombie, R. and D. Steadman. 1986. The lizards of Rarotonga and Mangaia, Cook Island group, Oceania. *Pacific Science* 40(1–4):44–57.

Czechura, G. 1999. Museums, preserves of nature. *Wildlife Australia* 36(2):12–18.

Dalhuijsen, K., W. Branch, and G. Alexander. 2015. A comparative analysis of the diets of *Varanus albigularis* and *Varanus niloticus* in South Africa. *African Zoology* 49. https://doi.org/10.1080/15627020.2014.11407621

Dalrymple, G. 1979. On the jaw mechanism of the snail-crushing lizards, Dracaena Daudin 1802 (Reptilia, Lacertilia, Teiidae). *Journal of Herpetology* 13(3):303–311.

D'Amore, D. 2015. Illustrating ontogenetic change in the dentition of the Nile monitor lizard, *Varanus niloticus:* A case study in the application of geometric morphometric methods for the quantification of shape-size heterodonty. *Journal of Anatomy* 226(5):403–419.

Daniel, J. C. 2002. *The book of Indian reptiles & amphibians*. Bombay Natural History Society and Oxford University Press, Mumbai.

Darevsky, I. 1958. Natural parthenogenesis in some subspecies of rock lizards (*Lacerta saxicola* Eversmann) distributed in *Armenia. Proceedings of the Academy of Science of the Armenian SSR* 122(4):730–732.

Darevsky, I. and S. Kadarsan. 1964. Zur Biologie des grosen Indonesischen Warens (*Varanus komodoensis* Ouwens). *Zoological Journal (Moscow)* 43:1263–1269.

Davis, R., R. Darling, and A. Darlington. 1986. Ritualized combat in captive Dumeril's monitors, *Varanus dumerilii. Herpetological Review* 17:85–88.

Darwin, C. 1859. *On the origin of species*. John Murray, London.

Daudin, F. 1802. *Histoire naturelle, reptiles. Tome 3*. Paris. 452 pp.

Dawson, W. and J. Templeton. 1963. Physiological responses to temperature in the lizard *Crotaphytus collaris. Physiological Zoology* 36:219–236.

Dedelmar, A. 2007. Keeping and breeding the blue tree monitor, *Varanus macraei. Reptilia* 50:25–27.

Deeming, D. 2004. Post-hatching phenotypic effects if incubation in reptiles. Pp. 229–251. IN: D. Deeming (ed.). *Reptilian incubation: Environment, evolution and behaviour.* Nottingham University Press, UK.

De Graf, H. 1886. Zur Anatomie unt Entwicklung der Epiphyse bei Amphibien und Reptilien. *Zoologisches Anzeiger* 9:191–194.

De Jong, J. and L. Brongersma. 1927. Die Bewegungen im Schädel von *Varanus komodoensis* Ouwens. *Zoologisches Anzeiger* 70:67–69.

Delfino, M. 2012. Synonymization of the Miocene varanid lizard *Iberovaranus* Hoffstetter, 1969 with *Varanus* Merrem, 1820. *Acta Palaeontologica Polonica*. https://doi.org/10.4202/app.2012.0025

Delfino, M., J.-C. Rage, A. Bolet, and D. Alba. 2013. Early Miocene dispersal of the lizard *Varanus* into Europe: Reassessment of vertebral Material from Spain. *Acta Palaeontologica Polonica* 58(4).

De Lisle, H. 1996. *Natural history of monitor lizards*. Krieger Publishing, Malabar, FL.

De Meester, G., K. Huyghe, and R. Van Damme. 2019. Brain size, ecology and sociality: A reptilian perspective. *Biological Journal of the Linnean Society* 126(3):381–391.

De Queiroz, K. 1998. The general lineage concept of species, species criteria, and the process of speciation. Pp. 57–75. IN: D. Howard and S. Berlocher (eds.). *Endless forms: Species and speciation*. Oxford University Press, Oxford, UK.

De Queiroz, K. 2007. Species concepts and species delimitation. *Systematic Biology* 56(6):879–886.

Dendy, A. 1899. On the development of the parietal eye and adjacent organs in Sphenodon (*Hatteria*). *Quarterly Journal of Microscopical Science* 42:111–153.

Deraniyagala, P. 1944. Four new races of the "kabaragoya" lizard, *Varanus salvator. Spoilia Zeylonica* 24:59–62.

Deraniyagala, R. 1958a. Pseudo-combat of the monitor lizard *Varanus bengalensis* (Daudin). *Spolia Zeylandica* 28:11–13.

Deraniyagala, P. 1958b. Reproduction in the monitor lizard *Varanus bengalensis. Spolia Zeylonica* 28(part 11):18–20, 165–166.

DeRooij, N. 1915. *The reptiles of the Indo-Australian archipelago. I. Lacertilia, Chelonia, Emydosauria*. E.J. Brill, Leiden. 384 pp.

Dick, T. and C. Clemente. 2016. How to build your dragon: Scaling of muscle architecture: From world's smallest to world's largest monitor lizard. *Frontiers in Zoology* 13(1). https://doi.org/10.1186/s12983-016-0141-5

Dick, T., J. Wakeling, and C. Clemente. 2016. Scaling of muscle architecture: From world's smallest to world's largest monitor lizard. *Integrative and Comparative Biology* 56(Supplement 1):E277.

Diehl, J., F. Baines, A. Heijboer, J. van Leeuwen, M. Kik, W. Hendriks, and D. Oonincx. 2018. A comparison of UVb compact lamps in enabling cutaneous vitamin D synthesis in growing bearded dragons. *Journal of Animal Physiology and Animal Nutrition (Berlin)*. 102(1):308–316. doi: 10.1111/jpn.12728. Epub 2017 Apr 27. PMID: 28452197

Ditmars, R. 1910. *Reptiles of the world*. Macmillan, New York.

Ditmars, R. 1956. *Reptiles of the world*. Macmillan, New York.

Dmi'el, R. 1967. Studies on reproduction, growth, and feeding in the snake *Spalerosophis cliffordi* (Colubridae). *Copeia* 1967:332–346.

Dobry, J., E. Wapstra, E. Stringer, B. Gruber, J. Deakin, and T. Ezaz. 2023a. Widespread chromosomal rearrangements preceded genetic divergence in a monitor lizard, *Varanus acanthurus* (Varanidae). *Chromosome Research* 31:1.

Dobry, J., Z. Zhu, Q. Zhou, E. Wapstra, J. Deakin, and T. Ezaz. 2023b. Unbalanced segmental duplication of W chromosomes in Australian ridge-tailed goannas. *Preprint.* https://doi.org/10.21203/rs.3.rs-2571045/v1

Dobry, J., Z. Zhu, Q. Zhou, E. Wapstra, J. Deakin, and T. Ezaz. 2023. Fixed allele differences associated with the centromere reveal chromosome morphology and rearrangements in a reptile (*Varanus acanthurus* Boulenger). *Molecular Biology and Evolution* 40(6):msad124. https://doi.org/10.1093/molbev/msad124

Dobson, J., C. Zdenek, C. Hay, A. Violete, R. Fourmy, C. Cochran, and B. Fry. 2019. *Toxins* 11(5):255. online at https://www.ncbi.nlm.nih.gov/pmc/articles/PMC6563220/

Dobzhansky, T. 1937. *Genetics and the origin of species.* Columbia University Press, New York, NY.

Doles, M. and W. Card. 1995. Delayed fertilization in the monitor lizard *Varanus gouldii. Herpetological Review* 26(4):196.

Dollo, L. 1923. *Saniwa orsmaelensis*, varanidé nouveau de Landénien supérieur d'Orsmael (Brabant). *Bulletin, Société Belge de Géologie de Paléontologie et d'Hydrologie* 33:76–82.

Domínguez-Guerrero, S., D. Esquerré, E. Burress, et al. 2024. Viviparity imparts a macroevolutionary signature of ecological opportunity in the body size of female *Liolaemus* lizards. *Nature Communications* 15:4966. https://doi.org/10.1038/s41467-024-49464-x

Dong, L., Y.-Q. Wang, Q., Zhao, D., Vasilyan, Y. Wang, and S. Evans. 2022. A new stem- varanid lizard (Reptilia, Squamata) from the early Eocene of China. *Philosophical Transactions of the Royal Society B* 377(1847):20210041. https://doi.org/10.1098/rstb.2021.0041

Donoghue, M., J. Doyle, J. Gauthier, A. Kluge, and T. Rowe. 1989. The importance of fossils in phylogeny reconstruction. *Annual Review of Ecology and Systematics* 20:431–460.

Doody, J., G. Burghardt, and V. Dinets. 2013. Breaking the social-non-social dichotomy: A role for reptiles in vertebrate social behavior research? *Ethology* 119:95–103. https://doi.org/10.1111/eth.12047

Doody, J. S., C. McHenry, L. Durkin, M. Brown, A. Simms, L. Coleman, C. Phizacklea, H. Jones, O. Phizacklea, and S. Clulow. 2018. Deep communal nesting by yellow-spotted monitors in a desert ecosystem: Indirect evidence for a response to extreme dry conditions. *Herpetologica* 74(4):306–310.

Doody, J. S., H. James, R. Ellis, N. Gibron, M. Raven, S. Mahony, D. Hamilton, D. Rhind, S. Clulow, and C. R. McHenry. 2014. Cryptic and complex nesting in the yellow-spotted monitor, *Varanus panoptes. Journal of Herpetology* 48:363–370.

Doody, J. S., H. James, K. Colyvas, C. R. Mchenry, and S. Clulow. 2015. Deep nesting in a lizard, deja vu devil's corkscrews: First helical reptile burrow and deepest vertebrate nest. *Biological Journal of the Linnean Society* 116:13–26.

Doody, J., K. Soennichsen, H. James, C. McHenry, and S. Clulow. 2021. Ecosystem engineering by deep-nesting monitor lizards. *Ecology* 102. https://doi.org/10.1002/ecy.3271

Doria, G. 1874. Enumerazione dei rettili raccolti dal Dott. O. Beccari in Amboina, alle isole Aru ed alle isole Kei. *Annali del Museo civico di storia naturale Giacomo Doria, Genoa* 6:330–352.

Downs, S. and R. Shine. 1998. Heat, safety or solitude? Using habitat selection experiments to identify a lizard's priorities. *Animal Behaviour* 55(5):1387–1396.

Dreisig, H. 1985. A time budget model of thermoregulation in shuttling ectotherms. *Journal of Arid Environments* 8:191–205.

Drivers, S. and M. Camus. 2020. Chapter 18. Lizards. IN: J. Heatley and K. Russell (eds.). *Exotic animal laboratory Diagnosis.* Wiley. Online Library at https://doi.org/10.1002/9781119108610.ch18

Druelle, F., J. Goyens, M. Vasilopoulou-Kampitsi, and P. Aerts. 2019. Compliant legs enable lizards to maintain high running speeds on complex terrains. *Journal of Experimental Biology* 222(6):jeb195511.

Dryden, G., B. Green, E. Wikramanayake, and K. Dryden. 1992. Energy and water turnover in two tropical varanid lizards, *Varanus bengalensis* and *V. salvator. Copeia* 1992:102–107.

Dubois, A. 1988. The genus in zoology: A contribution to the theory of evolutionary systematics. *Mémoires du Muséum National d'Histoire Naturelle. Série A* 140:1–122.

Dujsebayeva (Matveyeva), T. 1995. *Varanus griseus* (desert monitor lizard). Reproduction. *Herpetological Review* 26(2):100.

Duméril, M. and G. Bibron. 1836. *Erpetologie Generale. Tome 3.* Muséum National d'Histoire Naturelle, Paris. 517 pp.

Duméril, M. and A. Duméril. 1851. *Catalogue methodique de la collection des reptiles.* Muséum National d'Histoire Naturelle, Paris. 1–53 pp.

Dunn, E. and W. D. Burden. 1927. Results of the douglas burden expedition to the island of Komodo. 1. Notes on *Varanus komodoensis. American Museum Novitates* 286:1–10.

Dunson, W. 1974. Salt gland secretion in a mangrove monitor lizard. *Comparative Biochemistry and Physiology A* 47:1245–1255.

Dutt, K. 1968. A karyotype study of *Varanus monitor* Linn. *Caryologia* 21(1):1–10.

von Düring, M. and M. Miller. 1979. Sensory nerve endings of the skin and deeper structures. Pp. 407–441. IN: C. Gans, R. Northcutt, and P. Ulinski (eds.). *Biology of the reptilia, volume 9, neurology A.* Academic Press, New York., NY.

Dwyer, Q. 2008. Field observations on *Varanus spinulosus. Biawak* 2(4):162–168.

Dwyer, Q. and M. Perez. 2007. Husbandry and reproduction of the Black Water Monitor, *Varanus salvator komaini. BIAWAK* 1(1):13–20.

Eakin, R. M. 1973. *The third eye.* University of California Press, Berkeley.

Earley, B., O. Attum, and P. Eason. 2002. Varanid combat: Perspectives from game theory. *Amphibia-Reptilia* 22:469–485.

Earll, C. 1982. Heating, cooling, and oxygen consumption rates in *Varanus bengalensis. Comparative Biochemistry and Physiology* 72A:377–381.

Ebbesson, S. and R. Northcutt. 1976. Neurology of anamniotic vertebrates. Pp. 115–146. IN: R. Masterton, M. Bitterman, and C. Campbell (eds.). *Evolution of brain and behavior in vertebrates.* Erlbaum.

Ehmann, H., G. Swan, G. Swan, and B. Smith. 1991. Nesting, egg incubation and hatching by the Heath Monitor *Varanus rosenbergi* in a termite mound. *Herpetofauna* 21(1):17–24.

Eidenmüller, B. 2007. *Monitor lizards: Natural history, captive care and breeding.* Edition Chimaira, Frankfurt. 94 pp.

Eidenmüller, B. 2016. *Keeping and breeding emerald monitors: The Varanus prasinus group.* Edition Chimaira, Frankfurt.

Eidenmüller, B. 1999. Observations on the keeping and breeding of *Varanus mertensi* Glauert, 1951. *Salamandra* 26(2–3):132–139.

Eidenmüller, B. 1997. *Warane: Lebensweise, Pflege, Zucht*. Herpeton, Offenbach, Germany. 157 pp.

Eidenmüller, B. 1994. Bemerkungen zur Haltung und Zucht von *Varanus acanthurus* Boulenger, 1885, *V. storri* Mertens, 1966 und *V. gilleni* Lucas and Frost, 1895. *Herpetofauna (Germany)* 16(88):6–12.

Eidenmüller, B. 1989. Beobachtungen bei der Haltung und Nachzucht von *Varanus (Odatria) tristis orientalis*. *Salamandra* 25(3/4): 265–271.

Eidenmüller, B. and H.-D. Philippen. 2008. *Varanoid lizards: Warane und Krustenechsen. Terralog Band 6*. Chimaira Buchhandelsgesellschaft mbH, Frankfurt.

Eidenmüller, B. and R. Wicker. 1991. Einige Beobachtungen bei der Pflege und Nachzucht von *Varanus (Odatria) timorensis similis* Mertens, 1958. *Salamandra* 27(3):187–193.

Eidenmüller, B. and R. Wicker. 1995. The successful breeding of Mertens' monitor lizards, *Varanus mertensi*, Glauert 1951. *Herpetofauna (Australia)* 25(2):4–7.

Eidenmüller, B. and R. Wicker. 1998. Observations on *Varanus mertensi* clutches incubated under different conditions. *Herpetofauna (Germany)* 20(116):30–34.

Eldredge, N. and S. Gould. 1972. Punctuated equilibria: An alternative to phyletic gradualism. Pp. 82–115. IN: T. Schopf (ed.). *Models in paleobiology*. W. H. Freeman, San Francisco

Engbretson, G.A. and V.H. Hutchison 1976. Parietalectomy and thermal selection in the lizard *Sceloporus magister*. *Journal of Experimental Zoology* 198:29–38.

Engelmann, W.-E. and H.-G. Horn. 2003. First breeding of the monitor lizard, *Varanus jobiensis*, in Leipzig Zoo. *Zoological Garten* 736:353–358.

Espinoza, R., J. Wiens, and R. Tracy. 2004. Recurrent evolution of herbivory in small, cold-climate lizards: Breaking the ecophysiological rules of reptilian herbivory. *Proceedings of the National Academy of Sciences* 101(48):16819–16824.

Estes, R. 1983a. *Encyclopedia of paleoherpetology. Part 10A. Sauria terrestria, Amphisbaenia*. Gustav Fischer Verlag, Stuttgart. 398 pp.

Estes, R. 1983b. The fossil record and early distribution of lizards. Pp. 365–398. IN: A. Rhodin and K. Miyata (eds.). *Advances in herpetology and evolutionary biology: Essays in honor of Ernest E. Williams*. Museum of Comparative Zoology, Cambridge, MA.

Estes, R. 1984. Fish, amphibians, and reptiles from the Etadunna Formation, Miocene, of South Australia. *Australian Zoologist* 21: 335–343.

Estes, R., K. de Queiroz, and J. Gauthier. 1988a. Phylogenetic relationships within Squamata. Pp. 119–281. IN: R. Estes and G. Pregill (eds.). *Phylogenetic relationships of the lizard families*. Stanford University Press.

Estes, R. and G. Pregill (eds.). 1988b. *Phylogenetic relationships of the lizard families*. Stanford University Press. 119–281 pp.

Evans, S. 1984. The classification of the Lepidosauria. *Zoological Journal of the Linnean Society* 82:87–100.

Evans, S. 1988. The early history and relationships of the Diapsida. Pp. 221–253. IN: M. J. Benton (ed.). *The phylogeny and classification of the tetrapods*. Oxford University Press, Oxford.

Evans, S. 1994. A new anguimorph lizard from the Jurassic and Lower Cretaceous of England. *Palaeontology* 37:33–49.

Evans, S. 1996. *Parviraptor* (Squamata: Anguimorpha) and other lizards from the Morrison Formation at Fruita, Colorado. Pp. 243–248. IN: M. Morales (ed.). *The continental Jurassic: Museum of Northern Arizona Bulletin*, vol. 60.

Evans, S. 2007. At the feet of the dinosaurs: The early history and radiation of lizards. *Biological Reviews* 78:513–551.

Evans, S. 2008. The skull of lizards and tuatara. Pp. 1–347. IN: C. Gans (ed.). *Biology of the reptilia, volume 20, morphology H*. Society for the Study of Amphibians and Reptiles. 755 pp.

Falconer, H. 1868. XXI. On a fossil species of Emys from the Sewalik Hills, referable to the existing species Emys tecta (Bell). Pp. 382–388, pl. 32, figs. 4–7. IN: C. Murchison (ed.). *Palaeontological memoirs and notes of the late Hugh Falconer, A.M., M.D. with a biographical sketch of the author. Vol. I. Fauna antique Sivalensis*. Robert Hardwicke Publisher, London.

Fedak, M. and H. Seeherman. 1979. Reappraisal of energetics of locomotion shows identical cost in bipeds and quadrupeds including ostrich and horse. *Nature* 13;282(5740):713–716. doi: 10.1038/282713a0.

Fejérváry, G. 1918. Contributions to a monography on fossil Varanidae and Megalanidae. Annales historico-naturales Musei Nationalis Hungarici 16, *pars zoologie*:431–467.

Fejérváry, G. 1935. Further contributions to a monograph of the Megalanidae and fossil Varanidae—with notes on recent varanians. Annales historico-naturales Musei Nationalis Hungarici 29, *pars zoologie*:1–130.

Ferguson, G., A. Brinker, W. Gehrmann, S. Bucklin, F. Baines, and J. Mackin. 2010. Voluntary exposure of some western-hemisphere snake and lizard species to ultraviolet-B radiation in the field: how much ultraviolet-B should a lizard or snake receive in captivity? *Zoo Biology* 29:317–334.

Ferguson, G., W. Gehrmann, K. Karsten, S. Hammack, M. McRae, T. Chen, N. Lung, and M. Holick. 2003. Do panther chameleons bask to regulate endogenous vitamin D3 production? *Physiological and Biochemical Zoology* 76:52–59.

Ferguson, G., W. Gehrmann, B. Peavy, C. Painter, R. Hartigen, T. Chen, M. Holick, and J. Pinder, III. 2009. Restoring vitamin D in monitor lizards: Exploring the efficacy of dietary and UVB sources. *Journal of Herpetological Medicine and Surgery* 19(3):81–88.

Farris, J. 1986. *Hennig86*. Privately published software and documentation, IBM compatible version.

Finnegan, N., M. Lima, and J. Lynch. 2025. Mitochondrial DNA for phylogeny building: Assessing individual and grouped mtGenes as proxies for the mtGenome in platyrrhines. *American Journal of Primatology* 87(3):e70017. https://doi.org/10.1002/ajp.70017. PMID: 40059324; PMCID: PMC11891386.

Firth, B. and J. Turner. 1982. Sensory, neural, and hormonal aspects of thermoregulation. Pp. 213–274. IN: C. Gans and F. Pough (eds.). *Biology of the Reptilia, vol. 12. Physiology C. Physiological ecology*. Academic Press, New York.

Firth, T., and I. Belan. 1998. Daily and seasonal rhythms in selected body temperatures in the Australian lizard *Tiliqua rugosa* (Scincidae): Field and laboratory observations. *Physiological Zoology* 71(3):303–311.

Fitch, H. and A. Fitch. 1967. Preliminary experiments of physical tolerances of eggs of lizards and snakes. *Ecology* 48:160–165.

Flannery, T. 1994. *The future eaters: An ecological history of the Australasian lands and people*. Reed International Books, Kew, Victoria, New South Wales.

Fleishman, L., E. Loew, and M. Leal. 1993. Ultraviolet vision in lizards. *Nature* 365:397.

Fleishman, L., E. Loew, and M. Whiting. 2011. High sensitivity to short wavelengths in a lizard and implications for understanding the evolution of visual systems in lizards. *Proceedings of the Royal Society of London B: Biological Sciences* 278:2891–2899.

Foá, A. 1991. The role of the pineal and the retinae in the expression of circadian locomotor rhythmicity in the ruin lizard, *Podarcis sicula*. *Journal of Comparative Physiology A* 169:201–207.

Foá, A., F. Basaglia, G. Beltrami, M. Carnacina, and C. Bertolucci. 2018. Orientation of lizards in a Morris water-maze: Roles of the sun compass and the parietal eye. *Journal of Experimental Biology* 212:2918–2924.

Folwell, M., K. Sanders, P. Brennan, and J. Crowe-Riddell. 2022a. First evidence of hemiclitores in snakes. *Proceedings of the Royal Society B: Biological Sciences* 289(1989): 20221702

Folwell, M, K. Sanders, and J. Crowe-Riddell. 2022b. The squamate clitoris: A review and directions for future research. *Integrative and Comparative Biology* 62:559–568.

Forey, P., C. Humphries, I. Kitching, R. Scotland, D. Siebert, and D. Williams. 1992. *Cladistics — A practical course in systematics*. Systematics association publication no. 10. Oxford Science Publications, Oxford, UK.

Fost, M. 1995. Dumeril's monitors hatched at Zoo Atlanta. *AZA Communique* July 26.

Frappell, P., T. Schultz, and K. Christian. 2002. Oxygen transfer during aerobic exercise in a varanid lizard *Varanus mertensi* is limited by the circulation. *Journal of Experimental Biology* 205(17):2725–2736.

Frazzetta, T. 1962. A functional consideration of cranial kinesis in lizards. *Journal of Morphology* 111:287–320.

Freitas, S., S. Rocha, J. Campos, F. Ahmadzadeh, C. Corti, N. Sillero, Ç Ilgaz, Y. Kumlutaş, M. Arakelyan, D. Harris, and M. Carretero. 2016. Parthenogenesis through the ice ages: A biogeographic analysis of Caucasian rock lizards (genus *Darevskia*). *Molecular Phylogenetics and Evolution* 102:117–127. https://doi.org/10.1016/j.ympev.2016.05.035.

Freycinet, A. 1825. Voyage autour du monde : exécuté par ordre du roi, sur la corvette de Sa Majesté, la Coquille, pendant les années 1822, 1823, 1824, et 1825. Firmin-Didot (Firm), France. Ministère de la marine et des colonies. Paris. 712 pp.

Fry, B., B. Fry, N. Vidal, J. Norman, F. Vonk, H. Scheib, S. Ramjan, S. Kuruppu, K. Fung, S. Hedges, M. Richardson, W. Hodgson, V. Ignjatovic, V. Ignjatovic, R. Summerhayes, R. Summerhayes, and E. Kochva. 2006. Early evolution of the venom system in lizards and snakes. *Nature* 439:584–588.

Fry, B. G. and H. Scheib. 2007. Venom breathing dragons: The molecular diversity of protein toxins in varanid venoms. *Mertensiella* 16:10–24.

Fry, B., S. Wroe, W. Teeuwisse, M. J. P. van Osch, K. Moreno, J. Ingle, C. McHenry, T. Ferrara, P. Clausen, H. Scheib, K. L. Winter, L. Greisman, K. Roelants, L. van der Weerd, C. J. Clemente, E. Giannakis, W. C. Hodgson, S. Luz, P. Martelli, K. Krishnasamy, E. Kochva, H. F. Kwok, D. Scanlon, J. Karas, D. M. Citron, E. J. C. Goldstein, J. E. Mcnaughtan, and J. A. Norman. 2009. A central role for venom in predation by *Varanus komodoensis* (Komodo Dragon) and the extinct giant *Varanus (Megalania) priscus*. *Proceedings of the National Academy of Sciences* 106(22):8969–8974.

Frýdlova, P. and D. Frynta. 2010. A test of Rensch's rule in varanid lizards. *Biological Journal of the Linnean Society* 100:293–306.

Fuchs, K. and M. Fuchs. 2003. *Die Reptilhaut. The reptile skin*. Edition Chimaira, Frankfurt.

Fuller, S., P. Baverstock, and D. King. 1998. Biogeographic origins of goannas (Varanidae): A molecular perspective. *Molecular Phylogenetics and Evolution* 9(2):294–307.

Fujita, M. and C. Moritz. 2009. Origin and evolution of parthenogenetic genomes in lizards: Current state and future directions. *Cytogenetics and Genome Research* 127:261–272.

Furbinger, M. 1900. Beitrag zur Systematik und Genealogie der Reptilien. *Abdr. Jena. Zeitschrift Naturwissenschaft* XXXIV(XXVII):1–91.

Gadow, H. 1901. *Amphibia and reptiles*. Cambridge Natural History Series, vol. VIII.

Gaalema, D. 2011. Visual discrimination and reversal learning in rough-necked monitor lizards (*Varanus rudicollis*). *Journal of Comparative Psychology* 125(2):246–249.

Garberoglio, F., R. Gómez, T. Simões, M. Caldwell, and S. Apesteguía. 2019. The evolution of the axial skeleton intercentrum system in snakes revealed by new data from the Cretaceous snakes *Dinilysia* and *Najash*. *Nature Scientific Reports* 9:1276. https://doi.org/10.1038/s41598-018-36979-9

Gaikhorst, G., J. McLaughlin, B. Larkin, and M. McPHarlin. 2009. Successful captive breeding of Mitchell's Water Monitor, *Varanus mitchelli* (Mertens 1958), at Perth Zoo. *Zoo Biology* 28:1–11.

Galli, G., H. Gesser, E. Taylor, H. Shiels, and T. Wang. 2006. The role of the sarcoplasmic reticulum in the generation of high heart rates and blood pressures in reptiles. *Journal of Experimental Biology* 209:1956–1963.

Galli, G., D. Warren, and H. Shiels. 2009. Ca^{2+} cycling in cardiomyocytes from a high- performance reptile, the varanid lizard (*Varanus exanthematicus*). *American Journal of Regular, Integrative, and Comparative Physiology* 297(6):R1636–R1644.

Gans, C. and W. Dawson (eds.). 1976. *Biology of the reptilia, vol. 5. Physiology A*. Academic Press, London, New York, and San Francisco.

Gans, C. and D. Tinkle (eds.). 1977. *Biology of the reptilia, vol. 7. Ecology and behavior A*. Academic Press, New York, NY. 720 pp.

Gao, K. and M. Norell. 2000. Taxonomic composition and systematics of Late Cretaceous lizard assemblages from Ukhaa Tolgod and adjacent localities, Mongolian Gobi Desert. *Bulletin of the American Museum of Natural History* 249:118.

Garrick, D. 2008. Body surface temperature and length in relation to the thermal biology of lizards. *Bioscience Horizons* 1(2):136–142.

Gaulke, M. 1989. Zur Biologie des Bindenwaranes, unter Beruchsichtigung der palaogeographischen Verbreitung und der phylogenetischen Entwicklung der Varanidae. *Courier Forschung. Senckenberg* 112:1–242.

Gaulke, M. 2010. Overview of the present knowledge on *Varanus mabitang* (Gaulke and Curio 2001), including new morphological and meristic data. *Biawak* 4(2):50–58.

Gaulke, M., A. V. Altenbach, A. Demegillo, and U. Struck. 2007. On the diet of *Varanus mabitang*. *Mertensiella* 16:228–239.

Gauthier, J., R. Estes, and K. De Queiroz. 1988. A phylogenetic analysis of Lepidosauromorpha. Pp. 15–98. IN: R. Estes and G. Pregill (eds.). *Phylogenetic relationships of the lizard families*. Stanford University Press, Palo Alto, CA.

Gauthier, J., M. Kearney, J. Maisano, O. Rieppel, and A. Behlke. 2012. Assembling the squamate tree of life: Perspectives from the phenotype and the fossil record. *Bulletin of the Peabody Museum of Natural History* 53(1):3–308.

Georgalis, G., M. Gawad, S. Hassan, …, and M. Hamdan. 2020. Oldest co-occurrence of *Varanus* and Python from Africa: First record of squamates from the early Miocene of Moghra Formation, Western Desert, Egypt. *PeerJ.* https://doi.org/10.7717/peerj.9092

Georgalis, G., A. Villa, and M. Delfino. 2017. The last European varanid: Demise and extinction of monitor lizards (Squamata, Varanidae) from Europe. *Journal of Vertebrate Paleontology.* https://doi.org/10.1080/02724634.2017.1301946

Getsy, P., C. Ritzler, K. Becka, C. Gertiser, and W. Smith. 2022. The effects of multiple enrichment strategies on the behavior and space use of a zoo-housed Emerald Tree Monitor (*Varanus prasinus*). *BIAWAK* 15(1):22–32.

Ghimire, H. R. and K. B. Shah. 2014. Status and habitat ecology of the Yellow Monitor, *Varanus flavescens*, in the Southeastern part of Kanchanpur District, Nepal. *Herpetological Conservation and Biology* 9(2):387–393. http://www.herpconbio.org/Volume_9/Issue_2/Ghimire_Shah_2014.pdf

Gilmore, C. 1922. A new description of *Saniwa ensidens* Leidy, an extinct varanid lizard from Wyoming. *Proceedings of the U.S. National Museum* 60:1–28.

Gilmore, C. 1928. The fossil lizards of North America. *Memoirs of the National Academy of Sciences* 22:1–197.

Gilmore, C. 1943. Fossil lizards of Mongolia. *Bulletin of the American Museum of Natural History* 81(4):361–389.

Gleeson, T. 1981. Preferred body temperatures, aerobic scope, and activity capacity in the monitor lizard, *Varanus salvator. Physiological Zoology* 54:423–429.

Gluesing, E. A. 1983. Collared lizard predation: The effects of conspicuous morphology and movement. *Copeia* 1983(3):835–837.

Good, D., A. Bauer, and R. Günther. 1993. An annotated type catalogue of the anguimorph lizards (Squamata: Anguidae, Helodermatidae, Varanidae, Xenosauridae) in the Zoological Museum, Berlin. *Mitteilungen aus dem Zoologischen Museum in Berlin* 69(1993):45–56.

Good, D. 1994. Species limits in the genus *Gerrhonotus. Herpetological. Monographs* 8:180–202.

Goolsby, E., J. Bruggeman, and C. Ané. 2017. Rphylopars: Fast multivariate phylogenetic comparative methods for missing data and within species variation. *Methods in Ecology and Evolution* 8:22–27.

Gorman, D. 1993. Breeding the Bengal monitor in captivity. *Varanews* 3(4):2.

Gould, S. and R. Lewontin. 1979. The spandrels of San Marcos and the Panglossian paradigm: A critque of the adaptationist programme. *Proceedings of the Royal Society B Biological Sciences* 205(1161):581–598.

Gow, G. 1981. Herpetofauna of Groote Eylandt, Northern territory. *Australian Journal of Herpetology* 1(2):62–70.

Grabbe, J. and A. Koch. 2014. First and repeated cases of parthenogenesis in the varanid subgenus *Euprepiosaurus* (*Varanus indicus* species group) and the first successful breeding of *V. rainerguentheri* in captivity. *Biawak* 8(2):79–87.

Grace, M. and J. Beshare. 1994. Melatonin deacetylase activity in the pineal gland and brain of the lizards *Anolis carolinensis* and *Sceloporus jarrovi. Neuroscience* 62(2):615–623.

Grafen, A. 1990. Biological signals as handicaps. *Journal of Theoretical Biology* 144:517–546.

Grant, P. and B. R. Grant. 2014. *40 years of evolution: Darwin's finches on daphne major island.* Princeton University Press, Princeton, NJ.

Gray, J., E. Sherratt, M. Hutchinson, and M. Jones. 2019. Evolution of cranial shape in a continental-scale evolutionary radiation of Australian lizards. *Evolution.* 73(11):2216–2229

Gray, J. 1831. Synopsis reptilia. IN: E. Pidgeon (ed.). *Animal kingdom* vol. 9:1-1-21. Whittaker, Treacher & Co., London.

Gray, J. 1838. Catalogue of the slender-tongued sauriens ….

Gray, J. 1845a. *Catalogue of the specimens of lizards in the collection of the British museum.* Order of the Trustees of the British Museum by Edward Newman, London.

Gray, J. 1845b. Reptiles. Pp. 1-8, pl. 1-4, 8-9, 12-14, and 20. IN: J. Richardson and J. Gray (eds.). *The zoology of the voyage of HMS Erebus & Terror, under the command of Captain Sir James Clark Ross, R.N., F.R.S., during the years 1839 to 1843*, vol. 2. Longman, Brown, Green, and Longmans, London.

Green, B. and A. Christian. 2007. Field metabolic rates (FMRs) and water fluxes in free-living varanid lizards: A review. *Mertensiella* 16:240–246.

Green, B., G. Dryden, and K. Field. 1991. Field energetics of a large carnivorous lizard, *Varanus rosenbergi. Oecologia* 88:547–551.

Green, B. and D. King. 1993. *Goanna.* University of New South Wales Press. Armidale, NSW. 102 pp.

Green, B., D. King, and H. Butler. 1986. Water, sodium, and energy turnover in free-living perenties, *Varanus giganteus. Australian Wildlife Research* 13:589–595.

Green, B., D. King, M. Braysher, and A. Saim. 1991. Thermoregulation, water turnover, and energetics of free-living Komodo dragons, *Varanus komodoensis. Comparative Biochemistry and Physiology* 99A:97–101.

Green, B. and P. King. 1978. Home range and activity patterns of the sand goanna, *Varanus gouldii. Australian Wildlife Research* 5:417–424.

Greene, H. 1986. Diet and arboreality in the emerald monitor, *Varanus prasinus*, with comments on the study of adaptation. *Fieldiana Zoology, New Series* (31):1–12.

Greer, A. 1989. *The biology and evolution of Australian lizards.* Surrey Beatty & Sons, Chipping Norton, New South Wales.

Gregorovicova, M., M. Bartos, B. Jensen, J. Janacek, B. Minne, and D. Sedmera. 2022. Anguimorpha as a model group for studying the comparative heart morphology among Lepidosauria: Evolutionary window on the ventricular septation. *Ecology and Evolution* 12(11): e9476.

Gregorovicova, M., O. Zahradnicek, A. Tucker, P. Velensky, and I. Horacek. 2012. Embryonic development in the monitor lizard, *Varanus indicus. Amphibia-Reptilia* 33(3–4):451–468. https://doi.org/10.1163/15685381-00002849

Grismer, J., A Bauer, L. Grismer, K., Thirakhupt, A. Aowphol, J. Oaks, P. Wood, C. Onn, N. Thy, M. Cota, and T. Jackman. 2014. Multiple origins of parthenogenesis, and a revised species phylogeny for the Southeast Asian butterfly lizards, *Leiolepis. Biological Journal of the Linnean Society* 113:1080–1093. https://doi.org/10.1111/bij.12367

Gunawardena, S. 2016. Forensic significance of monitor lizard scavenging activity on human corpses. *Biawak* 10(2):45–47.

Gunderson, A. R and M. Leal. 2016. A conceptual framework for understanding thermal constraints on ectotherm activity with implications for predicting responses to global change. *Ecology Letters* 19:111–120. PMID:26647860

Günther, A. 1877. On a collection of reptiles and fishes from Duke of York Island, New Ireland, and New Britain. *Proceedings of the Zoological Society of London*:127–132..

Gutzke, W. and G. Packard. 1986. Sensitive periods for the influence of the hydric environment on eggs and hatchlings of painted turtles (*Chrysemys picta*). *Physiological Zoology* 59:9–17.

Hagen, I., S. Donnellan, and M. Bull. 2012. Phylogeography of the prehensile-tailed skink *Corucia zebrata* on the Solomon Archipelago. *Ecology and Evolution* 2(6):1220–1234.

Hall, R. 2013. The palaeogeography of Sundaland and Wallacea since the Late Jurassic. *Journal of Limnology* 72:1–17.

Halliwell, B., T. Uller, B. Holland, and G. While. 2017. Live bearing promotes the evolution of sociality in reptiles. *Nature Communications* 8:2030.

Hamilton, S., I. Paparella, P. Bell, N. Campione, F. Fanti, D. Larson, R. Sissons, M. Vavrek, M. Balsai, and C. Sullivan. 2023. New lizard specimens from the Campanian Wapiti Formation of Alberta, Canada. *Geosciences* 13(11):337.

Hanemaaijer, J., M. Gregorovicova, J. Nielsen, A. Moorman, F. Antoon, T. Wang, N. Planken, V. Christoffels, D. Sedmera, and B. Jensen. 2019. Heart development in the lizards (Varanidae) with the greatest extent of ventricular septation. *BioRxiv*. https://doi.org/10.1101/563767.

Hanemaaijer, J., M. Gregorovicova, J. Nielsen, A. Moorman, T. Wang, R. Planken, V. Christoffels, D. Sedmera, and B. Jensen. Identification of the building blocks of ventricular septation in monitor lizards (Varanidae). *The Company of Biologists* 146: dev177121. https://doi.org/10.1242/dev177121

Harrison, R. and E. Larson. 2014. Hybridization, introgression, and the nature of species boundaries. *Journal of Heredity* 105:795–809.

Hartdegen, R. 1998. Black rough-necked monitors (*Varanus rudicollis*). *Reptiles* 6(10):68–75.

Harvey, M. and D. Barker. 1998. A new species of blue-tailed monitor (genus *Varanus*) from Halmahera Island, Indonesia. *Herpetologica* 54(1):34–44.

Hawlitschek, O., Z. Nagy, and F. Glaw. 2012. Island evolution and systematic revision of Comoran snakes: Why and when subspecies still make sense. *PLoS One* 7:e42970.

Head, J., P. Barrett, and E. Rayfield. 2009. Neurocranial osteology and systematic relationships of *Varanus* (*Megalania*) *prisca* Owen, 1859 (Squamata: Varanidae). *Zoological Journal of the Linnean Society* 155:445–457.

Heatwole, H. and J. Taylor. 1987. *Ecology of reptiles*. Surrey Beatty & Sons, Chipping Norton. 325 pp.

Heatwole, H. and J. Vernon. 1977. Vital limit and evaporative water loss in lizards (Reptilia, Lacertilia): A critique and new data. *Journal of Herpetology* 11:341–348.

Hecht, M. 1975. The morphology and relationships of the largest known terrestrial lizard, *Megalania prisca* Owen, from the Pleistocene of Australia. *Proceedings of the Royal Society of Victoria* 87:239–250.

Hediger, H. 1934. Beitrag zur Herpetologie und Zoogeographie Neu Britanniens und einiger umliegender Gebiete. *Zoologisches Jahrbuch Syst* 65:441–582.

Heisler, N., P. Neumann, and G. Maloly. 1983. The mechanism if intracardiac shunting in the lizard *Varanus exanthematicus*. *Journal of Experimental Biology* 105:15–31.

Hendrickx, C, R. Ara´ujo, and O. Mateus. 2015. The non-avian theropod quadrate I: Standardized terminology with an overview of the anatomy and function. *PeerJ* 3:e1245. https://doi.org/10.7717/peerj.1245

Hennig, W. 1966. *Phylogenetic systematics*. University of Illinois Press, Champaign, IL.

Hermes, N. 1981. Merten's water monitor feeding on trapped fish. *Herpetofauna* 13:34.

Hernández, C. A., A. Bolet, A. Torices, and A. Perez-García. 2022. Identification of a large Anguimorph lizard (Reptilia, Squamata) by an articulated hindlimb from the upper Maastrichtian (Upper Cretaceous) of Basturs-1 (Lleida, Spain). *Cretaceous Research* 131.

Hertz, P., R. Huey, and T. Garland, Jr. 1988. Time budgets, thermoregulations, and maximal locomotor performance: Are reptiles Olympians or Boy Scouts? *American Zoologist* 28:927–938.

Hijmans, R., S. Cameron, J. Parra, P. Jones, and A. Jarvis. 2005. Very high resolution interpolated climate surfaces for global land areas. *International Journal of Climatology* 25:1965–1978.

Hillenius, W. 2020. Septomaxilla of nonmammalian synapsids: Soft-tissue correlates and a new functional interpretation. *Journal of Morphology* 245(1):29–50.

Hocknull, S., R. Lewis, L. Arnold, T. Pietsch, R. Joannes-Boyau, G. Price, P. Moss, R. Wood, A. Dosseto, O. Louys, J. Olley, and R. Lawrence. 2020. Extinction of eastern Sahul megafauna coincides with sustained environmental deterioration. *Nature Communications* 11:1–14.

Hocknull, S., P. Piper, G. van den Bergh, R. Due, M. Morwood, and I. Kurniawan. 2009. Dragon's paradise lost: Palaeobiogeography, evolution and extinction of the largest-ever terrestrial lizards (Varanidae). *PLoS One* 4:e724.

Hofer, H. 1960. Vergleichende Untersuchungen am Schädel von *Tupinambis* und *Varanus*, mit besonderer Berücksichtigung ihrer Kinetik. *Morphologisches Jahrbuch* 100:706–746.

Hoffmann, C. 1890. *Dr. H. G. Bronn's Klassen und Ordnungen des Their-Reichs, in Wort und Bild. Sechster Band, III Abtheilung. Reptilien. II Eidechsen und Wasserechsen*. C. F. Winter'sche Verlagshandlung, Leipzig. 443–1399 pp, plates XLIX-CVII.

Hoffstetter, R. 1969. Présence de Varanidae (Reptilia, Sauria) dans le Miocèrne de Catalogne. Considérations sur l'Histoire de la Famille. *Bulletin du Museum National d'Histoire Naturelle 2ᵉ serie* 40(5):1054–1064.

Hoffstetter, R., and J-P. Gasc. 1969. Vertebrae and ribs of modern reptiles. Pp. 201–310. IN: C. Gans (ed.). *Biology of the reptilia volume 1 - Morphology A*. Academic Press, London.

Holmes, R., M. King, and D. King. 1975. Phenetic relationships among varanid lizards based upon comparative electrophoretic data and karyotypicanalyses. *Biochemical Systematics and Ecology* 3:257–262.

Holmes, R., A. Murray, Y. Attia, E. Simons, and P. Chatrath. 2010. Oldest known *Varanus* (Squamata: Varanidae) from the Upper Eocene and Lower Oligocene of Egypt: Support for an African origin of the genus. *Palaeontology* 53(5):1099–1110.

Hooijer, D. 1972. Varanus (Reptilia, Sauria) from the pleistocene of timor. *Zoologische Mededelingen* 47:445–448.

Hörenberg, T. 2013. Parthenogenese bei *Varanus glauerti*. *Draco* 53:29–30.

Hörenberg, T. and A. Koch. 2013. Die Zwergwarane der Untergattung *Odatria* Gray, 1838. *Draco* 13(53):6–19.

Horn, H. G. 1977. Notizen zu Systematik, Fundortangaben und Haltung von *Varanus* (*Varanus*) *karlschmidti* (Reptilia: Sauria: Varanidae). *Salamandra* 13:78–88.

Horn, H. G. 2007. Transmission and reflection spectra of shedded skin of various varanids in the UV, VIS and IR and their intra and interspecific meaning. *Mertensiella* 16:291–303.

Horn, H. G. and G. Visser. 1989. Review of reproduction in monitor lizards. *International Zoo Yearbook* 28:140–150.

Horn, H. G. and G. Visser. 1997. Review of reproduction of monitor lizards *Varanus* spp. in captivity II. *International Zoo Yearbook* 35:227–246.

Horn, H. G., M. Gaulke, and W. Böhme. 1994. New data on ritualized combat in monitor lizards (Sauria: Varanidae), with remarks on their function and phylogenetic implications. *Zool. Garten N. F.* 64(1994):265–280.

Horn, H. G., S. Sweet, and K. Philipp. 2007. On the distribution of the Papuan monitor (*Varanus salvadorii* Peters & Doria, 1878) in New Guinea. *Mertensiella* 16:24–43.

Horowitz, A. 2009. *Inside of a dog: What dogs see, smell, and know*. Scribner, New York. 352 pp + addendum.

Houston, T. 1998. *Dragon lizards and goannas of South Australia*. South Australian Museum, Adelaide.

Howard, D. and M. Freeman, M. 2022. Overlooked and under-studied: A review of evidence-based enrichment in varanidae. *Journal of Zoological and Botanical Gardens* 3(1):32–43.

Hsieh, S. 2003. Three-dimensional hindlimb kinematics of water running in the plumed basilisk lizard (*Basiliscus plumifrons*). *Journal of Experimental Biology* 206:4363–4377.

Huang, W. 2006. Parental care in the long-tailed skink, *Mabuya longicaudata*, on a tropical Asian island. *Animal Behaviour* 72:791–795.

Huey, R. and J. Kingsolver. 2019. Climate warming, resource availability, and the metabolic meltdown of ectotherms. *The American Naturalist* 2019(194):E140–E150. PMID31738103.

Huo, X., K. Kunya, N. Ritthijan, P. Lohtongkam, C. Sukwan, S. Wongma, P. Kusolrat, and P. Kumroptanasan. 2024. A preliminary study of the effects of enrichment on the activity and spatial distribution of captive monitor lizards in Nakhon Ratchasima Zoo, Thailand. *Veterinary Integrative Sciences* 22(1):193–205.

Hutchinson, V. 1974. Thermoregulatory function of the parietal eye in the lizard *Anolis carolinensis*. *Oecologia* 16:173–177.

Iannucci, A., M, Altmanová, C., Ciofi, M., Ferguson-Smith, M. Massimo, J., Pereira, J., Pether, I, Rehák, M., Rovatsos, R., Stanyon, P., Velenský, P., Ráb, L., Kratochvil, and M, Pokorna. 2019. Conserved sex chromosomes and karyotype evolution in monitor lizards (Varanidae). *Heredity* 123:215–227.

Innocenti, A., L. Minutini, and A. Foà. 1994. Seasonal changes of locomotor activity patterns in ruin lizards *Podarcis sicula*. II. Involvement of the pineal. *Behavioral Ecology and Sociobiology* 35:27–32.

International Commission on Zoological Nomenclature. 2014. Zoological nomenclature and electronic publication—A reply to Dubois *et al.* (2013). *Zootaxa* 3779(1):3–5.

International Commission on Zoological Nomenclature. 2025. Declaration 47 – Addition of Examples to Article 13.1.1.

International Commission on Zoological Nomenclature. 2017. Declaration 45 – Addition of Recommendations to Article 73 and of the term "specimen, preserved" to the Glossary. https://www.iczn.org/the-code/declaration-45-addition-of-recommendations.

International Commission on Zoological Nomenclature. 2007. Declaration 44 – Amendment of Article 74.7.3. https://www.iczn.org/the-code/declaration-44-amendment-of-article-74-7-3/

International Commission on Zoological Nomenclature. 1999. *International code of zoological nomenclature*. 4th Ed. International Trust for Zoological Nomenclature, London. http://dx.doi.org/10.5962/bhl.title.50608

International Commission on Zoological Nomenclature. 1959. ICZN opinion 540. Protection under the plenary powers of the specific names "*bengalensis*" Daudin, [1802], as published in the combination "*Tupinambis bengalensis*, and "*salvator*" Laurenti, 1768, as published in the combination "*Stellio salvator*" (Class Reptilia). Volume 20, part 7:77–86.

Iordansky, N. 1970. Structure and biomechanical analysis of function of the jaw muscles in lizards. *Anatomischer Anzeiger* 127:383–413.

Irwin, K., A. Rasys, T. Roldan, D. Menke, and J. Lauderdale. 1994. The lizard third eye. *Investigative Ophthalmology & Visual Science, Annual Meeting Abstract Issue* 59(9):577.

Irwin, S. 1994. Notes on behaviour and diet of *Varanus teriae* Sprackland, 1991. *Memoirs of the Queensland Museum* 35(1):128.

Ishimatsu, A., J. Hicks, and N. Heisler. 1988. Analysis of intracardiac shunting in the lizard, *Varanus niloticus*: A new model based on blood oxygen levels and microsphere distribution. *Respiratory Physiology* 71:83–100.

Ivanov, M., M. Ruta, J. Klembara and M. Böhme. 2017. A new species of *Varanus* (Anguimorpha: Varanidae) from the early Miocene of the Czech Republic, and its relationships and palaeoecology. *Journal of Systematic Palaeontology*. https://doi.org/10.1080/14772019.2017.1355338

Iyai, D. and F. Pattiselanno. 2006. Diversitas dan ekologi Biawak (*Varanus indicus*) di Pulau Pepaya Taman Nasional Teluk Cenderawasih, Irian Jaya Barat. *Biodiversitas* 7(2):181–186.

Jacobson, E. 1909. De nieuwe fauna van Krakatau. *Jaarverslag van den Topografischen Dienst in Nederlandsch-Indië* 102;192–197.

James, C., J. Losos, and D. King. 1992. Reproductive biology and diets of goannas (Reptilia: Varanidae) from Australia. *Journal of Herpetology* 26(2):128–136.

Janik, D. and M. Menaker. 1990. Circadian locomotor rhythms in the desert iguana; the roles of the eyes and the pineal. *Journal of Comparative Physiology A* 166:803–810.

Jenison, G. and J. Nolte. 1980. An ultraviolet-sensitive mechanism in the reptilian parietal eye. *Brain Research* 194:506–510.

Janssen, J. 2018. Valuable varanoids: Surveys of reptile traders in Japan reveal monitor lizards without import records. *BIAWAK* 12(2):84–90.

Jessop, T. S., A. Ariefiandy, D. M. Forsyth, D. Purwandana, C. R. White, Y. J. Benu, T. Madsen, H. J. Harlow, and M. Letnic. 2020. Komodo dragons are not ecological analogs of apex mammalian predators. *Ecology* 101(4):e02970. https://doi.org/10.1002/ecy.2970. Epub 2020.

Johnson, C. 1972. Head-body temperature differences in *Varanus gouldii* (Sauria: Varanidae). *Comparative Biochemistry and Physiology* 43A:1025–1029.

Jollie, M. T. 1960. The head skeleton of the lizards. *Acta Zoologica* 41:1–64.

Joly, S., P. McLenachan, and P. Lockhart. 2009. A statistical approach for distinguishing hybridization and incomplete lineage sorting. *American Naturalist* 174(2):E54–E70.

Joshi, R. 2018. *Varanus bengalensis* predation by *Ophiophagus hannah* in Rajaji Tiger Reserve, India. *BIAWAK* 12(1):34–36.

Kalken, C van. 1994. *Varanus indicus. Aquarien und Terrarien Hobbyist*, Oct.: 42–44.

Kane, D., A. Davis, and C. Michaels. 2019. Play behaviour by captive tree monitors, *Varanus macraei* and *Varanus prasinus. The Herpetological Bulletin* 149:28–31.

Karin, B., A. Stubbs, U. Arifin, L. Bloch, G. Ramadhan, D. Iskandar, et al. 2018. The herpetofauna of the Kei Islands (Maluku, Indonesia): Comprehensive report on new and historical collections, biogeographic patterns, conservation concerns, and an annotated checklist of species from Kei Kecil, Kei Besar, Tam, and Kur. *Raffles Bulletin of Zoology* 66:704–738. https://escholarship.org/uc/item/14d553xm

Kaufman, J., G. Burghardt, and J. Phillips. 1996. Sensory cues and foraging decisions in a large carnivorous lizard, *Varanus albigularis. Animal Behavior* 52:727–736.

Keith, M. and A. Ginsberg. 2010. Active hearding of fish by the Nile Monitor (*Varanus niloticus*) in Kruger National Park. *African Herp News, Newsletter of the Herpetological Association of Africa* 51:19–21.

Kerr, T. 1992. An examination of the trigeminal nerve foramina of vertebrates. Unpublished B.Sc. thesis, Department of Anatomy, University College London. 202 pp.

Khan, A. 1987. *Bangladesher Bonnya Prani, Part-1 (= Wildlife of Bangladesh, Part-1).* Bangla Academy, Dhaka.

Khoirunnisa, I. 2020. Histological structure of the *Varanus salvator* kidney. *Proceedings of the International Conference on Science and Engineering* 3:117–119.

King, D. and B. Green. 1979. Notes on the diet and reproduction of the Sand Goanna, *Varanus gouldii rosenbergi. Copeia* 1979(1):64–70.

King, D. and B. Green. 1999. *Monitors: The biology of varanid lizards.* 2nd Ed. Krieger Publishing, Malabar, FL. 116 pp.

King, D., M. King, and P. Baverstock. 1991. A new phylogeny of the Varanidae. *Mertensiella* 2:211–219.

King, D. and E. Pianka. 2007. Ecology of the Pygmy monitor, *Varanus brevicauda*, in Western Australia. *Mertensiella* 16:304–311.

King, M., G. Mengden, and D. King. 1982. A pericentric-inversion polymorphism and s ZZ/ZW sex-chromosome system in *Varanus acanthurus* Boulenger analyzed by G-and C-banding and Ag staining. *Genetica* 58:39–45.

King, M. 1993. *Species evolution: The role of chromosome change.* Cambridge University Press, Cambridge, UK. 336 pp.

King, M. and D. King. 1975. Chromosomal evolution in the lizard genus *Varanus* (Reptilia). *Australian Journal of Biological Science* 28:89–108.

Kinghorn, J. 1923. A new *Varanus* from Coquet Island, Queensland. *Records of the Australian Museum* 14(2):135–137.

Kirkish, P., J. Fobes, and A. Richardson. 1979. Spatial reversal learning in the lizard *Coleonyx variegatus. Bulletin of Psychonomic Science* 13:265–267.

Kirschner, A., T. Muller, and H. Seufer. 1996. *Faszination warane.* Kirschner & Seufer Verlag, Keltern, Weiler.

Kis, A., L. Huber, and A. Wilkinson. 2015. Social learning by imitation in a reptile (*Pogona vitticeps*). *Animal Cognition* 18:325–331.

Kishida, K. 1929. A new monitor from the island of Saipan, South Sea Islands. *Lansania* 1:13–16.

Klabacka, R., et al. 2020. Rivers of Indochina as potential drivers of lineage diversification in the spotted flying lizard (*Draco maculatus*) species complex. *Molecular Phylogenetics and Evolution* 150:106861.

Klembara, J., K. Dobiašová, M. Hain, and O. Yaryhin. 2017. Skull anatomy and ontogeny of legless lizard *Pseudopus apodus* (Pallas, 1775): Heterochronic Influences on Form. *The Anatomical Record* 300(3):460–502.

Kluge, A. 1993. *Aspidites* and the phylogeny of pythonine snakes. *Records Australian Museum, Supplement* 19:1–77.

Koch, A. 2016. Der Blaue Baumwaran (*Varanus macraei*) von Batanta: Wunderschön und stark bedroht. *Reptilia* 21(122):38–41.

Koch, A., E. Arida, and W. Böhme. 2007. Zwischenbericht über der Herpetofauna Sulawesis unter besonderer Berücksichtigung der Gattung *Varanus*: phylogeographische Beziehungen zu angrenzenden Gebieten. *Elaphe* 15(3):42–52.

Koch, A., E. Arida, A. Schmitz, W. Böhme, and T. Ziegler. 2009. Refining the polytypic species concept of mangrove monitors (Squamata: *Varanus indicus* group): A new cryptic species from the Talaud Islands, Indonesia, reveals the underestimated diversity of Indo-Australian monitor lizards. *Australian Journal of Zoology* 57:29–40.

Koch, A., M. Auliya, A. Schmitz, U. Kuch, and W. Böhme. 2007. Morphological studies on the systematics of south East Asian water monitors (*Varanus salvator* Complex): Nominotypic populations and taxonomic overview. *Mertensiella* 16:109–180.

Koch, A., T. Ziegler, W. Böhme, E. Arida, and M. Auliyla. 2013. Pressing problems: Distribution, threats, and conservation status of the monitor lizards (Varanidae: *Varanus* spp.) of Southeast Asia and the Indo-Australian archipelago. *Herpetological Conservation and Biology* 8(Monograph 3):1–62.

Köhler, G. 2005. *Incubation of reptile eggs.* Krieger Publishing Company, Malabar. 214 pp.

Kok, R. 1993. Gelungene Zucht des Pazifik-Warans *(Varanus indicus). Monitor* 2(2):31.

Kopstein, F. 1926. Reptilien von den Molukken und den Benachbarten Inseln. *Zoologische Mededeelingen* IX(2–3):71–112.

Krebs, U. 1979. Der Dumeril-waran (*Varanus dumerilii*), ein spezialistieter krabbenfresser? *Salamandra* 15(3):146–157.

Krekorian C., V. Vance, and A. Richardson. 1968. Temperature-dependent maze learning in the desert iguana *Dipsosaurus dorsalis. Animal Behaviour* 16:429–436.

Kubary, F. 1872. Hartlaub and Finsch, untitled report. *Proceedings of the Zoological Society of London* 1872:107.

Kuhl, H. 1820. *Beitrage zur Zoologie und vergleichenden Anatomie.* Frankfurt.

Kukal, Z. 1993. Mangrove monitor: Captive *V. indicus* egg deposition. *VaraNews* 3(3):5.

Kushwaha, A., A. Pathak, and N. Kushwaha. 2015. Fatal monitor lizard bite: Case report. *International Journal of Nursing and Health Science* 2(6):70–72.

Kverková, K., L. Marhounová, A. Polonyiová, and P. Němec. 2022. The evolution of brain neuron numbers in amniotes. *Proceedings of the National Academy of Science* 119(11):e2121624119.

Lacépède, B. 1839. Histoire Naturelle de Lacépède comprenant les CétacVes, les Quadrupèdes, les Serpents, et les Poissons. Nouvelle Edition, PrécédVe de l'Éloge de Lacépède, Par Cuvier, Avec des Notes et la nouvelle Classification de M. A.-G. Desmarest. Tome I. Furne et Ce, Libraires-Éditeurs, Paris.

LaDage L., T. Cobb Irvin, and V. Gould. 2017. Assessing spatial learning and memory in small squamate reptiles. *Journal of Visual Experiments* 119:e55103.

Laing, C. and D. Fraser. 1999. The vitamin D system in iguanian lizards. *Comparative Biochemistry and Physiology Part B: Biochemistry and Molecular Biology* 123:373–379.

Lambert, H., G. Carder, and N. D'Cruze. 2019. Given the cold shoulder: A review of the scientific literature for evidence of reptile sentience. *Animals (Basal).* 9(10):821. https://doi.org/10.3390/ani9100821

Langerwerf, B. 2009. Preferred hatching time. P. 82. IN: R. Sprackland (ed.). *Giant lizards.* 2nd Ed. TFH Publications, Neptune, NJ.

Langkilde, T. and R. Shine. 2006. How much stress do researchers inflict on their study animals? A case study using a scincid lizard, *Eulamprus heatwolei. Journal of Experimental Biology* 209:1035–1043.

De Lapparent, F., S. Bailon, M. Augé, and J.-C. Rage. 2020. Amphibians and reptiles from the Neogene of Afghanistan. *Geodiversitas* 42(22). https://doi.org/10.5252/geodiversitas2020v42a22

Law, S., S. de Kort, D. Bennett, and M. van Weerd. 2016. Morphology, activity area, and movement patterns of the frugivorous monitor lizard *Varanus bitatawa. Herpetological Conservation and Biology* 11:467–475.

Law, S., R. Selvino, R. De Kort, D. Bennett, and M. Van Weerd. 2018. Diet and habitat requirements of Philippine endemic frugivorous monitor lizard *Varanus bitatawa. BIAWAK* 12(1):12–22.

Lawton, M. 2019. Ophthalmology. Pp. 721–735. IN: S. Divers and S. Stahl (eds.). *Mader's reptiles and amphibian medicine and surgery.* 3rd Ed. Elsevier. New York, NY.

LeBlanc, A., A. Morrell, S. Sirovica, M. Al-Jawad, D. Labonte, D. D'Amore, C. Clemente, S. Wang, et al. 2024. Iron-coated Komodo dragon teeth and the complex dental enamel of carnivorous reptiles. *Nature Ecology and Evolution* 8(9):1711–1722. https://doi.org/10.1038/s41559-024-02477-7. Epub 2024 Jul 24. PMID: 39048730; PMCID: PMC11383799.

Ledesma, D. and S. Scarpetta. 2018. The skull of the gerrhonotine lizard *Elgaria panamintina* (Squamata: Anguidae). *PLoS One* 13(6):e0199584.

Lee, H.-J., Y.-N. Lee, A. Fiorello, and J. Lü. 2018. Lizards ran bipedally 110 million years ago. *Scientific Reports* 8: 2617.

Lee, M. 1995. Species concepts and the recognition of ancestors. *Historical Biology* 10(4):329–339.

Lee, M. 1996. Possible affinities between *Varanus giganteus* and *Megalania priscus. Memoirs of the Queensland Museum* 39:232.

Lee, M. 1997. The phylogeny of varanoid lizards and the affinities of snakes. *Philosophical Transactions of the Royal Society, Series B* 352:53–91.

Lee, M. 2005. Squamate phylogeny, taxon sampling, and data congruence. *Organisms, Diversity and Evolution* 5:25–45.

Lee, M. and S. Scanlon. 2001. On the lower jaw and intramandibular septum in snakes and anguimorph lizards. *Copeia* 2001:531–535.

Leitch, D. and K. Catania. 2012. Structure, innervation and response properties of integumentary sensory organs in crocodilians. *Journal of Experimental Biology* 215(23):4217–4230.

Lemus, A. 1967. Contribucion al studio de la embryologia de reptiles chilenos. II. Tabla de desarollo de la largatija vivipara *Liolaemus gravenhorsti. Biologia Chili* 40:39–61.

Lenk, P., B. Eidenmüller, H. Stauder, R. Wicker, and M. Wink. 2005. A parthenogenetic *Varanus. Amphibia-Reptilia* 26:507–514.

Leslie, R. and W Grant. 1994. Meristic and morphometric variation among anglerfish of the genus *Lophius. Journal of Zoology (London)* 232(4):565–584.

Lesson, R. 1830. Zoologie. IN: M. Duperrey (ed.). *Voyage autour du monde… Sur la Corvette de sa Majeste.* La Coquille. Tome 2. Zoologie, Paris.

Lewin, R. 1987. *Bones of Contention.* Simon & Schuster. New York, NY.

Leydig, F. 1890. Das Parietalorgan der Amphibien und Reptilien. *Abhandlungen der Senckenberg Naturforschungistitut Gesellschaft* 16:441–550.

Lillywhite, H. 2008. *Dictionary of herpetology.* Krieger Publishing, Malabar, FL. 376 pp.

Lillywhite, H. 2023. Physiology and functional anatomy. Pp. 7–44. IN: C. Warwick, P. Arena, and G. Burghardt (eds.). *Health and welfare of captive reptiles.* Springer International Publishing AG. New York, NY.

Lin, L.-H. and J. Wiens. 2016. Comparing macroecological patterns across continents: Evolution of climatic niche breadth in varanid lizards. *Ecography.* https://doi:10.1111/ecog.02343

Linkem, C., et al. 2013. Stochastic faunal exchanges drive diversification in widespread Wallacean and Pacific Island lizards (Squamata: Scincidae: *Lamprolepis smaragdina). Journal of Biogeography* 40:507–520.

Linne, C. 1758. *Systema Naturae per regne tria naturae, secundum: Classes, ordines, genera, species. Cum characteribus synonymis, locus. Tomus I. Editio Decima, Reformata.* Duodecima, Reformata. Holmiae, Impensis Direct. Laurentii Salvii. 1758–1759 pp.

Linne, C. 1766. *Systema Naturae per regne tria naturae, secundum: classes, ordines, genera, species. Cum characteribus synonymis, locus. Tomus I. Editio Duodecima, Reformata.* Holmiae, Impensis Direct. Laurentii Salvii.

Locy, W. 1894. The derivation of the pineal eye. *Anatomischen Anzeiger* 9:169–180.

Longrich, N., B.-A. Bhullar, and J. Gauthier. 2012. Mass extinction of lizards and snakes at the Cretaceous-Paleogene boundary. *Proceedings of the National Academy of Science Early Edition.* https://doi.org/10.1073/pnas.1211526110

Loop, M. 1976. Autoshaping a simple technique for teaching a lizard to perform a visual discrimination task. *Copeia* 1976(3): 574–576.

Losos, J. 2011. *Lizards in an evolutionary tree: Ecology and adaptive radiation of anoles.* Organisms and Environments, vol. 10. University of California Press, Berkeley.

Losos, J. and H. Greene. 1988. Ecological and evolutionary implications of diet in monitor lizards. *Biological Journal of the Linnean Society* 35:379–407.

Losos, J. and R. Rickefs. 2010. *The theory of island biogeography revisited.* Princeton University Press, Princeton, NJ.

Lunney, D. and D. Ayers. 1993. *Herpetology in Australia, a diverse discipline: Transactions of the Royal Zoological Society of New South Wales*. Surrey Beatty & Sons. 414 pp.

MacArthur, R. and E. Wilson. 1967. *The theory of island biogeography*. Princeton.

Malhotra, A. and R. Thorpe. 1991a. Microgeographic variation in *Anolis oculatus*, on the island of Dominica, West Indies. *Journal of Evolutionary Biology* 4:321–335.

Malhotra, A. and R. Thorpe. 1991b. Experimental detection of rapid evolutionary response in natural lizard populations. *Nature* 353(6342):347–348.

Main, A. and M. Bull. 1996. Mother-offspring recognition in two Australian lizards, *Tiliqua rugosa* and *Egernia stokesii*. *Animal Behaviour* 52(1):193–200.

Mallet, J. 2005. Hybridization as an invasion of the genome. *Trends in Ecology and Evolution* 20:229–237.

Mallet, J., N. Besansky, and M. Hahn. 2015. How reticulated are species? *Bioessays* 2016 Feb;38(2):140–149. https://doi.org/10.1002/bies.201500149. Epub 2015 Dec 28. PMID: 26709836; PMCID: PMC4813508.

Manríquez-Morán, N., F. Cruz, and R. Murphy. 2014. Genetic variation and origin of parthenogenesis in the *Aspidoscelis cozumela* complex: Evidence from mitochondrial genes. *Zoological Science* 31:14–19. https://doi.org/10.2108/zsj.31.14

Manrod J., R. Hartdegen, and G. Burghardt. 2008. Rapid solving of a problem apparatus by juvenile black-throated monitor lizards (*Varanus albigularis albigularis*). *Animal Cognition* 11:267–273.

Marcellini, D. and S. Davis. 1982. Effects of handling on reptile egg hatching. *Herpetological Review* 13:43–44.

Markland, R. and C. Brown. 2009. Keeping and breeding *Varanus albigularis*. Pp. 220–221. IN: R. Sprackland (ed.). *Giant lizards, 2nd edition*. TFH Publications, Neptune, N.J.

Matsubara, K., S. Sarre, A. Georges, Y. Matsuda, J. Marshall Graves, and T. Ezaz. 2014. Highly differentiated ZW sex microchromosomes in the Australian *Varanus* species evolved through rapid amplification of repetitive sequences. *PLoS One* 9(4):e95226.

Mayes, P. J. 2007. The use of burrows and burrow characteristics of the semi-aquatic *Varanus mertensi* (Reptilia: Varanidae). *Mertensiella* 16:312–321.

Mayes, P. J., S. D. Bradshaw, and F. J. Bradshaw. 2007. Reproductive seasonality in the semi-aquatic monitor *Varanus mertensi* (Reptilia: Varanidae). *Mertensiella* 16:322–335.

Mayr, E. 1963. *Animal species and evolution*. Belknap Press, Harvard.

Mayr, E. and P. Ashlock. 1991. *Principles of systematic zoology*. 2nd Ed. McGraw Hill, New York, NY.

Mayr, E. and J. Diamond. 2001. *The birds of northern melanesia; speciation, ecology, and biogeography*. Oxford University Press. Oxford, UK.

Mays, S. 2007. Husbandry and captive propagation of the Crocodile Monitor, *Varanus salvadorii*, at the Houston Zoo. *Mertensiella* 16:422–429.

McCoid, M. 1993. Reproductive output in captive and wild mangrove monitors *(Varanus indicus)*. *VaraNews* 3(3):4.

McCoid, M. and R. Hensley. 1991. Mating and combat in *Varanus indicus*. *Herpetological Review* 22(1):16–17.

McCoid, M. and R. Hensley. 1993. Observations of *Varanus indicus* in the Mariana Islands. *VaraNews* 3(6):4–5.

McCoid, M. and G. Witteman. 1993. *Varanus indicus* (Mangrove monitor). Diet. *Herpetological Review* 24(3):105.

McCoy, M. 1980. *Reptiles of the Solomon Islands*. Wau Ecology Inst., Hndbk, vol. 7. Wau, Papua New Guinea.

McCurry, M., M. Mahony, P. Clausen, M. Quayle, C. Walmsley, T. Jessup, S. Wroe, H. Richards, and C. McHenry. 2015. The relationship between cranial structure, biomechanical performance and ecological diversity in varanoid lizards. *PLoS One* 10(6): e0130625. https://doi.org/10.1371/journal.pone.0130625, 27.

McDowell, S. 1972a. The extracolumella and tympanic cavity of the "Earless" monitor lizard, *Lanthanotus borneensis*. *Copeia* 1972(1):154–159.

McDowell, S. 1972b. The evolution of the tongue of snakes, and its bearing on snake origins. IN: T. Dobzhansky, M. Hecht, and W. Steere (eds.). *Evolutionary biology*. Springer, New York, NY. https://doi.org/10.1007/978-1-4684-9063-3_8

McDowell, S. and C. Bogert. 1954. The systematic position of *Lanthanotus* and the affinities of the anguinomorphan lizards. *Bulletin of the American Museum of Natural History* 105(1).

McGuire, J., et al. 2023. Species delimitation, phylogenomics, and biogeography of Sulawesi flying lizards: A diversification history complicated by ancient hybridization, cryptic species, and arrested speciation. *Systematic Biology*. https://doi.org/10.1093/sysbio/syad020

McMahan, C., L. Freeborn, W. Wheeler, and B. Crother. 2015. Forked tongues revisited. Molecular apomorphies Support Morphological Hypotheses of Squamate Evolution. *Copeia* 2015(3):525–529.

McNab, B. and W. Auffenberg.1976. The effect of large body size on the temperature regulation of the Komodo Dragon, *Varanus komodoensis*. *Comparative Biochemistry and Physiology* 55A:345–359.

McNaughtan, J. E. and J. A. Norman. 2009. A central role for venom in predation by *Varanus komodoensis* (Komodo dragon) and the extinct giant *Varanus (Megalania) priscus*. *Proceedings of the National Academy of Sciences* 106(22):8969–8974.

McWilliams, D. 2005. Nutrition research on calcium homeostasis. I. Lizards (with recommendations). *International Zoo Yearbook* 39:69–77.

Mehaffey, D. and D. McGinnity. 1996. Taxon management account: Black roughneck monitor *Varanus rudicollis*. P. 8. IN: S. H. Hammack (ed.). *American zoo and aquarium association lizard advisory group, taxon management accounts*. Fort Worth Zoological Park, Fort Worth.

Meier, J., D. Marques, S. Mwaiko, C. Wagner, L. Excoffier, and O. Seehausen. 2017. Ancient hybridization fuels rapid cichlid fish adaptive radiations. *Nature Communications* 8:14363.

Meiri, S. 2007. Size evolution in island lizards. *Global Ecological Biogeography* 16:702–708.

Menaker, M. and S. Wisner. 1983. Temperature-compensated circadian clock in the pineal of *Anolis*. *Proceedings of the National Academy of Science* 80(19):6119–6121.

Mendyk, R. 2012. Reproduction of varanid lizards (Reptilia: Squamata: Varanidae) at the Bronx Zoo. *Zoo Biology* 31(3):374–389.

Mendyk, R. 2016. An annotated bibliography of captive reproduction in monitor lizards (Varanidae: *Varanus*). Part 1. Odatria. *Biawak* 10(2):54–71.

Mendyk, R. 2017. An annotated bibliography of captive reproduction in monitor lizards (Varanidae: *Varanus*). Part II. *Empagusia* and *Philippinosaurus*. *Biawak* 11(1):40–49.

Mendyk, R. 2020. An annotated bibliography of captive reproduction in monitor lizards (Varanidae: *Varanus*). Part V. Varanus. *Biawak* 14(1&2):60–79.

Mendyk, R. and L. Augustine. 2023. Controlled deprivation and enrichment. Pp. 287–321. IN: C. Warwick, P. Arena, and G. Burghardt (eds.). *Health and welfare of captive reptiles*. Springer International Publishing AG New York, NY..

Mendyk, R. and H.-G. Horn. 2011. Skilled forelimb movements and extractive foraging in the arboreal monitor lizard *Varanus beccarii* (Doria, 1874). *Herpetological Review* 42(3):343–349.

Merrem, B. 1820. *Versuch eines Systems der Amphibien*. Marburg.

Mertens, R. 1926. Uber die Rassen einiger indo-australischer Reptilien. *Senckenbergiana* 8:272–276.

Mertens, R. 1934. Die Insel-Reptilien, ihre Ausbreitung, Variation und Arthildung. *Zoologica (Stuttgart)* 84:1–209.

Mertens, R. 1937. Zwei Bemerkungen über Warane (Varanidae). *Senckenbergiana* 19:177–181.

Mertens, R. 1941. Zwei neue Warane des Australischen Faunengebeties. *Senckenbergiana* 23(4/6):266–272.

Mertens, R. 1942a. Die Familie der Warane (Varanidae). Erster Teil: Der Allgemeines. *Abhandlungen der Senckenbergischen Naturforschenden Gesellschaft* 462:1–116.

Mertens, R. 1942b. Die Familie der Warane (Varanidae). Zweiter Teil: Der Schädel. *Abhandlungen der Senckenbergischen Naturforschenden Gesellschaft* 465:117–234.

Mertens, R. 1942c. Die Familie der Warane (Varanidae). Dritter Teil: Taxonomie. *Abhandlungen der Senckenbergischen Naturforschenden Gesellschaft* 466:235–391.

Mertens, R. 1951. A new lizard of the genus *Varanus* from New Guinea. *Fieldiana Zoology* 31(43):467–471.

Mertens, R. 1958. Bemerkungen uber die Warane Australiens. *Senckenbergiana Biologica* 39(5/6):229–264.

Mertens, R. 1959. Liste der Warane Asiens und der Indo-australischen Inselweltmit systematischen Bemerkungen. *Senckenbergiana Biologica* 40(5/6):221–240.

Mertens, R. 1960. Der Papua-Waran. *Kosmos* 56:547–549.

Mertens, R. 1962a. *Papusaurus*, eine neue Untergattung von *Varanus*. *Senckenbergiana Biologica* 43(5):331–333.

Mertens, R. 1962b. Philippinosaurus, eine neue Untergattung von *Varanus*. *Senckenbergiana Biologica* 43(5):331–333.

Mertens, R. 1963. Lanthanotidae, Helodermatidae, & Varanidae. *Das Tierreich* 79:1–26.

Mertens, R. 1971. Über eine waransammlung aus dem östlichen Neuguinea. *Senckenbergiana Biologica* 52(1/2):1–5.

Mertens, R. 1997. Die Familie der Warane. Ertster Teil: Allgemeines; Zwitter Teil: der Schadel; Dritter Teil: Taxonomie. IN: *Abbhandlungen Senckenbergischen Naturforschenden Gesellellschaft 462, 465, 466, reprint of 1942 volumes*. Edition Chimaira, Frankfurt.

Meyer, A. B. 1874. Ubersicht der von mir auf Neu-Guinea und den Inseln Jobi, Mysore und Mafoor. *Monatsberichte der Königlichen Preussische Akademie des Wissenschaften zu Berlin* 4–12.

Milenkaya, O. and J. Mckay. 2016. Observation in the wild of the poorly-known *Varanus yuwonoi*. *BIAWAK* 10(1):13–17.

Minnich, J. E. 1979. Reptiles. Pp. 391–641. IN: G. Maloiy (ed.). *Comparative physiology of osmoregulation in animals*, vol. 1. Academic Press, London.

Mitchell, G., T. Gleeson, and A. Bennett. 1981. Ventilation and acid-base balance during graded activity in lizards. *American Journal of Physiology* 240(1):R29–R37.

Mitra, D. 2001. Captive breeding of yellow monitor lizard (*Varanus flavescens*). *Cobra* (Chennai) 46:36–37.

Mo, J.-Y., X. Xu, and S. Evans. 2012. A large predatory lizard (Platynota, Squamata) from the Late Cretaceous of South China. *Journal of Systematic Palaeontology* 10(2):333–339. https://doi.org/10.1080/14772019.2011.588254

Mohan, A., P. Swamy, and K. Shanker. 2018, Oct. Population structure in the Andaman keelback, *Xenochrophis tytleri*: Geographical distance and oceanic barriers to dispersal influence genetic divergence on the Andaman archipelago. *PeerJ* 9(6):e5752. https://doi.org/10.7717/peerj.5752

Moldovan, D. 2007. Haltung und Zucht von *Varanus (Odatria) similis* Mertens, 1958. *Sauria* 29(3):21–26.

Molnar, R. 2004. *Dragons in the Dust*. University of Indiana Press, Bloomington, IN. 209 pp.

Molnar, R. 1990. New cranial elements of a giant varanid from Queensland. *Memoirs of the Queensland Museum* 29:437–444.

Montiani-Ferreira, F., B. Moore, and G. Ben-Shlomo (eds.). 2022. *Wild and exotic animal ophthalmology: Volume 1: Invertebrates, fishes, amphibians, reptiles, and birds*. Springer International Publishing, New York, NY. 696 pp.

Montuelle, S., A. Herrel, P-A. Libourel, S. Daillie, and V. Bels. 2012. Flexibility in locomotor-feeding integration during prey capture in varanid lizards: Effects of prey size and velocity. *Journal of Experimental Biology* 215:3823–3835.

Montuelle, S., A. Herrel, V. Schaerlacken, K. Metzger, A. Mutuyeyezu, and V. Bels. 2011. Inertial feeding in the teiid lizard *Tupinambis merianae*: The effect of prey size on the movements of hyolingual apparatus and the cranio-cervical system. *Zoology (Jena)* 114(3):165–170.

Mulcahy, D., et al. 2012. Estimating divergence dates and evaluating dating methods using phylogenomic and mitochondrial data in squamate reptiles. *Molecular Phylogenetics and Evolution* 65:974–991.

Müller, S. and H. Schlegel. 1845. *De in den Indischen archipel levendehagedisachtige dieren van het geslacht Monitor*. Leiden.

Munns, S., L. Hartzler, A. Bennett, and J. Hicks. 2004. Elevated intra-abdominal pressure limits venous return during exercise in *Varanus exanthematicus*. *Journal of Experimental Biology* 207:4111–4120.

Murphy, J. 1972. Notes on Indo-Australian varanids in captivity. *International. Zoo Yearbook*, 12:199–202.

Murphy, J., C. Ciofi, C. Panouse, and T. Walsh (eds.). 2002. *Komodo dragon: Biology and conservation*. Smithsonian Washington, DC. 268 pp.

Murphy, R. and N. Lovejoy. 1998. Punctuated equilibrium or gradualism in the lizard *Sceloporus*? Lost in plesiograms and a forest of trees. *Cladistics* 14:95–103.

Murphy, J., R. Mendyk, K. Miller, and L. Augustine. 2019. Tales of monitor lizards and other perspectives. *Herpetological Review* 50(1):178–201.

Murphy, J. and L. Mitchell. 1974. Ritualized combat behavior of the Pygmy Mulga Monitor Lizard, *Varanus gilleni* (Sauria: Varanidae). *Herpetologica* 30(1):90–97.

Murray, P. and D. Megirian. 1992. Continuity and contrast on Middle and Late Miocene vertebrate communities from the Northern Territory. *The Beagle* 9:195–218.

Muzammil, S. 1949. Studies on the anatomy of the tail in Sauria and Rhynchocephalia. Part III. *Varanus monitor* (Linne). *Proceedings of the Indian Academy of Sciences* XXX:155–167.

Myers, E., et al. 2019. Environmental heterogeneity and not vicariant biogeographic barriers generate community-wide population structure in desert-adapted snakes. *Molecular Ecology* 28:4535–4548.

Natusch, D. and J. Lyons. 2017. Notes on the natural history of Blue-tailed Monitors (*Varanus doreanus*) in Australia. *BIAWAK* 11(1):8–14.

Neaves, W. and P. Baumann. 2011. Unisexual reproduction among vertebrates. *Trends in Genetics* 27:81–88.

Nelling, C. 1995, Jan.–Feb. Mangrove monitors. *Reptile and Amphibian Magazine* 30–35.

Neugebauer, W. 1976. Monitors. Pp. 323–331. IN: B. Grzimek (ed.). *Animal life encyclopaedia*, vol. 6. Van Nostrand, New York, NY.

Nijboer, J. 2020. *Nutrition in lizards*. Merck Veterinary Manual, Professional Version (online): https://www.merckvetmanual.com/management-and-nutrition/nutrition-exotic-and-zoo-animals/nutrition-in-lizards?gclid=Cj0KCQjwjLGyBhCYARIsAPqTz19M7WMfkTBGj2lMQGk4anF3ihKnMq5o6uq8gYfSVMZ0RTFTLs4KrgaAhulEALw_wcB&gclsrc=aw.ds

Nishimura, T. 2020. The parietal eye of lizards (*Pogona vitticeps*) needs light at a wavelength lower than 580nm to activate light-dependent magnetoreception. *Animals (Basel)* 10(3):489. https://doi.org/10.3390/ani10030489

Noble, D., R. Byrne, and M. Whiting. 2014. Age-dependent social learning in a lizard. *Biological Letters* 10:20140430.

Noonan, B., J. Pramuk, R. Bezy, E. Sinclair, K. de Queiroz, and J. Sites. 2013. Phylogenetic relationships within the lizard clade Xantusiidae: Using trees and divergence times to address evolutionary questions at multiple levels. *Molecular Phylogenetics and Evolution* 69(1):109–122.

Nopcsa, F. 1903a. Über die *Varanus*-artigen Lacerten Istriens. *Beitr. z. Pal. Geol. Österr.-Ung. u. d. Orients, Wien* 15:31–42, 2 Taf.

Nopcsa, F. 1903b. Origin of the Mosasaurs. *Geological Magazine, London* 119–121.

Nopcsa, F. 1908. Zur Kenntnis der fossilen Eidechsen. *Beitrage zur Palaontologie und Geologie Osterteich-Ungarns und des Orients* 21:33–62.

Nopcsa, F. 1912. Zur Kenntnis fossiler Eidechsen. *Beitrage zur Palaontologie und Geologie Osterteich-Ungarns und des Orients* 21:33–62.

Nopcsa, F. 1923a. *Eidolosaurus* und *Pachyophis*. Zwei neue Neocom-Reptilien. *Paleontographica* 65:99–154.

Nopcsa, F. 1923b. Die Familien der Reptilien. *Fortschritte der Geologie und Palaeontologie* 2:1–210.

Norell, M. and K. Gao. 1997. Braincase and phylogenetic relationships of *Estesia mongoliensis* from the Upper Cretaceous, Gobi Desert and the recognition of a new clade of lizards. *American Museum Novitates* 3211:25.

Norell, M., K. Gao, and J. Conrad. 2007. A new platynotan lizard (Diapsida: Squamata) from the Late Cretaceous Gobi Desert (*Ömnögov*), Mongolia. *American Museum Novitates* 3605:22.

Norell, M., M. McKenna, and M. Novacek. 1992. *Estesia mongoliensis*, a new fossil varanoid from the Late Cretaceous Barun Goyot formation of Mongolia. *American Museum Novitates* 3045:1–24.

O'Brien M., S. Valverde, S. Duran-Nebreda, B. Vidiella, and R. Bentley. 2024. Punctuated equilibrium at 50: Anything there for evolutionary anthropology? Yes; definitely. *Evolutionary Anthropology* 33:e22009. https://doi.org/10.1002/evan.22009

Oelrich, T. 1956. The anatomy of the head of *Ctenosaura pectinata* (Iguanidae). *Miscellaneous Publications of the Museum of Zoology, University of Michigan* 94:1–122.

Olberding, J., L. Brayer, and T. Higham. 2012. Performance and three-dimensional kinematics of bipedal lizards during obstacle negotiation. *Journal of Experimental Biology* 215(2):247–255.

Oliveira, P., T. Souza, M. Castañon, J. Dergam, and B. Sousa. 2021. Note First report of hemiclitores in female lizards of the genus *Ophiodes* Wagler, 1828 (Squamata: Anguidae). *Cuadernos de Herpetología* 35:325–328.

Oliver, J. 1956. Reproduction in the King Cobra, *Ophiophagus hannah* Cantor. *Zoologica: New York Zoological Society* 41:145–152 + 5 pl.

Oliver, P., K. Smith, R. Laver, P. Doughty, and M. Adams. 2014. Contrasting patterns of persistence and diversification in vicars of a widespread Australian lizard lineage (the *Oedura marmorata* complex). *Journal of Biogeography* 41:2068–2079.

Olsson, M., S. Anderson, and E. Wapstra. 2011. UV-deprived coloration reduces success in mate acquisition in male sand lizards (*Lacerta agilis*). *PLoS One* 6:e19360.

Olsson, M., D. Stuart-Fox, and C. Ballen. 2013. Genetics and evolution of colour patterns in reptiles. *Seminars in Cell Developmental Biology* 24:529–541.

Oonincx, D., Y. Stevens, J. van den Borne, J. van Leeuwen, and W. Hendriks. 2010. Effects of vitamin D3 supplementation and UVb exposure on the growth and plasma concentration of vitamin D3 metabolites in juvenile bearded dragons (*Pogona vitticeps*). *Comparative Biochemistry and Physiology B Biochemical Molecular Biology* 156(2):122–128. doi: 10.1016/j.cbpb.2010.02.008. Epub 2010 Mar 4. PMID: 20206712.

Oonincx, D. and J. van Leeuwen. 2017. Evidence-based reptile housing and nutrition. *Veterinary Clinics of North American Exotic Animal Practice* 20(3):885–898.

Openshaw, G., D. D'Amore, M. Vidal-Garcia, and J. Keogh. 2016. Combining geometric morphometric analysis of multiple 2D observation views improves interpretation of evolutionary allometry and shape diversification in monitor lizard (*Varanus*) crania. *Biological Journal of the Linnean Society*. https://doi.org/10.1111/bij.12899

Oppel, M. 1811. *Die Ordnungen, Familien, und Gattungen der Reptilien als Prodrom einer Naturgeschichte derselben*. J. Lindauer, München.

O'Shea, M. 1991. The reptiles of Papua New Guinea. *British Herpetological Society Bulletin* 37:15–32.

Ostrom, J. and P. Wellnhofer. 1990. *Triceratops*: An example of flawed systematics. Pp. 245–254. IN: K. Carpenter and P. Curry (eds.). *Dinosaur systematics: Perspectives and approaches*. Cambridge University PressCambridge< UK..

Owerkowitz, T., C. Farmer, J. Hicks, and E. Brainerd. 1999. Contribution of gular pumping to lung ventilation in monitor lizards. *Science* 284:1161–1163.

Owen, R. 1859. Description of some remains of a gigantic land-lizard (*Megalania prisca*, Owen) from Australia. *Philosophical Transactions of the Royal Society of London* 149:43–48, pl. 7–8.

Packard, G. and J. Phillips. 1994. The importance of the physical environment for the incubation of reptilian eggs. Pp. 195–208. IN: J. B. Murphy, K. Adler, and J. T. Collins (eds.). *Captive management and conservation of amphibians and reptiles*. Society for the Study of Amphibians and Reptiles, Ithaca.

Packard, M. and V. DeMarco. 1991. Eggshell structure and formation in eggs of oviparous reptiles. Pp. 53–69. IN: D. C. Deeming and M. W. J. Ferguson (eds.). *Egg incubation: Effects on embryonic development in birds and reptiles*. Cambridge University Press, Cambridge.

Packard, M., G. Packard, and T. Boardman. 1982. Structure of eggshells and water relations of reptilian eggs. *Herpetologica* 38(1):136–155.

Page, R. and M. Charleston. 1997. From gene to organismal phylogeny: Reconciled trees and the genetree/species tree problem. *Molecular Phylogenetics and Evolution* 7:231–240.

Palci, A., M. Caldwell, M. Hutchinson, T. Konishi and M. Lee. 2020. The morphological diversity of the quadrate bone in squamate reptiles as revealed by high-resolution computed tomography and geometric morphometrics. *Journal of Anatomy* 236:210–227.

Palci, A., T. Konishi, and M. Caldwell. 2021. A comprehensive review of the morphological diversity of the quadrate bone in mosasauroids (Squamata: Mosasauroidea), with comments on the homology of the infrastapedial process. *Journal of Vertebrate Paleontology* e1879101:11.

Paluh, D., K. Olgun, and A. Bauer. 2018. Ontogeny, but not sexual dimorphism, drives the intraspecific variation of quadrate morphology in *Hemidactylus turcicus* (Squamata: Gekkonidae). *Herpetologica* 74(1):22–28.

Pandha, S. and J. Thapliyal. 1976. Egg laying and development in the garden lizard. *Calotes versicolor. Copeia* 1967:121–125.

Parker, F. 1970. Collecting reptiles and amphibians in New Guinea. *Australian Natural History* 16:309–314.

Paranjape, S. 1974. *The anatomy of the garden lizard (Calotes versicolor, Bouln.)*. University of Poona, India.

Pathak, S., M. Farooqui, A. Tripathi, and S. Chaturvedi. 2015. Morphological characterization of the tongue of Bengal monitor lizard (*Varanus bengalensis*). *Animal Science Reporter* 9(2):72–74.

Pavón-Vázquez, C., D. Esquerré, A. Fitch, B. Maryan, P. Doughty, S. Donnellan, and S. Keogh. 2022. Between a rock and a dry place: Phylogenomics, biogeography, and systematics of ridgetailed monitors (Squamata: Varanidae: *Varanus acanthurus* complex). *Molecular Phylogenetics and Evolution*. https://doi.org/10.1016/j.ympev.2022.107516

Payseur, B. and L. Riesberg. 2016. A genomic perspective on hybridization and speciation. *Molecular Ecology* 25(11):2337–2360.

Pellegrino, K., M. Rodrigues, J. Harris, Y., Yonenaga-Yassuda, and J. Sites. 2011. Molecular phylogeny, biogeography and insights into the origin of parthenogenesis in the Neotropical genus *Leposoma* (Squamata: Gymnophthalmidae): Ancient links between the Atlantic Forest and Amazonia. *Molecular Phylogenetics and Evolution* 61:446–459. https://doi.org/10.1016/j.ympev.2011.07.010.

Pérez i de Lanuza, G. and E. Font. 2014. Ultraviolet vision in lacertid lizards: Evidence from retinal structure, eye transmittance, SWS1 visual pigment genes and behaviour. *Journal of Experimental Biology* 217(16):2899–2909. https://doi.org/10.1242/jeb.104281

Peters, W. 1876. Uber die von *SMS Gazelle* mitgebrachten Amphibien. *Mber. Akad. Wiss. Berlin* 528–534.

Peters, W. and G. Doria. 1878. Catalogo dei rettili e dei batraci raccolti da O.Beccari, L. M. D'Albertis A. A. Bruijn. *Museum Civico Storia Natural Genova* XIII:323–450.

Pfeffer, P. 1965. *Auf den Inseln des Drachen*. Schwabenverlag, Stuttgart. 158 pp.

Philipp, K., S. Bowring, and T. Ziegler. 1999. The identity of *Varanus indicus*: Redefinition and description of a sibling species coexisting at the type locality (Sauria: Varanidae, *Varanus indicus* group). *Spixiana* 22(3):273–287.

Philipp, K. M. and D. P. Philipp. 2007. The monitor lizards of papua. Pp. 617–636. IN: A. J. Marshall and B. M. Beehler (eds.). *The ecology of papua – The ecology of Indonesia series volume VI part one*. Periplus Editions, Singapore.

Philipp, K. M., T. Ziegler and W. Böhme. 2007. Preliminary investigations on the natural diet of six monitor lizard species of the *Varanus (Euprepiosaurus) indicus* group. *Mertensiella* 16:336–345.

Philippen, H.-D. and B. Eidenmüller. 2007. *Terralog Vol. 6: Varanoid Lizards of the World*. Edition Chimaira, Frankfurt.

Phillips, J. and K. Howes. 1988. The pineal complex, aggressive behavior and thermoregulation in curly-tailed lizards, *Leiocephalus carinatus. Physiology and Behavior* 42(1):103–108.

Pianka, E. 1968. Notes on the biology of *Varanus eremius. Western Australian Naturalist* 11:39–44.

Pianka, E. 1969. Notes on the biology of *Varanus caudolineatus* and *Varanus gilleni. Western Australian Naturalist* 11:76–82.

Pianka, E. 1970a. Notes on *Varanus brevicauda. Western Australian Naturalist* 11:113–116.

Pianka, E. 1970b. Notes on the biology of *Varanus gouldii flavirufus. Western Australian Naturalist* 11:141–144.

Pianka, E. 1971. Notes on the biology of *Varanus tristis. Western Australian Naturalist* 11:180–183.

Pianka, E. 1972. Zoogeography and speciation of Australian desert lizards. *Ecology* 52:1024–1029.

Pianka, E. 1982. Observations on the ecology of *Varanus* in the Great Victoria desert. *Western Australian Naturalist* 15:37–44.

Pianka, E. 1986. *Ecology and natural history of desert lizards*. Princeton, NJ. 208 pp.

Pianka, E. 1994a. Evolution of body size: Varanid lizards as a model system. *American Naturalist* 146:398–414.

Pianka, E. 1994b. Comparative ecology of *Varanus* in the Great Victoria Desert. *Australian Journal of Ecology* 19:395–408.

Pianka, E. and D. King. 2000. *Varanoid lizards of the world*. Indiana University Press, Bloomington, IN.

Pianka, E. and L. Vitt. 2003. *Lizards: Windows to the evolution of diversity*. University of California Press, Berkeley.

Piskovská, A. 2019. *Clinical cardiology of reptiles*. Unpublished diploma thesis, University of Veterinary and Pharmaceutical Sciences Brno. 1–73 pp.

Platel, R. 1979. Brain weight-body weight relationships. Pp. 343–388. IN: C. Gans, R. Northcutt, and P. Ulinski (eds.). *Biology of the reptilia volume 9: Neurology A*. Academic Press, New York, NY.

Pledge, N. 1990. The upper fossil fauna of the Henschke fossil cave, Naracoorte, South Australia. *Memoirs of the Queensland Museum* 28:247–262.

Poder Le, J.-M. 2007. Notes on breeding *Varanus albigularis* in captivity. *Biawak* 1(2):73–76.

Pregill, G., J. Gauthier, and H. Greene. 1986. The evolution of helodermatid squamates, with description of a new taxon and an overview of Varanoidea. *Transactions of the San Diego Society of Natural History* 21(11):167–202.

Price, E. 2017. The physiology of lipid storage and use in reptiles. *Biological Review of the Cambridge Philosophical Society* 92(3):1406–1426. https://doi.org/10.1111/brv.12288. Epub 2016 Jun 27. PMID: 27348513.

Price, G., J. Louys, J. Cramb, Y. Feng, J. Zhao, S. Hocknull, G. Webb, A. Nguyen, and R. Joannes-Boyau. 2015. Temporal overlap of humans and giant lizards (Varanidae; Squamata) in Pleistocene Australia. *Quaternary Science Review* 125:98–105.

Purwandana, D., A. Ariefiandy, M. Imansyah, A. Seno, C. Ciofi, M. Letnic, and T. Jessop. 2016. Ecological allometries and niche use dynamics across Komodo dragon ontogeny. *The Science of Nature* 103(3–4):1–11.

Pyron, A. 2017. Novel approaches for phylogenetic Inference from morphological data and total-evidence dating in squamate reptiles (Lizards, Snakes, and Amphisbaenians). *Systematic Biology* 66(1):38–56.

Pullan, S. and P. Smith. 1987. Identification of hybrids between koi (*Cyprinus carpio*) and goldfish (*Carassius auratus*). *New Zealand Journal of Marine and Freshwater Research* 21(1):41–46. https://doi.org/10.1080/00288330.1987.9516198

Radford, L. and F. Paine. 1989. The reproduction and management of the Duméril's monitor *Varanus dumerilii* at the Buffalo Zoo. *International Zoo Yearbook* 28:153–155.

Ralph, C., B. Firth, and J. Turner. 1979a. The role of the pineal body in ectotherm thermoregulation. *American Zoologist* 19:273–293.

Ralph, C., B. Firth, and W. 1979b. The pineal complex and thermoregulation. *Biological Reviews* 54:41–72.

Ramsay, C. 2014. *Mate choice and hybridization: Comparing the Endler's Guppy (Poecilia wingei) and the Common Guppy (Poecilia reticulata).* Undergraduate honors thesis in ecology and evolutionary biology, University of Colorado Boulder.

Randow, H. 1932. Fauna und Flora von Dehiwala auf Ceylon. *Wochenschrift für Aquarien und Terrarien Kunde* 29:471–473.

Rannala, B. and Z. Yang. 2003. Bayes estimation of species divergence times and ancestral population sizes using DNA sequences from multiple loci. *Genetics* 164:1645–1656.

Rasmussen, A. 1994. A cladistic analysis of *Hydrophis* subgenus *Chitulia* (McDowell, 1972) (Serpentes: Hydrophiidae). *Zoological Journal of the Linnean Society* 111(2):161–178.

Rawlinson, P., A. Widjoya, M. Hutchinson, and G. Brown. 1990. The terrestrial vertebrate fauna of the Krakatau Islands, Sunda Strait, 1883–1986. *Philosophical Transactions of the Royal Society of London, B* 328:3–28.

Recchio, I. 2015. Reproduction of the rare Frugivorous Monitor Lizard *Varanus olivaceus* (Hallowell, 1857), at the Los Angeles Zoo and Botanical Gardens. Pp. 93–103. IN: M. Cota (ed.). *Proceedings of the 2015 interdisciplinary world conference on monitor lizards.* Suan Sunandha Rajabhat University, Bangkok. https://www.academia.edu/28631452/Reproduction_of_the_Rare_Frugivorous_Monitor_Lizard_Varanus_olivaceus_Hallowell_1857_at_the_Los_Angeles_Zoo_and_Botanical_Gardens

Read, J. and K. Moseby. 2006. Vertebrates of Tetepare Island, Solomon Islands. *Pacific Science* 60(1):69–79. http://muse.jhu.edu/journals/pacific_science/v060/60.1read.pdf

Reilly, S., L. McBrayer, and D. Miles (eds.). 2007. *Lizard ecology.* Cambridge University Press.

Reilly, S., L. McBrayer, and D. Miles (eds.). 2007. *Lizard ecology.* Cambridge University Press.

Reilly, S., et al. 2021. Phylogenomic analysis reveals dispersal-driven speciation and divergence with gene flow in lesser Sunda flying lizards (Genus *Draco*). *Systematic Biology* 71:221–241.

Reilly, S., et al. 2023. Bewildering biogeography: Waves of dispersal and diversification across southern Wallacea by bent-toed geckos (genus: *Cyrtodactylus*). *Molecular Phylogenetics and Evolution* 186:107853.

Reinholz, H. 1923. Über die Befestigung der Zähne von *Varanus niloticus*, ein Beitrag zur Frage nach der Herkunft des Zements. *Jena Zeitschrift fur Naturwissenschaften* 59:155–170.

Rieppel, O. 1978. The phylogeny and mechanics of feeding apparatus in *Varanus* (Reptilia, Lacertilia). Inaugural dissertation zur Erlangung der Würde eines Doktors der Philosophie, Universität Basel 1–32.

Rieppel, O. 1980. *The phylogeny of anguimorph lizards.* Birkhauser Verlag, Basel. 1–86 pp.

Rieppel, O. and L. Grande. 2007. The anatomy of the fossil varanid lizard *Saniwa ensidens* Leidy, 1870, based on a newly discovered complete skeleton. *Journal of Paleontology* 81(4):643–665.

Rieppel, O. and H. Zaher. 2000. The braincase of mosasaurs and *Varanus*, and the relationships of snakes. *Zoological Journal of the Linnean Society* 129:489–514.

Retes, F. and D. Bennett. 2001. Multiple generations, multiple clutches, and early maturity in four species of monitor lizards (Varanidae) bred in captivity. *Herpetological Review* 32(4):244.

Revell, L. J. 2009. Size-correction and principal components for interspecific comparative studies. *Evolution* 63:3258–3268.

Rhind, D., C. Jackson, N. Pezaro, and J. Doody. 2016. A nest of *Varanus mertensi* (Glauert, 1951) in northern Australia. *BIAWAK* 10(1):18–21.

Rich, T. and B. Hall. 1979. Rebuilding a giant. *Australian Natural History* 19:310–313.

Riley J., D. Noble, R. Bryne, and M. Whiting. 2017. Does social environment influence learning ability in a family-living lizard? *Animal Cognition* 20:449–458.

Rismiller, P., M. Mckelvey, and B. Green. 2010. Breeding phenology and behavior of Rosenberg's Goanna (*Varanus rosenbergi*) on Kangaroo Island, South Australia. *Journal of Herpetology* 44(3):399–408.

Rodda, G., T. Fritts, and J. Reichel. 1991. The distributional patterns of reptiles and amphibians in the Marshall Islands. *Micronesica*, 24(2):195–210.

Roger, O. 1898. Wirbelthierreste aus dem Dinotheriensande, II. Theil. Bericht des Naturwissenschaftlichen Vereins für Schwaben und Neuburg in Ausburg 33:385–396.

Romer, A. S. 1956. *Osteology of the reptiles.* Oxford University Press, New York, NY.

Romero, R. and F. de Souza. 2025. Evolution of pineal nonvisual opsins in lizards and the tuatara and identification of lepidopsin: A new opsin gene. *Genome Biology and Evolution* 17(5). https://doi.org/10.1093/gbe/evaf058.

Roscito, J., K. Sameith, B. Kirilenko, N. Hecker, S. Winkler, A. Dahl, M. Rodriguez, and M. Hiller. 2022. Convergent and lineage-specific genomic differences in limb regulatory elements in limbless reptile lineages. *Cell Reports* 38(2):110280. https://doi.org/10.1016/j.celrep.2021.110280

Rossi, N., M. Chiaraviglio, and G. Cardozo. 2022. Relationships among behavior, chromatism, and morphology in male aggressive encounters in *Tropidurus spinulosus*. *Ichthyology and Herpetology* 110:340–349.

Rossi, N., M. Chiaraviglio, and G. Cardozo.2023. Behavioural plasticity in activity and sexual interactions in a social lizard at high environmental temperatures. *PLoS One* 18(7):e0285656. https://doi.org/10.1371/journal.pone.0285656

Roth, T., A. Krochmal, and L. LaDage. 2019. Reptilian cognition: A more complex picture via integration of neurological mechanisms, behavioral constraints, and evolutionary context. *Bioessays* ;41(8):e1900033. https://doi.org/10.1002/bies.201900033

Rotter, J. 1963. *Die Warane*. A. Ziemsen Verlag, Wittenberg. 74 pp.

Rubinoff, D. and S. Brenden. 2005. Between two extremes: Mitochondrial DNA is neither the panacea nor the nemesis of phylogenetic and taxonomic inference. *Systematic Biology* 54(6):952–961. https://doi.org/10.1080/10635150500234674

Rubinoff, D., and S. Brenden. 2005. Between Two Extremes: Mitochondrial DNA is neither the Panacea nor the Nemesis of Phylogenetic and Taxonomic Inference, *Systematic Biology* 54(6): 952–961. https://doi.org/10.1080/10635150500234674

Sanderson, M. 1990. Estimating rates: A bias due to homoplasy. *Cladistics* 6(4):387–391.

Sannolo, M., R. Ponti, and M. Carretero. 2019. Waitin' on a sunny day: Factors affecting lizard body temperature while hiding from predators. *Journal of Thermal Biology* 84:146–153.

Santos, M., M. Oliveira, L. Verrastro, and A. Tozetti. 2010. Playing dead to stay alive: Death-feigning in *Liolaemus occipitalis* (Squamata: Liolaemidae). *Biota Neotropica* 10(4). http://www.biotaneotropica.org.br/v10n4/pt/abstract?short-communication+bn03110042010

Säve-Söderbergh, G. 1947. Notes on the brain-case in *Sphenodon* and certain Lacertilia. *Zoologiska Bidrag från Uppsala* 25:489–516.

Säve-Söderbergh, G. 1946. On the fossa hypophyseos and the attachment of the retractor bulbi group in *Sphenodon, Varanus*, and *Lacerta. Arkiv for Zoologi (Stockholm)* 38A(11):1–24.

Säve-Söderbergh, G. 1947. Notes on the braincase of *Sphenodon* and certain Lacertilia. *Särtryck ur Festrskrift tillägnad Professor Nils von Hofsten, Zoologiska Bidrag Från Uppsala* 25:489–516.

Schaerlaeken, V., S. Montuelle, P. Aerts, and A. Herrel. 2012. Jaw and hyolingual movements during prey transport in varanid lizards: Effects of prey type. *Journal of Experimental Biology* 215(part 21):3823–3835.

Schlegel, H. 1872. *De direntium van het Koninklijk zoologisch Genootschap Natura Artis Magistra te Amsterdam, Reptilia*. Amsterdam. 64 pp.

Schlegel, H. 1837–1841. *Abbildungen neuer oder unvollständig bekannter Amphibien, nach der Natur oder dem Leben entworfen und mit einem erläuternden Texte begleitet*. Arne & Co., Düsseldorf.

Schmida, G. 2017. *Australian monitors*. Gunther Schmida-Nature Shots, Gold Coast, Queensland.

Schmidt, K. 1919. Article II—Contributions to the herpetology of the Belgian Congo based on the collection of the American Museum Congo Expedition, 1909–1915. *Bulletin of the American Museum of Natural History* XXXIX:385–624.

Schmidt, K. 1932. Solomon Island reptiles. *Field Museum Natural History Zoology Series* 18: 1–45.

Schuett, G. W., R. S. Reiserer, and R. L. Earley. 2009. The evolution of bipedal postures in varanoid lizards. *Biological Journal of the Linnaen Society* 97(3):652–663.

Schwenk, K. 1984. *Evolutionary morphology of the lepidosaur tongue*. Unpublished PhD thesis, University of California, Berkeley.

Schwenk, K. 1985. Occurrence, distribution and functional significance of taste buds in lizards. *Copeia* 1985(1):91–101.

Schwenk, K. 1995. Of tongues and noses: Chemoreception in lizards and snakes. *Tree* 10(1):7–12.

Seba, A. 1734–1765. Locupletissimi rerum naturalium thesauri accurata descriptio, et iconibus artificiosissimis expressio, per universam physices historiam: opus, cui, in hoc rerum genere, nullum par exstitit. Amstelaedami, Apud J. Wetstenium, & Gul. Smith, & Janssonio-Waesbergios.

Secor, S. and J. Phillips. 1997. Specific dynamic action of a large carnivorous lizard, *Varanus albigularis. Comparative Biochemistry and Physiology* 117A:515–522.

Seebacher, F. and G. Grigg. 2001. Changes in heart rate are important for thermoregulation in the varanid lizard *Varanus varius. Journal of Comparative Physiology B* 171:395–400.

Sheren, A., E. Nasr, and S. Hassan. 2018. Light and scanning electron microscopic observations on the tongue of Nile monitor, *Varanus niloticus niloticus. International Journal of Advanced Research in Biological Sciences* 5(4):1–11.

Shine, R. 1986. Food habits, habitats, and reproductive biology of four sympatric species of varanid lizards in tropical Australia *Herpetologica* 42:346–360.

Shrivastava, R., 1963. The structure and the development of the chondrocranium of *Varanus*. I. The development of the ethmoidal region. *Okajimas Folia Anatomica of Japan* 55–83.

Shrivastava, R. K. 1964a. The structure and development of the chondrocranium of *Varanus*. II. The development of the orbito-temporal region. *Journal of Morphology* 115:97–107.

Shrivastava, R. K., 1964b. The structure and development of the chondrocranium of *Varanus*. III. The otic and occipital regions, basal plate, viscerocranium and certain features of the osteocranium of a juvenile. *Gegenbaurs Morphologisches Jahrbuch* 106:147–187.

Simões, T. and R. Pyron. 2021. The squamate tree of life. *Bulletin of the Museum of Comparative Zoology* 163(2):47–95. https://bioone.org/journals/bulletin-of-the-museum-of-comparative-zoology/volume-163/issue-2

Simon, J., F. Delmotte, C. Rispe, and T. Crease. 2003. Phylogenetic relationships between parthenogens and their sexual relatives: The possible routes to parthenogenesis in animals. *Biological Journal of the Linnean Society* 79:151–163. https://doi.org/10.1046/j.1095-8312.2003.00175.x

Sinclair, A. 1983. *The mechanics of feeding in reptiles*. Unpublished Ph.D. thesis, University of Leeds.

Sinervo, B., F. Méndez-De-La-Cruz, D. B. Miles, B. Heulin, E. Bastiaans, M. Villagran-Santa Cruz, R. Lara-Resendiz, et al. 2010. Erosion of lizard diversity by climate change and altered thermal niches. *Science* 328:894–899.16 of 17ARIANO-SANCHEZET AL.

Singhal, S. 2023. *Why species delimitation matters*. Invited Speaker Presentation at Smithsonian Institution, PowerPoint.

Sirimanna, D., W. Kanthika, J. Manamperi, W. Suranga, and H. Herath. 2020. Cannibalism and other ecological observations of *Varanus bengalensis* (Daudin, 1802) in Lowland Wet Zone, Sri Lanka. *Sri Lanka Naturalist* 10(1–2):75–78.

Slowinski, J. 1994. A phylogenetic analysis of *Bungarus (Elapidae)* based on morphological characters. *Journal of Herpetology* 28(4):440–446.

Smissen, P., J. Melville, J. Summer, and T. Jessop. 2013. Mountain barriers and river conduits: Phylogeographical structure in a large mobile lizard (Varanidae: *Varanus varius*) from eastern Australia. *Journal of Biogeography* 40:1729–1740.

Smith, K. 2006. A diverse new assemblage of Late Eocene squamates (Reptilia) from the Chadron Formation of North Dakota, USA. *Palaeontographica Electronica* 9(5A):44.

Smith, K. 2009. A new lizard assemblage from the earliest Eocene (zone Wa0) of the Bighorn Basin, Wyoming, USA: Biogeography during the warmest interval of the Cenozoic. *Journal of Systematic Palaeontology* 7(3):299–358.

Smith, K. 2011. Morphology and function of the tongue and hyoid apparatus in *Varanus* (Varanidae, Lacertilia). *Journal of Experimental Biology* 212(part 16):2501–2510.

Smith, K., B.-A. Bhullar, G. Köhler, and J. Habersetzer. 2018. The only known jawed vertebrate with four eyes and the Bauplan of the pineal complex. *Current Biology* 28:1101–1107.

Smith, K. and J. Habersetzer. 2021. The anatomy, phylogenetic relationships, and autecology of the carnivorous lizard "*Saniwa*" *feisti* Stritzke, 1983 from the Eocene of Messel, Germany. Pp. 441-506. In: A. Folie, E. Buffetaut, N. Bardet, A. Houssaye, A. Gheerbrant, and M. Laurin(eds). *Palaeobiology and palaeobiogeography of amphibians and reptiles: An homage to Jean-Claude Rage. Comptes Rendus Palevol* 20(23):441–506. https://doi.org/10.5852/cr-palevol2021v20a23

Smith, K. and W. L. Hylander. 1985. Strain gage measurements of mesokinetic movement in *Varanus exanthematicus*. *Journal of Experimental Biology* 114:53–70.

Smith, L., S. Sweet, and D. King. 2004. 7.19. *Varanus scalaris*. IN: E. Pianka and D. King (eds.). *Varanoid lizards of the world*. Indiana University Press, Bloomington, IN.

Solessio, E. and G. Engbretson. 1999. Electroretinogram of the parietal eye of lizards: Photoreceptor, glial, and lens cell contributions. *Visual Neuroscience* 16(5):895–907.

Somma, L. and J. Fawcett. 1989. Brooding behaviour of the prairie skink, *Eumeces septentrionalis*, and its relationship to the hydric environment of the nest. *Zoological Journal of the Linnean Society* 95(3):245–256.

Soule, M. and W Kerfoot. 1972. On the climatic determination of scale size in a lizard. *Systematic Zoology* 21(1):97–105.

Spencer W. 1887. On the presence and structure of the parietal eye in Lacertilia. *Quarterly Journal of Microscopical Science* 27:165–238.

Sprackland, R. 1970. Further notes on *Lanthanotus*. *Sarawak Museum Journal* 1970:412–413.

Sprackland, R. 1972a. A summary of observations on the earless monitor, *Lanthanotus borneensis*. *Sarawak Museum Journal* 1972:323–327.

Sprackland, R. 1972b. Habits of *Lygosoma Rooije* (say). *Bulletin of the Philadelphia Herpetological Society* 20:23–24.

Sprackland, R. 1980. Some notes on Storr's dwarf spiny-tailed monitor. *Kansas Herpetological Society Newsletter* 40(1):7–9.

Sprackland, R. 1982a. *Varanus "monitor"*, an invalid name for *Varanus bengalensis*. *Herpetological Review* 13(4):117.

Sprackland, R. 1982b. Feeding and nutrition of monitor lizards in captivity and in the wild. *Kansas Herpetological Society Newsletter* 47(1):15–18.

Sprackland, R. 1989a. Notes on Duméril's monitor lizard, *Varanus dumerilii* (Schlegel). *Sarawak Museum Journal* 1976:287–291.

Sprackland, R. 1989b. Captive maintenance of green tree monitors (*Varanus prasinus*) and their kin. *Northern California Captive Propagation and Husbandry Conference Proceedings* 1989:49–56.

Sprackland, R. 1989c. Mating and waiting: A status report on reproduction in captive monitor lizards (Sauria: Varanidae). *Northern California Captive Propagation and Husbandry Conference Proceedings* 1989:57–63.

Sprackland, R. 1989d. A preliminary study of food discrimination in monitor lizards (Reptilia: Lacertilia: Varanidae). *Bulletin of the Chicago Herpetological Society* 25(10):18–23.

Sprackland, R. 1991a. The origin and zoogeography of monitor lizards of the subgenus *Odatria* Gray (Sauria: Varanidae): A re-evaluation *Mertensiella* 2:240–252.

Sprackland, R. 1991b. Taxonomic review of the *Varanus prasinus* group with descriptions of two new species. *Memoirs of the Queensland Museum* 30(3):561–576.

Sprackland, R. 1992. *Giant lizards*. TFH Publications, N.J. 1–288 pp.

Sprackland, R. 1993a. The taxonomic status of the monitor lizard *Varanus dumerilii heteropholis* Boulenger, 1892 (Reptilia: Varanidae). *Sarawak Museum Journal* 45(65):113–121.

Sprackland, R. 1993b. Rediscovery of a Solomon Islands monitor lizard (*Varanus indicus spinulosus*). *Reptiles* 4(5):25–27.

Sprackland, R. 1993c. Etymology of varanid names. *VaraNews* 3(4):3.

Sprackland, R. 1994a. Rediscovery and taxonomic review of *Varanus indicus spinulosus* Mertens, 1941. *Herpetofauna (Australia)* 24(2):33–39.

Sprackland, R. 1994b. De mangrove varan groep. *Dutch Varanid Assoc. Nieuwsbrief* 2:10–12.

Sprackland, R. 1994c. Emerald monitors: Lean, green, and rarely seen. *Reptiles* 1(5):16–23.

Sprackland, R. 1995a. Duméril's monitor lizard. *Reptiles* 3(7):56–69.

Sprackland, R. 1995b. *Evolution, systematics, and variation in Pacific mangrove monitor lizards*. Unpublished Ph.D. thesis, Department of Anatomy, University College London.

Sprackland, R. 1995c. Taxonomy: The making of scientific names. *The Vivarium* 6(4):14–17.

Sprackland, R. 1995d, Nov./Dec. The French origins of herpetology. *Reptile and Amphibian* 82–89.

Sprackland, R. 1995e, Mar./Apr. The rocky origins of British herpetology. *Reptile and Amphibian* 42–49.

Sprackland, R. 1995f. Halloween herptiles (about warning coloration). *The Vivarium* 6(2):26–29.

Sprackland, R. 1996. Taxon management account for mangrove monitor/Pacific monitor, *Varanus indicus* (Daudin, 1802). IN: S. Hammack (ed.). *Lizard advisory group taxon management accounts*. American Association of Zoological Parks and Aquariums.

Sprackland, R. 1998. Mangrove monitor lizards. *Reptiles* 5(3):48–63.
Sprackland, R. 1999a. A new species of monitor from Indonesia. *Reptile Hobbyist* 4(6):20–27.
Sprackland, R. 1999b. Character displacement among sympatric varanid lizard species: A study of natural selection and systematics in action (Reptilia: Squamata: Varanidae). *Mertensiella* 11:113–120.
Sprackland, R. 1999c. Are monitors speaking with forked tongues? *Herpetofauna (Australia)* XV:37–49.
Sprackland, R. 1999d. Sarawak's earless monitor lizard (*Lanthanotus borneensis*). *Reptiles* (11):72–79.
Sprackland, R. 1999e. The emerald monitor. *Reptile & Amphibian Hobbyist* 4(11):8–16.
Sprackland, R. 2001a. Herpetofauna of Fitzroy Island, Australia. *Reptiles* 9(9):8–16.
Sprackland, R. 2001b. Dwarf monitors. *Reptile & Amphibian Hobbyist* 6(10):8–16.
Sprackland, R. 2000a. Pronunciation of scientific names. *Reptile & Amphibian Hobbyist* 5(8):44–47.
Sprackland, R. 2000b. Magnificent millennial monitor maintenance: A beginner's guide to five easy species. *Reptiles USA* 54–63.
Sprackland, R. 2000c. Snakes with legs jog (theorists' minds, that is). "Reptile Road." *Reptile & Amphibian Hobbyist* 5(11):82–83.
Sprackland, R. 2003. Sampling Yathong's reptiles. *Reptiles* 11(8):42–51.
Sprackland, R. 2007a. Peach-throat monitors. *Reptiles* 15(2):40–47.
Sprackland, R. 2007b, February 17. Climate change: Whatever the cause, we are in hot water. *EzineArticles*. http://ezinearticles.com/?Climate-Change:-Whatever-the-Cause,-We-Are-in-Hot-Water&id=458683
Sprackland, R. 2009. *Giant lizards: The definitive guide to the natural history, care, and breeding of monitors, iguanas, and other large lizards*. TFH Publications., Neptune, NJ. 1–334 pp.
Sprackland, R. 2010a. *The pocket expert's guide to lizards*. TFH Publications, Neptune, NJ.
Sprackland, R. 2010b. True blue: Care for the rare and challenging blue-spotted monitor. *Reptiles* 18(7):32–41.
Sprackland, R. 2012. Varanid origins: The first monitors. *Scales & Tails* 26(Nov):28–32.
Sprackland, R. 2015. *I am King*. Online at https://www.academia.edu/37088956/I_AM_KING
Sprackland, R., H. M. Smith, and P. D. Strimple. 1997a. *Hydrosaurus gouldii* Gray, 1838 (currently *Varanus gouldii*) and *Varanus panoptes* Storr, 1980 (Reptilia: Squamata): Proposed conservation of the specific names by the designation of a neotype for *H. gouldii*. Case 3042. *Bulletin of Zoological Nomenclature* 54(2):95–99.
Sprackland, R., H. M. Smith, and P. D. Strimple. 1997b. *Varanus teriae* Sprackland, 1991 (Reptilia: Squamata): Proposed conservation of the specific name. Case 3043. *Bulletin of Zoological Nomenclature* 54(2):100–103.
Sprackland, R., T. Sprackland, and D. Diessner. 2004. Reptiles and mammals of Fitzroy Island, Queensland. *Memoirs of the Queensland Museum* 49(2):733–739.
Srikulnath, K., Y. Uno, C. Nishida, and Y. Matsuda. 2013. Karyotype evolution in monitor lizards: Cross-species chromosome mapping of cDNA reveals highly conserved synteny and gene order in the Toxicofera clade. *Chromosome Research* 21:805–819.
Stamps, J. and A. Krishnan. 1994. Territory acquisition in lizards: II. Establishing social and spatial relationships. *Animal Behavior* 47:745–753.
Starck, D. 1979. Cranio-cerebral relations in recent reptiles. Pp. 1–38. IN: C. Gans (ed.). *Biology of the reptilia, volume 9, neurology A*. Academic Press, New York, NY.
Stebbins, G. 1959. The role of hybridization in evolution. *Proceedings of the American Philosophical Society* 103:231–251.
Stebbins, R., J. Lowenstein, and N. Cohen. 1967. A field study of the lava lizard (*Tropidurus albermarlensis*) in the Galapagos Islands. *Ecology* 48(5):839–851.
Steindachner, F. 1877. Über zwei neue Eidechsen-arten aus Süd-Amerika und Borneo. *Denkschriften der Kaiserlichen Akademie der Wissenschaften, Mathematisch- Naturwissenschaftliche Classe*, XXXVIII, Bild I, Abhteilung 4: 1-5, Taf. 2, 5 Juli 1877. Preprint.
Sternfeld, R. 1920. Zur Tiergeographie Papuasiens und der pazifischen Inselwelt. *Abhandlungen der Senckenbergischen Naturforschenden Gesellschaft* 36:375–436.
Stirling, E. C. 1912. Observations on the habits of the large Central Australian monitor (*Varanus giganteus*), with a note on the "fat bodies" of this species. *Transactions and Proceedings of the Royal Society of South Australia* 36:26–33.
Storr, G. 1980. The monitor lizards (genus *Varanus* Merrem, 1820) of Western Australia. *Records of the Western Australian Museum* 8(2):237–293.
Storr, G., L Smith, and R. Johnstone. 1983. *Lizards of western Australia 2, dragons and monitors*. Western Australian Museum Press, Perth.
Srikulnath, K., Y. Uno, C. Nishida, and Y. Matsuda.2013. Karyotype evolution in monitor lizards: Cross-species chromosome mapping of cDNA reveals highly conserved synteny and gene order in the Toxicofera clade. *Chromosome Research* 21:805–819.
Streicher, J. and J. Wiens. 2017. Phylogenomic analyses of more than 4000 nuclear loci resolve the origin of snakes among lizard families. *Biology Letters* 13(9).
Stritzke, R. 1983. *Saniwa feisti* n. sp., ein Varanide (Lacertilia, Reptilia) aus dem Mittel-Eozan von Messel bei Darmstadt. *Senckenbergiana Lethaea* 64(5/6):497–508.
Surahya, S. 1989. *Atlas Komodo: An anatomical study of Komodo Dragon and its position in animal systematic*. Gadjah Mada University Press, Yogyakarta, Indonesia. 382 colour figures, unpaginated, in Indonesian and English.
Suwala, G., M. Altmanová, S. Mazzoleni, E. Karameta, P. Pafilis, L. Kratochvíl, and M. Rovatsos. 2020. Evolutionary variability of W-linked repetitive content in lacertid lizards. *Genes (Basel)* 11(5):531.
Suzuki, N., K. Igarashi, and T. Hamada. 1991. Bipedal standing facility of *Varanus komodoensis* (Reptilia) in a wild habitat. *Scientific Papers of the College of Arts and Sciences, the University of Tokyo* 41(2):95–105.
Sweet, S. 1999. Spatial ecology of *Varanus glauerti* and *V. glebopalma* in Northern Australia. *Mertensiella* 11:317–366.
Sweet, S. 2016. Chasing flamingos: Toxicofera and misinterpretation of venom in varanid lizards. Pp. 123–149. IN: M. Cota (ed.). *Proceedings of the 2015 interdisciplinary world conference on monitor lizards*. Institute for Research and development, Sun Snandha Rajabhat University, Bangkok, Thailand. 250 pp.
Sy, E. 2023. *A naturalist's guide to the reptiles of the Philippines*. John Beaufoy Publishing, Oxford. 176 pp.

Szabo, B., S. Hoefer, and M. Whiting. 2020. Are lizards capable of inhibitory control? Performance on a semi-transparent version of the cylinder task in five species of Australian skinks. *Behavioural Ecology and Sociobiology* 74:118.

Szabo, B., D. Noble, and M. Whiting. 2021. Learning in non-avian reptiles 40 years on: Advances and promising new directions. *Biological Reviews* 96:331–356.

Szabo, B., D. Noble, R. Byrne, D. Tait, and M. Whiting. 2019. Precocial juvenile lizards show adult level learning and behavioural flexibility. *Animal Behaviour* 154:75–84.

Taylor, S. and E. Larson. 2019. Insights from genomes into the evolutionary importance and prevalence of hybridization in nature. *Nature Ecology and Evolution* 3:170–177.

Ten Donkelaar, H. 1987. Evolution of the red nucleus and rubrospinal tract. *Behavior and Brain Research* 28(1/2):9–20.

Ten Donkelaar, H. and G. Bangma. 1987. The cerebellum. IN: C. Gans and R. Northcutt (eds.). *Biology of the reptilia, volume 17: neurology C*. Academic Press, New York, NY.

ten Donkelaar, H. J., G. C. Bangma, H. A. Barbas-Henry, R. de Boer-van Huizen, and J. G. Wolters. The brain stem in a lizard, *Varanus exanthematicus*. *Advances in Anatomy, Embryology and Cell Biology* 107:1–168.

Ten Donkelaar, H., G. Bangma, H. Barbas-Henry, R. Huizen, and J. Wolters. 1987. *The brain stem in a lizard, Varanus exanthematicus*. Springer-Verlag, Berlin.

Ten Donkelaar, H., G. Bangma, and R. Huizen. 1985. The fascisulus longitudinalis medialis in the lizard *Varanus exanthematicus*. 2. Vestibular and internuclear components. *Anatomy and Embryology* 172:205–215.

Ten Donkelaar, H. and R. Huizen. 1984. The fascisulus longitudinalis medialis in the lizard *Varanus exanthematicus*. 1. Interstitiospinal, reticulospinal and vestibulospinal components. *Anatomy and Embryology* 169:177–184.

Thakur, M. 2020. Notes on eggs and hatchlings of the Bengal Monitor (*Varanus bengalensis*). *IRCF Reptiles & Amphibians* 27(2):310–311.

Thangadurai, D. and J. Sangeetha (eds.). 2015. *Genomics and proteomics: Principles, technologies, and applications*. Routledge.

Thissen, R. 1992. Breeding the spiny-tailed Monitor (*Varanus acanthurus* Boulenger). *The Vivarium* 3(5):32–34.

Thompson, A., V. Kapsanaki, H. Liwang, P. Pafilis, I. Wang, and K. Brock. 2023. Some like it hotter: Differential thermal preferences among lizard color morphs. *Journal of Thermal Biology Volume* 113:7.

Thompson, G. 1977. Comparative morphology … Western Australian varanids. *Journal of Morphology* 233:127–152.

Thompson, G. 1999. Goanna metabolism: Different to other lizards, and if so, what are the ecological consequences? Pp. 79–90. IN: H.-G. Horn and W. Boehme (eds.). *Advances in monitor research II. Mertensiella 11*, DGHT, Bonn.

Thompson, G., N. Heger, T. Heger, and P. Withers. 1995. Standard metabolic rate of the largest Australian lizard, *Varanus giganteus*. *Comparative Biochemistry and Physiology* 111A:603–608.

Thompson, G. and E. Pianka. 2001. Allometry of clutch and neonate sizes in monitor lizards (Varanidae: *Varanus*). *Copeia* 2001(3):443–458.

Thompson, G. and P. Withers. 1992. Effects of body mass and temperature on standard metabolic rates for two Australian varanid lizards (*Varanus gouldii* and *V. panoptes*). *Copeia* 1992:343–350.

Thompson, G. and P. Withers. 1994. Standard metabolic rates of two small Australian varanid lizards (*Varanus caudolineatus* and *V. acanthurus*) *Herpetologica* 50:494–502.

Thompson, G., P. Withers, and S. Thompson, S. 1992. The combat ritual of two monitor lizards, *Varanus caudolineatus* and *Varanus gouldii*. *Western Australian Naturalist* 19:21–24.

Thompson, M. and K. Russell. 1998. Metabolic cost of development in one of the world's smallest lizard eggs. *Copeia* 1998:1016–1020.

Thorpe, R. and R. Brown. 1989. Testing hypothesized causes of within-island geographic variation in the colour of lizards. *Experientia* 45:397–400.

Tinkle, D. 1972. *The role of environment in the evolution of life history differences within and between lizard species*. University of Arkansas Occasional Papers No. 4:77–100.

Tollis, M., E. Hutchins, J. Stapley, S. Rupp, W. Eckalbar, I. Maayan, E. Lasku, C. Infante, …, and K. Kusumi. 2018. Comparative genomics reveals accelerated evolution in conserved pathways during the diversification of anole lizards. *Genome Biology and Evolution* 10(2):489–506. https://doi.org/10.1093/gbe/evy013. PMID: 29360978; PMCID: PMC5798147.

Torres-Cervants, R., X. Hernandez-Ibarra, and A. Ramirez-Bautista. 2004. *Anelytropsis papillosus* (Mexican blind lizard). Death feigning and autotomy. *Herpetological Review* 35(4):384.

Tosini, G. 1997. The pineal complex of reptiles: Physiological and behavioral roles. *Ethological and Ecological Evolution* 9:313–333.

Tosini, G. and R. A. Avery. 1996. Dermal photoreceptors regulate basking behavior in the lizard *Podarcis muralis*. *Physiology & Behavior* 59(1):195–198.

Townsend, T., A. Larson, E. Louis, and R. Macey. 2004. Molecular phylogenetics of Squamata: The position of snakes, amphisbaenians, and dibamids, and the root of the squamate tree. *Systematic Biology* 53(5):735–757.

Tracholt, C. 1997. Effect of masking the parietal eye on the diurnal activity and body temperature of two sympatric species of monitor lizards, *Varanus s. salvator* and *Varanus b. nebulosus*. *Journal of Comparative Physiology B* 167:177–184.

Trainor, C., O. Hidayat, and S. Cooleman. 2022. New records and observations of the Peacock Monitor (*Varanus auffenbergi*) from Rote Island, East Nusa Tenggara, Indonesia, with an analysis of extent of occurrence. *BIAWAK* 15(1):8–13.

Tsellarius, A. 1994. Behaviour and mode of life of the monitor in sand desert. *Priroda* 5:26–35.

Tsellarius, A. and Y. Men'shikov. 1994. Indirect communications and its role in the formation of social structure in *Varanus griseus* (Sauria). *Russian Journal of Herpetology* 1:121–132.

Tsellarius, A. and E. Tsellarius. 1997. Behaviour of *Varanus griseus* during encounters with conspecifics. *Asiatic Herpetological Research* 7:108–130.

Turvey, S., J. Crees, J. Hansford, et al. 2017. Quaternary vertebrate faunas from Sumba, Indonesia: Implications for Wallacean biogeography and evolution. *Proceedings of the Biological Sciences B* 284(1861):20171278. https://doi.org/10.1098/rspb.2017.1278

Tyler, M. J. 1989. A new species of *Lechriodus* (Anura: Leptodactylidae) from the Tertiary of Queensland, with a redefinition of the ilial characteristics of the genus. *Transactions of the Royal Society of South Australia* 113(1):15–22.

Uchida, T. 1966. Observations on the monitor lizard, *Varanus indicus*, as a rat-control agent on Ifaluk, western Caroline Islands. *Bulletin of the World Health Organization* 35:976–980.

Uchida, T. 1967. Observations on the monitor lizard, *Varanus indicus* (Daudin) as a rat control agent on Ifaluk, western Caroline Islands. *Micronesica* 3(1):17–18.

Underwood, G. 1957. *Lanthanotus* and the anguinomorphan lizards: A critical review. *Copeia* 1957(1):20–30.

Underwood, G. 1970. The eye. Pp. 1–97. IN: C. Gans and T. Parsons (ed.). *Biology of the Reptilia*, vol. 2. Morphology B. Academic Press, New York, NY.

U.S.D.A. 2018. *FoodData Central*. U.S. Department of Agriculture, Agricultural Research Service. https://fdc.nal.usda.gov/fdc-app.html#/food-details/1099155/nutrients

Valdecantos, S. and F. Lobo. 2015 First report of hemiclitores in females of South American Liolaemid lizards. *Journal of Herpetology* 49:291–294.

Valenzuela, N. 2004. Temperature-dependent sex determination. Pp. 211–227. IN: D. Deeming (ed.). *Reptilian incubation: Environment, evolution and behaviour.* Nottingham University Press, Nottingham, UK.

Van Damme, R., D. Bauwens, F. Brana, and R. Verheyen. 1992. Incubation temperature differentially affects hatching time, egg survival, and hatchling performance in the lizard *Podarcis muralis*. *Herpetologica* 48(2):220–228.

van den Ende C., M. Puttick, A. Urrutia, and M. Wills. 2023. Why should we compare morphological and molecular disparity? *Methods in Ecology and Evolution* 14, 2390–2410. https://doi.org/10.1111/2041-210X.14166.

Van der Lee, F. 1996. Breeding the monitor lizard *Varanus exanthematicus*. *Aquarium (Hilversum)* 669:213–215.

Van Wassenbergh, S. and P. Aerts. 2013. In search of the pitching momentum that enables some lizards to sustain bipedal running at constant speeds. *Journal of the Royal Society Interface* 10:20130241. https://doi.org/10.1098/rsif.2013.0241

Vasilyan, D. and M. Bukhsiandidze. 2020. The fossil record of the genus *Varanus* from the southern Caucasus (Armenia, Georgia). *PeerJ* 8(3–4). https://doi.org/10.7717/peerj.8322

Venczel, M. 2006. Lizards from the late Miocene of Polgardi (W-Hungary). *Nymphaea* XXXIII:25–38.

Vickers, M. and L. Schwarzkopf. 2016. A simple method to predict body temperature of small reptiles from environmental temperature. *Ecology and Evolution* 6:3059–3066. PMID: 27252829.

Vidal, N. and S. Hedges. 2005. The phylogeny of squamate reptiles (lizards, snakes, and amphisbaenians) inferred from nine nuclear protein-coding genes. *Comptes Rendus. Biologies* 328:1000–1008. https://doi.org/10.1016/j.crvi.2005.10.001

Vidal, N., M. Maldonado, and L. Antonieta. 2008. *Herpetologia de Chile*. Science Verlag, Santiago. 593 pp.

Vidal, N., J. Marin, J. Sassi, F. Battistuzzl, S. Donnellan, A. Fitch, B. Fry, F. Vonk, R. Rodriguez de la Vega, A. Couloux, and S. B. Hedges. 2012. Molecular evidence for an Asian origin of monitor lizards followed by Tertiary dispersals to Africa and Australasia. *Biology Letters* 8:853–855.

Vikrant, S. and B. Verma. 2014. Monitor lizard bite-induced acute kidney injury—A case report. *Renal Failure* 36(3):444–446.

Vila-Pouca, C., S. Vedder, and A. Kotrschal. 2022. Hybridization may promote variation in cognitive phenotypes in experimental Guppy Hybrids. *American Naturalist* 200(4):607–619.

Villa, A., J. Abella, D. M. Alba, S. Alme´cija, A. Bolet, G. D. Koufos, et al. 2018. Revision of *Varanus marathonensis* (Squamata, Varanidae) based on historical and new material: morphology, systematics, and paleobiogeography of the European monitor lizards. *PLoS One* 13(12):e0207719. https://doi.org/10.1371/journal.pone.0207719

Vincent, M. and S. Wilson. 1999. *Australian goannas*. New Holland Publishers, Sydney.

Visser, G. 1985. Notizen zur Brutbiologie des Gelbwarans *Varanus (Empagusia) flavescens* (Hardwicke & Gray, 1827) im Zoo Rotterdam. *Salamandra* 21(2/3):161–168.

Vitt, L. and W. Cooper, Jr. 1989. Maternal care in skinks (*Eumeces*). *Journal of Herpetology* 23(1):29–34.

Vitt, L. and E. Pianka (eds.). 2014. *Lizard ecology: Historical and experimental perspectives*. Princeton University Press, Princeton, NJ.

Von Euw, S., Y. Wang, G. Laurent, C. Drouet, F. Babonneau, N. Nassif, and T. Azaïs. 2019. Bone mineral: New insights into its chemical composition. *Scientific Reports* 9:8456. https://doi.org/10.1038/s41598-019-44620-6

Voris, H. 2000. Maps of Pleistocene sea levels in southeast Asia: Shorelines, river systems and time durations. *Journal of Biogeography* 27:1153–1167.

Walsh, T., R. Roscoe, and G. Birchard. 1993. Dragon tales: The history, husbandry and breeding of Komodo monitors at the National Zoological Park. *The Vivarium* 4(6):23–26.

Warwick, C., P. Arena, and G. Burghardt (eds.). 2023. *Health and welfare of captive reptiles*. 2nd Ed. Springer Cham, Springer International Publishing AG. New York, NY.

Waterman, J., R. McNally, D. Harrold, M. Cook, G. Garcia, A. Fidgett, and L. Holmes. 2021. Evaluating environmental enrichment methods in three zoo-housed varanidae lizard species. *Journal of Zoological and Botanical Gardens* 2(4):716–727.

Waltering, J. 2012. From lizard to snake: Behind the evolution of an extreme body plan. *Current Genomics* 13:289–299.

Watkinson, G. 1906. The cranial nerves of *Varanus bivittatus*. *Gegenbaurs Morphologisches Jahrbuch* 35:450–472, Taf. XI-XIII.

Watts, P., K. Buley, S. Sanderson, W. Boardman, C. Ciofi, and R. Gibson. 2006. Parthenogenesis in Komodo dragons. *Nature* 444(7122):1021–1022. https://doi.org/10.1038/4441021a

Wang, T., A. Smits, and W. Burggren. 1998. The cardiovascular system: Comparative aspects of heart morphology. IN: C. Gans (ed.). *Biology of the Reptilia* 19:297–374.

Webb, G., H. Heatwole, and J. Bavay. 1971. Comparative cardiac anatomy of the Reptilia I. the chambers and septa of the varanid ventricle. *Journal of Morphology* 134:335–350.

Weavers, B. 2011. Solar absorptance, reflectance and emissivity of the skin of Lace Monitors (*Varanus varius*) in south-eastern Australia. *BIAWAK* 15(1/2):10–16.

Weijola, V. 2010. Geographical distribution and habitat use of monitor lizards of the north Moluccas. *Biawak* 4:7–23.

Weijola, V. 2015. *Tupinambis indicus* Daudin, 1802 (currently *Varanus indicus*; Reptilia, Squamata): proposed conservation of usage of the specific name by replacement of the neotype. *Bulletinof Zoological Nomenclature* 72:134–141.

Weijola, V., S. Donnellan, and C. Lindqvist. 2016. A new blue-tailed Monitor lizard (Reptilia, Squamata, *Varanus*) of the *Varanus indicus* group from Mussau Island, Papua New Guinea. *ZooKeys* 568:129–154. https://doi.org/10.3897/zookeys.568.6872

Weijola, V. and S. Sweet. 2010. A new melanistic species of monitor lizard (Reptilia: Squamata: Varanidae) from Sanana Island, Indonesia. *Zootaxa* 2434:17–32. [Google Scholar].

Weijola, V. and S. Sweet. 2015. A single species of mangrove monitor (*Varanus*) occupies Ambon, Seram, Buru and Saparua, Moluccas, Indonesia. *Amphibian and Reptile Conservation* 9:14–23.

Weijola, V., V. Vahtera, A. Koch, A. Schmitz, and F. Kraus. 2020, May. Taxonomy of micronesian monitors (Reptilia: Squamata: *Varanus*): Endemic status of new species argues for caution in pursuing eradication plans. *Royal Society Open Science* 7(5). https://doi.org/10.1098/rsos.200092

Weiner, J. 1995. Evolution made visible. *Science* 267(5194):30–33.

Wells, R. and R. Wellington. 1983. A synopsis of the class Reptilia in Australia. *Australian Journal of Herpetology* 1(3–4):73–129.

Wells, R. and R. Wellington. 1985. A synopsis of the Amphibia and Reptilia of New Zealand. *Australian Journal of Herpetology, Supplemental Series* 1:62–64.

Wellnhoffer, P. 1993. Das siebte Exemplar von *Archaeopteryx* aus den Solnhofener Schichten. *Archaeopteryx* 11(1):1–47.

Welton, L., C. Siler, C. Linkem, A. Diesmos, M. Diesmos, E. Sy, and R. Brown. 2013. Dragons in our midst: Phyloforensics of illegally traded Southeat Asian monitor lizards. *Biological Conservation* 159:7–15.

Welton, L., C. Siler, J. Oaks, A. Diesmos, and R. Brown. 2013. Multilocus phylogeny and Bayesian estimates of species boundaries reveal hidden evolutionary relationships and cryptic diversity in Southeast Asian monitor lizards. *Molecular Ecology* 22:3495–3510. https://doi.org/10.1111/mec.12324

Welton, L., S. Travers, C. Siler, and R. Brown. 2014. Integrative taxonomy and phylogeny-based species delimitation of Philippine water monitor lizards (*Varanus salvator* Complex) with descriptions of two new cryptic species. *Zootaxa* 3881(3):201–227.

Werneburg, I., K. Polachowski, and M. Hutchinson. 2015. Bony skull development in the Argus Monitor (Squamata, Varanidae, *Varanus panoptes*) with comments on developmental timing and adult anatomy. *Zoology* 118:255–280.

Werner, F. 1900. Die Reptilien- und Batrachienfauna des Bismarck-Archipels. *Mitt. Zoologischen Samml. Museums Naturk. Berlin* 1(4):1–132.

Wesiak, K. 1993a. Eine gelungene Nachzucht von *Varanus indicus* (Daudin 1802). *Monitor* 2(2):10.

Wesiak, K. 1993b. Uber haltung und nachtzucht von *Varanus (Euprepiosaurus) indicus indicus* (Daudin, 1802). *Herpetofauna (Germany)* 15(87):21–25.

Wesiak, K. 2006. Zur Pflege und Nachzucht des *westafrikanischen* Steppenwarans *Varanus (Polydaedalus) exanthematicus* (Bosc, 1792). *Elaphe* 14(3):21–34.

Wesiak, K. 2008. Unterbringung, Pflege und Nachzucht von *Varanus (Odatria) auffenbergi* Sprackland, 1999. *Elaphe* 16(1):37–47.

Wesiak, K. and A. Koch. 2009. Successful husbandry and first breeding of *Varanus juxtindicus* Böhme et al., 2002, with remarks on the development of juveniles of this "rarely- kept" endemic Solomon monitor species. *Biawak* 3:106–121.

Wever, E. 1967. The tectorial membrane of the lizard ear: Types of structure. *Journal of Morphology* 122(4):307–320.

White, F. 1969. Functional anatomy of the heart of reptiles. *American Zoologist* 8:211–219.

Whiteside, D., S. Chambi-Trowell, and M. Benton. 2022. A Triassic crown squamate. *Science Advances* 8(2022):eabq8274. https://doi.org/10.1126/sciadv.abq8274

Whiteside, D., S. Chambi-Trowell, and M. Benton. 2024. Late Triassic †*Cryptovaranoides microlanius* is a squamate, not an archosauromorph. *Royal Society Open Science* 11(11):231874. https://doi.org/10.1098/rsos.231874. PMID: 39606587; PMCID: PMC11597406.

Whitford, W. 1998. Contribution of pits dug by goannas (*Varanus gouldii*) to the dynamics of banded mulga landscapes in eastern *Australia. Journal of Arid Environments* 40:453–457.

Whiting, M., D. Stuart-Fox, D. O'Connor, D. Firth, N. Bennett, and S. Blomberg. 2006. Ultraviolet signals ultra-aggression in a lizard. *Animal Behaviour* 72:353–363.

Whiting, M., and G. While. 2017. Sociality in lizards. Pp. 390–426. IN: D. R. Rubenstein and P. Abbott (eds.). *Comparative Social Evolution*. Cambridge University Press, Cambridge.

Whiting, M., X. Feng, K. Fonti, J. Riley, R. Byrne, and D. Noble. 2018. Evidence for social learning in a family living lizard. *Frontiers of Ecology and Evolution* 6. https://doi.org/10.3389/fevo.2018.00070

Wickramasinghe, M., B. Kekulandala, I. Peabotuwage, and S. Karunarantha. 2010. A remarkable feeding behavior and a new distribution record of *Varanus salvator salvator* (Laurenti, 1768) in eastern Sri Lanka. *Biawak* 4(3):93–98.

Wiechmann, R. 2012. Observations on parthenogenesis in monitor lizards. *Biawak* 6(1):11–21.

Wiens, J., M. Brandley, and T. Reeder. 2006. Why does a trait evolve multiple times within a clade? Repeated evolution of snakelike body form in squamate reptiles. *Evolution* 60:123–141.

Wiens, J. and M. Servedio. 2000. Species delimitation in systematics: Inferring diagnostic differences between species. *Proceedings of the Royal Society of London B* 267:631–636.

Wikipedia Contributors. 2025. Wells and Wellington affair. *Wikipedia, The Free Encyclopedia*. Retrieved 21:53, June 3, 2025, from https://en.wikipedia.org/w/index.php?title=Wells_and_Wellington_affair&oldid=1292103606

Wikramanayake, E. and G. Dryden. 1988. The reproductive ecology of *Varanus indicus* on Guam. *Herpetologica* 44(3):338–344.

Wilén, M. 2011. *Experimental thermoregulation in the Wedge-snouted Lizard (Meroles cuneirostris)*. Unpublished Paper, Elective Course in Neuroscience, Uppsala University. 34 pp.

Wiles, G., A. Amerson, and R. Beck. 1989. Notes on the herpetofauna of Tinian, Marshall Islands. *Micronesica* 22(1):107–118.

Wiley, E. 1980. *Phylogenetic systematics*. J. Wiley Publ., New York, NY. 439 pp.

Wiley, E., D. Siegel-Causey, D. Brooks, and V. Funk. 1991. *The compleat cladist: A primer of phylogenetic techniques*, vol. 19. University of Kansas Museum of Natural History Special Publications. 158 pp.

Wilken, A., K. Middleton, K. Sellers, I. Cost, and C. Holliday. 2019. The roles of joint tissues and jaw muscles in palatal biomechanics of the savannah monitor (*Varanus exanthematicus*) and their significance for cranial kinesis. *Journal of Experimental Biology* 222:jeb201459. https://doi.org/10.1242/jeb.201459

Wilkinson, A. and L. Huber. 2012. Cold-blooded cognition: reptilian cognitive abilities. Pp. 129–141. IN: J. Vonk and T. Shackelford (eds.). *The Oxford handbook of comparative evolutionary psychology*. Oxford University Press, Oxford, UK.

Wilson, P. 1992. On inferring hybridity from morphological intermediacy. *Taxon* 41:11–23.

Wilson, S. and D. Knowles. 1988. *Australia's reptiles, a photographic reference to the terrestrial reptiles of Australia*. Collins Publ., Sydney. 447 pp.

Winston, J. 1999. *Describing species: Practical taxonomic procedure for biologists*. Columbia University Press, New York, NY.

Wonosasmito, B. 2009. *Varanus nebulosus* (Gray, 1831). *Bulletin de liason du G.E.V.* 1(1):21.

Wood, S., K. Johansen, M. Glass, and G. Maloiy. 1978. Aerobic metabolism of the lizard *Varanus exanthematicus*: Effects of activity, temperature, and size. *Comparative Biochemistry and Physiology B* 127:331–336.

Worrell, E. 1966. *Reptiles of Australia: Crocodiles, turtles, tortoises, lizards, snakes*. Angus and Robertson, Sydney.

Wunderlich, S., T. Griffiths, and F. Baines. 2024. UVB-emitting LEDs for reptile lighting: Identifying the risks of nonsolar UV spectra. *Zoo Biology* 43:61–74.

Wüster, W, S. Otsuka, A. Malhotra, and R. Thorpe. 1992. Population systematics of Russell's viper: A multivariate study. *Biological Journal of the Linnean Society* 47:97–113.

Wüster, W. and R. Thorpe. 1989. Population affinities of the Asiatic cobra *(Naja naja)* species complex in southeast Asia: Reliability and random resampling. *Biological Journal of the Linnean Society* 36:391–409.

Yadaz, A. and K. Rana. 1988. Sexual dimorphism in the Indian Monitor lizard, *Varanus bengalensis* (Daudin). *Geobios* 15(5):210–214.

Yeager, V. 1973. The vascular anatomy of the liver of the monitor lizard (genus *Varanus*). *American Journal of Anatomy* 136(4):441–453.

Yoshida, M., Y. Kanto, M. Tsuboi, and Y. Sakai. 2013. Rapid acquisitioned response in an intertidal fish, *Tridentiger trigonocephalus* (Gobiidae) using an ethologically relevant conditioning paradigm. *Behaviour* 150(6):585–598.

Young, B. 1990. Is there a direct link between the ophidian tongue and Jacobson's organ? *Amphibia-Reptilia* 11:263–276.

Young, B. 1988. The subclavian loop of *Varanus salvator. Copeia* 1988(4):1029–1030.

Young, B. 1997. On the absence of taste buds in monitor lizards (*Varanus*) and snakes. *Journal of Herpetology* 31(1):130–137.

Yuyek, M. 2004. Only in the Philippines: The natural history and reproductive husbandry of the Gray's monitor, *Varanus olivaceus*, with emphasis on the world's first successful rearing of a neonate bred at Avilon Zoo (Part One). *Animal Scene* 4(5):80–82.

Yuyek, M. 2004. Only in the Philippines: The natural history and reproductive husbandry of the Gray's monitor, *Varanus olivaceus*, with emphasis on the world's first successful rearing of a neonate bred at Avilon Zoo (Part Two). *Animal Scene* 4(6):78–81.

Zamudio, K., R. Bell, and N. Mason. 2016. Phenotypes in phylogeography: Species' traits, environmental variation, and vertebrate diversification. *Proceedings of the National Academy of Sciences* 113:8041–8048.

Zdunek, P., G. Harold, and S. Karunaratha. 2024. Additional records of intrageneric predation and cannibalism in monitor lizards. *Herpetology Notes* 17:749–755.

Zerova, G. and V. Čhkhikvadze. 1983. [*Varanus tyrasiensis* G. Zerova et V. Čhkhikvadze sp. nov.]. Pp. 417–420 (in Russian). IN: A. Lungu, G. Zerova, and V. Čhkhikvadze (eds.) *[Primary evidence on the Miocene Varanus of the north Black Sea littoral.] Bulletin of the Academy of Sciences of the Georgian SSR*, vol. 110.

Zerova, G. and V. Čhkhikvadze. 1986. Neogene varanids of the USSR. Pp. 689–694. IN: Z. Roček (ed.). *Studies in herpetology: Proceedings of the European herpetological meeting*, Charles University, Prague.

Zhang, Y.-P., S.-R. Li, J. Ping, S.-W. Li, H.-B. Zhou, B.-J. Sun, and W.-G. Du. 2016. The effects of light exposure during incubation on embryonic development and hatchling traits in lizards. *Scientific Reports* 6:38527. https://doi.org/10.1038/srep38527.

Zheng, Y. and J. Wiens. 2016. Combining phylogenomic and supermatrix approaches, and a time-calibrated phylogeny for squamate reptiles (lizards and snakes) based on 52 genes and 4162 species. *Molecular Phylogenetics and Evolution* 94(Pt B):537–547. https://doi.org/10.1016/j.ympev.2015.10.009

Zhue, X.-M., Y. Du, Y.-F. Qu, and L.-H. Lin. 2020. The geographical diversification in varanid lizards: The role of mainland versus island in driving species evolution. *Current Zoology* 66(2). https://doi.org/10.1093/cz/zoaa002

Ziegler, T., W. Böhme, and K. Philipp. 1999. *Varanus caerulivirens* sp. n., a new monitor lizard of the *V. indicus* group from Halmahera, Moluccas, Indonesia (Squamata: Sauria Varanidae). *Herpetozoa* 12(1/2):45–56.

Ziegler, T., A. Rauhaus, and I. Gill. 2016. A preliminary review of monitor lizards in zoological gardens. *Biawak* 10(1):26–35.

Zietz, F. 1920. Catalogue of Australian Lizards. *Records of the Australian Museum* 1:181–228.

Zimmermann, E. 1986. *Breeding terrarium animals*. TFH Publications, Neptune City. 384 pp.

Zweifel, R. 1952. Pattern variation and evolution of the Mountain Kingsnake, *Lampropeltis zonata*. *Copeia* 1952(3):152–168.

Zweifel, R. 1980. Results of the Archbold expeditions. No. 103. Frogs and lizards from the Huon Peninsula, Papua New Guinea. *Bulletin of the American Museum of Natural History* 165(5):387–434.

Zweifel, R. 1981. Color pattern morphs of the kingsnake (*Lampropeltis getulus*) in southern California: Distribution and evolutionary status. *Bulletin of the Southern California Academy of Sciences* 80(2):70–81.

INTERNET CITATIONS

Varanus prasinus using dog feeder toy: https://www.youtube.com/watch?v=-eooVUYh8kg

Varanus rudicollis using dog feeder toy: https://www.youtube.com/watch?v=z41BbVMWsjc

Index

A

Acclimatized, to handling 149–151
†*Aiolosaurus*, genus 231–232
Albino 62, 193–194
Angle of snout 31, 72, 120
Anglerfish, Ceratioidei 23
Anguimorpha 18, 24, 28, 32, 50, 52, 55, 120, 135, 155, 199, 200, 206, 211, 215, 221, 223, 234, 235, 239, 241, 242, 250
Antebrachial scales 54
Arch, postorbital 29, 103, 200, 206, 208, 222–223, 225, 235, 238–241
†*Archaeovaranus*, genus 231–232
Auffenberg, Walter 18

B

Basioccipital 30, 71, 91–93, 95–99, 101–102, 103, 119, 122, 231, 234, 242
Basisphenoid 30, 71, 92, 93, 96–99, 138, 234, 242
Behavior 5–8, 10, 17, 19, 35, 54–55, 117, 122, 137, 145, 157, 158, 183, 185, 201
 aggressive 151, 156
 bipedal 151–153
 definition 145
 learned 3, 146, 148
 mating 135
 and systematic relationships 220
Bibron, Gabriel 14
Bile salts 130
Biogeography 5, 15, 18, 166, 199, 203, 212–213, 215, 249; *see also* Zoogeography
Biozones 109–110
Bipedality 55, 56, 151–153, 162
Bite 87, 91, 139, 153, 155, 160, 193
 force 83, 102, 137–138
 venomous 135–136, 160, 221
Böhme, Wolfgang 8, 10, 11, 16, 19, 42, 116, 129, 170, 179, 181, 204, 213, 218, 230, 248
Books about monitors 3
Boulenger, George Albert 14
Brain 92, 99, 120–125, 139, 146–147, 162, 163
 brain stem 122–125
 cerebellum 92, 99, 122–124, 187
 cerebrum 96, 124
 choroid plexus 124
 diencephalon 96, 120
 hypothalamus 124
 medulla oblongata 124
 neurohypophysis 123
 neuron number 123, 146
 olfactory stem 88, 92, 96, 99, 104, 123, 124, 155, 162, 227, 238
 optic chiasma 123
 pineal complex 124, 156–157, 237, 239
 pituitary, *see* Pituitary
 pons 124
 size, compared to other reptiles 122
 stem 92, 94, 122, 123
 tectum, optic 122–124
 telencephalon 122
Braincase 30, 71, 74, 88, 90, 92–99, 101, 119–120, 123, 125, 146, 200, 205, 224
 alar process of proötic 92, 94, 95, 98
 basioccipitals 99
 basipterygoid 93, 96, 98
 basisphenoid 93, 96, 98, 99
 crista cranii 74, 88, 104
 crista intranasalis 110, 225
 crista juxtafovealis 229
 crista proötica 30, 95, 122, 225, 229, 239
 cristae parietalii 227
 ectopterygoid 71, 85, 100, 101, 103, 224
 epipterygoid 90, 91, 96, 97, 99
 exoccipitals 95, 99
 foramina 27, 29, 69, 71, 74, 75, 94, 98, 99
 opisthotic 95, 99
 otic capsule 94–96, 122
 otoöccipitals 95
 parasphenoid 92, 96
 parietal, *see* Parietal bone
 proötic 95, 99, 108, 123
 pterygoid 30, 74, 87, 90, 95–97, 99, 101–104, 122
 quadrate, *see* Quadrate
 sella turcica 93, 95, 96, 98, 123
 sphenoid 93, 98
 supraoccipital 92, 95, 96
Burrows 3, 71, 118, 165
 spiral 3, 182, 185

C

Calcium: phosphorus ratios 167
Camp, Charles Lewis 12, 55, 137, 204, 223
Canthus rostralis 24, 27, 36, 37, 41, 64, 65, 74, 230
Capsule, otic 94–96, 99, 122; *see also* Nasal capsule
Caudal scales 33, 43, 48, 52, 54–55, 218
Character state 18, 23, 24–26, 28, 32, 37
Characters 6, 10, 12, 15, 17, 18, 23–25, 27, 30–31, 39
 analysis 7, 12, 14, 17–18, 24–25, 69, 111
 molecular 1, 5, 7, 17, 19, 98, 127, 175, 176, 192, 194, 200, 218–220, 222, 247–248
†*Cherminotus*, genus 239–241
Choanae 41, 113, 115
Cladistics 18, 23, 219, 222
Claws, *see* Digits
Cloaca 44–45
 scales 51
Clutch size 182, 184
Coloration
 body 32, 57, 60–62
 head 32, 57
 ontogenetic change 63, 64–65, 191
 pattern, *see* Pattern
 tail 32, 62–63
 tongue, *see* Tongue
 ventral 32
Combined data sets 219–220
Competition
 on islands 174–175
Complex, species 10
Conrad, Jack 12
Convergence 23, 24, 71, 90, 109–111, 113, 123, 161, 222, 249
Cope, Edward Drinker 10–11, 200
Cranial bones, *see* Skulls
Cranial organs, chart 108
Cranial scales 50–51
Crocodiles 3, 165, 176
 feeding rolls 112
 heart 126
 sensory organs, cutaneous 49
Cryptic taxa 17, 116
†*Cryptovaranoides microlanius* 199

D

Data, lack of 15
Darwin's finches 5, 174, 215
Daudin, François Marie 11–14, 245, 248
Defense 41, 46, 54–55, 136, 148, 155, 159, 161, 162, 221; *see also* Threat
Dentary 30–31, 71, 74–75, 85, 86, 104–107, 119, 124, 135–136, 138, 140, 207, 223–224, 227–229, 235–237, 240
Dentition, *see* Teeth
Dermis 16, 47, 119, 122; *see also* Epidermis
 α keratin layer 46
 β keratin layer (*Oberhautchen* or stratum corneum) 46
 stratum compactum 47
Deutsche Gesellshaft für Herpetologie and Terrarienkunde (DGHT) 18–19
Die Familie der Warane (monograph) 15–16, 69
Digestion 171
Digits 3, 32, 46, 56, 67, 147, 224–225, 236
Dimorphism, sexual 179
Dispersal, barriers to 176
Distribution
 Cretaceous 204–208
 over time 203–212
 tertiary 208–212
Divergence, morphological 109, 164, 175, 190
Diversity 4, 17, 30, 56, 69, 79, 89, 109, 125, 135, 175, 177, 192, 199, 203, 207, 208, 210, 215, 230, 242, 250
DNA 24, 165, 178, 190, 204, 218
Duméril, André Marie Constant 14
Dystocia (egg binding) 181
†***Dytikosaurus*** gen. n. 210–211, 229, 233–234, 237, 249, 250

E

Ear 42, 122, 124
Egg tooth 186
Eggs
 clutches 184
 communal 148, 151, 182
 development 181, 185, 189
 incubation time 184
 retention 181
 in termite mounds 148, 178, 182, 183
 water needs 188
Eidenmüller, Bernd 3, 153, 159, 174, 179–181, 183–187, 202, 230, 246

Embryo development 183, 185–187
Energy 54–55, 133, 163, 164, 167–168, 172
  oxygen consumption 167
  use 155, 159, 165, 167, 170
Epidermis 46, 47, 136
  β keratin layer (*Oberhautchen* or stratum corneum) 46
Estes, Richard 12
Evans, Susan E. 7, 69, 71, 108, 201, 204, 223, 231, 249
Evolution 4–6, 10, 12, 15, 18, 19, 23–24, 28, 67, 86, 101, 110, 134–136, 146, 171, 174–175, 185–186, 192–193, 197, 199–205, 208, 210, 213–215, 218, 220–222, 224, 246
Extinction 6, 165, 175, 208, 210, 212
Extracolumella 42, 96, 101–102, 122, 138
Eyes 118–120
  color vision 38, 120, 164
  and depth perception 120, 123, 141
  of parietal eye 121, 156
  pineal, *see* Parietal eye
  retina 120
  third, *see* Parietal eye

F

Feeding 69, 112, 145, 153–155, 167, 193
  and gape 104, 114, 137, 153
  specialized 105, 111, 171, 212
  and venom 221
  while gravid 181
Fejérváry, Baron Géza 7, 10, 11–12, 17–18, 54, 69–70, 75, 81, 84–85, 111, 201, 204, 212, 217, 228
Finches, *see* Darwin's finches
Fold 57, 80, 130, 137, 181
  of coprodaeum 134
  gular 27, 32, 35–36, 44, 137, 140, 224–225, 239
  lateral 137
  of liver 130
  nasal chamber 118, 170
  nuchal 52
  of stomach 130, 133
  transverse 137
Foramen 76–77, 92, 93, 99
  ethmoid 28, 76
  hyomandibular 30
  hypoglossal 30
  jugular 99
  lacrimal 29, 87
  magnum 51, 92, 98, 99
  nutrient 27, 75
  parietal 30, 74, 91, 99
  septomaxillary 80
Foramina 108–109
Fossa 113
  supratemporal 30
Fossil 11–12, 15–17, 20, 25, 28–30, 69–71, 75, 84–85, 87, 91, 100, 103, 112, 113, 116, 167, 177, 189, 197, 199–201, 203–213, 215–217, 219, 221–223, 225–228, 231–232, 234, 236–238, 248–250
Fossil history 199, 235

G

Gape 137, 171
Genetics 5, 7, 17, 136, 174, 179, 193, 194, 199, 201–203, 219, 222
Gilmore, Charles W. 12, 17, 87, 111, 201, 204–206, 210, 216–217, 237–238
Glands 134, 141, 170; *see also* Organs, visceral, glands
  cutaneous 47
  and water resorption 134
Gray, John Edward 10–12, 14, 219, 223, 230, 245

H

Habitat 3, 8, 10, 18, 35, 54–57, 66, 86, 111, 113, 118, 146–148, 156, 165–167, 169–171, 174, 176, 177, 193, 200, 203, 215, 219, 243, 247, 250
Hatching 42, 55, 65, 178, 180, 183, 185, 189, 190
  size correlated 183, 186–188, 207
  timing and prey 187
  and water 188–189
Head
  general morphology 35–42
  shape, ontogenetic change 63–64
Head-neck lengths 35–36
Hearing 41, 122, 125
Heart, *see* Organs, visceral
Heart rate 125, 171
Heat 46, 48, 54, 124, 141, 155, 163, 174, 200
  absorption 151, 156, 157, 159, 172
  dissipation 157, 159, 169
  and eggs 182–183
  and hatching 187
  production 171
  tolerance 164–166, 210
Hemibaculum 137
Hemiclitoris 181, 225
Hemipenis 36, 64, 125, 137, 178, 181
Hybrids 192–195
  in captivity 194–195
Hydrolysis, physiological 169
Hysteresis 171

I

Identification 62, 82–83
  from locality data 80
  sensory 125
  of species 7, 23, 32, 62, 80, 83, 116, 210, 228
  of vertebrae 80, 210, 216, 226
Incubation
  temperatures 183–184
  timing of 184, 186
Intelligence 122, 145–151
  concepts of 146
  and counting 149
  habituation versus 145–146
  and incubation temperature 148
  instinct versus 145–146
  and puzzle solving 149
  and tameness 150
  study limitations 145
  why study 146
Interocular scales 38
Intramandibular joint 105, 107
Intraspecific combat 158–161
Islands
  competition on 175
  and dwarfism 213
  and endemic varanids 176
  evolution on 175
  and extinction 177
  and gigantism 175, 213
  insularity 212
  and melanism 175, 176, 213–214
  types of 175
  variants 175–177

J

Jacobson's organ 74, 102, 117–118
Jugal 99–101

K

Kidneys 125, 132–134, 136, 141, 164, 169–170, 172
King, Denis 19
Komodo dragon 3, 10, 17, 21, 34, 77, 108, 151, 160
  skull cf. *Varanus caudolineatus* 68

L

Lacépède, compe de 12, 14, 52
Lanthanotidae 238–239
Lanthanotids 240–241
Lanthanotus, as snake ancestor 200, 239
Lateral recessus scala tympani (lrst) 30, 93–94, 98, 99, 122
Limbs 23, 25, 32, 35, 46, 55–57, 65–67, 125, 165, 200, 224–225, 236, 246–247
  antebrachial scales of 54, 56
  claws 46, 56–57
  digits 46, 56–57
  embryonic 183
  posturing 147, 156; *see also* Bipedality
  proportions 56, 190, 231, 239, 243
Linnaeus, Carolus 10

M

Mammals, and varanid evolution 174
Mandible (lower jaw) 74, 104
  angular 105
  articular 106–107
  coronoid 105
  dentary 104–105
  Meckelian canal (fossa) 74, 105
  prearticular 74–75, 104–105, 224
  splenial 105, 106–107
  surangular 105, 107
Mating 178–180
Maxilla, *see* Skulls
Maxillary morphotypes 81–83
  and biozone distribution 110
McDowell, Samuel Booker 7, 12, 18, 25, 28–30, 39, 48, 69–71, 111, 118, 136, 200–201, 204, 208, 240
†*Megalania* genus (=*Varanus*) 69, 208, 211, 218, 221, 244
Melanism 59, 175–176, 213–214
Memory 122; *see also* Intelligence
Merrem, Blasius (author of "*Varanus*") 10
Mertens, Robert Friedrich 1, 3, 6, 9, 11, 12, 14–18, 25, 28, 52, 68–69, 75, 76, 78, 84, 85, 107, 134, 155, 175, 194, 204, 213, 218–219, 222, 229, 230, 242, 243–247, 249
Metabolic rate 167, 172
Monitors 1, 3, 4, 6–8, 10, 12, 14, 15, 17, 18, 21, 25, 28, 35–38, 40, 42, 53, 56, 67, 85, 200, 248
  aquatic 39, 118, 131, 166

combat, male-male 161–162, 181
common names 10
heat-loving 210
invasive in North America 167
locomotion 151, 153
number of species 14
prey size, preferred 171; *see also* Scales
small 53, 54, 102, 183, 187, 213, 231
tree 52, 54, 57, 64, 148, 190, 202, 218, 248–249
uric acid secretion 133
use by humans 10, 177
variation in size 34
Monstersauria 237, 250
Morphology 1, 5–7, 15, 17, 18, 23, 29–30, 51–53, 56, 65, 69, 71, 74–76, 78–79, 109–111
macro 21, 34, 40, 41–42, 48, 55, 63, 81, 83–84, 91, 101, 219
micro 219
Morphometric comparisons 31, 69
Mosasaurs 10–11, 18, 85, 200–201, 230
Most recent common ancestor (MRCA) 25
Mouth 25, 26, 41–42, 49, 86–87, 102–103, 113–115, 124, 128, 135, 137, 139, 153–155, 157, 159, 169
Muscles 47, 64, 70, 71, 75, 81, 84, 90, 105–106, 138, 164, 224, 229
constrictor colli 139
costal 129–130
genioglossus 140
geniohyopideus 140
glenohyoideus 140
glossopharyngeal 116
hyolingual 116
intercostal 132
m. depressor mandibularis 137
m. hemipenus repulsors 181
m. pterygoideus 136–138
m. retractor penis 181
m. superficialis adductor mandibulae externus (MAME) 138
mylohyoideus 140
of orbit 119–120, 123
of throat 137, 139, 169
sternohyoideus 140
temporal 92, 138, 179
tympanus 122, 124
Mystery bones 233–234

N

Nasal capsule 115, 122, 169–170
Nasal chamber 115
Nasal salt glands 41, 79, 111, 170
Nasals 28, 77–78; *see also* Skulls, bones of, nares and nasal
Neck length 35
Neonates 189–190
Nerves
cranial nerves (CN)
CN I 96, 99, 122, 127
CN II 96, 99, 123
CN III 99, 125
CN IV 99, 123
CN V 95, 99, 108, 109, 123, 136; *see also* Nerve, trigeminal (CN V)
CN VI 99, 123
CN VII 99, 124
CN VIII 99, 125
CN IX 99, 116, 125
CN X 99, 116, 123, 125
CN XI 123, 125
CN XII 99, 116, 123, 125
Nerve, trigeminal (CN V)
mandibular branch 123
maxillary (cheek) branch 123
ophthalmic branch 123
ramus muscularis 123
New Caledonia, varanids 177, 248–249, 255
Nitrogen 132, 169
Nopcsa, Baron Franz 11–12, 17, 201, 204
Nostril 25
flap 41, 77, 118
and Jacobson's organ 117–118
nasal opening 117–118
position on snout 39, 41, 118
shape 39, 41
Nuchal scales 36, 44, 47, 219
shape 32, 43
size 32
structure 44, 48–50, 51

O

Oberhautchen, *see* Epidermis
Odatria, subgenus 230
Ontogenetic changes 17, 25, 63, 83–85, 89, 91, 102, 111, 190, 201, 218, 219, 237
color and pattern 64–67
head shape 28, 37, 63, 154
SVL: tail proportions 32, 64
Orbit, anatomy 119
Orbitosphenoid 104
Organs, visceral
cloaca 134
gall bladder 129, 130
gastrointestinal tract 130–132
glands
salt 134–135
venom system 135–136
heart 125–128
intestines 131–133
kidneys 132–134
liver 125, 129–130
lungs 125, 128–129
pancreas 125, 129–130
spleen 125, 130
stomach 130–131, 132
Origins, varanids 200–201
Otic capsules 95
Ovaries 123, 132, 136–137
†*Ovoo*, genus 233–234

P

Palate 103, 117, 140, 205, 238
hard (bony) 75, 96, 113
soft 113–114
Paraoccipital process 91, 92, 98–99, 207, 225, 227, 229–230, 238, 239, 241
Parasphenoid 30
†*Paravaranus* genus 241
Parental care 147, 183, 185, 189
Parietal bone
crista parietalis lateralis 89, 229
dorsal shape 29
eye 29, 119–122; *see also* Parietal eye
foramen 29
lateral processes 29
morphology 88–92
ontogeny 89–90
pteroticum 30
supratemporal process 29
Parietal eye 29, 119–122
and behavior 121
in compass orientation 157
function hypotheses 156
morphological types 120
pineal nerve 120
in regulating reproduction 157
structure 121
third eye 156
as ultraviolet dosimeter 156–157
Parietals and diet 91
Parsimony 23
Parthenogenesis, *see* Reproductive modes
Pattern (markings) 7, 17, 18, 28, 32–33, 45, 51, 57–67; *see also* Tongue
head 26, 27, 37, 58
sensilla 49
Phenotypes 175
Phylogenetics
cladistics 23
Hennigian 18
and links to feeding habits
Phylogenies 12, 15, 17, 23, 199, 201, 220, 221, 232, 249
Anguimorpha 220, 249
apomorphy based 23–24
branch based 23–24
complete data set 223, 249
chromosome based 19
node based 23–24
varanids 14, 220, 249
Physiology 5, 6, 7, 19, 54, 127, 164, 170, 178, 189, 219, 221
abiotic factors 165
behavioral 153
pathways 164
rates 171
respiratory 127, 129
and water 169
Platynota 199–201, 216, 223
Plesiomorphies 23
Pianka, Eric 18–19
Pilae 30, 92, 96, 106
Pituitary 98, 123, 124
hormones 123
†*Placovaranus* genus 17
Play 151
Populations 174
Postocular (temporal) scale rows 36
Preanal scales 32, 44, 45
Premaxilla
ascendens process 27
and divided nasals 27
dorsal ridge 28
dorsal texture 27
length 28
number of tooth sockets 28
Prey, of monitors 3, 25, 55, 74, 75, 91, 96, 97, 102, 131, 133, 134, 164, 166–167, 170–171, 174, 176, 184, 190; *see also* Feeding
capture 54, 86, 91, 123, 136, 138–139, 141, 148, 153–155, 162, 165, 169
counting 149
and hatching timing 187
hearing 122
ontogenetic shift 65, 85
tracking 114, 120, 154, 162

Proötic 30
alar process angle 30
Punctuated equilibrium 214–215
Pythonomorpha 10, 200

Q

Quadrate
contacts surangular 107
kinesis 100
morphology 100–101
and stapes 102

R

Reproduction
mating 178–181
nests 182–185
parthenogenesis 178, 179
Reproductive modes
hybrid speciation
parthenogenesis 178, 192
steps in 178
Reproductive systems 136–137
Respiration 127–129, 141, 153, 165, 169–170
Response to being called by name 150–151
Reversal learning 149
Rictal scales 31, 32, 38
Roach (posture) 159–160

S

Salt glands, *see* Nasal salt glands
Salvaguardia xi, 11
†*Saniwa* genus 234–238
†*Saniwides* genus 242
Scales
antebrachial 54
dorsals 32, 43–44, 50–52, 53
labial 27, 31, 36, 43, 48–50, 57, 64, 77, 224–225, 239, 246
nuchals 32, 43–44, 47, 49–52, 66
parietals 29, 36, 88
scale organs 43; *see also* Sensilla, dermal
tail (caudal) 33–34, 43–45, 48, 52–55, 62, 65–66, 218, 231
vertebrae 111–112, 204, 217, 224
temporal 36, 50–51
ventrals 32, 43–45, 48, 51–53, 66, 224, 225
Schlegel, Hermann 11, 216, 242
Seba, *Thesaurus* 12
Sensilla, dermal 43, 48–51, 65–67
Septomaxilla 29, 71, 74, 77–80, 83–85, 102, 113, 115–118, 135, 207–208, 219, 225, 231, 234
Sex determination 190, 192
Sibling species 18, 21, 116
Size 3, 4, 8, 10, 21, 34, 35, 43, 54, 56, 58, 64, 69, 151
Skin 7, 19, 20, 29, 43, 44–51, 87, 115, 118, 145, 156, 170, 172; *see also* Dermis; Epidermis; Sensilla
Skull, bones of
angle of snout 72
ectopterygoid 71, 100–101, 103, 224
epipterygoid 90–91, 96–97, 99, 103
frontals 87–88, 104
jugal 27, 50, 71, 74, 81, 84–85, 99–101, 103, 105, 119, 137–138, 223, 225–226, 231, 234–236, 238, 239, 241, 242
kinesis between 71, 74–75, 91, 97, 105, 200, 224
lacrimal 87
maxilla 71–74, 80
nares and nasal 24, 27, 28, 30, 41, 71, 74–81, 83, 87, 88, 140, 170, 200, 205, 207, 215, 218–219, 223, 225–227, 230–236, 239–243
occipital bones 92–99; *see also* Braincase
opisthotic 95, 99
orbitosphenoid 30, 93, 101, 104
parietal, *see* parietal bone
premaxilla 75–77
proötic 30, 71, 90–92, 94–96, 98–99, 101, 108, 122–123
crista proötica 95, 99
pterygoid 207, 208, 224, 231, 234–236, 241, 242; *see also* Braincase
teeth 87, 101, 238, 234, 235, 238, 239
septomaxilla 29, 78–86
stapes 96, 102
vomer 28, 41, 77, 79, 80, 87, 102–103, 113, 141, 223, 232, 234, 235, 238–241
teeth 236, 215, 235
Skulls
convergence within Varanidae 109–111
shared characters in *Varanus* 71–75
Snakes
varanids as mimics of 64, 161, 190
varanoid ancestry of 10, 18, 24, 30, 55, 69, 123, 141, 200–201, 240
Snout
angle of 100–101
and depth of field
tip, depressed 39
width 37
Sodium (Na) 169
Speciation 176, 192–194, 201–203, 213, 226, 248, 250
branching 202–203
chronospecies 203, 215
hybrid 178, 192, 193
punctuated equilibrium 215
reticulate 203
Species complexes
acanthurus group 40, 42, 50, 53, 56, 59, 88, 243
Odatria tristis/O. scalaris group 51, 55, 59, 63, 70, 218, 243–247
unresolved 242–243
Varanus indicus group 70, 248–249
Varanus prasinus group 249
Varanus varius 24, 31, 42, 54, 58, 61–62, 84, 141, 203, 248
Species definitions 201
Sphenodon 23, 38, 54, 120
Spines, cloacal 44
Splenial 30, 74, 104–107, 224, 231, 235, 239
Stratum corneum, *see* Epidermis
Subgenera 10, 16, 218–219, 230
Subgenus 8, 9, 70, 76–78, 80, 177, 194–195, 207, 215, 218–219, 230–231, 243, 248
Subspecies 8–9, 10, 15–17, 63, 69, 80, 84, 166, 179, 193–194, 212, 214, 218–219, 222, 230, 243–246
Supratemporal 91, 92, 207, 219
arch 30, 71, 89–90, 227, 238, 239, 241
fenestra/fossa 30, 74, 89, 91, 92
process 91, 237
Supraocular scales 7, 26, 35–38
Swallowing 71, 104, 139
Sweet, Sam 8, 38, 52, 129, 135, 155, 160, 221, 248
Symplesiomorphy 24, 218
Synapomorphy 23, 24, 28–30, 32, 54, 60, 205–207
Synonyms 8–10, 14–15, 17, 24, 63, 71, 202, 204, 210–212, 216, 218–219, 222, 225, 230, 237, 243–246, 249
Systematics 1, 6, 18, 23–24, 71, 197, 208, 218–219, 221, 231, 246, 248–250
cladograms 24
of Lace Monitors 248
molecular 1
principles 218–219

T

Tail
annuli 33
autotomy, loss of 54
color 64–65
lateral cf. ventral scales 33
ontogeny 64
proportion to total length 54, 55–56
scalation 32
shape, in section 32, 52
uses 54–55
Taxonomic considerations 8–9
Taxonomic history 10–14
Taxonomy 6–7, 9, 10, 12, 14–16, 18, 23, 52, 70, 78, 85, 151, 201, 204, 213, 215, 218–219, 221–222, 237, 243, 245, 247
Teeth 86–87
palatal 75, 235, 236
plicidentine 74, 86, 205, 218, 223, 224, 225, 226, 229, 231, 235
†*Telmasaurus* genus 238
Temperament of monitors 161
Temperature, *see also* Thermoregulation
extremes 166
variation 156
Temporal arch 90, 92, 139
Temporal jaw muscles 138–139
Termite mounds 3
temperature control in 183–185
Testes 123, 132–133, 136–137, 141
Thanatosis (playing dead) 159
Thermoregulation 121, 155, 157–159, 147
behavioral 147, 157, 159, 174
effects on body 155
heliotherms 155–156
nasal chamber and 170
physical actions 147
poikilotherms 155
thigmotherms 156
Threat postures 3, 162, 163, 174
Toe length 32
Tongue 30, 41, 113, 115–117
color 31, 40, 41, 116
color, species specific 41–42
structure 40
Total evidence phylogenies 7
Toxicofera 220–221
Trabeculae cranii 98–99, 101, 126

Transformation series 24
Tympanum 42, 46, 96, 101–102, 122, 124; *see also* Ear

U

Ultraviolet light 159, 174–175
  from LED sources 157
  needed in captivity 157–158
Ultraviolet radiation (UVR)
  and calcium levels 172–173
  dangers of 174
  parietal eye detection of 121, 156
  recognition by varanids 116
  and skin 46
Uric acid ($C_5H_4N_4O_3$) 130, 132–134, 169

V

Varanidae, family 5–6, 7, 10, 15, 23, 24, 28, 56, 82–83, 85–87, 118, 123, 126, 128, 141, 152, 153, 155, 168, 171, 178, 199–201, 206–208, 211, 215, 218, 220– 223, 234, 235, 237, 241–242, 248–250
Varanoidea 24, 54, 199–201, 204, 207, 208, 223
Varanoids 3, 6, 10, 25, 29, 80, 86–87, 102, 111–112, 201, 128, 154, 167, 199–201, 204–206, 208, 210, 215, 219, 221–222
Varanus
  in captivity 17–19, 35, 62, 148, 151, 153, 155, 159, 161, 166, 171, 174, 179, 185
  genus 225
  number of species described by year 11
  reproduction in captivity 179–183, 185, 187–189, 193
Venom
  effects 136, 160
  system 135–136, 155
Vertebrae 111–112, 215–217
Vidian canal 93, 95
Vision 31, 38, 67, 120, 141, 156, 162, 164
Vitamin D 130, 158
  and calcium levels 182
  effected by UVB 163, 173
  and egg binding 182
  levels in monitors 173
  synthesis 158
Vomer 102–103; *see also* Skull, bones of, vomer
Vomeronasal organ 41, 78, 113, 114, 117, 122, 140, 223, 234
VO_2, in metabolic oxygen consumption 167–168

W

Water
  dynamics in varanids 169, 170
  from food 118
  and hatching 189
  hearing in 122
  in Kansas 248
  loss 157, 170
  refusing as sign of illness 182
  resorption 132–133, 134, 164, 169, 172
  retention by skin 46, 55, 164
  running on 152
  and sensory detection 49
  submerged in 5, 41, 54, 114, 118
  and thermoregulation 169

Z

Zoogeography 7, 212–214

For Product Safety Concerns and Information please contact our
EU representative GPSR@taylorandfrancis.com Taylor & Francis
Verlag GmbH, Kaufingerstraße 24, 80331 München, Germany